COAGULATION
AND
FLOCCULATION

SURFACTANT SCIENCE SERIES

COAGULATION AND FLOCCULATION

Second Edition

Edited by
Hansjoachim Stechemesser
Technische Universitat
Bergakademie Freiberg
Freiberg, Germany

Bohulav Dobiáš
University of Regensburg
Regensburg, Germany

CRC Press
Taylor & Francis Group
Boca Raton London New York

CRC Press is an imprint of the
Taylor & Francis Group, an **informa** business
A TAYLOR & FRANCIS BOOK

CRC Press
Taylor & Francis Group
6000 Broken Sound Parkway NW, Suite 300
Boca Raton, FL 33487-2742

First issued in paperback 2019

ISBN-13: 978-1-57444-455-1 (hbk)
ISBN-13: 978-0-367-39314-4 (pbk)

Library of Congress Card Number 2004059308

Library of Congress Cataloging-in-Publication Data

Coagulation and flocculation: theory and applications / edited by Hansjoachim Stechemesser, Bohuslav Dobiás.-- 2nd ed.
    p. cm. -- (Surfactant science series ; v. 126)
    Includes bibliographical references and index.
    ISBN 1–57444–455–7 (alk. paper)
    1. Coagulation. 2. Flocculation. I. Stechemesser, Hansjoachim. II. Dobiás, B. III. Series.

QD547.C63 2004
541'.3415--dc22
                                                         2004059308

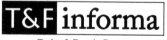

Taylor & Francis Group
is the Academic Division of T&F Informa plc.

Visit the Taylor & Francis Web site at
http://www.taylorandfrancis.com

and the CRC Press Web site at
http://www.crcpress.com

Preface

Following its publication in 1993, the first edition of *Coagulation and Flocculation* has been well received around the world by researchers in this very interesting field, that of stabilization and destabilization of fine solid dispersions. Since then, there have been significant advances in this field. This second edition, intended to capture these advances, will be a useful reference work.

The chapters have been significantly updated and expanded to include advances in both theoretical and application aspects. For different reasons — work on another branch or retirement — some contributors to the first edition did not choose to update their chapters, so that the original Chapter 6, "Structure Formation in Disperse Systems," Chapter 8, "Floc Stability in Laminar and Turbulent Flow," Chapter 9, "Measurement of the Size of Aggregates in Suspension," and Chapter 14, "The Effect of Coagulation and Flocculation on the Filtration Properties of Suspensions Incorporating a High Content of Fines" have been deleted completely.

However, famous scientists in corresponding topics were enlisted to participate in their place. For this reason, more than half of the information from the first edition of the book is new or updated.

Chapter 1, "Thermodynamics of Adsorption from Solution," has been extended to reflect the dissociation from a solid surface with independent sites, and Chapter 2, "Surface Charge and Surface Potential" has been increased by the description of improvements of the Gouy-Chapman model. Chapter 3, "Coagulation Kinetics" has been rewritten. The chapter by Luuk K. Koopal called "Ion Adsorption on Homogeneous and

Heterogeneous Surfaces," has been extended essentially to review a more comprehensive theoretical treatment concerning the electrical double layer and the surface ionization. A special section is devoted to surface heterogeneity. "Adsorption of Surfactant" in the first edition was deleted, and Jonas Addai-Mensah and Clive A. Prestidge were enlisted as new authors for Chapter 4, "Structure Formation in Dispersed Systems." Chapter 5, "Modeling Polymer Adsorption, Steric Stabilization, and Flocculation" has been completely rewritten. The new chapter "Fundamentals of Homopolymers at Interfaces and their Effect on Colloid Stability" gives the best available comparison of exact numerical data and analytical descriptions and contains many figures and equations that have never before been published. Chapter 7, "Flotation as a Heterocoagulation Process" has been revised and extended by the following topics: "Rupture of the Intervening Thin Liquid Film" and "Modeling of a Semi Batch Process on the Basis of the Microprocesses Probabilities." New sections on "Intracrystalline Reactions" and "Applications" have been added to Chapter 8, "From Clay Mineral Crystals to Colloidal Clay Mineral Dispersions."

The first edition's section "Aggregation of Clay Minerals" has been extended to "Aggregation of Clay Minerals and Gelation" and has been newly written, and Chapter 13, "Mineral Aggregate Formation and the Measurement of Aggregate Size" has been replaced completely by the more interesting subject "Flocculation and Dispersion of Colloidal Suspensions by Polymers and Surfactants" with the concentration on the stability of suspensions in the presence of surfactants, polymers and their mixtures, as well as the population balance model for flocculation. The chapter "Flocculation and Dewatering of Fine-Particle Suspensions" by Richard Hogg replaces Chapter 14, "The Effect of Coagulation and Flocculation on the Filtration Properties of Suspensions Incorporating a High Content of Fines." The emphases are: "Process Evaluation and Floc Characterization," "Destabilization of Fine-Particle Suspension," "Floc Formation and Growth," "Floc Structure," "Polymer-Induced Flocculation," and "Application to Dewatering by Sedimentation and Filtration."

The authors hope that, with the update and expansion, the second edition will continue to serve as a useful and handy reference for the researchers in the field of the stabilization and the destabilization of fine solid dispersions.

We would like to thank all the authors for their contributions.

Unfortunately, we have to mourn the deaths of two contributors. Hans Sonntag died after a long illness on February 2nd, 1997. He led the Department of Colloid Chemistry at the Institute of Physical Chemistry of the Academy of Sciences of the GDR. After the reunification of Germany, Sonntag was an active and competent member of the founding committee of the new Max Planck Institute of Colloids and Interfaces. In him we have lost an internationally known scientist who dedicated his work to colloid chemistry. He belonged to a group of outstanding colloid scientists who inspired many colleagues around the world.

Hans Joachim Schulze died on September 4th, 2003, after a gallant fight with cancer. He was leader of the Max Planck Research Group for Colloids and Interfaces at the Freiberg University of Mining and Technology in Germany. His activities embraced the broad field of colloids science, ranging from the fundamental research of flotation and paper de-inking to wetting of building materials and interfacial biophysics.

In these two we have lost not only excellent scientists but also friends. It was our privilege to have learned from and worked with these great men of science. They will remain in our collective memory forever. May their work continue to stimulate the minds of a new generation of researchers.

Hansjoachim Stechemesser
Bohuslav Dobiáš

List of Contributors

Jonas Addai-Mensah
Ian Wark Research Institute
University of South Australia
Adelaide, Australia

Bohuslav Dobiáš
Institute of Physical and
Macromolecular Chemistry
University of Regensburg
Regensburg, Germany

G.J. Fleer
Laboratory of Physical
Chemistry and Colloid Science
Wageningen University
Wageningen, The Netherlands

R. Hogg
Department of Energy and Geo-
Environmental Engineering
The Pennsylvania State
University
University Park, PA

P.C. Kapur
Tata Research Development
and Design Centre
Pune, India

H.-H. Kohler
Institute of Analytical
Chemistry, Chemo- and
Biosensors
University of Regensburg
Regensburg, Germany

Luuk K. Koopal
Laboratory of Physical
Chemistry and Colloid Science
Wageningen University
Wageningen, The Netherlands

Gerhard Lagaly
Institute of Inorganic Chemistry
University of Kiel
Kiel, Germany

F.A.M. Leermakers
Laboratory of Physical
Chemistry and Colloid Science
Wageningen University
Wageningen, The Netherlands

Clive A. Prestidge
Ian Wark Research Institute
University of South Australia
Adelaide, Australia

Venkataramana Runkana
Langmuir Center for Colloids
and Interfaces
Columbia University, New York
and
Tata Research Development and
Design Centre
Pune, India

Hans Joachim Schulze
Max-Planck-Research Group for
Colloids and Interfaces at the
Technische Universität
Bergakademie Freiberg
Freiberg, Germany

H. Sonntag
Max-Planck-Institut
für Kolloid- und
Grenzflächenforschung
Potsdam, Germany

H. Stechemesser
Arbeitsgruppe Kolloide und
Grenzflächen am Institut für
Keramik, Glas- und
Baustofftechnik, Technische
Universität Bergakademie
Freiberg, Germany

Christian Simon
Functional Ceramics Group
SINTEF
Oslo, Norway

P. Somasundaran
Langmuir Center for Colloids
and Interfaces
Columbia University
New York

Werner Stöckelhuber
Max-Planck-Research Group for
Colloids and Interfaces at the
Technische Universität
Bergakademie Freiberg
Freiberg, Germany
and
Institute for Polymer Research
Dresden, Germany

S. Woelki
Institute of Analytical
Chemistry, Chemo- and
Biosensors
University of Regensburg
Regensburg, Germany

Contents

1

Thermodynamics of Adsorption from Solution

H.-H. KOHLER

Institute of Analytical Chemistry, Chemo- and Biosensors, University of
Regensburg, D-93040 Regensburg, Germany

I. INTRODUCTION

Adsorption processes often dominate processes of aggregation
and flocculation in solution. The aim of this chapter is to give
a concise description of the thermodynamic background of
adsorption including its relation to interfacial tension.

In the author's experience, the implications and limitations of a theoretical result often remain obscure, unless the
theoretical line of arguments is traced back to the fundamental equations and assumptions. Therefore, an attempt is
made to establish a reasonable balance between theoretical
foundations and derivations, on the one hand, and results, on
the other hand. The fundamental thermodynamic relations
will be presented in Section II. Subsequent sections are devoted to special types of single-layer adsorption advancing
from one-component adsorption to two-component and to electrolyte adsorption. A similar approach worth reading is presented in Aveyard and Haydon [1].

II. FUNDAMENTAL RELATIONS

A. The System

Adsorption is the exchange of matter between a volume phase and an interface. Therefore, the thermodynamics of adsorption are closely related to the thermodynamic description of interfaces. We assume that adsorption takes place between a liquid solution and a macroscopically homogeneous and plane phase boundary between the solution and another solid or fluid phase. Assuming, for convenience, that the composition of the second phase is practically fixed and that all changes of the interfacial properties are due to exchange with the solution phase, we will (mostly) refer to a simplified thermodynamic system consisting of the solution and the interfacial phase only. We further assume that temperature T and pressure p are uniform all over the system.

The thermodynamic description of the system starts from the first law of thermodynamics.

Reversible differential changes of the internal energy U are given by

$$dU = dQ^r + dW^r, \tag{1}$$

where dQ^r and δW^r are heat and work reversibly exchanged with the surroundings (the greek δ is used to denote differential changes of path-dependent variables). We assume that reversible work can be exchanged by variation of the volume V, giving rise to volume work $dQ^r_{vol} = -p\,dV$, by variation of the amount n_i of component i, giving rise to the chemical work $dW^r_{chem} = \sum \mu_i\,dn_i$, where μ_i is the chemical potential of component i and by variation of the interfacial area A of the phase boundary, giving rise to interfacial work $dW^r_{surf} = \sigma\,dA$, where σ is the interfacial tension. Hence, with $dQ^r = T\,dS$, Equation (1) transforms into Gibbs' fundamental equation:

$$dU = T\,dS - p\,dV + \sum \mu_i\,dn_i + \sigma\,dA. \tag{2}$$

It is convenient to introduce the Gibbs free energy

$$G = U + pV - TS = H - TS. \tag{3}$$

Inserting Equation (2), $\mathrm{d}G = \mathrm{d}U + p\,\mathrm{d}V + V\,\mathrm{d}p - T\,\mathrm{d}S - S\,\mathrm{d}T$ can be expressed by

$$\mathrm{d}G = -S\,\mathrm{d}T + V\,\mathrm{d}p + \sum \mu_i\,\mathrm{d}n_i + \sigma\,\mathrm{d}A. \tag{4}$$

We use c_i to denote the molar concentration of component i in the bulk solution. Component 1 is always the solvent. Since equilibrium is assumed between the interface and the solution, the interfacial tension σ is determined by the intensive state of the bulk solution, which in turn is a function of T, p and the solute concentrations $c_2, \ldots, c_n$. Thus

$$\sigma = \sigma(T, p, c_2, \ldots, c_n). \tag{5}$$

The system may be built up continuously by increasing the $n'_i s$ and the interfacial area A in such a way that the intensive state of the system keeps constant (which implies $\mathrm{d}T$, $\mathrm{d}p = 0$ and μ_i, $\sigma = $ constant) Then, according to Equation (4), G can be expressed by

$$G = \sum \mu_i \int_0^{n_i} \mathrm{d}n_i + \sigma \int_0^A \mathrm{d}A \quad \text{or} \quad G = \sum \mu_i n_i + \sigma A \tag{6}$$

From the last equation

$$\mathrm{d}G = \sum \mu_i\,\mathrm{d}n_i + \sum n_i\,\mathrm{d}\mu_i + \sigma\,\mathrm{d}A + A\,\mathrm{d}\sigma, \tag{7}$$

which if equated with Equation (4), leads to the Gibbs–Duhem equation

$$S\,\mathrm{d}T - V\,\mathrm{d}p + \sum n_i\,\mathrm{d}\mu_i + A\,\mathrm{d}\sigma = 0. \tag{8}$$

B.   The Interface

We define the amount of substance i contained in the bulk phase by

$$n_i^{\mathrm{b}} = c_i V. \tag{9}$$

The amount of substance n_i^σ contained in the interfacial phase, often called the interfacial excess of i, is $n_i - n_i^{\mathrm{b}}$.

The interfacial concentration (or the interfacial excess concentration) Γ_i is defined by $\Gamma_i = n_i^\sigma/A$. Thus

$$n_i = n_i^b + n_i^\sigma = c_i V + \Gamma_i A. \tag{10}$$

The definition of n_i^b in Equation (9) is not complete as long as the volume V is not defined precisely. In our thermodynamic description, an interface is a two-dimensional plane separating homogeneous bulk phases. In physical reality interfacial effects are spatially distributed and extend into the neighboring volume phases. Therefore the precise location of the interface and, consequently, the precise value of the volume of the bulk phase is a matter of definition.

Taking into account the special role of substance 1, we define the position of the dividing interface by

$$V = \frac{n_1}{c_1}. \tag{11}$$

This implies $n_1^b = n_1$ and $n_1^\sigma = 0$ or $\Gamma_1 = 0$. (See Defay [2] for more details, including the distinction between the *interface of tension* and the *dividing interface*.) With volume V given by Equation (11), the bulk value of any extensive variable X, related to the volume, is given by

$$X^b = x^b V, \tag{12a}$$

where x^b is the bulk concentration. As a generalization of Equation (10), the interfacial contribution to the extensive variable X, X^σ, and the interfacial concentration, x_A^σ, are given by

$$X^\sigma = X - X^b, \qquad x_A^\sigma = \frac{X^\sigma}{A}. \tag{12b}$$

According to Equation (11), we have $V = V^b$. Hence from Equation (12b)

$$V^\sigma = 0. \tag{12c}$$

With Equation (12b), dG and G now can be written as

$$dG = dG^b + dG^\sigma, \qquad G = G^b + G^\sigma, \tag{13a}$$

where G^b is the Gibbs free energy of an ordinary bulk phase. Accordingly

$$dG^b = -S^b\,dT + V\,dp + \sum \mu_i\,dn_i^b,$$
$$G^b = \sum \mu_i n_i^b. \tag{13b}$$

On subtracting these equations from Equations (4) and (6), we find

$$dG^\sigma = -S^\sigma\,dT + \sigma\,dA + {\sum}' \mu_i\,dn_i^\sigma, \quad G^\sigma = {\sum}' \mu_i n_i^\sigma + \sigma A. \tag{13c}$$

Because of $n_1^\sigma = 0$, summation over i here can be restricted to the range $i = 2$ to $i = n$, as indicated by ${\sum}'$. For the Helmholtz free energy, $F = U - TS$, we have, correspondingly

$$dF = dF^b + dF^\sigma, \quad F = F^b + F^\sigma \tag{14a}$$

with

$$dF^b = -S^b dT - p\,dV + \sum \mu_i\,dn_i^b, \quad F^b = \sum \mu_i n_i^b - pV, \tag{14b}$$

and

$$dF^\sigma = -S^\sigma\,dT + \sigma\,dA + {\sum}' \mu_i\,dn_i^\sigma, \quad F^\sigma = {\sum}' \mu_i n_i^\sigma + \sigma A. \tag{14c}$$

Note that, according to Equations (13c) and (14c), there is no difference between Gibbs and Helmholtz interfacial free energies, i.e.

$$dG^\sigma = dF^\sigma, \quad G^\sigma = F^\sigma. \tag{14d}$$

This also implies

$$H^\sigma = U^\sigma \tag{14e}$$

The Gibbs–Duhem equation of the bulk phase is

$$S^b\,dT - V\,dp + \sum n_i^b\,d\mu_i = 0. \tag{15}$$

We subtract this from Equation (8) and obtain

$$S^\sigma \, dT - A \, d\sigma + {\sum}' n_i^\sigma \, d\mu_i = 0. \tag{16}$$

This is the Gibbs–Duhem equation of the interface.

C. Gibbs' Adsorption Equation, Adsorption Isotherm, and Interfacial State Equation

Dividing Equation (16) by A gives

$$d\sigma = -s_A^\sigma \, dT - {\sum}' \Gamma_i \, d\mu_i, \tag{17a}$$

which is known as the Gibbs' adsorption equation [3]. Note that σ does not explicitly depend on p. At constant temperature we obtain

$$d\sigma = - {\sum}' \Gamma_i \, d\mu_i. \tag{17b}$$

For an ideal solution under constant pressure (by ideal solution we always mean an ideally diluted solution) this simplifies to

$$d\sigma = -RT {\sum}' \Gamma_i \, d \ln c_i. \tag{17c}$$

Hence, the dependence of the interfacial tension on the bulk concentrations c_i is regulated by the interfacial concentration Γ_i. If Γ_i is positive, substance i is called surface-active (or a surfactant), otherwise it is surface-inactive. In view of Equation (17a), σ can be written as

$$\sigma = \sigma(T, \mu_2, \ldots, \mu_n). \tag{18}$$

According to Equation (17a) Γ_i is the partial derivative of σ with respect to μ_i. Therefore it depends on the same independent variables as σ. Thus

$$\Gamma_i = \Gamma_i(T, \mu_2, \ldots, \mu_n), \qquad i = 2, 3, \ldots \tag{19a}$$

The set of equations given by Equation (19a) can be inverted to

$$\mu_i = \mu_i(T, \Gamma_2, \ldots, \Gamma_n), \qquad i = 2, 3, \ldots \tag{19b}$$

This result shows that μ_i, taken as a function of interfacial variables, can be written as a function of T and Γ_2 through Γ_n alone and does not depend on p. In bulk solution thermodynamics μ_i is written as

$$\mu_i = \mu_i(T, p, c_2, \ldots, c_n). \tag{19c}$$

According to Equations (19b) and (19c), chemical equilibrium between the interface and the bulk solution now can be expressed by

$$\mu_i(T, \Gamma_2, \ldots, \Gamma_n) = \mu_i(T, p, c_2, \ldots, c_n). \tag{19d}$$

On the left-hand side there are n independent variables, on the right-hand side, however, $n + 1$. Obviously, something is wrong. Recall that the interface is part of a two-phase system. Although, in our context, we (try to) ignore the second phase, it is still there. Gibbs' phase rule states that, at equilibrium, the number of independent intensive variables of the two-phase system is smaller by one than the number of the one-phase system. So actually one of the variables on the right-hand side of Equation (19d) is a function of the remaining ones and therefore should be removed from the list of independent intensive bulk variables of our system. Arbitrarily, we omit the pressure. Equations (19c) and (19d) now become

$$\mu_i = \mu_i(T, c_2, \ldots, c_n), \tag{19e}$$

$$\mu_i(T, \Gamma_2, \ldots, \Gamma_n) = \mu_i(T, c_2, \ldots, c_n). \tag{19f}$$

In a shorter notation we write the last equation as

$$\mu_i^{\sigma} = \mu_i^{b}, \tag{19g}$$

where μ_i^{σ} and μ_i^{b} are introduced to denote the chemical potential as a function of the intensive variables of the interfacial phase and the bulk phase, respectively. Due to Equation (19e), Equations (19a) and (19b) can be transformed into

$$\Gamma_i = \Gamma_i(T, c_2, \ldots, c_n), \qquad i = 2, 3, \ldots \tag{20a}$$

$$c_i = c_i(T, \Gamma_2, \ldots, \Gamma_n), \qquad i = 2, 3, \ldots \tag{20b}$$

These equations relate bulk concentrations to interfacial concentrations and are general forms of the adsorption isotherms of the system, the term isotherm reflecting that, in practical use, the Γ_i s and c_i's are usually treated as variables (while T is used as a parameter).

Inserting Equation (19b) in Equation (18) we obtain

$$\sigma = \sigma(T, \Gamma_2, \ldots, \Gamma_n) \tag{21}$$

which is the general form of the interfacial equation of state.

Changes of the interfacial tension due to adsorption can be expressed in terms of the interfacial pressure π defined by

$$\pi = \sigma_0 - \sigma, \tag{22}$$

where σ_0 is the interfacial tension of the "clean interface" (more precisely, at $\Gamma_2, \ldots, \Gamma_n = 0$). If interfacial concentrations are proportional to bulk concentrations, so that

$$\Gamma_i = a_i(T)c_i \tag{23a}$$

then Equation (17c) gives

$$\pi = RT \sum{}' \Gamma_i \tag{23b}$$

which is the two-dimensional equivalent of the ideal gas equation.

Note that, as a result of Equations (10) and (11), Γ_i can also be expressed as

$$\Gamma_i = \frac{\partial n_i}{\partial A}\bigg|_{c_i, V} = \frac{\partial n_i}{\partial A}\bigg|_{c_i, c_1, n_1} \tag{24a}$$

But c_1 will be constant if T and $c_2, \ldots, c_n$ are constant (again we omit the pressure as an independent variable, see discussion following Equation (19b)). Therefore, Equation (24a) can be rewritten as

$$\Gamma_i = \frac{\partial n_i}{\partial A}\bigg|_{T, n_1, c_{j \neq 1}}. \tag{24b}$$

This shows that Γ_i is the amount of substance i ($i \neq 1$) to be added to the system per change of interfacial area to preserve the intensive state of the system at constant n_1.

D. Sustained Equilibrium and Enthalpy of Adsorption

Because of $G^\sigma = H^\sigma - TS^\sigma$, Equation (13c) yields

$$\mu_i^\sigma = \left.\frac{\partial G^\sigma}{\partial n_i^\sigma}\right|_{T,A,n_{j\neq i}^\sigma} = h_i^\sigma - Ts_i^\sigma, \tag{25a}$$

where (cf. Equation (14e))

$$h_i^\sigma = \frac{\partial H^\sigma}{\partial n_i^\sigma} = \frac{\partial U^\sigma}{\partial n_i^\sigma}, \qquad s_i^\sigma = \frac{\partial S^\sigma}{\partial n_i^\sigma}. \tag{25b}$$

The partial derivatives are taken under the conditions specified in Equation (25a). Accordingly

$$\mu_i^b = \left.\frac{\partial G^b}{\partial n_i^b}\right|_{T,p,n_{j\neq 1}^b} = h_i^b - Ts_i^b. \tag{26}$$

Introducing the partial molar adsorption enthalpy

$$\Delta h_i^{ad} = h_i^\sigma - h_i^b, \tag{27a}$$

and the partial molar adsorption entropy

$$\Delta s_i^{ad} = s_i^\sigma - s_i^b, \tag{27b}$$

and using Equations (25a), (26), (27a), and (27b), the equilibrium condition of Equation (19g) can be written as

$$\Delta \mu_i^{ad} = \mu_i^\sigma - \mu_i^b = \Delta h_i^{ad} - T\Delta s_i^{ad} = 0. \tag{28}$$

Reversible adsorption processes therefore require $d(\Delta \mu_i^{ad}) = 0$ or

$$d\mu_i^b = d\mu_i^\sigma. \tag{29a}$$

If the bulk solution is ideal and $\Gamma_2, \ldots, \Gamma_n$ are kept constant, then, with Equation (19b), we obtain from Equation (29a) (under the neglect of pressure effects in the bulk solution):

$$-s_i^b dT + RT \, d\ln c_i = \frac{\partial \mu_i^\sigma}{\partial T}\bigg|_{\Gamma_2, \ldots, \Gamma_n} dT. \tag{29b}$$

By applying Schwarz' equality to Equation (13c) (with μ_i now denoted by μ_i^σ) one finds

$$\frac{\partial \mu_i^\sigma}{\partial T}\bigg|_{\Gamma_2, \ldots, \Gamma_n} = \frac{\partial^2 G^\sigma}{\partial T \partial n_i^\sigma} = -\frac{\partial S^\sigma}{\partial n_i^\sigma}\bigg|_{T, A, n_{j \neq i}^\sigma} = -s_i^\sigma,$$

where the definition of s_i^σ given in Equation (25b) is used With Equation (27b), Equation (29b) thus yields $\Delta s_i^{\text{ad}} \, dT = -RT \, d \ln c_i$ or, with Equation (28),

$$\Delta h_i^{\text{ad}} = -RT^2 \frac{\partial \ln c_i}{\partial T}\bigg|_{\Gamma_2, \ldots, \Gamma_n}. \tag{30}$$

By analogy with a quite similar equation relating the temperature derivative of the (partial) vapor pressure p_i to the (partial) heat of condensation, Equation (30) is called Clausius–Clapeyron equation. Often, the value obtained from the right-hand side is called the isosteric heat of adsorption.

The total differential enthalpy change produced by multicomponent adsorption can be written as

$$\Delta h^{\text{ad}} = \sum{}' h_i^{\text{ad}} \frac{d\Gamma_i}{d\Gamma_2}, \tag{31a}$$

where component 2 is used as a reference component. Δh^{ad} is called the differential molar enthalpy of adsorption (or the differential heat of adsorption). The integral specific enthalpy of adsorption now is

$$\Delta h_A^{\text{ad}} = \sum{}' \int_0^{\Gamma_i} \Delta h_i^{\text{ad}} \, d\Gamma_i = \int_0^{\Gamma_2} \Delta h^{\text{ad}} \, d\Gamma_2. \tag{31b}$$

For a more comprehensive treatment of adsorption enthalpies see Wittrock [4] and Seidel [5]

E. Nonequilibrium Interfaces and the Gibbs Free Energy of Adsorption

We define the Gibbs free energy of adsorption, ΔG^{ad}, as the difference between the actual free energy of the system, G, and the hypothetical value, G^{ref}, obtained at the same values of T, p, $n_1, \ldots, n_n$ and A without adsorption. In this reference case the interface is "clean," that is $\Gamma_2, \Gamma_3, \ldots \Gamma_n = 0$. Denoting the interfacial tension and the chemical potentials in the reference case by σ^{ref} and μ_i^{ref}, we obtain, with Equation (6):

$$\Delta G^{ad} = G - G^{ref} = \sum (\mu_i - \mu_i^{ref})n_i + (\sigma - \sigma^{ref})A. \qquad (32a)$$

Although σ^{ref} is related to vanishing interfacial excess concentrations, it is not necessarily the same as σ_0, the interfacial tension of the pure solvent. For σ_0 is an equilibrium tension, while σ^{ref} is related to a nonequilibrium situation. At this point some properties of interfacial phases should be pointed out. An interfacial phase is not conceivable without extended neighboring phases (it is nonautonomous). Consequently, the intensive interfacial state variables, such as σ and μ_i^σ, do not depend on temperature T and the interfacial concentrations $\Gamma_2, \ldots, \Gamma_n$ alone, but on the bulk concentrations $c_2, \ldots, c_n$ as well. The dependence on $c_2, \ldots, c_n$, however, does not become explicit if bulk and interface are in equilibrium, since the Γ_i's and the c_i's are related by Equations (20a) and (20b). In the preceding sections, we dealt with such equilibrium situations and, therefore, could ignore any explicit dependence on the bulk concentrations. As an example of a dynamic interfacial state, imagine an interface freshly created from a solution. In the first moment the composition of the interfacial region will not differ from that of the bulk liquid, thus $\Gamma_i = 0$ for all i. Yet, the composition of the bulk phase is different from the pure solvent and so the interfacial tension of the freshly created interface will be different from the equilibrium tension σ_0 of the pure solvent. So σ^{ref}, in general, will be different from σ_0. However in the case of an ideal solution, the composition of both the solution and the reference interfacial layer is nearly that of the pure solvent. Hence, the difference between σ^{ref} and σ_0 can be neglected. Let us further assume that the bulk phase is large ($n_i \gg n_i^\sigma$), then the sum $\sum (\mu_i - \mu_i^{\text{ref}}) n_i$ on the right-hand side of Equation (32a) can be shown to be vanishingly small. For ideal solutions Equation (32a) thus simplifies to

$$\Delta G^{\text{ad}} = (\sigma - \sigma_0)A. \tag{32b}$$

Thus, defining the specific Gibbs free energy of adsorption by $\Delta g_A^{\text{ad}} = \Delta G_{\text{ad}}/A$, we finally obtain

$$\Delta g_A^{\text{ad}} = \sigma - \sigma_0 = -\pi. \tag{32c}$$

According to the Gibbs' adsorption equation, Equation (17b), we further have at constant T (a tilde here is used to denote integration variables)

$$\Delta g_A^{\text{ad}} = (\sigma - \sigma_0) = \int_{\sigma_0}^{\sigma} d\tilde{\sigma} = -\sum{}' \int_0^{\mu_i^b} \tilde{\Gamma}_i d\tilde{\mu}_i. \tag{32d}$$

Integration by parts yields

$$\Delta g_A^{\text{ad}} = -\sum{}' \Gamma_i \mu_i^b + \sum{}' \int_0^{\Gamma_i} \tilde{\mu}_i d\tilde{\Gamma}_i = \sum{}' \int_0^{\Gamma_i} (\tilde{\mu}_i - \mu_i^b) d\tilde{\Gamma}_i. \tag{32e}$$

Thus, Δg_A^{ad} can be interpreted as the specific change of the Gibbs free energy when the amounts Γ_i are exchanged between a bulk phase with constant chemical potentials μ_i^b and the interface. At the beginning of the exchange process all interfacial concentrations are zero, at the end the interface has reached equilibrium with the bulk phase. Although this process involves nonequilibrium states, the above argument shows that, in the case of ideal solutions, the chemical potentials $\tilde{\mu}_i$'s of the interface can be identified with the equilibrium potentials given, for example, by Equation (19b).

In view of the close relation found in Equation (32d) between the energy of adsorption and the interfacial tension the latter is a variable of major importance for all adsorption processes, including adsorption at solid interfaces, where it may be of minor interest in its original meaning.

Interfacial tensions are often calculated on the basis of (imaginary) electrostatic charging and discharging processes. If, per interfacial area, a differential charge $d\gamma^\sigma$ is exchanged at a potential difference φ^σ between interface and bulk, then $d\sigma$ is given by the Lippmann equation

$$\mathrm{d}\sigma = -s_A^\sigma \mathrm{d}T - {\sum}' \Gamma_i \,\mathrm{d}\mu_i - \gamma^\sigma \,\mathrm{d}\varphi^\sigma \qquad (32f)$$

which is a generalization of Equation (17a).

III. ONE-COMPONENT ADSORPTION

A. General Relations

In this chapter, the general theory is applied to two-component ideal solutions. Substance 2, which might be a nonionic surfactant, is the only adsorbate. Recall that substance 1 is the solvent. For simplicity, we omit subscript 2. An inert substance 3 is tacitly assumed to keep the pressure constant (see discussion of independent intensive variables in Section II.B). The pressure is considered to be constant throughout the rest of this chapter and will no longer be used as a variable. From Equations (17c) and (22):

$$-\mathrm{d}\sigma = \mathrm{d}\pi = RT\Gamma \mathrm{d}\ln c. \qquad (33)$$

Accordingly, Γ can be determined from the slope of the σ (or π) versus $\ln(c/c_0)$ curve (where c_0 is a normalization concentration).

From Equation (30) and Equation (31a):

$$\Delta h^{\mathrm{ad}} = -RT^2 \frac{\partial \ln c}{\partial T}\bigg|_\Gamma. \qquad (34)$$

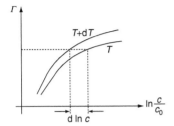

Figure 1.1 Schematic drawing illustrating the relation between $d \ln c$ and dT given by Equation (34). Shown are adsorption isotherms for neighboring temperatures T and $T + dT$.

Thus, Δh^{ad} can be obtained from adsorption isotherms determined for neighboring temperatures, see Figure 1.1.

It should be emphasized that knowledge of Δh^{ad} alone provides little information about the precise form of the adsorption isotherm. In fact, irrespective of Δh^{ad}, the relation between c, T and Γ (cf. Equation (20b)) can be freely chosen at some temperature T_0. This is seen more clearly by integrating Equation (34) at a given Γ between T_0 and T:

$$\ln c \left| \frac{c(\Gamma, T)}{c(\Gamma, T_0)} \right. = - \int_{T_0}^{T} \frac{\Delta h^{\text{ad}}}{RT^2} \, dT,$$

where $\Delta h^{\text{ad}} = \Delta h^{\text{ad}} \, (\Gamma, T)$, or

$$c(\Gamma, T) = c(\Gamma, T_0) \exp \left(- \int_{T_0}^{T} \frac{\Delta h^{\text{ad}}}{RT^2} \, dT \right). \tag{35}$$

As claimed above, the isotherm $c(\Gamma, T_0)$, playing the role of an *integration constant*, can be chosen at will.

According to Equation (28), the equilibrium of one-component adsorption is given by

$$\Delta \mu^{\text{ad}} = \mu^{\sigma} - \mu^{\text{b}} = \Delta h^{\text{ad}} - T \Delta s^{\text{ad}} = 0. \tag{36}$$

In the subsequent paragraphs we will investigate different types of one-component adsorption with the restriction to single-layer adsorption.

B. Langmuir Adsorption

The Langmuir isotherm describes adsorption to identical independent sites and is given by (cf. Equations (20a) and (20b)):

$$\Theta = \frac{\Gamma}{\hat{\Gamma}} = \frac{c}{c+K} \tag{37a}$$

or

$$c = K\frac{\Theta}{1-\Theta}, \tag{37b}$$

where $\hat{\Gamma}$ is the maximum interfacial concentration, K is the
dissociation constant (identical with the concentration of half
saturation), and $\Theta = \Gamma/\hat{\Gamma}$ is the degree of occupation. A convenient linear representation of the Langmuir isotherm is the
Γ^{-1} versus c^{-1} plot (Lineweaver–Burk plot):

$$\Gamma^{-1} = \hat{\Gamma}^{-1}(1 + Kc^{-1}) \tag{37c}$$

which is often used to test the agreement between Langmuir
isotherm and experimental data and also to determine the
parameters $\hat{\Gamma}$ and K by linear regression.

Assume that $\hat{\Gamma}$ is independent of temperature, while the
dependence of K is given by

$$K = K_0 e^{A/RT} \tag{38a}$$

with constant values of K_0 and A. Using Equation (34), one
can easily verify

$$A = \Delta h^{\mathrm{ad}}. \tag{38b}$$

The chemical potential of the adsorbate, in terms of the intensive variables of the bulk solution, is (cf. Equation (19e))

$$\mu = \mu^{\mathrm{b}}(T,c) = \mu_0(T) + RT\ln\frac{c}{c_0}. \tag{39a}$$

With Equation (37b) this gives (cf. Equation (19b)):

$$\mu = \mu^{\sigma}(T,\Gamma) = \mu_0 + RT\ln\frac{K}{c_0} + RT\ln\frac{\Theta}{1-\Theta}. \tag{39b}$$

Introducing the interfacial standard chemical potential

$$\mu_0^\sigma(T) = \mu_0 + RT \ln \frac{K}{c_0} \tag{40a}$$

and the interfacial activity coefficient

$$f^\sigma = \frac{1}{1 - \Theta}. \tag{40b}$$

Equation (39b) can also be written as

$$\mu = \mu^\sigma = \mu_0^\sigma + RT \ln(f^\sigma \Theta). \tag{41}$$

With μ^b and μ^σ given by Equations (39a) and (39b), Equation (36) is equivalent to the Langmuir isotherm. With K given by Equation (38a), Equation (36) yields

$$\Delta s^{\mathrm{ad}} = R \ln \left(\frac{c}{K_0} \frac{1 - \Theta}{\Theta} \right) \tag{42a}$$

or, in the standard case, $f^\sigma \Theta = 1$ (or $\Theta = 1/2$)

$$\Delta s_0^{\mathrm{ad}} = R \ln \frac{c_0}{K_0}, \tag{42b}$$

where K_0 is seen to have a purely entropic meaning. Note that the standard adsorption entropy increases with decreasing value of K_0.

The effect of Langmuir adsorption on the interfacial pressure can be calculated from Equation (33). With Equation (37a) and Equation (37b):

$$\sigma_0 - \sigma = RT \hat{\Gamma} \int_0^c \frac{dc}{c + K} = RT \hat{\Gamma} \ln \left(1 + \frac{c}{K} \right)$$

$$= -RT \hat{\Gamma} \ln(1 - \Theta). \tag{43a}$$

This equation is known as the Szyskowski equation. According to Equations (32d) and (32e), $\sigma_0 - \sigma$ can also be obtained from

$$\Delta g_A^{\mathrm{ad}} = (\sigma - \sigma_0) = \int_0^\Gamma (\tilde{\mu} - \mu^b) d\tilde{\Gamma}, \tag{43b}$$

where μ^b and $\tilde{\mu} = \tilde{\mu}^\sigma$ are given by Equations (39a) and (39b), respectively. The reader will find that calculation of $\sigma_0 - \sigma$ from Equation (43b) is much more complicated than the calculation presented above.

Equation (43a) is plotted in Figure 1.2. The interfacial tension σ becomes zero for $\pi = \sigma_0$, the corresponding monomer concentration is $c^* = K \cdot (\exp(\sigma_0/(RT\hat{\Gamma})) - 1)$. For aqueous surfactant solutions in contact with air experience shows that c^* is far above the critical micelle concentration (cmc). In contact with oil, however, $\sigma \approx 0$ may be reached. This is the basis of microemulsions between water and oil forming at extremely small interfacial tensions. With respect to suspensions of solid particles it should be noted that adsorption, in

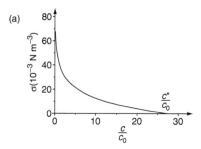

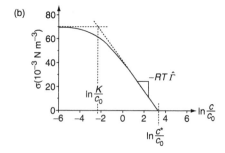

Figure 1.2 Relationship between interfacial tension and concentration in the case of Langmuir adsorption (Equation (43a)). $\sigma_0 = 70 \times 10^{-3}$ N/m, $\hat{\Gamma} = 5 \times 10^{-6}$ mol/m^2, $K = 0.1$ mol/m^3, $T = 300$ K, $c_0 = 1$ mol/m^3. (a) Linear concentration axis. (b) Logarithmic concentration axis.

lowering σ, reduces the driving force of Ostwald ripening and thus helps to stabilize suspensions of solid particles.

C. Temkin Adsorption

The isotherms known under the names of Temkin, Frumkin, and Fowler–Guggenheim are essentially the same. Although often mentioned, the physicochemical background of these isotherms mostly remains obscure [6–9]. They can be obtained from the Langmuir isotherm by introducing a molar interaction energy $\mu_{\text{inter}}^\sigma$, which is proportional to the degree of occupation Θ:

$$\mu_{\text{inter}}^\sigma = \hat{\mu}(T)\Theta. \tag{44}$$

The chemical potential μ^σ for Langmuir adsorption given by Equation (39b), now becomes

$$\mu^\sigma(T, \Gamma) = \mu_0^\sigma + RT\ln\frac{\Theta}{1 - \Theta} + \hat{\mu}\Theta \tag{45}$$

with μ_0^σ from Equation (40a). With the activity coefficient

$$f^\sigma = \frac{1}{1 - \Theta}\exp\frac{\hat{\mu}\Theta}{RT}.$$

Equation (45) can be given the form of Equation (41). On equating Equation (45) with Equation (39a) we obtain the adsorption isotherm

$$c = Ke^{\Theta\beta}\frac{\Theta}{1 - \Theta}, \tag{46a}$$

where

$$\beta = \frac{\hat{\mu}}{RT}. \tag{46b}$$

Let us write $\hat{\mu} = \hat{\mu}(T)$ as

$$\hat{\mu} = \hat{h} - T\hat{s}, \qquad \hat{h}, \hat{s} = const. \tag{47}$$

Using Equation (34), we find

$$\Delta h^{\text{ad}} = A + \hat{h}\Theta \tag{48a}$$

and from Equations (36) and (38a)

$$\Delta s^{ad} = R \ln\left(\frac{c}{K_0}\frac{1-\Theta}{\Theta}\right) + \hat{s}\Theta. \tag{48b}$$

Thus, with Equation (47), we are able to model purely ener-getic ($\hat{s} = 0$), purely entropic ($\hat{h} = 0$), and mixed interactions.

Figure 1.3a and b shows Temkin isotherms for different choices of $\hat{\mu}$. The Θ versus $\ln(c/K)$ isotherms are symmetric with respect to $\Theta = 0.5$. The critical values of Θ and β, where the slope becomes infinite, are

$$\Theta_c = 0.5 \tag{49a}$$

and

$$\beta_c = -4 \tag{49b}$$

corresponding to a critical temperature

$$T_c = -\frac{1}{4}\frac{\hat{\mu}}{R}. \tag{49c}$$

At $\beta < \beta_c$ parts of the isotherms have negative slope. A homo-geneously occupied interface then is unstable and phase separation will occur at some concentration c_{pc} (pc = phase change). To determine c_{pc}, we assume that T and p are con-stant and that no work is exchanged with the surroundings but volume work. Then the state of the materially closed system changes in such a way that

$$dG \leq 0. \tag{50a}$$

Now assume that the system has uniform chemical potential and that the interfacial area A is constant. If there are two interfacial phases I and II occupying the areas A_I and $A_{II} = A - A_I$, then according to Equations (4) and (50a), dG is given by

$$dG = \sigma_I dA_I + \sigma_{II} dA_{II} = (\sigma_{II} - \sigma_I)dA_{II} \leq 0. \tag{50b}$$

It follows that the interfacial phase with the lower interfacial tension is stable ($\sigma_{II} < \sigma_I$ implies $dA_{II} > 0$) and that equilib-rium between the interfacial phases requires

$$\sigma_{II} - \sigma_I = 0. \tag{51}$$

So both phases must have the same interfacial tension. According to Equation (36) they must also have the same chemical potentials. According to Equation (33) the interfacial tension is obtained from

$$\sigma = \sigma_0 - RT\,\hat{\Gamma} \int_{c=0} \Theta\,\mathrm{d}\ln c. \qquad (52)$$

Hence, because of Equation (51), c_{pc} must satisfy the condition (cf. Figure 1.3a)

$$\int_{B'}^{B''} \Theta\,\mathrm{d}\ln\frac{c}{K} = 0, \qquad (53a)$$

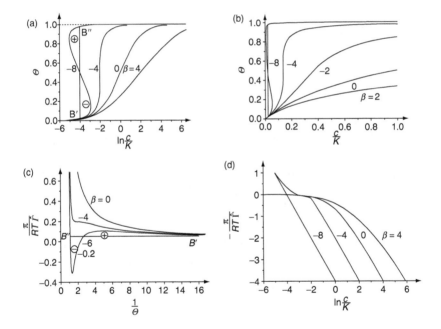

Figure 1.3 Temkin isotherms for different values of the interaction parameter $\beta = \hat{\mu}/(RT)$. (a) Θ versus $\ln(c/K)$; phase change at $\beta = -8$ along the vertical line. (b) Θ versus c/K; phase change at $\beta = -8$ along the vertical line. (c) $\pi/(RT\,\hat{\Gamma})$ versus $1/\Theta$; phase change at $\beta = -6$ along the horizontal line. (d) $-\pi/(RT\,\hat{\Gamma})$ versus $\ln(c/K)$; phase change at $\beta = -8$ at the intersection between the flat and the steep branch of the curve.

where integration from B' to B'' has to be carried out along the isotherm. Geometrically, Equation (53a) requires that the two areas labeled $+$ and $-$ in Figure 1.3a are equal. Therefore, the phase change line intersects the isotherm at $\Theta = \Theta_c = 0.5$. With Equation (46a) we obtain

$$c_{\mathrm{pc}} = K e^{\beta/2} \quad \text{for } \beta < -4. \tag{53b}$$

Equation (52) can also be used to calculate σ as a function of Θ (using Equation (46a) and $d \ln c = (d \ln c/d\Theta) \, d\Theta$). The result is

$$\pi = \sigma_0 - \sigma = RT \hat{\Gamma}\left(-\ln(1 - \Theta) + \frac{1}{2}\beta\Theta^2 \right). \tag{54a}$$

Inserting the critical values of the degree of occupation and the temperature (Equations (49a) and (49c)), the critical value of the interfacial pressure becomes

$$\pi_c = RT_c\hat{\Gamma}(\ln 2 - 0.5) = 0.193 \, RT_c\hat{\Gamma}. \tag{54b}$$

Figure 1.3c shows π versus Θ^{-1} curves obtained from Equation (54a). Θ^{-1} characterizes the interfacial area per molecule. These curves correspond to the p versus v curves ($v = $ molar volume) of a three-dimensional gas. The critical isotherm ($\beta = -4$) has a terrace point. This is the critical point. According to our previous result, Equation (51), the phase transition lines of the isotherms in Figure 1.3c must be drawn at constant π. The following argument will show that, irrespective of the particular type of isotherm, the areas labeled $+$ and $-$, here again, must be equal. According to Equation (17b) isothermal changes of the chemical potential are described by

$$d\mu = -\frac{d\sigma}{\Gamma} = \frac{1}{\hat{\Gamma}}\Theta^{-1}d\pi.$$

We have seen that interfacial phases coexisting under equilibrium conditions have identical values of the chemical potentials and the interfacial tension. Therefore the interfacial pressure of coexisting phases must satisfy the condition

$$\int_{B''}^{B'} d\mu = \frac{1}{\hat{\Gamma}} \int_{B''}^{B'} \Theta^{-1}d\pi = 0,$$

where integration is along the isotherm. This proves that the areas labeled $+$ and $-$ in Figure 1.3c must be equal.

Figure 1.3d shows π versus $\ln c/K$ isotherms calculated from Equations (54a) and (46a) (with Θ used as an auxiliary variable). Note that in the case $\beta = -8$ there is a phase change at the intersection between the flat and the steep branch of the curve. According to Equations (50b) and (51) these two branches are thermodynamically stable while the gusset contains unstable states. It can be shown that the states on the prolongation of the stable branches are metastable while those on the upper branch are absolutely unstable. More complete sets of Temkin isotherms, which might be used for curve fitting, are presented in Figure 1.4.

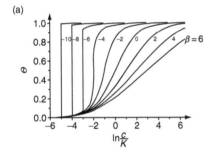

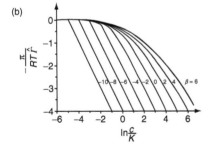

Figure 1.4 Set of Temkin isotherms; only thermodynamically stable states are shown. (a) Θ versus $\ln(c/K)$ (cf. Figure 1.3a). The slope at $\Theta = 0.5$ equals $1/(\beta + 4)$ (for $\beta > -4$). (b) $-\pi/(RT\,\hat{\Gamma})$ versus $\ln(c/K)$ (cf. Figure 1.3d).

D. van der Waals Adsorption

We start from the interfacial state equation

$$(\pi + \alpha \Gamma^2)\left(\frac{1}{\Gamma} - b\right) = RT, \tag{55a}$$

which is the two-dimensional analog of the van der Waals equation. The coefficients a and b are assumed to be constant and b to be nonnegative. With

$$\hat{\Gamma} = \frac{1}{b} \tag{55b}$$

and

$$\Theta = \frac{\Gamma}{\hat{\Gamma}}. \tag{55c}$$

Equation (55a) can be written as

$$\pi = RT \hat{\Gamma}\left(\frac{\theta}{1-\theta} - \frac{\alpha \hat{\Gamma}}{RT}\theta^2\right). \tag{55d}$$

Figure 1.5a shows π versus $1/\theta$ isotherms calculated from this relation.

One of the isotherms has a terrace point, which is the critical point. We introduce the abbreviations

$$\hat{\mu} = -2\alpha\hat{\Gamma} \quad \text{and} \quad \beta = \frac{\hat{\mu}}{RT} \tag{56a}$$

so that Equation (55d) becomes (cf. Equation (54a))

$$\pi = RT \hat{\Gamma}\left(\frac{\theta}{1-\theta} + \frac{1}{2}\beta\theta^2\right). \tag{56b}$$

From the terrace point (that is from $d\pi/d(\theta^{-1}) = d^2\pi/d^2(\theta^{-1}) = 0$) we obtain the critical values

$$\beta_c = -\frac{27}{4}\left(\text{or } T_c = -\frac{4}{27}\frac{\hat{\mu}}{R}\right),$$

$$\theta_c = \frac{1}{3}, \; \pi_c = 0.125 \, RT_c\hat{\Gamma}, \tag{56c}$$

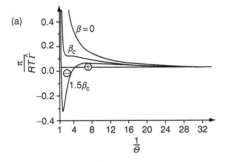

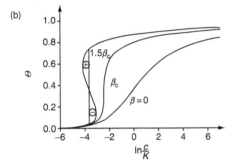

Figure 1.5 van der Waals adsorption. (a) $\pi/(RT\ \hat{\Gamma})$ versus $1/\Theta$ isotherms. (b) Θ versus $\ln(c/K)$ isotherms (Hill–de Boer). In both part figures a and b, the phase change line separates equal areas labeled + and −.

which should be compared with the values for Temkin adsorption, Equations (49a)–(49c) and (54b). As shown in Section III.C, the phase transition line must be drawn such that the areas labeled by + and − become equal.

For constant T Equation (55d) gives

$$-\mathrm{d}\sigma = \mathrm{d}\pi = \left(\frac{RT}{(1 - \Gamma/\hat{\Gamma})^2} - 2\alpha\Gamma\right)\mathrm{d}\Gamma.$$

On inserting this in Equation (33) one finds by integration (using partial fraction expansion)

$$c = K(T)e^{\beta\theta}\frac{\theta}{(1 - \theta)}e^{\theta/(1-\theta)}, \tag{57a}$$

where K is an integration constant, cf. Equation (46a). This adsorption isotherm is known by the name Hill–de Boer isotherm, see e.g., Leja [10], and is shown in Figure 1.5b. Using for K Equation (38a), we find, with Equation (34),

$$\Delta h^{\mathrm{ad}} = A + \hat{\mu}\theta \tag{57b}$$

which should be compared with Equation (48a). Note that $\hat{\mu}$, because of Equation (56a), is temperature-independent. Therefore the van der Waals equation implies $\hat{\mu} = \hat{h}$.

The above treatment shows that the van der Waals equation and the Temkin isotherm (Section III.C) give qualitatively similar results. Therefore, we omit further discussion of the van der Waals case.

E. Surface Tension of Micellar Solutions

We include a short discussion of the surface tension of an air–liquid interface of a micellar solution. Assume that a micelle M is formed from ν monomers of substance 2. The equilibrium condition is

$$\nu\,\mu_{\mathrm{mon}} = \mu_{\mathrm{M}}, \tag{58}$$

where the subscript "mon" is for the monomer and "M" for the micelle. The Gibbs' adsorption equation, taken at constant temperature, now reads

$$\mathrm{d}\sigma = -\Gamma_{\mathrm{mon}}\,\mathrm{d}\mu_{\mathrm{mon}} - \Gamma_{\mathrm{M}}\,\mathrm{d}\mu_{\mathrm{M}}. \tag{59a}$$

Due to Equation (58), we have $\mathrm{d}\,\mu_{\mathrm{M}} = \nu\,\mathrm{d}\mu_{\mathrm{mon}}$. Introducing the total interfacial concentration of component 2

$$\Gamma = \Gamma_{\mathrm{mon}} + \nu\Gamma_{\mathrm{M}}.$$

We obtain from Equation (59a)

$$\mathrm{d}\sigma = -\Gamma\mathrm{d}\mu_{\mathrm{mon}}. \tag{59b}$$

This shows that Γ is measurable, while Γ_{mon} and Γ_{M} are not measurable. Let us assume that the solution is ideal with respect to the surfactant found in the monomer state. Then Equation (59b) becomes

$$d\sigma = -RT\,\Gamma\,d\ln c_{\text{mon}}. \tag{59c}$$

At low total concentrations c, the monomer concentration c_{mom} equals c. This leads back to Equation (33). But at $c \gg cmc$, which implies $c_M \approx c/\nu$, Equation (58) yields $dc_{\text{mon}} \approx \nu^{-1}$ $d\ln c$, and, therefore, from Equation (59c)

$$d\sigma = -RT\frac{\Gamma}{\nu}d\ln c. \tag{59d}$$

Comparison with Equation (33) shows that the decrease of σ with increasing c now is reduced by a factor of $1/\nu$. If ν is known, Equation (59d) can be used to determine values of Γ above the cmc from interfacial tension versus concentration measurements.

IV. SOLUTION CONTAINING TWO NONIONIC ADSORBATES

We now consider three-component ideal solutions ($n = 3$) containing two nonionic adsorbates 2 and 3. The following treatment is focused on adsorption enthalpies. Section IV.A is devoted to general relations and Section IV.B to the special case of competitive Langmuir adsorption.

A. Differential Adsorption Enthalpy

According to Equation (31a), the differential molar adsorption enthalpy is

$$\Delta h^{\text{ad}} = \Delta h_2^{\text{ad}} + \Delta h_3^{\text{ad}}\frac{d\Gamma_3}{d\Gamma_2}. \tag{60a}$$

Let us consider here adsorption under the condition that c_2 is varied while c_3 and T are kept constant. Equation (60a) then becomes

$$\Delta h^{\text{ad}} = \Delta h_2^{\text{ad}} + \Delta h_3^{\text{ad}}\frac{\partial \Gamma_3}{\partial \Gamma_2}\bigg|_{T,\,c_3}. \tag{60b}$$

According to Equation (30), Δh_2^{ad} and Δh_3^{ad} are related to the adsorption isotherms by

$$\Delta h_i^{ad} = -RT^2 \frac{\partial \ln c_i}{\partial T}\bigg|_{\Gamma_2, \Gamma_3}. \tag{61}$$

To evaluate the right-hand side, the bulk concentrations must be varied as a function of temperature with both interfacial concentrations kept constant. Experimentally, this is a rather awkward condition. Therefore, let us investigate how the derivative

$$\frac{\partial \ln c_2}{\partial T}\bigg|_{\Gamma_2, c_3} \tag{62a}$$

is related to Δh^{ad} given by Equation (60b). Note that the derivative in Equation (62a) requires only a single interfacial concentration to be kept constant. From the theory of partial derivatives we have

$$\frac{\partial \ln c_2}{\partial T}\bigg|_{\Gamma_2, \Gamma_3} = \frac{\partial \ln c_2}{\partial T}\bigg|_{\Gamma_2, c_3} + \frac{\partial \ln c_2}{\partial \ln c_3}\bigg|_{T, \Gamma_2} \frac{\partial \ln c_3}{\partial T}\bigg|_{\Gamma_2, \Gamma_3}.$$

With Δh_i^{ad} from Equation (61) this transforms into

$$-RT^2 \frac{\partial \ln c_2}{\partial T}\bigg|_{\Gamma_2, c_3} = \Delta h_2^{ad} - \frac{\partial \ln c_2}{\partial \ln c_3}\bigg|_{T, \Gamma_2} \Delta h_3^{ad}. \tag{62b}$$

On the other hand, a Legendre transformation of the Gibbs adsorption equation, Equation (17c), gives

$$d\left(\sigma + RT\Gamma_2 \ln \frac{c_2}{c_0}\right) = RT\left(\ln \frac{c_2}{c_0} d\Gamma_2 - \Gamma_3 d\ln \frac{c_3}{c_0}\right),$$

which on applying Schwarz's equality to the right-hand side yields

$$-\frac{\partial \ln c_2}{\partial \ln c_3}\bigg|_{T, \Gamma_2} = \frac{\partial \Gamma_3}{\partial \Gamma_2}\bigg|_{T, c_3}. \tag{62c}$$

On inserting in Equation (62b) and comparing with Equation (60b), we find that Δh^{ad} is simply given by

$$\Delta h^{\mathrm{ad}} = -RT^2 \frac{\partial \ln c_2}{\partial T}\bigg|_{\Gamma_2,\, c_3}. \tag{63a}$$

This important result can be generalized to solutions containing n components: The adsorption enthalpy Δh^{ad} occurring under variation of c_2 at constant T and constant $c_3,..., c_n$ satisfies

$$\Delta h^{\mathrm{ad}} = -RT^2 \frac{\partial \ln c_2}{\partial T}\bigg|_{\Gamma_2,\, c_3,...,\, c_n}. \tag{63b}$$

Irrespective of n, the derivative on the right-hand side is obtained with a single interfacial concentration kept constant only.

B. Competitive Langmuir Adsorption

If two species 2 and 3 compete for the same sites, the Langmuir adsorption isotherms become

$$\Theta_2 = \frac{\Gamma_2}{\hat{\Gamma}} = \frac{c_2/K_2}{1 + (c_2/K_2) + (c_3/K_3)} \tag{64a}$$

and

$$\Theta_3 = \frac{\Gamma_3}{\hat{\Gamma}} = \frac{c_3/K_3}{1 + (c_2/K_2) + (c_3/K_3)}. \tag{64b}$$

With these relations the integration of Equation (17c) (along any path in the c_2, c_3 plane that starts at $c_2 = c_3 = 0$) yields (cf. Equation (43a))

$$\sigma_0 - \sigma = RT\hat{\Gamma} \ln\left(1 + \frac{c_2}{K_2} + \frac{c_3}{K_3}\right) \tag{65}$$
$$= -RT\hat{\Gamma} \ln(1 - \Theta_2 - \Theta_3).$$

Equations (64a) and (64b) can be easily solved for c_2 as a function of Γ_2 and c_3. Using this dependence and assuming that both K_2 and K_3 are given by Equation (38a), that is

$$K_2 = K_{20} e^{A_2/RT}, \qquad K_3 = K_{30} e^{A_3/RT} \tag{66}$$

we obtain, from Equation (63a), the differential adsorption enthalpy Δh^{ad} as defined by Equation (60b):

$$\Delta h^{ad} = \Delta h_2^{ad} - \frac{c_3/K_3}{1 + (c_3/K_3)} \Delta h_3^{ad}. \qquad (67)$$

Thus Δh^{ad} equals Δh_2^{ad} if c_3 is small, and equals $\Delta h_2^{ad} - \Delta h_3^{ad}$ if c_3 is large. As a consequence, Δh^{ad} easily changes sign as a function of c_3. The result for large c_3 reflects that, in the saturation range, an additional particle of species 2 can only adsorb if one of species 3 desorbs.

C. Dissociation from a Solid Surface with Independent Sites

Consider a soluble solid of composition AB (A, B electrically neutral) at constant temperature and pressure. In equilibrium, the concentrations of A and B in solution, c_A and c_B, are coupled by

$$c_A c_B = L, \qquad L = const. \qquad (68a)$$

When the area of the interface changes by dA, the concentrations in solution change by (materially closed system)

$$dc_A = -\frac{1}{V} \Gamma_A dA \quad \text{and} \quad dc_B = -\frac{1}{V} \Gamma_B dA. \qquad (68b)$$

Because of Equation (68a) these changes are coupled by

$$c_B d c_A + c_A d c_B = 0. \qquad (68c)$$

With Equation (68b), this leads to

$$\frac{\Gamma_B}{\Gamma_A} = \frac{dc_B}{d c_A} = -\frac{c_B}{c_A}. \qquad (68d)$$

Thus, the excess concentrations Γ_A and Γ_B are coupled to the concentrations of the solution and always have different signs. With Equation (68d) one obtains from the Gibbs adsorption isotherm, Equation (17c):

$$d\sigma = -RT\,\Gamma_A\left(1+\frac{c_B}{c_A}\right)d\ln c_A, \qquad (69a)$$

We now restrict considerations to the case $c_A \gg \sqrt{L}$, or $c_A \gg c_B$. Equation (69a) then simplifies to

$$d\sigma = -RT\,\Gamma_A\,d\ln c_A, \qquad (69b)$$

which is formally the isotherm for a one-component adsorption. Assume that there is a total concentration $\hat{\Gamma}$ of surface sites and that A has a stronger tendency to dissociate from the surface than B. According to Equation (68d) this means that Γ_A is negative. If adsorption of A is described by a Langmuir type isotherm, then Γ_A becomes

$$\Gamma_A = -\hat{\Gamma}+\hat{\Gamma}\frac{c_A}{c_A+K} = -\hat{\Gamma}\frac{K}{c_A+K}. \qquad (70a)$$

Obviously, Γ_A goes to zero (with interfacial tension $\sigma=\sigma_0$) when c_A becomes very large. On inserting Equation (70a) in the Gibbs adsorption equation (69b), we obtain

$$\sigma = \sigma_0 + RT\,\hat{\Gamma}\,\ln\frac{c_A}{K+c_A}. \qquad (70b)$$

So, this time, the interfacial tension increases with concentration. A graph is shown in Figure 1.6. Formally, at sufficiently low values of c_A, σ becomes negative. This means that low concentrations of A favor the formation of small particles (large interfacial area), while high concentrations favor Ostwald ripening. This is just the reverse effect of ordinary one-component adsorption. Using Equation (32c) we obtain from Equation (70b)

$$\Delta g_A^{ad} = RT\,\hat{\Gamma}\,\ln\frac{c_A}{K+c_A}. \qquad (70c)$$

so that, this time, Δg_A^{ad} increases with increasing c_A (cf. Equations (43a) and (43b)).

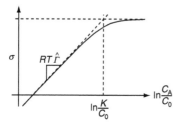

Figure 1.6 Surface tension of a solid, σ, as a function of $\ln(c_A/c_0)$; calculated from Equation (70b). The slope of the left asymptote is $RT\hat{\Gamma}$.

V. ADSORPTION FROM ELECTROLYTE SOLUTIONS

A. Symmetrical Electrolyte: General Relations

Bulk phases are electrically neutral. Therefore, if the system is electrically neutral as a whole, the interfacial phase must also be electrically neutral. Thus, adsorption from the solution of a single electrolyte resembles adsorption of a single neutral substance and can be treated as one-component adsorption. Yet a new aspect appears: since electrolytes are more or less dissociated, they are never ideally solved if taken as a single species. It is only the particles actually existing in solution that can exist in the state of ideal solution.

Dissociation equilibrium of the electrolyte is described by

$$\mu^{u} = \mu^{+} + \mu^{-}, \tag{71a}$$

where μ^{u}, μ^{+}, and μ^{-} are the chemical potentials of the undissociated electrolyte, the cation, and the anion, respectively. Let c^{u}, c^{+}, c^{-} and Γ^{u}, Γ^{+}, Γ^{-} denote the according bulk and interfacial concentrations and c the total electrolyte concentration. Due to electroneutrality

$$c^{+} = c^{-} \quad \text{and} \quad \Gamma^{+} = \Gamma^{-}. \tag{71b}$$

The total interfacial concentration of the electrolyte thus can be written as

$$\Gamma = \Gamma^{\mathrm{u}} + \Gamma^{+}. \tag{71c}$$

In terms of single-component adsorption, the Gibbs adsorption equation is

$$d\sigma = -\Gamma\, d\mu^{\mathrm{u}}. \tag{72a}$$

In terms of the particles actually existing in solution it becomes

$$d\sigma = -\Gamma^{\mathrm{u}}\, d\mu^{\mathrm{u}} - \Gamma^{+}\, d\mu^{+} - \Gamma^{-}\, d\mu^{-} = -\Gamma(d\mu^{+} + d\mu^{-}), \tag{72b}$$

where Equations (71a)–(71c) are used to establish the last equality. Of course, thermodynamically, Equations (72a) and (72b) are equivalent. If all components are ideally solved, Equations (72a) and (72b) give, with $d\mu^{+} = d\mu^{-} = RT\, d\ln c^{+}$ and $d\mu^{\mathrm{u}} = RT\, d\ln c^{\mathrm{u}}$:

$$d\sigma = -RT\,\Gamma\, d\ln c^{\mathrm{u}}, \tag{73a}$$

$$d\sigma = -2RT\,\Gamma\, d\ln c^{+} \tag{73b}$$

or, introducing the degree of dissociation α,

$$d\sigma = -RT\,\Gamma\, d\ln[(1-\alpha)c], \tag{74a}$$

$$d\sigma = -2RT\,\Gamma\, d\ln(\alpha c). \tag{74b}$$

For strong electrolytes ($\alpha = 1$) the last equation becomes

$$d\sigma = -2RT\,\Gamma\, d\ln c. \tag{75}$$

Compared with Equation (33), an additional factor of 2 has appeared on the right-hand side, which is due to the simultaneous adsorption of cation and anion (cf. Equation (72b)). Except for this factor, adsorption of strong electrolytes can be described by the same models as single-component adsorption (see previous chapter). On combining Equations (31), (30), and (71b) the differential heat of adsorption of a strong electrolyte becomes

$$\Delta h^{\mathrm{ad}} = -RT^{2}2\frac{\partial \ln c}{\partial T}\bigg|_{\Gamma}. \tag{76}$$

Again, there is an additional factor of 2 if compared with Equation (34).

B. Symmetrical Electrolyte: Gouy–Chapman–Stern Theory

Consider a plane interface with electrically neutral sites in contact with an ideal solution of a strong $(z, -z)$ electrolyte. Assume that the cation can directly adsorb to the sites located at $x = 0$. The positive charge accumulating in the interface modifies the electrical potential φ and the ionic concentrations c^+ and c^- in front of the interface. If the system is treated as one-dimensional, c^+, c^- and φ become functions of the normal coordinate x: $c^+ = c^+(x)$, $c^- = c^-(x)$, and $\varphi = \varphi(x)$ (see Figure 1.7). Assume that the cation exchange between the sites and the solution in the immediate neighborhood of the interface (at $x = 0$) is governed by a Langmuir isotherm. Then, denoting the interfacial concentration of the adsorbed cations by Γ_0^+, we obtain

$$\frac{\Gamma_0^+}{\hat{\Gamma}} = \Theta = \frac{c^+(0)}{K + c^+(0)}, \tag{77}$$

where $\hat{\Gamma}$ is the total interfacial concentration of the sites.

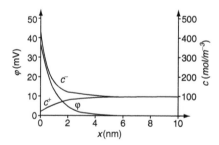

Figure 1.7 Profiles of cation and anion concentrations, c^+ and c^-, and of the electrical potential, φ, for an interfacial charge density $\gamma_0 = 0.03\,\text{As/m}^2$. Curves calculated from Gouy–Chapman theory for a 1, −1 electrolyte with bulk concentration 0.1 mol/l, $\varepsilon/\varepsilon_0 = 80$ ($\varepsilon_0 = $ vacuum permittivity) and $T = 298\,\text{K}$.

Under equilibrium conditions $c^\pm(0)$ is related to the bulk concentration $c(\infty)$ through the Boltzmann equation. Writing c instead of $c(\infty)$, we have

$$c^\pm(0) = e^{\mp zF\varphi(0)/(RT)}c \tag{78}$$

and Equation (77) can be rewritten as

$$\frac{\Gamma_0^+}{\hat{\Gamma}} = \Theta = \frac{c}{K\exp(zF\varphi(0)/(RT)) + c}, \tag{79a}$$

which is equivalent to

$$c = K\, e^{zF\varphi(0)/(RT)}\frac{\Theta}{1-\Theta}. \tag{79b}$$

Note the similarity with the Temkin isotherm in Equation (46a). The interfacial charge density is

$$\gamma_0 = zF\Gamma_0^+. \tag{80}$$

According to the Gouy–Chapman theory (see Chapter 2 in this volume) γ_0 is related to the interfacial potential, $\varphi(0)$, by

$$\gamma_0 = (8RT\varepsilon c)^{1/2}\sinh\frac{zF\varphi(0)}{2RT}, \tag{81}$$

where the dielectric constant of the solution, ε, is assumed to be independent of x. Combination of Equation (80) with Equations (79a) and (81) gives

$$zF\hat{\Gamma}\frac{c}{K\exp(zF\varphi(0)/(RT)) + c} = (8RT\varepsilon c)^{1/2}\sinh\frac{zF\varphi(0)}{2RT}, \tag{82}$$

which is an equation of the Gouy–Chapman–Stern type. With values of z, $\hat{\Gamma}$, K, ε, and $\varphi(0)$ given, this equation can be solved for c analytically. Then $c^\pm(0)$ and Γ_0^+ are obtained from Equations (78) and (79a). An example is shown in Figure 1.8.

As is seen from Figure 1.8b, the shape of the adsorption isotherm Γ_0^+ versus $\ln(c/K)$ does not differ too much from a simple Langmuir isotherm. It can be shown that for $\varphi(0) \gg RT/F$ the maximum of the $\varphi(0)$ versus $\ln(c/K)$ curve is at

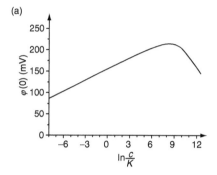

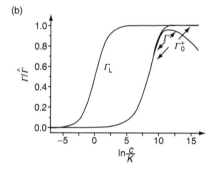

Figure 1.8 Potential-modified Langmuir adsorption: interfacial potential and concentrations caused by a cation adsorbing to interfacial sites from a symmetrical 1, -1 electrolyte solution of concentration c: $T = 298\,\text{K}$, $\varepsilon/\varepsilon_0 = 80$, $F\hat{\Gamma} = 0.5\,\text{As/m}^2$, $K = 10^{-6}\,\text{mol/l}$. (a) $\varphi(0)$ versus $\ln(c/K)$, calculated from Equation (82). (b) Γ_0^+ and Γ versus $\ln(c/K)$, calculated from Equations (79a) and (84). For comparison an ordinary Langmuir isotherm with the same parameter K, marked Γ_L, is shown.

$\Theta = 0.5$. As a consequence the adsorption isotherm is flatter than a Langmuir isotherm for $\Theta < 0.5$, where the negative cooperativity of charge accumulation prevails, and steeper for $\Theta > 0.5$, where screening by the ionic cloud is prevailing (high ionic strength).

It must be emphasized that the concentration Γ_0^+ differs from the interfacial excess concentration Γ^+ used in the thermodynamic theory. (The discrepancy can be immediately seen: Γ_0^+ and Γ_0^- are different (Γ_0^- is zero), while Γ^+ and Γ^-

must have the same value Γ (Equation (71b).) Γ can be calculated from the Gouy–Chapman theory using the relation

$$\Gamma = \int_0^\infty (c^-(x) - c)\mathrm{d}x, \tag{83}$$

which yields

$$\Gamma = \Gamma_0^+ \left(\frac{1}{2} + \frac{1}{2} \mathrm{tgh}\, \frac{zF\varphi(0)}{4RT} \right). \tag{84}$$

At low interfacial potentials, where the amount of counterion enrichment in the ionic cloud equals that of coion displacement, Γ is only one-half of Γ_0^+. At high interfacial potentials, however, the enrichment of counterions is much stronger than the displacement of coions (the latter is limited as concentrations cannot become negative). Therefore, in agreement with Equation (84), the ratio Γ/Γ_0^+ approaches 1. An adsorption isotherm Γ versus $\ln (c/K)$ is also shown in Figure 1.8.

C. Adsorption from Solutions with High Total Electrolyte Concentration

1. General Thermodynamic Relations

The essential thermodynamic features can be seen by considering a single directly adsorbing ionic species. We consider solutions of strong electrolytes with two cationic species, marked by prime and double prime, and a single anionic species. The cation marked by prime is assumed to adsorb strongly at the interface at low concentrations. All ions are assumed to be monovalent. The condition of electroneutrality of the bulk solution reads

$$c' + c'' = c^- = c, \tag{85}$$

where c is the total concentration of cations or anions. Assuming

$$c' \ll c'' \tag{86a}$$

we obtain with Equation (85)

$$c' \approx c^- = c \tag{86b}$$

and thus from Equation (17c)

$$d\sigma = -RT\,\Gamma'd\ln c' - RT(\Gamma'' + \Gamma^-)d\ln c. \tag{87a}$$

At constant c this simplifies to

$$d\sigma = -RT\,\Gamma'd\ln c'. \tag{87b}$$

At the same time we have from Equation (63b)

$$\Delta h^{\mathrm{ad}} = -RT^2\frac{\partial \ln c'}{\partial T}\bigg|_{\Gamma'}. \tag{88}$$

It can be inferred from Equations (87b) and (88) that adsorption at a constant ion concentration c and small adsorbate concentration $c' \ll c$ formally resembles that of a neutral species.

2.   Gouy–Chapman–Stern Adsorption

(a) Adsorption of a single ion

Following the same lines of argument as in Section V.B and using the notations and assumptions of the preceding Section V.C.1 (particularly Equation (86a)), we consider the case that adsorption of the cation prime follows a Langmuir isotherm relating Γ_0' to the concentration $c'(0)$. If the sites exist in a concentration $\hat{\Gamma}$, we obtain with $c' = c'(\infty)$ (cf. Equations (78)–(81))

$$c'(0) = e^{-zF\varphi(0)/(RT)}c', \tag{89}$$

$$\Theta = \frac{\Gamma_0'}{\hat{\Gamma}} = \frac{c'}{K\exp\left(F\varphi(0)/(RT)\right) + c'}, \tag{90}$$

$$\gamma_0 = F\Gamma_0', \tag{91}$$

$$\gamma_0 = (8RT\varepsilon c)^{1/2}\sinh\frac{F\varphi(0)}{2RT}. \tag{92}$$

The concentration c under the square root of Equation (92) now is the total ion concentration in the bulk (while it was the adsorbate concentration in Equation (81)). Instead of Equation (84) we have

$$\Gamma' = \Gamma'_0, \tag{93}$$

which emphasizes the similarity with the adsorption of a neutral species.

Isotherms calculated from Equations (89)–(92) with c kept constant are shown in Figure 1.9. (For given values of the interfacial potential $\varphi(0)$ the corresponding value of c can be easily calculated from the combination of Equations (90)–(92).) Note that, contrary to Figure 1.8, the curve Γ_0' versus $\ln(c'/K)$ is much flatter than a Langmuir isotherm, since, this time, the screening effect of an increasing ionic strength is missing. Accordingly, there is no maximum in $\varphi(0)$. The Γ_0' versus $\ln(c'/K)$ isotherm of Figure 1.9 can be approximated by a Temkin isotherm with interaction parameter $\beta \approx 3$ (cf. Figure 1.4a). Generally, in comparison with neutral adsorb-

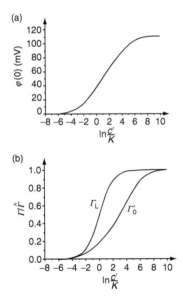

Figure 1.9 Same situation as in Figure 1.8, but total electrolyte concentration kept constant at $1\,\mathrm{mol/l}$. The concentration of the adsorbing cation, c', is varied.

ates, the negative cooperativity in the diffusive double-layer impedes the occurrence of phase changes in the interface as well as the formation of micellar aggregates in its immediate neighborhood.

In the examples of this section and the last, we have combined Langmuir adsorption with Gouy–Chapman theory. Replacing the Langmuir isotherm by a Temkin or some other isotherm, various types of interaction of adsorbing ionic particles can be modeled [11].

(b) *Simultaneous adsorption of a cation and an anion: the isoelectric point*

Particles suspended in a solution reach the isoelectric point if the electrical potential in their interface of shear becomes zero (see zeta potential in Chapter 2 in this volume). In terms of the present Gouy–Chapman–Stern model the isoelectric point coincides with the condition of zero interfacial potential. Consider competitive Langmuir adsorption of a monovalent cationic and a monovalent anionic species, marked by plus prime and minus prime (c_+', c_-'). Then (cf. Equations (64a) and (64b)):

$$\Gamma^+ = \hat{\Gamma} \frac{c_+'(0)/K_+'}{1 + (c_+'(0)/K_+') + (c_-'(0)/K_-')}, \tag{94a}$$

$$\Gamma^- = \hat{\Gamma} \frac{c_-'(0)/K_-'}{1 + (c_+'(0)/K_+') + (c_-'(0)/K_-')}. \tag{94b}$$

Denoting the bulk concentrations of the cation and the anion by c_+' and c_-', we have

$$c_\pm'(0) = e^{\mp F\varphi(0)/(RT)} c_\pm'. \tag{95}$$

If the empty sites are electrically neutral, the interfacial charge density γ_0 is

$$\gamma_0 = F(\Gamma^+ - \Gamma^-). \tag{96}$$

Under isoelectric conditions, i.e., when $\varphi(0)=0$, we have $c_\pm'(0)=c_\pm'$ (Equation (95)). But $\varphi(0)=0$ implies $\gamma_0=0$ (Equation (92)). From Equations (94a), (94b), and (96) we then find

$$\frac{c'_-}{c'_+} = \frac{K'_-}{K'_+}. \tag{97}$$

Thus, at the isoelectric point, the concentration ratio c'_-/c'_+ equals the ratio of the corresponding dissociation constants.

REFERENCES

1. R. Aveyard and D.A. Haydon, *An Introduction to the Principles of Surface Chemistry*, Cambridge University Press, Cambridge, 1973, pp. 1–30.
2. R. Defay, I. Prigogine, and A. Bellemans, *Surface Tension and Adsorption*, Longmans, London, 1966, pp. 1f, 71f.
3. H.T. Davis, *Statistical Mechanics of Phases, Interfaces and Thin Films*, VCH Publishers, New York, 1996, p. 358.
4. C. Wittrock, H.-H. Kohler, and J. Seidel, *Langmuir*, 12, 5550–5556, 1966.
5. J. Seidel, C. Wittrock, and H.-H. Kohler, *Langmuir*, 12, 5557, 1966.
6. A.W. Adamson, *Physical Chemistry of Surfaces*, John Wiley & Sons, New York, 1982, pp. 216, 530, and 575.
7. R.J. Hunter, *Foundations of Colloid Science*, Vol. II, Clarendon Press, Oxford, 1989, p. 712.
8. M.J. Jaycock and G.D. Parfitt, *Chemistry of Interfaces*, Ellis Horwood, Chichester, 1981, p. 192.
9. G.D. Parfitt and C.H. Rochester, Eds., *Adsorption from Solution at the Solid/Liquid Interface*, Academic Press, London, 1983, p. 256.
10. J. Leja, *Surface Chemistry of Froth Flotation*, Plenum Press, New York, 1982, p. 370.
11. A. Semmler and H.-H. Kohler, *Colloid Polym. Sci.*, 278, 911, 2000.

2

Surface Charge and Surface Potential

H.-H. Kohler and S. Woelki

Institute of Analytical Chemistry, Chemo- and Biosensors, University of
Regensburg, D-93040 Regensburg, Germany

I. INTRODUCTION

Surface charges play an important role in translocation and aggregation processes. The following treatment is restricted to uniformly charged particles immersed in ideally diluted electrolyte solutions. After treating the Gouy–Chapman theory of an isolated planar charged surface, we will consider the DLVO theory (named after Derjaguin, Landau, Verwey, and Overbeek) describing, in an approximate manner, the interaction between two identical charged surfaces. The theory incorporates electro-osmotic repulsion (which is due to the ionic cloud between the particles) and van der Waals attraction. After these thermodynamic considerations we turn to electrokinetics and consider the electrical potential of the surface of shear (called the zeta potential) of particles moving under the influence of a uniform external electrical field. An attempt is made to work out the theoretical background of the equations commonly used in practice.

II. THE DIFFUSE DOUBLE LAYER

A. General Features and Basic Equations of the Gouy–Chapman Model

The model system consists of a plane charged surface with a homogeneously distributed surface charge density γ_0 and an ideal electrolyte solution in front. The coordinate normal to the surface is x, where $x = 0$ denotes the location of the surface (Figure 2.1). The solution in front of the surface contains a positive excess of counterions and a negative excess of coions that compensate the surface charge. Due to thermal agitation the counterions are spatially distributed and give rise to a diffuse ionic cloud. Surface and ionic cloud form a diffuse electrical double layer.

Considering the problem as one-dimensional in x, we come to the Gouy–Chapman model of the diffuse double layer [1–4]. The Gouy–Chapman theory (set up in 1910) is the one-dimensional analog, in Cartesian coordinates, of the Debye–Hückel theory. Due to this one-dimensionality, the

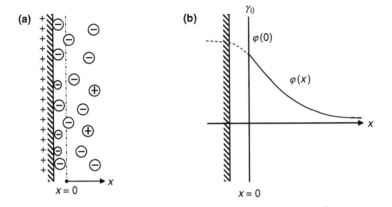

Figure 2.1 Schematic representation of charge and potential distribution in front of a (positively) charged surface. (a) Particle surface with (partially) dehydrated, small, and hydrated, large, counterions directly adsorbed to the surface and a diffuse cloud of ions in front. x is the normal coordinate. (b) One-dimensional potential profile $\varphi(x)$; γ_0 is the surface charge density.

Gouy–Chapman theory gets along without linearization. The electrical potential is denoted by φ, the electrical space charge density by ρ and the molar concentration of the ions of species i by c_i.

The entire system $(-\infty < x < \infty)$ is electrically neutral. Therefore we have

$$\int_{-\infty}^{\infty} \rho(x)\mathrm{d}x = 0. \tag{1}$$

We define the surface charge density γ_0 by

$$\gamma_0 = \int_{-\infty}^{0^+} \rho(x)\mathrm{d}x. \tag{2a}$$

According to this definition, the precise location of the charges contributing to γ_0 is of no importance as long as they are located at negative values of x. Surface charges located in the plane $x = 0$ itself are included in γ_0. Therefore the upper limit in Equation (2a) is denoted by 0^+. Because of Equation (1), γ_0 can also be expressed by

$$\gamma_0 = - \int_{0^+}^{\infty} \rho(x) dx. \tag{2b}$$

In the following the superscript $+$ of 0^+ will be omitted.

There are two basic equations to describe the electrochemical state in the solution range, $x \geq 0$. First, the Poisson equation, which, with the dielectric constant ε assumed to be independent of x, reads

$$\varphi''(x) = -\frac{\rho(x)}{\varepsilon}. \tag{3a}$$

The prime denotes derivation with respect to x. The dielectric constant may be written as $\varepsilon = \varepsilon_r \varepsilon_0$, where ε_0 is the vacuum permittivity and ε_r the relative dielectric constant of the solvent. The Poisson equation relates the curvature of the potential profile to the space charge density $\rho(x)$, which, in turn, is related to the ion concentrations by

$$\rho(x) = \sum_i z_i F c_i(x), \tag{3b}$$

where z_i is the valence and F is the Faraday constant. Inserting this relation in Equation (3a) we obtain

$$\varphi''(x) = -\frac{1}{\varepsilon} \sum_i z_i F c_i(x). \tag{3c}$$

The second basic relation is the thermodynamic equilibrium condition of constant electrochemical potentials. In the case of an ideal solution it is sufficient to consider the equilibrium of the ionic species, which, disregarding volume effects, can be expressed by the "Boltzmann equation"

$$c_i(x) = e^{-z_i F \varphi(x)/(RT)} c_i(\infty) \tag{4}$$

Actually, concentrations may become very high in the ionic cloud. Therefore the neglect of nonideality and volume effects is a severe shortcoming of the Gouy–Chapman theory.

We assume that the electrical potential of the bulk solution, $\varphi(\infty)$, is zero:

$$\varphi(\infty) = 0 \quad \text{with} \quad \varphi'(\infty) = 0. \tag{5a}$$

This is the first boundary condition. A second boundary condition is obtained by inserting Equation (3a) in Equation (2b) and subsequent integration. With Equation (5a), we find

$$-\varphi'(0) = \frac{1}{\varepsilon}\gamma_0. \tag{5b}$$

Thus, with ε given, the field strength in front of the surface is determined by the surface charge density alone.

According to the above relations, the position of the surface plane, $x = 0$, can be freely chosen, provided that for $x > 0$ the solution is ideal with constant ε. Hence, the electrostatic surface plane may be chosen some way out in the solution. Observe that, according to Equations (2a) and (2b), the value of γ_0 depends on the choice of $x = 0$.

B. General Solutions

On inserting Equation (4) in Equation (3c) a differential equation in φ of the second order is obtained:

$$\varphi''(x) = -\frac{1}{\varepsilon}\sum_i z_i F e^{-z_i F \varphi(x)/(RT)} c_i(\infty). \tag{6}$$

Hence, two integrations over x have to be carried out to obtain φ or any of the concentrations c_i as a function of x. With the identity

$$\varphi'' = \frac{d\varphi'}{dx} = \frac{d\varphi'}{d\varphi}\frac{d\varphi}{dx} = \frac{d\varphi'}{d\varphi}\varphi'. \tag{7}$$

Equation (6) transforms into

$$\varphi' d\varphi' = -\frac{1}{\varepsilon}\sum_i z_i F e^{-z_i F \varphi/(RT)} c_i(\infty) d\varphi.$$

With the boundary conditions of Equation (5a), one obtains, on integrating from infinity to x

$$\int_0^{\varphi'} \tilde{\varphi}' \, d\tilde{\varphi}' = -\frac{1}{\varepsilon} \sum_i z_i F c_i(\infty) \int_0^{\varphi} e^{-z_i F \tilde{\varphi}/(RT)} d\tilde{\varphi}. \tag{8a}$$

The result is

$$\frac{1}{2}(\varphi')^2 = \frac{RT}{\varepsilon} \sum_i c_i(\infty)(e^{-z_i \psi} - 1), \tag{8b}$$

where

$$\psi = \frac{F\varphi}{RT} \tag{9a}$$

is used. At $T = 298\,\mathrm{K}$ the relationship between ψ and φ becomes

$$\psi = \frac{\varphi}{25.7\,\mathrm{mV}}. \tag{9b}$$

For $x = 0$ Equation (8b) yields, with Equation (5b),

$$\gamma_0^2 = 2\,RT\varepsilon \sum_i c_i(\infty)(e^{-z_i \psi(0)} - 1) \tag{10a}$$

or alternatively, with Equation (4)

$$\gamma_0^2 = 2\,RT\varepsilon \sum_i (c_i(0) - c_i(\infty)). \tag{10b}$$

Equation (10a) relates the surface potential $\varphi(0)$ to the surface charge density γ_0, while Equation (10b) shows that the total ion excess concentration in front of the surface, $\sum(c_i(0) - c_i(\infty))$, is determined by the surface charge density and does not depend on the valences of the ions. At high absolute values of γ_0, the surface concentrations of counterions become large and Equation (10b) can be approximated by

$$\gamma_0^2 = 2\,RT\varepsilon \sum_i' c_i(0). \tag{10c}$$

The prime here denotes summation over all counterion species. Thus, if surface potentials are high, the total counterion concentration in front of the surface, is determined by γ_0 alone. This shows that a charged surface has a tendency to create an ionic environment of its own.

Let us consider in more detail the case of small surface potentials given by

$$\psi(0) \ll 1. \tag{11}$$

Then $\exp{(z_i\,\psi(0))}$ can be approximated by

$$e^{z_i\psi(0)} = 1 + z_i\psi(0) + \frac{1}{2}z_i^2\psi^2(0). \tag{12}$$

Inserting this in Equation (10a) we find

$$\gamma_0 = \varepsilon\frac{1}{\kappa}\varphi(0), \tag{13a}$$

where κ is the Debye–Hückel coefficient

$$\kappa = \left(\frac{F^2\sum z_i^2 c_i(\infty)}{RT\varepsilon}\right)^{1/2}. \tag{13b}$$

Formally, Equation (13a) resembles the equation relating charge density and voltage of a capacitor. This is the basis of the Helmholtz model of the diffuse double layer. In this model the countercharge, with a charge density $-\gamma_0$, is thought to be located in a plane at a distance $1/\kappa$ from the surface (which leads to Equation (13a)). Note that, according to Equation (11), the Helmholtz model is only valid for very small surface potentials.

So far, a first integration of Equation (6) has been achieved. To obtain φ as a function of x, a second integration, based on Equation (8b), must be carried out. Equation (8b) is a nonlinear first-order differential equation for $\varphi(x)$. In general it does not have a closed solution. A closed solution, however, does exist in the case of symmetrical electrolytes. This case will be considered next.

C. Symmetrical Electrolyte

Consider a single $(z, -z)$ electrolyte with bulk concentration $c(\infty)$. Equation (10a) than can be written as

$$\gamma_0 = (8\,RT\varepsilon c(\infty))^{1/2} \sinh\frac{z\psi(0)}{2}, \tag{14}$$

while Equation (8b) yields the differential equation

$$\varphi' = \left(\frac{8\,RTc(\infty)}{\varepsilon}\right)^{1/2} \sinh\frac{zF\varphi}{2\,RT}. \tag{15}$$

The solution of this equation is

$$\varphi = \left(\frac{2\,RT}{zF}\right)\ln\frac{1+y_0 e^{-\kappa x}}{1-y_0 e^{-\kappa x}}, \tag{16a}$$

where

$$y_0 = \frac{e^{z\psi(0)/2}-1}{e^{z\psi(0)/2}+1} = \mathrm{tgh}\frac{z\psi(0)}{4}, \tag{16b}$$

while κ, according to Equation (13b), now is

$$\kappa = \left(\frac{2F^2 z^2 c(\infty)}{RT\varepsilon}\right)^{1/2}. \tag{16c}$$

Note that y_0 tends to ± 1 when $\psi(0)$ approaches $\pm\infty$. Denoting the concentrations of cation and anion by c^+ and c^-, respectively, we obtain from Equation (4) with Equation (16a)

$$c^{\pm} = c(\infty)\left(\frac{1-y_0 e^{-\kappa x}}{1+y_0 e^{-\kappa x}}\right)^{\pm 2}. \tag{17}$$

Figure 2.2 shows examples of potential and concentration profiles calculated from Equations (16a), (16b), and (17) in combination with Equation (14).

In the validity range of the Helmholtz model $(\psi(0)\ll 1)$, Equation (16b) reduces to

$$y_0 = \frac{z\psi(0)}{4}. \tag{18a}$$

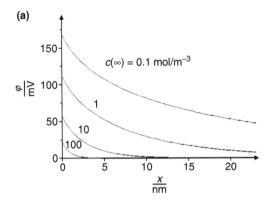

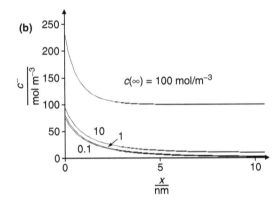

Figure 2.2 Potential and counterion concentration profiles in front of a surface with charge density $\gamma_0 = 0.016\,As/m^2$ (corresponding to one unit charge per $10\,nm^2$) in aqueous solutions of a $(1,-1)$ electrolyte with bulk concentrations $c(\infty)$ of 0.1, 1, 10, and $100\,mol/m^3$ ($T = 293\,K$, $\varepsilon_r = 78.61$). (a) Potential profiles: Note that all curves have the same slope at $x = 0$ (Equation (7b)). At sufficiently high values of $\varphi(x)$, the curves are displaced in parallel by $\Delta\varphi = (R \times 293\,K/F)\,\ln 10 = 58\,mV$. (b) Counterion concentration profiles: With decreasing $c(\infty)$, the surface concentration $c^-(0)$ falls to a limiting value given by Equation (11). In the scale chosen, differences between the concentration profiles for bulk concentrations of 1.0 and $0.1\,m\,mol/l$ can be hardly seen. This illustrates the tendency of a charged surface to create an ionic atmosphere of its own at low electrolyte concentrations.

Equations (16a) and (17) become

$$\varphi = \varphi(0)e^{-\kappa x} \tag{18b}$$

and

$$c^{\pm} = c(\infty)(1 \mp z\psi(0)e^{-\kappa x}). \tag{18c}$$

Thus, in the limit of small surface potentials, potential, and concentrations decline exponentially with x. In this range of low surface potentials, the Debye length $1/\kappa$ equals the mean distance $\langle x \rangle$ of the countercharge from the surface. This justifies the name of *thickness of the ionic cloud* given to $1/\kappa$. However, at higher surface potentials the mean distance $\langle x \rangle$ is different from $1/\kappa$. Generally it is given by

$$\langle x \rangle = \frac{\int_0^{\infty} x\rho(x)\mathrm{d}x}{\int_0^{\infty} \rho(x)\mathrm{d}x}. \tag{19}$$

With Equations (3b) and (17), we find, after some calculation (including partial integrations and repeated use of the addition formulas of hyperbolic functions)

$$\langle x \rangle = \frac{1}{\kappa}\frac{z\psi(0)/2}{\sinh(z\psi(0)/2)} = \frac{2\,RT\varepsilon}{zF\gamma_0}\mathrm{arcsinh}\frac{\gamma_0}{(8\,RT\varepsilon c(\infty))^{1/2}}. \tag{20a}$$

To obtain the last relation Equation (14) was used. Notice that, at a given value of $c(\infty)$, $\langle x \rangle$ decreases with increasing $|\psi(0)|$. Therefore, at very high values of the surface potential the mean distance of the countercharge is much smaller than $1/\kappa$ and approximately given by (cf. Figure 2.2b)

$$\langle x \rangle = \frac{1}{\kappa}z|\psi(0)|e^{-z|\psi(0)|/2}. \tag{20b}$$

At large distances, where $\psi(x) \ll 1$, Equation (16a) can be approximated by

$$\psi(x) = \frac{4}{z}y_0 e^{-\kappa x}. \tag{21}$$

Recall that $|y_0|$ approaches 1 for high values of $|\psi(0)|$. Equation (21) thus shows that, in the case of large surface potentials, the potential profile found far away from the surface does not provide information about the precise value of $\varphi(0)$.

D. Improvements of the Gouy–Chapman Model

There are two different approaches for the description of the electrical double layer: The classical local balance approach based on the thermodynamic condition of constant electrochemical potentials of the components and leading, in its simplest form, to the Gouy–Chapman theory; and the more fundamental statistical mechanics approach using pair correlation functions. An upto-date description of the latter is found in Refs. [5–7]. Because of a variety of mathematical and computational problems, the statistical mechanics approach is still rather academic and not too useful for practical purposes. On the other hand, the local balance approach widely ignores the discrete nature of matter, leading, for instance, to questionable results regarding the counterion concentration profiles in the immediate neighborhood of the interface. One may expect, however, that reasonable results are obtained for integral quantities like the electrostatic free energy per surface area.

Making use of the local balance approach, one may go beyond the Gouy–Chapman theory by choosing more appropriate expression for the electrochemical potentials of the components than the simple Boltzmann term. An entire spectrum of corrections of the ordinary Poisson–Boltzmann theory has been proposed by Bell and Levine [8], which include excluded volume effects, dependence of permittivity on composition and field strength, polarization effects, self-atmosphere and image forces, and curvature effects. A comprehensive local balance treatment of all these effects comprising theoretical considerations and numerical calculations can be found in Refs. [9,10]. At low surface charge densities (below $0.15\,As/m^2$ for a plane interface) small deviations from the Gouy–Chapman results are found. At higher surface charge densities, however, there is a strong synergism between excluded volume effects and dielectric saturation, leading to strong specific counterion effects (see Figure 2.3). This finding might be a clue to the problem, often unsolved, of

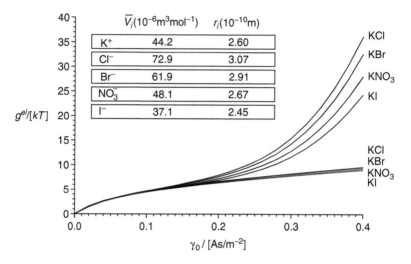

Figure 2.3 Electrostatic free energy g^{el} (free energy divided by number of interfacial elementary charges) at $T = 298\,K$ of a plane, positively charged surface separating an aqueous solution from a solid (with a relative dielectric constant of 2.0) as a function of surface charge density γ_0 for different sodium salts, from Woelki and Kohler [10]. Bulk electrolyte concentration $c = 0.01\,mol/l$. The upper curves are calculated from the extended local balance model in Ref. [9], the lower curves from an ordinary Gouy–Chapman–Stern model. $\bar{V}_i$ is the volume of the hydrated species, r_i the corresponding radius. Note the increase in energy and ion specificity in the extended model at higher surface charge densities.

counterion specificity. Another remarkable result is that at high surface charge densities small counterions displace larger ones from near the surface even if the latter have higher charge. This casts some doubt on the dominating role generally attributed to polyvalent ions in the neighborhood of highly charged surfaces.

III. REPULSION BETWEEN PLANAR SURFACES: DLVO THEORY

Figure 2.4 shows a schematic drawing of the potential distribution between two identical positively charged flat plates

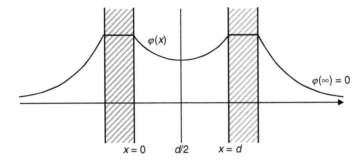

Figure 2.4 Schematic drawing of the potential profile between two positively charged plates in an electrolyte solution.

immersed in a symmetrical electrolyte solution. If the distance d between the plates is finite, there is an excess of anions (counterions) between the plates at $x = d/2$. Therefore, as seen from the Boltzmann equation, Equation (4), the potential $\varphi(d/2)$ is positive. Also, because of an excess in total ion concentration, there is an osmotic pressure given by

$$\Delta p = p(d/2) - p(\infty)$$
$$= RT(c^+(d/2) + c^-(d/2) - 2c(\infty)), \qquad (22a)$$

where $p(\infty)$ is the bulk pressure. With Equation (4) Δp becomes

$$\Delta p = RT2c(\infty)(\cosh(z\psi(d/2)) - 1). \qquad (22b)$$

Imagine that the system is cut into two at $x = d/2$. As at this point the electrical field strength is zero, the electrical force exerted on either half of the system is zero. The remaining force is due to the pressure difference Δp given by Equation (22b). Thus, the repulsive force between the plates is electroosmotic in nature and not electrostatic in the proper sense.

As seen from Equation (22b), the repulsive force can be calculated if $\psi(d/2)$ is known. The exact evaluation of $\psi(d/2)$ is difficult. Even in the simple case of a symmetrical electrolyte one ends up with elliptic integrals. A useful approximation is available under the condition $\psi(d/2) \ll 1$ (which is always fulfilled when d is sufficiently large). Then $\psi(d/2)$ can be obtained from the superposition of the potentials of the isolated plates (weak overlap approximation). With Equation (21) we obtain

$$\psi(d/2) = \frac{8}{z}y_0 e^{-\kappa d/2}. \tag{23}$$

At the same time, because of $\psi(d/2) \ll 1$, the approximation $\cosh(x) = 1 + x^2/2$ can be used in Equation (22b) which then simplifies to

$$\Delta p = RTc(\infty)z^2\psi^2(d/2)^2. \tag{24a}$$

With Equation (23)

$$\Delta p = 64\,RTc(\infty)y_0^2 e^{-\kappa d}. \tag{24b}$$

The electro-osmotic energy per surface area therefore is

$$w_{el}(d) = -\int_{\infty}^{d}\Delta p\,dx = 64\,RTc(\infty)y_0^2\frac{1}{\kappa}e^{-\kappa d}. \tag{25}$$

We see that, as a function of d, Δp and w_{el} decrease exponentially with a characteristic length of $1/\kappa$. Usually, the most important interaction energy besides w_{el} is the van der Waals energy (dispersion energy). For two plates separated by a distance d the van der Waals energy per surface area is

$$w_{vdW} = -\frac{A}{12\pi d^2}. \tag{26a}$$

The negative sign means that the force is attractive. The Hamaker constant A depends on the material of plates and solution. Typically, it amounts to something like 10^{-20} J. The total energy per surface is the sum of electro-osmotic and van der Waals contributions

$$w = w_{el} + w_{vdW} = 64\,RTc(\infty)\frac{1}{\kappa}y_0^2 e^{-\kappa d} - \frac{A}{12\pi d^2}. \tag{26b}$$

This is the basic equation of the DLVO-theory [11]. Notice that the van der Waals term is larger, in absolute value, than the electro-osmotic term for both very large and very small distances d. Thus repulsion, if effective at all, will be prevailing in a barrier region at intermediate distances. Repulsion can be reduced by increasing $c(\infty)$ (decrease of $e^{-\kappa d}$ outweighs the increase of the preexponential factor $c(\infty)/\kappa$) so that increasing electrolyte concentrations favor flocculation and coagulation (see Figure 2.5). On the far side of the barrier, there is a flat energy minimum, called the secondary minimum. If the surface area of the particles is large, the energy in this minimum may exceed kT in absolute value and cause a stabilization of the secondary minimum configuration. Such metastable coagulation is a characteristic feature of the process of creaming and similar reversible aggregation phenomena.

IV. THE ZETA POTENTIAL

There are only a small number of experimental possibilities to determine electrical surface potentials. One is to measure the potential of the surface of shear. As the symbol ζ is currently used to denote this potential, it is called the zeta potential. Experimental values of the zeta potential can be obtained

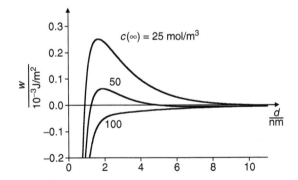

Figure 2.5 Energy per surface area of two identical charged plates with surface charge density $\gamma_0 = 0.016\,\text{As/m}^2$ as a function of their distance d, calculated from Equation (26b) for the indicated values of the concentration $c\ (\infty)$; $(1,-1)$ electrolyte, Hamaker constant $A = 2 \times 10^{-20}$ J, $T = 293\,\text{K}$, $\varepsilon_r = 78.6$, values of y_0 from Equation (16b) in combination with Equation (14).

from electrokinetic experiments, such as electrophoresis, electro-osmosis, streaming potential, and sedimentation potential measurements. Let us speak in terms of electrophoresis. The quantity primarily measured in electrophoresis is the velocity v_E of an isolated particle (which may be a solid particle, a droplet, or a bubble) relative to the surrounding solution caused by an external electric field of strength E_0. v_E is called the electrophoretic velocity of the particle. Using an appropriate theoretical model, it is possible to convert v_E into a value of the zeta potential, which, in the presence of the external field, is the difference between the actual potential φ of the surface of shear and the potential φ_0 caused by the external field alone (i.e., without disturbance by the particle).

The concept of a surface of shear implies some idealization. It suggests that the viscosity of the fluid flowing around the particle retains its bulk value η up to this surface and then, abruptly, becomes infinite. Moreover, the concept implies some averaging in time and, for rough particles, in space. In any case, a model is needed to calculate ζ from the experimental value

As pointed out in the above treatment of the Gouy–Chapman model, there is some freedom to choose the location of the *electrostatic* surface. Using this freedom, we now identify the electrostatic surface with the surface of shear (and both with the surface of the particle). Following the notation of the preceding sections we now have

$$\zeta = \varphi(0). \tag{27}$$

To establish a relationship between the electrophoretic velocity and the zeta potential the underlying set of partial differential equations must be solved. This set consists of the Poisson equation (electrostatics), the Navier–Stokes equation (hydrodynamic force equation = Newton's second law), the continuity equation (conservation of matter), and the Nernst–Planck equations for the ionic components (electrodiffusion). Generally, even if the zeta potential has the same value all over the surface, it is difficult to solve the set of these equations, and numerical methods must be used.

Yet, there are two important special cases where the relationship between v_E and ζ can be established analytically. The respective results are known as the Smoluchowski and the Hückel equation.

Let us consider the Smoluchowski equation first. It is of greater practical importance than the Hückel equation and reads

$$\zeta = \frac{\eta}{\varepsilon} \frac{v_E}{E_0}, \tag{28}$$

where ε and η are the (bulk) values of the dielectric constant and viscosity. It applies, among other cases, to insulating plates and to spheres provided the radius R_0 of the spheres satisfies the condition

$$R_0 \gg \frac{1}{\kappa}, \tag{29}$$

where $1/\kappa$ is the Debye length. To derive Equation (28), let us first consider a positively charged plate with homogeneous planar surfaces moving with velocity v_E in an electrolyte solution under the influence of an electrical field of strength E_0. We use an x,y,z-coordinate system. The electrical field has direction z, and the plate is assumed to be oriented parallel to this direction. The surface of shear is assumed to be located at $x = 0$, where x is the coordinate normal to the surface. Because of the positive value of the surface charge, there is an excess of negative ions in the adjacent liquid. Under the influence of the electric field these ions are driven in the minus z direction and, due to friction, draw the liquid with them. Under the neglect of edge effects, this produces a stationary velocity profile $v_z(x)$ described by the hydrodynamic force equation (Navier–Stokes equation)

$$f_z(x) = -\eta v_z'' (x), \tag{30a}$$

where f_z is the nonfrictional force per fluid volume in direction z (as above, the prime denotes derivation to x). The boundary conditions are

$$v_z(0) = v_E, \qquad v_z(\infty) = 0 \qquad \text{and} \qquad v_z'(\infty) = 0. \tag{30b}$$

With $f_z(x) = E_0\, \rho(x)$, where ρ is the electrical space charge density, Equation (30a) becomes

$$E_0 \rho(x) = -\eta v_z''(x). \tag{30c}$$

Due to the external field, the electrical potential φ depends on x and z: $\varphi = \varphi(x,z)$. In this two-dimensional case the Poisson equation reads

$$\frac{\partial^2 \varphi}{\partial x^2} + \frac{\partial^2 \varphi}{\partial z^2} = -\frac{\rho}{\varepsilon}. \tag{30d}$$

In our case φ can be split up into

$$\varphi = \Phi(x) - E_0 z. \tag{30e}$$

Further we have $\rho = \rho(x)$. Inserting this in Equation (30d) we come back to Equation (3a). More precisely we find

$$\varphi''(x) = \Phi''(x) = -\frac{\rho(x)}{\varepsilon}. \tag{31a}$$

Inserting the last result in Equation (30c) we obtain

$$\Phi'' = \frac{\eta}{\varepsilon E_0} v_z''. \tag{31b}$$

On integrating twice from x to infinity, with $\Phi(\infty) = \Phi'(\infty) = 0$ and the boundary conditions of Equations (30b), we obtain

$$\Phi(x) = \frac{\eta}{\varepsilon E_0} v_z(x). \tag{31c}$$

Thus $v_z(x)$ is proportional to $\Phi(x)$. With $\Phi(0) = \zeta$ and $v_z(0) = v_{\mathrm{E}}$ this yields

$$\zeta = \frac{\eta}{\varepsilon E_0} v_{\mathrm{E}}, \tag{31d}$$

which is identical to Equation (28). Note that, as a consequence of Equation (30c), shearing of the liquid is restricted to the space charge region of the ionic cloud in front of the surface.

Next, let us consider a positively charged insulating sphere in an incompressible liquid moving with velocity

v_E under the influence of an electric field of strength E_0 in direction z. Assume that Equation (29) is satisfied. As in the case of the plate, shearing primarily occurs in the thin space charge region near the surface. The liquid, forced to flow around the particle, will cause some shearing outside the space charge region as well, but shearing in that region will be very weak in comparison with the space charge region. Therefore, outside the space charge region the liquid can be approximately treated as ideal (i.e., as a liquid with zero viscosity). Under rather general conditions (which are fulfilled in our case) the flow of an incompressible ideal liquid is characterized by a vanishing curl of the fluid velocity $\vec{v}$ (irrotational flow, rot $\vec{v} = 0$ [12]). The curl-free flow is called *ideal flow*.

Consider the hypothetical case that the liquid flow around the sphere is ideal everywhere. Ideality implies complete absence of adhesive forces between the surface and the liquid so that the liquid is free to slip along the particle surface. Because of rot $\vec{v} = 0$ the spatial velocity distribution of ideal flow can be written as

$$\vec{v} = -\text{grad } \varphi_f, \tag{32a}$$

where φ_f is the flow potential. With Equation (32a) the continuity equation, div $\vec{v} = 0$, transforms into the *potential equation*

$$\text{div grad } \varphi_f = 0. \tag{32b}$$

Since the sphere is moving through the liquid, $\vec{v}$ is time-dependent, and so is φ_f. Seen from the sphere itself, the flow, however, is stationary. Therefore, to get rid of the time dependence, we introduce a moving coordinate system fixed to the center of the sphere. The radius r and the angle ϑ relative to the z-axis are used as spherical coordinates (Figure 2.6). (Due to cylindrical symmetry we do not need the third coordinate.) From now on, both $\vec{v}$ and φ will be described in this coordinate system. This implies that the differentiations of the operators grad and div in Equations (32a) and (32b) are carried out in the new system. (The velocity $\vec{v}$, however, remains the quantity defined in the original, nonmoving system.)

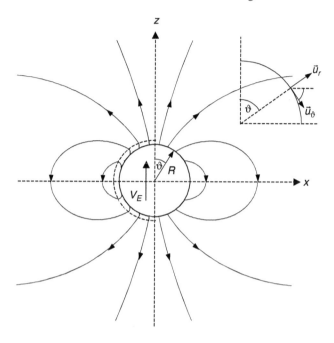

Figure 2.6 Streamlines produced by a sphere moving with velocity v_E through a resting liquid. Right half of the figure: streamlines of an ideal liquid not adhering to the sphere. Left half of the figure: streamlines obtained from the boundary layer approach. Because of adherence of the liquid to the sphere all streamlines have vertical direction in the immediate neighborhood of the sphere. Due to the electrostatic forces acting in the space charge region the streamlines merge into the streamlines of the ideal liquid within a very short distance from the surface. Insert: Relation between ϑ and the tangential and normal unit vectors, $\vec{u}_\vartheta$ and $\vec{u}_r$.

Equations (32b) and (32a) allow us to determine $\vec{v}$ without any reference to the force equation. The normal component of the liquid velocity at the particle surface, v_n^{id}, equals the normal component of the electrophoretic velocity $v_E\,\vec{u}_z$, ($\vec{u}_z$ is the unit vector in direction z). Thus

$$v_n^{id} = v_E \cos\vartheta. \tag{33a}$$

Standard methods of potential theory now can be used to calculate the spatial profile of the liquid velocity $\vec{v}$. A convenient way is to assume a flow dipole in the center of the sphere.[1] The tangential component of the velocity at the surface, v_t^{id}, then is found to be

$$v_{t^{id}} = 0.5 v_E \sin \vartheta. \tag{33b}$$

In the real case, the liquid does not slip but sticks to the surface and therefore the normal and tangential components of the velocity at the surface are linked to v_E by

$$v_n^{real} = v_E \cos \vartheta, \quad v_t^{real} = -v_E \sin \vartheta. \tag{33c}$$

The difference between the surface velocities in the two cases is

$$\Delta \vec{v}(R_0, \vartheta) = \vec{v}^{real}(R_0, \vartheta) - \vec{v}^{id}(R_0, \vartheta) = \Delta v_t(\vartheta)\vec{u}_\vartheta$$

$$= -\frac{3}{2} v_E \sin \vartheta \vec{u}_\vartheta. \tag{33d}$$

Apart from the negative sign, this difference is just the slipping velocity of the ideal fluid relative to the sphere. To reconcile the case of slipping with the real case of sticking, $\vec{v}^{id}$ must be transformed into $\vec{v}^{real}$ in the thin space charge region of the diffuse ionic cloud surrounding the particle. The corresponding velocity difference across the layer is given by Equation (33d). Since Δv_t is of the order of magnitude of v_E, shearing within the thin space charge layer of the ionic cloud

---

[1] The potential of a flow dipole located in the centre of the sphere, with its dipole moment oriented in direction z, is given by Nayfeh and Brussel [17]:

$$\varphi_f = A \frac{\cos \vartheta}{r^2},$$

where r is the distance from the center. The value of A varies with the dipole strength. (Note the analogy with an electrostatic dipole.)
    With $\vec{v} = \vec{v}(r, \vartheta) = -\mathrm{grad}\, \varphi_f$, see Equation (32a), and

$$\mathrm{grad}\, \varphi_f = \frac{\partial \varphi_f}{\partial r} \vec{u}_r + \frac{1}{r} \frac{\partial \varphi_f}{\partial \vartheta} \vec{u}_\vartheta$$

$\vec{u}_r$ and $\vec{u}_\vartheta$ are unit vectors) the boundary condition of Equation (33a) can be satisfied by $A = v_E R_0^3/2$. This leads to the tangential component given by Equation (33b). (Use $r = R_0$ and $\vec{v}(R_0, \vartheta) = v_n^{id}(\vartheta) \vec{u}_r + v_t^{id}(\vartheta) \vec{u}_\vartheta$.)

is strong. Therefore, liquid flow within this region is treated as purely viscous flow.[2] To relate it to the external field $\vec{E}_0 = E_0 \vec{u}_z$, we must determine the electrical field strength $\vec{E}$ in the neighborhood of the surface. In an electrolyte solution with specific conductivity κ_{el} the electrical field $\vec{E}$ is related to the electrical current density $\vec{i}$ by $\vec{E} = \kappa_{el} \vec{i}$. Further, Kirchhoff's law states div $\vec{i} = 0$. With $\vec{E} = -\text{grad } \varphi$, and the approximation $\kappa_{el} = $ constant, the two relations combine to div grad $\varphi = 0$. Thus, again, the potential field method can be used to determine $\vec{E}$. By superposition of the electrical field of a current dipole located in the center of the sphere and of the external field $\vec{E}_0$ the boundary condition of a vanishing normal component of the current density (and thus of the field) at the surface of the sphere can be satisfied. This treatment is quite similar to that of the fluid velocity above. In this way, the electrical field near the particle surface with tangential component E_t is found to be related to E_0 by

$$\vec{E}(R_0, \vartheta) = E_t(\vartheta)\vec{u}_\vartheta = -\frac{3}{2} E_0 \sin \vartheta \vec{u}_\vartheta. \tag{33e}$$

Note that for ϑ near $\pm \pi/2$ E_t is larger in absolute value than E_0. This reflects the compression of the electrical field lines near the surface of the sphere seen in Figure 2.7.

Condition (29) implies that, with respect to the ionic cloud, the surface can be considered as planar. Therefore, the tangential component of the electrical field at the surface, $E_t = E_t(\vartheta)$, can be related to the tangential velocity difference across the boundary layer, $\Delta v_t = \Delta v_t(\vartheta)$, by the analog of Equation (31d):

$$\zeta(\vartheta) = \frac{\eta}{\varepsilon} \frac{\Delta v_t(\vartheta)}{E_t(\vartheta)}. \tag{34}$$

---

[2] Our present strategy to combine nonviscous potential flow at distances far away from the particle, where shearing is weak, with purely viscous flow in a layer near the surface, where shearing is strong, is a standard approach of hydrodynamic theory known as the boundary layer approach [18,19]. Note that potential flow neglects viscosity, while purely viscous flow neglects inertial forces. It seems that the boundary layer approach, so far, has not been used in the theoretical treatment of the zeta potential.

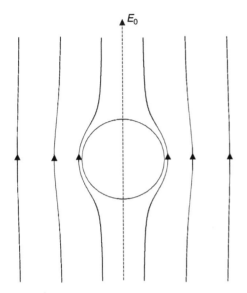

E_0

Figure 2.7 Field lines caused by an external field of strength E_0 in an electrolyte solution surrounding an insulating sphere, calculated from Equation (36c).

On inserting Equations (33d) and (33e), the angle ϑ drops out and we come back to Equation (28) with a zeta potential independent of ϑ. Conversely we conclude: If the electrical potential of the surface of shear of a sphere has a uniform value ζ and if, at the same time, Equation (29) is satisfied, then v_E and E_0 are related by Equation (28).

Due to the adherence of the liquid to the particle, the streamlines shown on the left-hand side of Figure 2.6 have direction z all over the surface of the sphere. Outside the ionic cloud, the streamlines are identical with those of the potential flow profile shown on the right-hand side of the figure.

Since the electric field strength and the liquid velocity developing outside the boundary layer are described by analogous potential field equations, the spatial profiles of field and velocity are closely connected, irrespective of whether the particle is spherical or not. As a consequence, the slipping velocity Δv_t and the tangential field strength E_t are related by $\Delta v_t/E_t = v_E/E_0$ at any point of the surface of an insulating particle, provided the radii of curvature are large all over the surface in comparison with $1/\kappa$. Equation (28) therefore holds for nonconducting particles in general, provided R^{min}, the minimal radius of curvature of the particle, satisfies the condition

$$1/\kappa \ll R^{min}. \tag{35a}$$

Equation (29) is just a special case. A planar plate satisfies this condition with $R^{min} = \infty$.

In the above derivation of the Smoluchowski equation a uniform value of the specific conductivity κ_{el} was assumed. But, due to the ionic cloud surrounding the charged particle, the specific conductivity near the surface is different from that of the bulk solution. For spherical particles in symmetrical electrolyte solutions this effect is in fact negligible if, instead of condition (29), the more stringent condition

$$\frac{1}{\kappa} \exp\left(|zF\zeta|/(2RT)\right) \ll R_0 \tag{35b}$$

is satisfied [13]. As one would expect intuitively, the impact of surface conductivity increases with increasing (absolute value of the) zeta potential.

The above boundary layer approach provides a comparatively simple derivation of the Smoluchowski formula and, at the same time, an analytical description of the velocity field around a spherical particle (cf. Figure 2.6). Using Equation (33e) for the tangential electrical field strength and interpreting x and z in Equation (31c) as the normal and the tangential coordinate of a surface element, the velocity profile in the space charge layer, at a given angle ϑ, can be obtained from Equation (31c) if the corresponding potential profile $\Phi(x)$ is known. In the case of a symmetrical electrolyte, for example, this profile is given by Equation (16a) (with φ now taken as Φ and $\Phi(0) = \zeta$). The velocity profile in the potential flow region can be calculated from Equation (32a) and the formulas in footnote 1 and is found to be

$$\frac{\vec{v}}{v_E} = v_r\vec{u}_r + v_\vartheta\vec{u}_\vartheta = \frac{1}{(r/R_0)^3}(\cos\vartheta\vec{u}_r + 0.5\sin\vartheta\vec{u}_\vartheta). \tag{35c}$$

Due to the dipole character of the velocity field, $|v|$ decreases pretty fast with increasing r. For example, at $r = 3\,R_0$ $|v|$ has dropped to less than 5% of its value at R_0.

In the case of rotational symmetry the streamlines of the flow field generally follow from

$$\frac{\mathrm{d}r}{v_r} = \frac{r\,\mathrm{d}\vartheta}{v_\vartheta}. \tag{36a}$$

In the potential flow region of the dipole this leads to

$$\frac{r}{x_{\max}} = \sin^2\vartheta, \tag{36b}$$

where $x_{\max}$ is the x-value of the intersection of the individual streamline with the x-axes. In the x, z-plane Equation (36b) becomes

$$z^2 = x_{\max}^{2/3}\,x^{4/3} - x^2. \tag{36c}$$

This relationship is used to calculate the streamlines in Figure 2.6.

Now let us consider the second special case. Assume a sphere with radius R_0 much smaller than the Debye length

$$1/\kappa \gg R_0. \tag{37a}$$

This condition is the reverse of Equation (29). In the limit, the countercharge is infinitely far away from the particle. Then the zeta potential and the charge Q_{el} of the particle are simply related by Coulomb's law

$$\zeta = \frac{Q_{\mathrm{el}}}{4\pi\varepsilon R}. \tag{37b}$$

The electrostatic force acting on the particle must be balanced by the frictional force. According to Stoke's law this requires

$$E_0 Q_{\mathrm{el}} = 6\pi\eta R v_{\mathrm{E}}. \tag{38}$$

On inserting Equation (37b), we immediately get the Hückel equation

$$\zeta = \frac{3}{2}\frac{\eta}{\varepsilon}\frac{v_{\mathrm{E}}}{E_0}, \tag{39}$$

which differs by a factor of 3/2 from the Smoluchowski equation (28).

Applicability of the Hückel equation is limited to spherical particles satisfying condition (37a). The Smoluchowski equation, on the other hand, can be applied if condition (35a) is satisfied which is much more flexible than condition (37a). Therefore, zeta potential measurements are preferentially made under conditions where the Smoluchowski equation can be used ($1/\kappa$ can be made small by the addition of salt). Observe, that the condition (35a) ensures the applicability of the Gouy–Chapman theory and of related adsorption models (see Chapter 1 in this volume).

The theoretical relationship between the electrophoretic velocity and the zeta potential becomes rather involved if neither conditions (35a) and (35b) nor condition (37a) are satisfied. For zeta potentials below 150 mV this regards, roughly, the range $0.01 < \kappa R < 10^3$. In this intermediate range, the relationship between zeta potential and electrophoretic velocity is very sensitive to the distortion of the ionic cloud by the applied electric field (relaxation effect). For more details the reader is referred to Refs. [13–16].

REFERENCES

1. P.C. Hiemenz, *Principles of Colloid and Surface Chemistry*, Marcel Dekker, New York, 1986, pp. 677–735.
2. R.J. Hunter, *Foundations of Colloid Science*, Vol. I, Clarendon Press, Oxford, 1989, pp. 329–394.
3. R. Aveyard and D.A. Haydon, *An Introduction to the Principles of Surface Chemistry*, Cambridge University Press, Cambridge, 1973, pp. 31–57.
4. J.N. Israelachvili, *Intermolecular and Surface Forces*, Academic Press, London, 1985, pp. 144–187.
5. L. Blum and D. Henderson, in *Fundamentals of Inhomogeneous Fluids*, D. Henderson, Ed., Marcel Dekker, New York, 1992, pp. 239–276.
6. P. Attard, *Adv. Chem. Phys.*, 92, 1–159, 1996.
7. H.T. Davis, *Statistical Mechanics of Phases, Interfaces and Thin Films*, VCH Publishers, New York, 1996, pp. 381–423

8. G.M. Bell and S. Levine, in *Chemical Physics of Ionic Solutions*, B.E. Conway, Ed., John Wiley & Sons, New York, 1966, pp. 409–461.
9. S. Woelki and H.-H. Kohler, *Chem. Phys.*, 261, 411–419, 2000.
10. S. Woelki and H.-H. Kohler, *Chem. Phys.*, 261, 421–438, 2000.
11. R.J. Hunter, *Foundations of Colloid Science*, Vol. I, Clarendon Press, Oxford, 1989, p. 418f.
12. L.D. Landau and E.M. Lifschitz, *Lehrbuch der Theoretischen Physik*, Bd. 6 Hydrodynamik, Akademie Verlag, Berlin, 1991, pp. 1–27.
13. R.J. Hunter, *Foundations of Colloid Science*, Vol. II, Clarendon Press, Oxford, 1989, pp. 787–826.
14. R.J. Hunter, *Zeta Potential in Colloid Science*, Academic Press, London, 1981
15. J.L. Anderson, *Ann. Rev. Fluid Mech.*, 21, 61–99, 1989.
16. A. Moncho, F. Martinez-Lopez, and R. Hidalgo-Alvarez, *Colloids and Surfaces A*, 192, 215–226, 2001.
17. M.H. Nayfeh and M.K. Brussel, *Electricity and Magnetism*, John Wiley & Sons, New York, 1985.
18. R.B. Bird, W.E. Stewart, and E.N. Lightfoot, *Transport Phenomena*, 2nd ed., John Wiley & Sons, New York, 2002, p. 133.
19. L.M. Milne-Thomson, *Theoretical Hydrodynamics*, Macmillan, London, 1968.

3

Coagulation Kinetics

H. Stechemesser and H. Sonntag*

Arbeitsgruppe Kolloide und Grenzflächen am Institut für Keramik, Glas- und Baustofftechnik, Technische Universität Bergakademie Freiberg, Freiberg, Germany

Max-Planck Institut für Kolloide- und Grenzflächenforschung, Potsdam, Germany

* Deceased

I. INTRODUCTION

The translational and rotational motions of particles dispersed in liquids and the interaction forces between them during their approach and collisions determine the properties of the dispersions with respect to their stability, their rheological behavior, and their structure formation.

Much of colloid science and especially that dealing with the stability of dispersions are based on perikinetic phenomena, i.e., those originating from Brownian translation and rotation of the dispersed particles or their aggregates. These motions follow random statistics.

The orthokinetic phenomena result from the particle movement caused by flow of the fluid medium in which they are dispersed. In a laminar flow the particle motion is nonrandom.

Coagulation kinetics is classified as *rapid* or *fast* if every collision is effective, or *slow* if not. The latter is caused either by an energy barrier of some k_BT (thermal energy) or a low shallow primary minimum or secondary minimum also comparable to k_BT, where k_B is the Boltzmann constant and T is temperature.

From slow primary minimum coagulation, one can obtain information on the height of the energy barrier and on the depth of the energy minimum.

II. DIFFUSION-CONTROLLED COAGULATION KINETICS: PERIKINETIC COAGULATION

A. Diffusion of Colloidal Particles and their Aggregates

1. Brownian Motion of Single Particles and their Aggregates

Brownian motion can be considered to be a random walk in which particles or particle aggregates make random walks of characteristic length. The average square of the displacement (x^2) is given by the following equation

$$x^2 = 2D_1 t, \tag{1}$$

where D_1 is the translational diffusion coefficient of a single particle or a single aggregate, and t is time.

The chaotic motion of a single spherical particle treated theoretically by Einstein [1]. The diffusion coefficient is related to the frictional coefficient f_s through the following relationship

$$D_1 = \frac{k_B T}{f_s}. \tag{2}$$

For spherical particles where the particle radius a is much greater than the radius of the solvent molecules and under condition of laminar flow in a liquid of dynamic viscosity η one obtains

$$D_1 = \frac{k_B T}{6\pi \eta a}. \tag{3}$$

The frictional term becomes higher for aggregates and dependent on the orientation. For a doublet of spherical particles the translational diffusion coefficient in the direction of the line joining the centers of the particles [2,3] is given by

$$D_1 = \frac{k_B T}{6\pi \eta a \times 1.29} \tag{4}$$

and for directions perpendicular to the axis is given by η

$$D_1 = \frac{k_B T}{6\pi \eta a \times 1.449}. \tag{5}$$

With increasing number of single particles in aggregates, the number of possible morphological states in the aggregates increases and therefore the value of the translational diffusion coefficient becomes dependent on the direction of the movement. The greater the number of primary particles in the aggregate, the greater the number of principle modes of motion.

Beside the translational diffusion we have to consider the rotational diffusion of particles and aggregates. The Brownian rotational diffusion coefficient for a single sphere [4] is

$$D_{\text{rot}} = \frac{k_B T}{8\pi\eta a^3} \tag{6}$$

and for a doublet of spheres

$$D_{\text{rot}} = \frac{k_B T}{29.9\pi\eta a^3}. \tag{7}$$

It is clear that for aggregates containing more than two particles there exist more than two configurations. The most extreme configurations are a linear aggregate, in which the particles are oriented in a line, and a compact or clustered aggregate. A clustered aggregate is defined as one in which the coagulated particles form an aggregate of spherical symmetry.

2.   Measurement of Single Particle and Aggregate
     Diffusion

The Brownian motion of colloidal particles is a fundamental process of considerable interest owing to, among other things, its role in the coagulation and rheology of dispersions. As a result, many techniques are available to determine the diffusion coefficient of particles [5], including spin-echo NMR [6,7], and photo-correlation spectroscopy [8,9].

Krystev et al. [10] used the barycentric method for studying the Brownian diffusion of colloidal particles without any assumptions about the shape of the diffusant particles. The Taylor–Aris dispersion technique can be used to measure the diffusivities of molecular solutes, mutual diffusion coefficients in binary solutions, as well as diffusivity of colloidal particles in aqueous suspensions up to $0.3\,\mu m$ in diameter [5]. The electron paramagnetic resonance is applicable for diffusion coefficients within the range from 10^{-21} to $10^{-18}\,m^2\,s^{-1}$ for particle sizes between $50\,nm$ and $1\,\mu m$ [11]. Tawari et al. [12] used the dynamic light scattering (DLS) to study the Brownian translational diffusion of Laponite clay particles at low electrolyte concentrations. The rotational diffusion coefficient of hard-sphere colloidal particles is measured by studying the time dependence of the depolarized field scattered in the forward direction (zero angle) [13].

Koenderink et al. [14] investigated the generalization of the Stokes–Einstein–Debye (SED) relation to tracer diffusion in suspensions of neutral and charged colloidal host spheres. Rotational diffusion coefficients are measured with dynamic light scattering and phosphorescence spectroscopy, and calculated including two- and three-particle interactions. The rational tracer diffusion is always faster than predicted by the SED relation.

The particle movement was studied by Vadas et al. [15,16] using the microtube technique. A similar improved apparatus was later used to measure the translational diffusion coefficient of single particles and aggregates [17,18]. Vadas et al. [15] investigated the translational diffusion coefficient of two different polystyrene lattices and compared the results with the calculated diffusion coefficient according to Equation (2). Similar results were obtained by Cornell et al. [17] and by Reynolds and Goodwin [18]. The results are summarized in Table 3.1.

There is a good agreement between calculated and measured diffusion coefficients. The diffusion coefficient of doublets was measured in line of the axis and perpendicular to it for two kinds of latexes by Vadas et al. [15]. The results are presented in Table 3.2. $D_{\parallel}$ and $D_{\perp}$ represent the diffusion coefficients in line of the axis and perpendicular to it.

Reynolds and Goodwin [18] measured translational diffusion coefficients for single particles and aggregates up to sextets. The results are summarized in Table 3.3. The values of the translational diffusion coefficients for the aggregates

Table 3.1 Translational diffusion coefficients of single latex particles at 298 K compared with calculated values according to Equation (2)

Particle radius (μm)	$D_{singlet}$ (10^{-13} m^2/sec)	D_{calc} (10^{-13} m^2/sec)	Ref.
0.595	4.26[a]	4.12	[6]
1.035	2.46[a]	2.43	[6]
0.458	3.60 + 0.45[b]	4.00	[7]
1.000	2.00 + 0.25[b]	1.91	[7]
1.88	0.91 + 0.10[b]	1.09	[7]
1.735	1.36[c]	1.41	[8]
1.000	2.43[c]	2.45	[8]

[a] Measured in distilled water.
[b] Measured in 10^{-5} mol NaCl/l.
[c] Measured in a mixture of 52% D_2O and 48% H_2O and viscosity 1.002×10^{-3} Pa s and multiplied by the viscosity of the D_2O/H_2O mixture to that of water to give the values in pure water at 298 K.

Table 3.2 Translational diffusion coefficients of doublets in line of the axis $D_{\parallel}$ and perpendicular to it $D_{\perp}$ [15]

Particle radius (μm)	$D_{\parallel}$ (10^{-13} m^2/sec)		$D_{\perp}$ (10^{-13} m^2/sec)	
	Measured	Calculated	Measured	Calculated
0.595	3.18	3.18	2.98	2.83
1.035	1.91	1.83	1.60	1.63

show that there is a rapid decrease in the mobility of the aggregates with increasing aggregate size. Linear, chain-like aggregates have higher translational diffusion coefficients and are therefore more mobile than clustered aggregates.

Will and Leipertz [19] used a novel optical scattering cell for the measurement of particle diffusion coefficients by dynamic light scattering. A major feature of the cell is that allows measurements without knowledge of the refraction index of the dispersing liquid. This improvement is achieved by the use of a symmetrical set-up. Silica particles of 222.8 nm ± 0.6% showed a translational diffusion coefficient in water of 1.989×10^{-12} m^2 s^{-1} ± 0.53%.

Table 3.3 Translational diffusion coefficients of aggregates of polystyrene latex particles (2.00 μm diameter) in water at 298 K [18]

Number of primary particles	Diffusion coefficient	
	10^{-9} cm^2/sec	Error (%)
1	2.43	1.0
2	1.76	1.5
3 linear	1.56	1.5
3 cluster	1.47	3.0
4 linear	1.34	5.0
4 cluster	1.09	9.0
5 linear	0.98	7.5
5 cluster	0.80	18.0
6 linear	0.87	7.0
6 cluster	0.76	17.0

Depolarized Fabry–Perot interferometry (FPI) is a less common dynamic light-scattering technique that is applicable to optically anisotropic particles. The translational diffusion coefficient measured by dynamic light-scattering techniques and the rotational diffusion coefficient measured by depolarized FPI may be combined to obtain the dimensions of nonspherical particles [20].

3. Relative Diffusion of Colloidal Particles

The simplest case is the relative diffusion of two spherical particles with radius a_1 and a_2, respectively, which are independent of each other until their contact (Smoluchowski approximation [21]). Under this condition the relative diffusion coefficient is described by

$$D_1 + D_2 = D_{12} = \frac{k_B T}{6\pi\eta}\left(\frac{1}{a_1} + \frac{1}{a_2}\right). \tag{8}$$

This equation holds when the particles are far from each other relative to their radii.

The steady-state flux I_{12} of two particles according to von Smoluchowski is given by the following equation:

$$I_{12} = 4\pi(a_1 + a_2)D_{12}z_0 = \frac{2}{3}z_0\frac{k_BT}{\eta}\frac{(a_1 + a_2)^2}{a_1a_2}, \tag{9}$$

where z_0 is the initial number of particles per cubic centimeter.

It has been shown by Derjaguin et al. [22], Spielman [23], and Honig et al. [24] that the assumption of the additivity of D_1 and D_2 is not correct at least at small distances. The relative diffusion coefficient has to be corrected by the viscous drag dependent on the distance between the particles. Because the viscous friction becomes infinitely large with decreasing distance, the viscous drag would hinder the particles to come in contact.

Independent of the knowledge of hydrodynamic interaction between approaching particles, it was shown by McGown and Parfitt [25] that rapid coagulation in the primary minimum could take place only under the action of van der Waals attraction forces. Instead of Equation (9) the following equation must be used for the diffusion flux of the aggregated particles to the surface of one of them

$$I_1 = \frac{8\pi aD_1z_0}{\int_0^\infty (1/(u+2)^2)\exp\left(V_A(u)/k_BT\right)du}, \tag{10}$$

where $u = h/a$ the dimensionless distance of closest approach between two spherical particles on the condition that no impenetrable layer surrounds the particles, i.e., $\delta = 0$ (see Figure 3.1).

If hydrodynamic interaction is introduced, the following equation is obtained

$$I_1 = \frac{8\pi aD_1z_0}{\int_0^\infty (\beta(u)/(u+2)^2)\exp\left(V_A(u)/k_BT\right)du}, \tag{11}$$

where $\beta(u)$ is the coefficient of hydrodynamic retardation for two equal spherical particles, reflecting the influence of the hydrodynamic interaction of the approaching particles on their diffusivity, given by Honig et al. [24]:

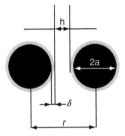

Figure 3.1 Geometrical quantities of two approaching particles.

$$\beta(u) = \frac{6(u)^2 + 13(u) + 2}{6(u)^2 + 4(u)}. \tag{12}$$

Let us now compare the values of the dimensionless rate constants $(I_1/8\pi a D_1 z_0)$ without and with a hydrodynamic interaction and with different values of the Hamaker constant, A:

A (Joule)	$(I_1/8\pi a D_1 z_0)$	
	Without	With
5.0×10^{-20}	2.292	1.210
2.5×10^{-20}	3.005	1.936

It follows from this example that the coagulation velocity with hydrodynamic interaction reaches 60% and 97% of the Smoluchowski coagulation velocity depend on Hamaker constant.

The theoretical computations by Honig et al. [24] predict that hydrodynamic interaction changes the rate of fast coagulation by a factor of about 0.4 to 0.6 in general. When a repulsive force is acting between the particles, owing to electric double-layer interaction, the rate of coagulation is diminished. Under these conditions the interaction–distance curves have a maximum, which prevents every encounter from being effective, i.e., the coagulation will be *slow*.

The stability ratio W (sometimes called the delay factor) is a criterion of the stability of the colloidal system. It is defined as the ratio of the diffusion fluxes for slow to rapid coagulation and given by the following equation

$$W = \frac{(I_1)_{V_{el=0}}}{I_1} = \frac{\int_0^\infty (\beta(u)/(u+2)^2)\exp\left[V(u)/k_BT\right]du}{\int_0^\infty (\beta(u)/(u+2)^2)\exp\left[V_A(u)/k_BT\right]du},$$

(13)

where $V(u)$ is the energy of interaction of two spherical particles of the radius a and $(I_1)_{V_{el}=0}$ the diffusion flux in the absence of electrostatic repulsion.

According to this definition, W cannot be less than 1. Only a fraction, $1/W$, of the collisions leads to aggregation. Since analytical calculations of W are not possible, Honig et al. [24] published a number of numerical calculations.

Dukhin and coworkers [26,27] made an attempt to incorporate the double-layer dynamics into the theory of stability. For spherical particles the mutual diffusion coefficient is given by the following equation

$$D_1 = \frac{k_BT}{6\pi\eta a + k_{rel}(u)},$$

(14)

where k_{rel} is the relaxation-bound resistance coefficient.

It was shown by these authors that the nonequilibration of the double layer reduced the rate of coagulation in the same order as the hydrodynamic interaction. The same result was described by Muller [28]. It should be noticed that with this correction, slow coagulation becomes independent of the particle size. This is in agreement with experimental results (see Section II.B.3).

The assumption that coagulation in the primary minimum is always irreversible has not been confirmed experimentally. Either due to the hydration of surface groups or due to the adsorption of solvated counterions, the primary minimum energy may become much lower than the calculated values from the superposition of van der Waals attraction and double-layer repulsion forces. The consequence is the reversibility of coagulation.

B. Coagulation Kinetics

1. Theory of Rapid Coagulation: von Smoluchowski Approach

The classical understanding of coagulation kinetics is given by the von Smoluchowski theory [21,29], which follows from the assumption that the collisions are binary and that fluctuations in density are sufficiently small, so that the collisions occur at random. Computer simulations [30–35] can serve as a means to test the validity of the mean field approach. A simple comparison of the observed and predicted size distribution does not test the basic assumptions of the theory. The size distribution is only a coarse reflection of the basic processes.

An aggregate formed from i identical single particles is called an i-mer. The average number of i-mers per unit volume is the particle concentration z_i. As the unit volume, 1 cm^3 is used.

The coagulation of two clusters of the kind i and j is given by the following relation:

$$i\text{-mer} + j\text{-mer} \xrightarrow{\ k_{ij}\ } (i+j)\text{-mer} = k\text{-mer}, \tag{15}$$

where k_{ij} is the concentration-independent coagulation constant or kernel. Smoluchowski was able to show that all the coagulation rates between all kinds of i-mers and j-mers is identical, i.e., they have approximately the same value. For dilute dispersions with volume fractions less than 1% only two-particle collisions need be considered, since the probability of three-particle collisions is small.

The equation describing the temporal evolution of the cluster of kind k is as follows:

$$\frac{dz_k}{dt} = \frac{1}{2} \sum_{i=1}^{k-1} k_{i(k-i)} z_i z_{k-i} - z_k \sum_{i=1}^{\infty} k_{ik} z_i. \tag{16}$$

The first term in Equation (16) describes the increase in z_k owing to coagulation of an i-mer and a j-mer and the second term describes the decrease of z_k owing to the coagulation of a k-mer with other aggregates.

The coagulation constant k_{ij} depends on details of the collision process between i-mers and j-mers $= (k-i)$-mer. This kernel embodies the dependence on i and j of the meeting of an i-mer and a j-mer including the mutual diffusion coefficient (D_{ij}) and the so-called sphere of interaction (or collision radius) R_{ij}, i.e., the distance at which the van der Waals attraction becomes dominant. By this means the coagulation constant k_{ij} is

$$k_{ij} = 4\pi R_{ij} D_{ij}. \tag{17}$$

von Smoluchowski made the following assumptions for two spherical particles of radii a_i and a_j

$$D_{ij} = D_i + D_j = \frac{k_{\rm B} T}{6\pi\eta} \left(\frac{1}{a_i} + \frac{1}{a_j} \right), \tag{18}$$

$$R_{ij} = R_i + R_j. \tag{19}$$

The coagulation constant for spherical symmetry equals

$$k_{ij} = \frac{2 k_{\rm B} T}{3\eta} \frac{(a_i + a_j)^2}{a_i a_j}. \tag{20}$$

The following connection is valid depending on the rate of the radii:

$$\frac{(a_i + a_j)^2}{a_i a_j} = \begin{cases} 4 & \text{for } a_1 = a_2 \\ > 4 & \text{for } a_1 \neq a_2 \end{cases}. \tag{21}$$

From this it follows that the collision probability increases with increasing polydispersity. The change of the total number of particles z_i with time is obtained by summing over all aggregate types

$$\frac{d \sum_1^\infty z_i}{dt} = \frac{1}{2} \sum_{i=1}^\infty k_{ij} z_i^2 + \sum_{\substack{i=1 \\ j=i+1}}^\infty k_{ij} z_i z_j - \sum_{i=1}^\infty \sum_{j=1}^\infty k_{ij} z_i z_j$$

$$= -\frac{1}{2} k_{ij} \left(\sum_{i=1}^\infty z_i \right)^2. \tag{22}$$

Therefore the total number of particles decreases according to a bimolecular reaction, which can be easily integrated. One obtains

$$\sum_1^\infty z_i = \frac{z_0}{1 + (1/2)k_{ij}z_0 t}, \tag{23}$$

where

$$\frac{1}{(1/2)k_{ij}z_0} = T_{\text{ag}} \tag{24}$$

is called coagulation time, at which the total number of particles is reduced to $z_0/2$.

Equation (16), which determines the number of k-mers, can now be solved by means of Equation (23); one obtains for singlets, doublets, and k-mers

$$z_1 = \frac{z_0}{(1 + t/T_{\text{ag}})^2}, \tag{25}$$

$$z_2 = \frac{z_0 t/T_{\text{ag}}}{(1 + t/T_{\text{ag}})^3}, \tag{26}$$

$$z_k = \frac{z_0 (t/T_{\text{ag}})^{k-1}}{(1 + t/T_{\text{ag}})^{k+1}}, \tag{27}$$

respectively.

In Figure 3.2, the decrease in the total number of particles and the change of the number of singlets, doublets, and triplets, is shown as a function of time.

In general, the collision probability among particles is depending upon their sizes and their polydispersity. In this case the coagulation constant of particles (or aggregates) can be calculated from the following kinetic equation [36–40]:

$$\frac{\delta n(v,t)}{\delta t} = \frac{1}{2} \int_0^v \beta(\tilde{v}, v - \tilde{v}) \cdot n(\tilde{v}, t) \cdot n(v - \tilde{v}, t) \mathrm{d}\tilde{v}$$

$$- \int_0^\infty \beta(\tilde{v}, v) \cdot n(\tilde{v}, t) \cdot n(v, t) \mathrm{d}\tilde{v}, \tag{28}$$

where $n(v, t)$ is the particle size distribution function, $\beta(v, \tilde{v})$ is the collision probability of particles of volume v with particles of volume $\tilde{v}$.

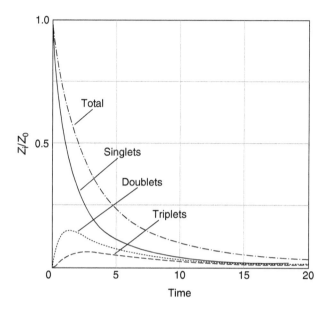

Figure 3.2 The decrease in the normalized number of total particles, singlets, doublets, and triplets according to Smoluchowski theory as a function of time.

In this equation the first integral describes the growth of the number of particles with volume v by collision of particles of volume $(v - \tilde{v})$, and $\tilde{v}$, and the second one the decrease in particle number of volume v by coagulation with particles of the kind $\tilde{v}$.

In the case of Brownian coagulation, the normalized collision probability is given by

$$\beta(v, \tilde{v}) = \frac{2k_{\mathrm{B}}T}{3\eta}(v^{1/3} + \tilde{v}^{1/3})\left(\frac{1}{v^{1/3}} + \frac{1}{\tilde{v}^{1/3}}\right). \tag{29}$$

A general analytic solution of Equation (28) does not exist due to the particle size dependency of the collision coefficient β and the complexities involved in the equation itself. Some analytic solutions have been investigated on the assumption that the collision coefficient is constant and a certain assumption regarding the form of size distribution is employed [39,40] or an analytic work includes a pursuit of an asymptotic solution [37,38]. The calculated decay of the total number of particles shows that the polydispersity has a pronounced effect on coagulation kinetics. But these calculations have also shown that von Smoluchowski's theory (Equation (17)) may be used with reasonable accuracy for initially monodisperse systems and for dispersion having low polydispersity initially.

2. Reversible Coagulation

In Section II.B.1 the rapid coagulation was calculated with the assumption that the approaching spheres come into direct contact, i.e., $h = 0$ (see Figure 3.1). In this case the coagulation is irreversible. As described in Section II.A.3, coagulation in the primary minimum is not necessarily irreversible. Retardation of coagulation kinetics may be caused by deaggregation.

In many cases, a surface layer of immobilized liquid exists around the particles and prevents a direct contact of there surfaces. This layer acts as a barrier and is caused by

- Macromolecules or tensides, which generate a sterical barrier [41–50]
- Layers of adsorbed solvent especially water which generate a structural barrier [51–62]
- Thickness of the Stern layer [63].

Denoting the thickness of such a layer by δ, the distance of the closest approach is $h = r - 2\,(a + \delta) = 2\,\delta$ (Figure 3.1). The minimum energy becomes comparable with few $k_B T$, and therefore, deaggregation will occur, as illustrated in Figure 3.3.

The effect of deaggregation was first considered by Martynov and Muller [64] and independently by Frens and Overbeek [65,66]. According to Martynov and Muller [64], the change of the number of k-mers by disaggregation is given by the following equation:

$$\frac{dz_k}{dt} = z_k \sum_{i=1}^{k-1} b_{i,k-i} - \sum_{i=1}^{\infty} (1 + \delta_{ik}) b_{ik} z_{i+k}, \qquad (30)$$

$$\delta_{i,k} = \begin{cases} 1 & \text{for } i = j \\ 0 & \text{for } i \neq j \end{cases},$$

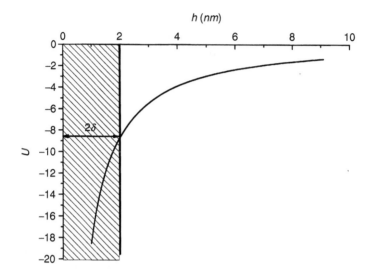

Figure 3.3 Interaction energy–distance curve between two particles enclosed with an immobilized layer of thickness δ.

where b_{ik} is the disaggregation probability of aggregates of kind $i + j$ in aggregates of the kind i and kind j. The first term describes the decrease in z_k by disaggregation into smaller aggregates and the second term, the increase in z_k by disaggregation of higher aggregates.

In analogy to the coagulation time (Equation (24)), the disaggregation time may be introduced:

$$T_{\text{dis}} = \frac{1}{b_{ik}}. \tag{31}$$

The total change of the number of k-mers is obtained by summing up Equations (16) and (30):

$$\frac{dz_k}{dt} = \frac{1}{2} \sum_{i=1}^{k-1} k_{i,k-i} z_i z_{k-i} - z_k \sum_{i=1}^{\infty} k_{ik} z_i$$

$$+ \sum_{i=1}^{\infty} (1 + \delta_{ik}) b_{ik} z_{i+k} - \frac{z_k}{z} \sum_{i=1}^{k-1} (1 + \delta_{i,k-i}) \, b_{i,k-i}. \tag{32}$$

Odriozola et al. [167,168] simulated reversible aggregation processes for systems of freely diffusing particles. The clusters were considered to move due to free Brownian motion. The authors studied the time evolution of aggregating systems where all collisions lead to the formation of bonds that are then allowed to disintegrate with a break-up probability to be the same for all bonds. This situation models systems characterized by interparticle potentials with a finite primary minimum and no repulsive energetic barrier. In order to describe the kinetics of such aggregation–fragmentation processes, a fragmentation kernel was developed and then used together with the Brownian aggregation kernel for solving the corresponding kinetic master equation. The simulated cluster-size distributions could be described only by introducing the concept of fragmentation effectiveness, i.e., it became necessary to consider as cluster fragmentation only those events where a bond breaks and the obtained fragments are separate enough so that they will not stick again during the following time steps. In a further paper, Odriozola et al. [167,168] focused their theoretical investigations on the study of reversible aggregation processes of more realistic systems characterized by interparticle potentials showing primary and secondary minima separated by an energetic barrier. Hence, two different kinds of bonds, primary and secondary, can be formed and should be treated separately, i.e., this behavior was implemented by considering bonds with different break-up probabilities. The agreement between the simulations and the kinetic description was found to be quite satisfactory. Moreover, studying the time evolution of the bond population showed that cluster ageing appears as a natural consequence of the employed model.

3. Slow Coagulation

In this section, the mutual diffusion of colloidal particles is considered, when electrostatic repulsion forces are not fully suppressed. Under this condition the interaction energy distance curves have a maximum that prevents every encounter from being effective. Only a fraction, $1/W$, of collisions lead to aggregation. W is the stability ratio, which is defined as the ratio of the diffusion-controlled rapid coagulation rate to the slow or reaction-limited coagulation rate.

For the slow coagulation of an i-mer and a j-mer to a k-mer we introduce

$$W_{ij} = \frac{k_{ij}(\text{rapid})}{k_{ij}(\text{slow})}. \tag{33}$$

The temporal evolution of the clusters of kind k is given in analogy to Equations (16) and (32), respectively

$$\frac{dz_k}{dt} = \frac{1}{2}\sum_{i=1}^{k-1}\left(\frac{k_{i(k-i)}}{W_{i(k-i)}}\right)z_i z_{k-i} - z_k \sum_{i=1}^{\infty}\frac{k_{ik}}{W_{ik}}z_i. \tag{34}$$

The total change of the number of k-mers under the conditions of slow coagulation is obtained by summing up Equations (16) and (30).

$$\frac{dz_k}{dt} = \frac{1}{2}\sum_{i=1}^{k-1}\left(\frac{k_{i,k-i}}{W_{i,(k-i)}}\right)z_i z_{k-i} - z_k \sum_{i=1}^{\infty}\frac{k_{ik}}{W_{ik}}z_i$$
$$+ \sum_{i=1}^{\infty}(1+\delta_{ik})b_{ik}z_{i+k} - \frac{z_k}{z}\sum_{i=1}^{k-1}(1+\delta_{i,k-i})\,b_{i,k-i}. \tag{35}$$

Equation (36) defines the aggregation time for slow coagulation

$$T_{\text{ag}} = \frac{W_{ij}}{(1/2)k_{ij}z_0} \tag{36}$$

and the disaggregation time by Equation (37)

$$T_{\text{dis}} = \frac{R_{ij}^2}{D_{ij}}. \tag{37}$$

III. EXPERIMENTS ON COAGULATION KINETICS

A. Methods for Measuring Particle Concentration and Aggregate Size

The determination of the particle size and the particle size distribution is the requirement for the study of kinetic processes, which govern the aggregation of colloidal particles and the stability of colloidal suspensions.

A wide range of different measuring methods is available, each with its own limitations. Some of these techniques give only global information on the state of aggregation; others give a detailed picture of the particle and floc size distribution. Ideally the monitoring technique should be suited to online application without a sample pretreatment such as dilution.

1. Bulk Techniques

Turbidity

A widely used procedure to follow coagulation kinetics is to monitor the turbidity as a function of time. One of the problems is that the turbidity of a coagulating dispersion may not increase monotonously with time.

In the Rayleigh region a $< \lambda/5$, (λ-wavelength), the turbidity (τ) is defined as

$$\tau = \frac{1}{L} \ln \frac{I_0}{I}, \tag{38}$$

where L is the scattering path length, I_0 is the incident light intensity, and I is the transmitted light intensity.

For a given dispersion the turbidity depends on the particle concentration and the light-scattering properties of the dispersed particles. If the dispersion contains only identical particles, then the turbidity is given by

$$\tau = C v_0^2 z_0, \tag{39}$$

where C is the optical constant, v_0 is the particle volume, and z_0 is the initial particle concentration.

One problem is that the turbidity may not increase monotonously with the degree of coagulation; this makes the interpretation difficult. By measuring the turbidity in the initial stages of coagulation one can obtain an average value of the coagulation constant. In measuring the turbidity the scattering factor should be low, i.e., $q \ll 1/a$.

The scattering factor q is given by

$$q = \frac{4\pi n_0}{\lambda_0} \sin \frac{\theta}{2}, \tag{40}$$

where θ is the scattering angle that should be less than $1°$, n_0 is the refractive index of the medium and λ_0 is the wavelength.

Suitable instruments have been described by many authors [67–69].

One technique is based on fluctuations of turbidity [70–72]. The number of particles in the observed volume is not constant but fluctuates in time. These variations follow the Poisson distribution so that the variance is equal to the mean value and the standard deviation of the latter is equal to the square root of the mean. The lower the observation volume, the higher is the standard deviation. This method is sensitive to the beginning of coagulation. The principle is that a laser beam is focused on the centre of a flow-through cell. Particles passing the focus produce an attenuation of the laser beam by absorption and scattering, which is measured photoelectrically. Within the last years the turbimetry as a method that is not disturbed by multiple light scattering was developed as an investigative tool for the analysis of the stability of dispersions [73–75].

Static and dynamic light scattering

The most used bulk technique to measure coagulation kinetics is volume scattering [76,76a,78]. By the static light-scattering method one measures the angular dependence of the scattered light from the whole ensemble of aggregates and single particles. By this way one obtains only information on the coagulation constant at the beginning of coagulation, when mainly single–single collisions determine the coagulation rate. However, one gains little information about the distribution of the aggregate size.

With dynamic light scattering [78–80] one measures the temporal autocorrelation of the scattered intensity. From detailed analysis of the decay of the autocorrelation function caused by the diffusion of particles and aggregates, one deduces the average hydrodynamic radius of the ensemble of clusters. Because the measurements can be performed quickly, the growth of the aggregates can be followed through R_H (hydrodynamic radius) as a function of time.

Due to technical progress over the past few years, new measurement methods for measuring particle concentration and cluster size as well as coagulation kinetic studies have been developed [20,81–90]. Fibre-optic-based multiangle light scattering setups enable simultaneous time-resolved static and dynamic light-scattering measurements over a wide range of scattering vectors [81–83]. This method leads to a significant decrease of the recording time for scattering curves and is therefore suitable for routine coagulation measurements. Further the measurement of light-scattering intensity at different angles by using a fibre-optic-based setup avoids the dust and bubble problems of low-angle light scattering [84].

Pecora [20] showed that the depolarized Fabry–Perot interferometry is applicable to optically anisotropic nanosized particles and is used for measurement of rotational diffusion coefficients. By combination with photon correlation spectroscopy one can obtain the dimensions of nonspherical particles.

Methods using a variety of probing radiations

Other solution sizing techniques applicable to sizing nanoparticles in liquids include static scattering of radiation whose wavelength is comparable to the size of the particle. For nanoparticles, this includes mainly small-angle x-ray and neutron scattering. Chu and Liu [89] characterized nanoparticles by using of small-angle x-ray scattering and small-angle neutron scattering. By choosing a suitable scattering technique or a combination of different techniques for nanoparticle characterization, the particle's molecular weight, radius of gyration, hydrodynamic radius, size distribution, shape, and internal structure as well as interparticle interactions of nanoparticles, can be determined.

2. Single-Particle Techniques

In contrast to bulk techniques, single-particle techniques measure the size of single particles or individual aggregates.

Coulter counter

A Coulter counter is able to detect and size particles in an electrical conducting medium by changes of the electrical resistance, when they pass through a short capillary the diameter of which may vary from 20 to 400 μm. At the orifice are two electrodes. If the particle concentration is between 10^6 and 10^7 particles per cm^3, they can be detected individually. The sensitivity of the apparatus depends much on the particle radius or capillary radius ratio. If the ratio is less than 0.02, the method becomes rather insensitive. The count rate is 10^4 particles per minute. Coulter counters require the presence of an electrolyte to provide electrical conductivity, which can influence the stability of the dispersion. The original Coulter counters were restricted to particles larger than 1 μm. Deblois et al. [91] developed a counter that can detect submicrometer particles.

Flow cytometry

Flow cytometry is mainly used to detect and size dispersed animal and human cells by fluorescence or light-scattering detection [92]. The count rate is 10^5 particles per minute. This apparatus works like a single particle sizer. By staining the cells or latex particles with a fluorescent dye, individual particles can be determined. The dye molecules can be excited. The particle radius should be larger than 1 μm.

Single Particle Light Scattering

The single particle light-scattering technique (SPLS) is a static scattering method based on measurements of individual particles or aggregates. The advantage of the SPLS in contrast to volume light-scattering photometry is that only one particle (or aggregate) is located in the light-scattering volume with high probability.

Light-scattering pulses are therefore measured by only one particle. It is in principle easier to analyze the scattering of light by a single particle to draw conclusions from the scattering of light by a greater ensemble of particles on the single units if it concerns polydisperse or aggregated samples.

The predecessors of today's single particle scattering light photometers are ultramicroscopes of Siedentopf and Zsigmondy [93,94]. They used the dark-field microscopy to prove discreet gold particle in ruby glasses and to investigate colloidal sols, especially coagulation of gold sols [95]. A stationary volume was used for counting of the particles. Flow cells are better suited for the examination of dispersions as they were used in the flow ultramicroscopy [96–105].

The flow ultramicroscope was first developed for individual particle counting in aerosols [96–99]. The number of particles was determined by visual counting of light flashes within the scattering volume. The measuring times are shorter and there are fewer difficulties in determining the scattering volume. Despite this innovation, the distinction of different size fractions proved to be often time-consuming and imperfect.

Reasons for include the following:

- Insufficiently defined scattering volume (edge effects)
- Insufficient receivers for the quantitative recording of the scattering light
- Time consuming data processing
- Observation of the scattering light at 90°.

With the use of lasers [106] and hydrodynamical focusing [107,108] a qualitative improvement was reached. The dispersion flux is now restricted so that every particle reaches the illuminated zone. This zone is completely located within the image of the reception optics. Therefore, every particle can be registered.

The breakthrough for automatic counting of particles or aggregates and measuring their size was the development of the single particle light-scattering technique [108–119,166]. Several key features of this technique are the use of a spatial filter and an elliptical focus to obtain an almost homogeneous light intensity in a small detection volume, a low-angle detection of the scattered light, and an accurate hydrodynamic focusing with relatively small shear forces.

Particle size distributions can be measured in the diameter range of 0.1–$5\,\mu m$ and particle concentrations in the range of 10^7–10^8 particles/cm^3.

The technique is based on measurements of the small-angle light scattering by individual particles or aggregates as they pass through the volume defined by the diameter of the laser beam and the diameter of the particle stream. The scattered light impulses are collected according to their pulse heights by a multichannel analyzer. The latter is coupled with a computer. The optics of several instruments differs only in details. Buske et al. [109] employed a spatial filter, a diaphragm, and a tilted lens. Cummins and coworkers [110,111] used a diaphragm and a spherical lens, Pelssers and others [112,113] used a spatial filter and a combination of cylindrical and spherical lenses, and Broide and coworkers [114,115] used two crossed cylindrical lenses.

Two kinds of flow cells are used. Cummins et al. [110], Pelssers et al. [113], and Fernández-Barbero et al. [116,117] modified a flow cell used by Cahill et al. [114], who improved a cell described by Hershberger et al. [118].

The construction by Pelssers and others [112,113] is shown in Figure 3.4. The dispersion flows through the capillary (1) of $0.75\,mm$ diameter, which ends in a tip of $0.2\,mm$ diameter. The inner water flow (2) is much faster than the dispersion velocity; therefore the dispersion stream is focused and accelerated. The typical value of the dispersion flow is $3.05 \times 10^{-2}\,cm^3/s$. At the tip of this nozzle the inner flow enters the outer flow. The purpose of the latter is to avoid contaminations of the cell window and to enable the separation of reflections from the scattered light of the particles. The lower nozzle stabilizes the dispersion stream spatially because all flows are forced to leave the cell through this nozzle. The flow must be laminar and spatially stable.

The particles are entirely concentrated in the center of the inner water flow, and their velocity is $54\,cm/s$. The

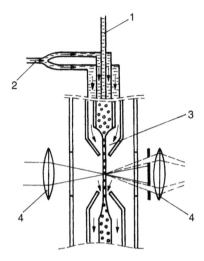

Figure 3.4 Scheme of the flow cell [97,110–113,116]. (1) Particle stream, (2) focusing liquid, (3) hydrodynamic focusing, and (4) optical system.

dispersion flux is calculated by multiplying this velocity with the cross-sectional area. The cross-sectional area is calculated as $5.65\,\mu m^2$, corresponding to a diameter of the dispersion stream in the upper nozzle of $2.68\,\mu m$.

The diameter of the laser focus is $30\,\mu m$ and the diameter of the particle stream in the laser focus is $5.7\,\mu m$, so that the detection volume is a vertical cylinder with these dimensions. The particle number concentration should be chosen such that at any one moment, only one particle is present in this detection volume. For a random distribution of particles the probability distribution for the number of particles in a given volume element is a Poisson distribution. For a coincidence error of 0.5% the following relation is obtained between the maximum particle concentration z_{max} and the scattering volume V: $z_{max} = 0.1\,V$. From this it follows that for a concentration of 10^8 particles/cm^3, the scattering volume may not exceed $10^{-9}\,cm^3$.

Another kind of flow cell (Figure 3.6) is used in the labor-made single particle light-scattering apparatus (Figure 3.5)

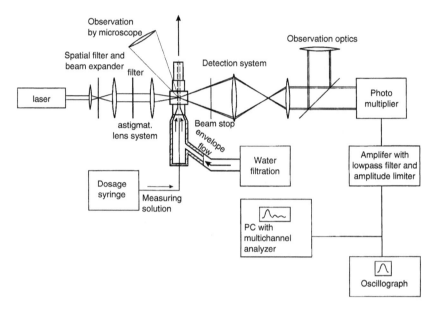

Figure 3.5 The scheme of the single particle light-scattering equipment [119].

by the authors [119] based on their first version described in Buske et al. [109]. The advantages of forward scattering are used in this equipment. The particle size is a monotonous function of the scattered light intensity for a relatively large size interval. The form factor is close to one under the conditions of the Rayleigh–Gans–Debye approximation [120]. This is particularly important for coagulation kinetic investigations. Neither the structure nor the orientation of small aggregates has measurable influence on the scattered light intensity.

As shown in Figure 3.5, linear polarized light (wavelength: 532 nm) from a low noise Nd:YVO$_4$ laser (Verdi^{TM} V-2, Coherent) is passed through a spatial filter (Newport) in order to remove random fluctuations from the intensity profile of the laser beam by dust particles. This beam is collimated into an elliptical focus by an astigmatic lens system. In this way, a flat beam is generated broad enough to ensure an almost constant intensity across the particle stream without giving an undesirably large illuminated volume. The scattering volume is determined by the cylinder of the dispersion flow and is restricted at these two fronts by the surface of the laser focus. The dispersion is fed into a platinum tube with 12 cm of length and an inside diameter of 1.5 mm by means of a motor-driven syringe. The volume flow rate is 4×10^{-4} ml/ min for typical operating conditions. A narrow part with a 0.08-mm hole is in the upper end of the tube like illustrated in Figure 3.6.

The dispersion drawing out into the cell through the hole where it meets an envelope stream with the same dispersion liquid provided by a gravitational feed. The envelope stream flows in the same direction with a slightly higher velocity than the central dispersion. The envelope stream is pressed through a filter (pore size: $0.05 \, \mu m$, Sartorius) by means of a compressor to remove disturbing dust particles.

The dispersion flow is hydrodynamically focused by the conical restriction in the cell head. Thus, drawing out the particles into a stream with a diameter of about $15 \, \mu m$ (for typical operating conditions). In this way, the scattering volume amounts to $1.6 \times 10^{-9} \, cm^3$ as calculated from the number of the coincidences by means of the Poisson distribution and the particle velocity in the scattering volume is 2 cm/sec.

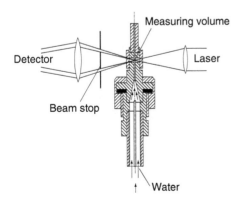

Figure 3.6 Scheme of the flow cell [119].

With this velocity 10^3 particles/sec are analyzed. The scattered light pulses are monitored in a detection system consisting of a beam stop, a lens, a diaphragm, and a photomultiplier (R2801HA, Hamamatsu Photonics).

The beam-limiting device (beam stop) includes a circular slit that allows the transmission of only the light scattered at the intervals between 5° and 10°. The diaphragm with a diameter of 0.3 mm blocks most of the reflections eliminates the scattered light of the dust particles in the envelope stream. The scattered light pulses of particles can be seen on an oscilloscope, sorted according to their pulse height by a multichannel analyzer (Trump-PCI-card, PerkinElmer™ instruments), and processed by means of a computer (software: Maestro-32, Perkin-Elmer). Results can be displayed on the screen or by printer.

In the case of reception angles below the range of 5° to 10° the resolution is higher but consequently there is generally higher scattering volume. In this case, the detection of individual particles is only possible at relatively low particle concentrations and the influence of foreign particles is higher. Moreover, a small particle concentration decelerates the coagulation kinetics in coagulation experiments and the measuring time is unacceptably high for the determination of the velocity constant. For these reasons we have decided to retain the relatively high angular aperture of 5° to 10° [121]. Thus we are still able to investigate dispersions with particle concentrations up to (2 to 3) $\times 10^8 \, \text{cm}^{-3}$, provided the influence of coincidences is taken into account. The precise location of this peak at twice the location of the monomer peak indicates incoherent scattering from two monomers passing simultaneously through different regions of the scattering volume. Under the present conditions the lower measuring limit for PS latex was achieved with a diameter of 115 nm (Figure 3.7).

B. Experimental Results: Rapid Coagulation

First quantitative experiments on rapid coagulation were performed with gold particles by slit ultramicroscopy and individual counting [93,122–124]. The average value of the

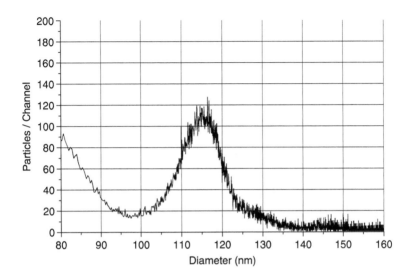

Figure 3.7 Scattering intensity distribution of 115 nm — PS latex [119].

coagulation constant of gold particles was found to be $k_{ij} = (12.64 \pm 1.45) \times 10^{-18}$ m^3/sec at 298 K. These are the only values described in the literature that confirm apparently the theoretical value.

The coagulation of gold particles was also investigated later by streaming ultramicroscopy with visual counting by Derjaguin and Kudravzeva [125] with different electrolytes. For sodium chloride $k_{ij} = 8.24 \times 10^{-18}$ m^3/sec, for magnesium sulfate $k_{ij} = 8.9 \times 10^{-18}$ m^3/sec, and for lanthanum nitrate $k_{ij} = 6.5 \times 10^{-18}$ m^3/sec was obtained. Selenium sols were investigated spectrophotometrically by Watillon et al. [126]. The rapid rate constant was 4×10^{-18} m^3/sec.

Silver iodide sols were coagulated with barium nitrate and lanthanum nitrate by Ottewill and Rastogi [127]. The coagulation constants were 8.78×10^{-18} and 10.2×10^{-18} m^3/sec, respectively.

The influence of the particle size on rapid coagulation was investigated with hematite particles by Penners [128] and Penners and Koopal [129]. The coagulation constant was dependent on the particle radius for perikinetic and orthokinetic coagulations. The results for the perikinetic coagulation are shown below (Table 3.4). The rate constant increased with the exception of the smallest particles with increasing particle size.

Using the single particle light-scattering technique (Figure 3.5), the authors investigated the rapid coagulation of monodisperse Fe_2O_3 particles (hematite, diameter: 220 nm) induced by $5 \times 10^{-1} M$ KNO$_3$. It was revealed that the value of the velocity constant of coagulation, $k_{11} = 11.0 \times 10^{-18}$ m^3/sec, was significantly closer to predictions by Smoluchowski's theory than to theories which allows for hydrodynamic and van der Waals interactions between spherical particles [130]. In order to give a theoretical interpretation of this result Shilov et al. [131] considered the influence of close hydrodynamic and van der Waals interaction on the basis of a model in which the surfaces of interacting particles have areas with variable surface curvature.

If two nearly spherical particles approach, microscopic curvature radii determine the interaction instead of macroscopic particle radii at the moment before they contact each other. The following relation could be derived [131]:

$$k_{11} = \frac{8\pi D_0 a}{(1/2) + (\langle a_l \rangle / a_T)\left\{ \int_0^\infty \left(\exp\left[V_A(u)/k_B T\right]\beta(u)/(2+u)^2\right) du - (1/2)\right\}},$$
$$a_l = \frac{a_{l1} * a_{l2}}{a_{l1} + a_{l2}}, \tag{41}$$

where a_T is the *macroscopic* particle radius and a_{li} is the *microscopic* curvature radius, $\langle a_l \rangle$ is the value after having been averaged based on all possible mutual orientations of the particles.

Table 3.4 Rapid rate constant for doublet formation of hematite particles of different sizes

Radius (nm)	37	86	134	173	225	282	352
Rate constants (m^3/sec) $\times 10^{-18}$	4	2.6	6.8	3.8	5.6	6.0	7.5

In the case of nearly spherical particles with a highly variable radius of curvature the velocity constant of coagulation merges with that of Smoluchowski according to the simplified assumptions $V = 0$ and $\beta(u) = 1$.

Walz [132] gives a review of the investigations on the effect of surface heterogeneities on colloidal forces. Some of major findings are:

- The effect of surface roughness on the van der Waals interaction between surfaces becomes insignificant when the surface-to-surface separation distance becomes much larger that the characteristic size of the roughness.
- The effect of roughness on particle adhesion force depends on the characteristic size of the roughness features on both particles and substrate. When the substrate contains hills and valleys that are comparable in size to the particle, the adhesion force can be increased or decreased relative to a smooth surface, depending on where the adhesion takes place (i.e., in a valley or on a hill). When the characteristic size of the roughness is much smaller than the particle size, the adhesion force is substantially reduced (e.g., by orders of magnitude).
- Surface roughness lowers the height of the primary potential energy barrier to coagulation by orders of magnitude, leading to significant coagulation in systems where no coagulation would be predicted assuming smooth surfaces.

Predictions of interaction energy indicate, in general, that the DLVO interaction energy profiles for rough surfaces deviate significantly from those derived assuming perfectly smooth surfaces, particularly at very short distances [133–135].

Many experiments have been carried out using polystyrene lattices because of their monodispersity and their ideal spherical shape. They are commercially available but may be easily prepared in the laboratory. According to their size they coagulate either in the primary or secondary minimum. However, they also have their disadvantages, e.g., the surface potential calculated from surface charge density is always higher or even much higher than the potential calculated from the mobility of the particles in an electric field. In addition, the electrokinetic potential at first increases with increasing electrolyte concentration up to a concentration of $10^{-2} M$ of potassium nitrate and then decreases. Some of the older results on latex coagulation are summarized in Ref. [63].

Monodisperse polystyrene particles of $a = 915$ nm and at an initial particle concentration of 5.2×10^7 parts/cm^3 were coagulated with 1% hydrochloric acid [136]. The time evolution of singlets, doublets, and triplets was measured with the Coulter counter. From the initial slope of the time evolution of singlets, doublets, and triplets, the coagulation constants $k_{11} = 7 \times 10^{-18}$ m^3/sec and $k_{22} = 3 \times 10^{-18}$ m^3/sec were calculated. The rate constant for the higher aggregates was lower than for singlets. One would expect just the opposite behavior. For different electrolytes such as magnesium sulfate, magnesium chloride, and sodium sulfate, a same rapid rate constant of 7×10^{-18} m^3/sec was found.

Holthoff et al. [81] performed the measurements on a multiangle static and dynamic light-scattering instrument using a fiber-optic-based detection system that permits simultaneous time-resolved measurements of different angles. The absolute coagulation rate constants are determined from the change of the scattering light intensity as well as from the increase of the hydrodynamic radius at different angles. The combined evaluation of static and dynamic light-scattering results permits the determination of coagulation rate constants without the explicit use of light scattering form factors for the aggregates. For different electrolytes fast coagulation rate constants of 215-nm-polystyrene latex were estimated, as shown in Table 3.5. Taking into account the van der Waals attraction and retarding effect of hydrodynamic interaction [23,24], the estimate value of the fast coagulation rate constant is found to be 40–65% of the theoretical Smoluchowski value with 12.2×10^{-18} m^3/sec at 25°C.

van Zanten and Elimelech [84] reported on the use of simultaneous, multiangle static light-scattering technique for the determination of absolute coagulation rate constants

Table 3.5 Comparison of the fast coagulation rate constants, k_{11} by using different electrolytes [81]

Electrolyte	Fast coagulation rate constant, k_{11} (10^{-18} m^3/sec)
NaClO$_4$	3.2 ± 0.4
NaNO$_3$	3.5 ± 0.1
KCl	4.3 ± 0.4
CaCl$_2$	3.7 ± 0.2

of two particle suspensions, as shown in Table 3.6. For both particle suspensions it was found that rapid coagulation is obtained at CaCl$_2$ concentrations larger than about 20 mM. The absolute coagulation rate constants are about half the theoretical value given by Equation (20). The retarding effect of hydrodynamic interaction [23,24] and deviations from Smoluchowski kinetics [21] were discussed as possible reasons of this finding. The rate constants were determined from analysis of the change in the angular dependence (15 fixed angles ranging from 23° to 128°) of the scattered light intensity with time, at the early stage of the aggregation process. They are obtained from the slope of the curve describing the initial normalized intensity change at each scattering angle as a function of the corresponding dimer form factor.

Matthews and Rhodes [137] investigated the coagulation of Dow lattices of $a = 357$ and 600 nm by particle counting

Table 3.6 Coagulation rate constants, k_{11} of polystyrene latex for different salt concentrations [84]

Suspension (nm)	N_0 (cm^{-3})	CaCl$_2$ (mM)	k_{11} (m^3/sec)
202	2.636×10^8	133	2.28×10^{-18}
202	1.318×10^8	133	2.25×10^{-18}
202	6.590×10^7	133	2.40×10^{-18}
202	1.318×10^8	20	2.50×10^{-18}
202	1.318×10^8	15	8.99×10^{-19}
202	1.318×10^8	10	3.79×10^{-19}
98	2.636×10^8	133	2.37×10^{-18}
98	2.636×10^8	15	2.98×10^{-18}
98	2.636×10^8	5	5.80×10^{-19}

with Coulter counter. The coagulant was $AlCl_3$. The values obtained were $k_{11}(357) = 8.1 \times 10^{-18}\,m^3/sec$ and $k_{11}(600) = 5.4 \times 10^{-18}\,m^3/sec$. The lower rate constant of the bigger particles is probably due to reversible coagulation in the secondary minimum. The same authors described an increase in the rate constant with increasing particle concentration.

Hatton et al. [138] investigated rapid coagulation of different polystyrene lattices of the radii 185, 250, and 950 nm with streaming ultramicroscopy. The coagulant was $0.1M$ $MgSO_4$ at pH 5.5. The rate constant was independent of the particle size but strongly dependent on the initial particle number, namely, at $z_0 = 10^7$, 10^8, and 10^9 particles/cm^3, k_{11} was 3.1×10^{-18}, 3.37×10^{-18}, and $5.5 \times 10^{-18}\,m^3/sec$, respectively.

Holthoff et al. [117] compared the results of absolute coagulation rate constants of polystyrene latex (diameter: 683 nm resp. 580 nm) in aqueous suspension measured by single particle light scattering (SPLS) and simultaneous static and dynamic light scattering (SLS + DLS) at the early stages of the coagulation process. The results are summarized in Table 3.7 and are close to the experimental mean value for diffusion-limited coagulation, $(6 \pm 3) \times 10^{-18}\,m^3/sec$, reported by Sonntag and Strenge [63]. This comparative study shows that the two different techniques lead to very similar results for the absolute coagulation rate constants. Due to the in situ character of the multiparticle technique it is indicated that no alteration of the cluster-size distribution during early stages of the coagulation process is caused by the hydrodynamic shear forces involved in the single particle light-scattering technique. This finding is supported by

Table 3.7 Comparison of coagulation rate constants, k_{11} ($10^{-18}\,m^3/sec$) obtained by single (SPLS) and multiparticle light-scattering (SLS + DLS) techniques [117]

Latex diameter (nm)	SPLS	SLS + DLS
580	7.3 ± 0.4	6.9 ± 0.4
683	7.1 ± 0.4	6.1 ± 0.4

the investigations of Fernández-Barbero et al. [116]. On the other hand, the corresponding results show that the simple Smoluchowski kinetic model with constant rate kernel does describe the aggregation process in its early stages to a good degree of approximation. This observation is in turn helpful in the validation of the results obtained from the multiparticle light-scattering technique, since the data analysis relies on the rates of change of the static and dynamic scattering signal over short time.

Lips et al. [76,76a,139] investigated the rapid coagulation of polystyrene lattices ($a = 178\,\mathrm{nm}$) using low-angle light-scattering technique. From $z_0 = 10^6$ to $3.1 \times 10^7\,\mathrm{cm}^3$ the rate constant was independent of the initial particle number $6.4 \times 10^{-18}\,\mathrm{m}^3/\mathrm{sec}$.

Cahill et al. [141] determined the initial rate constant for polystyrene particles of different sizes and for different initial particle concentrations by using the classical treatment of Smoluchowski for the change in the total particle concentration z_0 (Equation (22)). Table 3.8 shows the results obtained.

Measurements of polystyrene and PVAC by Lichtenfeld et al. [141] showed that the coagulation and deaggregation constants (for values see Table 3.9) were independent of the initial particle number in the range from 0.30×10^8 to $2.63 \times 10^8\,\mathrm{cm}^{-3}$. Recent investigations of silica by the authors also show an independence of the coagulation and deaggregation constants (for values see Table 3.9) at the isoelectric point (pH. 2.1) both of the particle size (radii: 125, 250, 275 nm) and of the initial particle number 0.5×10^8, 1×10^8, and $2 \times 10^8\,\mathrm{cm}^{-3}$.

From the experiments on rapid coagulation described above and summarized in Ref. [63], with the exception of

Table 3.8 Initial rate constant for polystyrene particles using Smoluchowski treatment [140]

z_0 (cm^{-3})	5.8×10^8	1.3×10^8	3.66×10^8
Radius (μm)	0.26	0.35	0.5
k_{11} (m^3/sec)	5.16×10^{-18}	7.6×10^{-18}	6.54×10^{-18}

Table 3.9 Coagulation and deaggregation constants as well as the barrier thickness at the surface of different particles

	PVAC (Figure 3.8)	Polystyrene (Figure 3.9)	Fe_2O_3 (Figure 3.10)	Silica (Figure 3.11)
Radius (nm)	300	305	110	125
Electrolyte (m)	$0.5\,KNO_3$	$0.5\,KNO_3$	$0.5\,KNO_3$	IEP — pH 2.1
Hamaker constant (J)	5×10^{-21}	5×10^{-21}	5×10^{-20}	1.2×10^{-20}
k_{11}, $10^{-18}\,m^3/sec$	6.1	5.8	11	1.6
b_{11}, $10^{-5}\,s^{-1}$	6.0	14.0	3.0	8.0
δ (nm)	—	0.8	1.99	0.7

Zsimondy's results [93,95] with gold particles, the coagulation rates are substantially lower than the theoretical value than the theoretical value of von Smoluchowski even when the hydrodynamic interaction was taken into consideration. In the same cases, this discrepancy can be explained by the interplay between the van der Waals forces and the hydrodynamic interactions, which typically reduces the theoretically predicted rate constant by a factor of about 2, in comparison with the von Smoluchowski value. In other cases, however, the consideration of the van der Waals forces and the hydrodynamic interactions leads to theoretical rate constants which exceed the experimental values systematically [83].

There are different ways to explain this behavior. The first one is the introduction of reversibility into the coagulation process. This was suggested by Martynov and Muller [142], Muller [143], Frens and Overbeek [65], and Frens [66]. Under this assumption the coagulation kernel can remain constant. On the other hand, one can postulate that the coagulation kernel is not constant but depends on the aggregate size. After all, since the absolute values of the rate constants depend on the type of counterion [81], additional short-range forces represent an explanation for the discrepancy.

A widely propagated opinion exists, which is that the information about the surface forces obtained from the measurements of coagulation kinetics is only the information about parameters that determine the threshold of rapid coagulation. In other words the only information about parameters that determine the point, where potential barrier on the curve of inter-particle energy disappears may be extracted from the data about kinetics of colloid coagulation.

It looks like close to reality, but only for the case if only the rate of particle aggregation is the measured value. But it is nolonger valid in the case where not only the rate of aggregation but also the rate of disaggregation of arising aggregates occurs to be experimentally determined value. Such detailed information about coagulation process may be obtained with the measurement of single-particle light scattering by the flowing coagulated suspension. Previous investigations [130,141,144] and the work presented here on the rapid coagulation process led to the conclusion that the deaggregation constants of hematite, latex, and silica suspensions are finite values. The theoretical calculations coincide with the experimental results if deaggregation is only considered. For this reason the idea of reversibility seems sensible.

In the mentioned investigations the time evolution of singlets, doublets, and triplets was measured by means of the authors' single particle light-scattering apparatus (Figure 3.5) and the results were directly analyzed by a computer. Coagulation and deaggregation were considered in the description of the time evolution of singlets and higher aggregates. The starting point for the determination of the coagulation constants is given by Equation (32) that describes the change of the number of particles or aggregates of kind j. As a result of the fitting of the kinetic parameters, k_{ij} which characterize the junction of two aggregates each of which contains, correspondingly, i and j single particles to the one containing $(i+j)$ particles and b_{ij}, which characterizes the disaggregation of large aggregate having $(i+j)$ particles onto two smaller ones containing i and j particles can be determined. To allow for the influence of higher aggregates, all factors k_{11} up to k_{99} and b_{11} up to b_{45} have been used in the calculations.

Lichtenfeld et al. [130,141,144] measured the influence of the initial number of particles in the range of 0.3×10^8 up to $2 \times 10^8 \, \mathrm{cm}^{-3}$ on coagulation kinetics with hard-sphere polystyrene (surface charge density: $6.3 \times 10^{-2} \, \mathrm{As/m}^2$), soft-sphere polyvinyl acetate (PVAC) (surface charge density: $1.53 \times 10^{-2} \, \mathrm{As/m}^2$), Fe_2O_3 (hematite) as well as silica particles. For all these experiments a rather good fit was obtained between the experimental and calculated curves with the same coagulation and deaggregation constants corresponding to the kind of particles (Table 3.9).

Deaggregation becomes possible when, as described in Section II.B.2, the approach of two particles resulting in aggregation does not lead to a direct contact of the surface. There is a "wall" of the thickness δ resulting from the Born forces of repulsion or forces of interaction of thin solvate layers (structural barrier) at the particle surface, which is not destroyed by the high electrolyte concentration (Figure 3.1). In this case, the only long-range force of interparticle interaction is the van der Waals force. The fact that the coagulation contact between the surfaces is not a zero contact

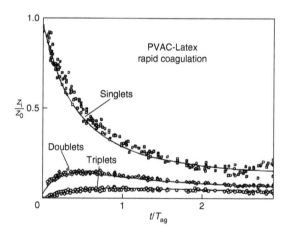

Figure 3.8 Time dependence of the scaled number of singlets, doublets, and triplets for PVAC latex: $z_0 = 0.75 \times 10^8 \, \mathrm{cm}^{-3}$, $t_{max} = 240 \, \mathrm{min}$, $k_{ij} = 6.1 \times 10^{-18} \, \mathrm{m}^3/\mathrm{sec}$, $b_{11} = 6.0 \times 10^{-5} \, \mathrm{s}^{-1}$, $0.5 \, M$ KNO$_3$ [142].

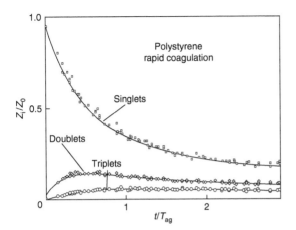

Figure 3.9 Time dependence of the scaled number of singlets, doublets, and triplets for polystyrene latex: $z_0 = 1.816 \times 10^8 \, \text{cm}^{-3}$, $t_{\max} = 99 \, \text{min}$, $k_{ij} = 5.8 \times 10^{-18} \, \text{m}^3/\text{sec}$, $b_{11} = 1.4 \times 10^{-4} \, \text{s}^{-1}$, $0.5 \, M$ KNO_3 [142].

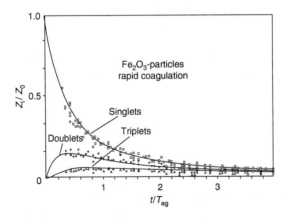

Figure 3.10 Time dependence of the scaled number of singlets, doublets, and triplets for α-Fe_2O_3: $z_0 = 1.63 \times 10^8 \, \text{cm}^{-3}$, $t_{\max} = 90 \, \text{min}$, $k_{ij} = 11 \times 10^{-18} \, \text{m}^3/\text{sec}$, $b_{11} = 3 \times 10^{-5} \, \text{s}^{-1}$, $0.5 \, M \, KNO_3$ [130].

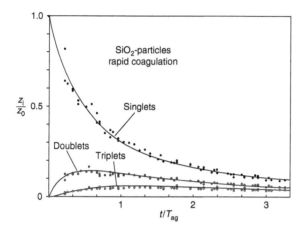

Figure 3.11 Time dependence of the scaled number of singlets, doublets, and triplets for silica particles: $z_0 = 2.0 \times 10^8 \, \text{cm}^{-3}$, $t_{\text{max}} = 90 \, \text{min}$, $k_{ij} = 1.6 \times 10^{-18} \, \text{m}^3/\text{sec}$, $b_{11} = 8 \times 10^{-5} \, \text{s}^{-1}$, isoelectric point — pH = 2.1.

leads to the logical conclusion that the minimum energy is reduced to values of some $k_B T$ units (Figure 3.3). Low value of bond energy makes disintegration of an aggregate possible under the effect of thermal molecular motion [143].

In the following, the value of the distance between particle surfaces, which corresponds to the coagulation contact, is estimated. This will be performed on the basis of the interpretation of measured values of aggregation k_{11} and deaggregation constants b_{11} in the case of rapid coagulation. This is the simplest way to do this because only the van der Waals energy and the corresponding Hamaker constant are used.

The statement of the problems for theoretical calculations of the kinetic coefficients k_{11} and b_{11} is based on the consideration of the initial stage of coagulation process in a very diluted suspension. In such an ideal suspension almost all the particles are single, and the doublets are present in much smaller concentration (as their concentration is quadratic with respect to small concentration of single particles, see Equations (16) and (30)). The existence of the triplets and the aggregates of higher order is negligible in our ideal system such as their concentration values are of third or higher orders with respect to small parameter, i.e., the concentration of single particles. It is very important to note that in as much as booth kinetic parameters k_{11} and b_{11} do not depend on the initial concentration and on the time, their values, obtained on the basis of the consideration of ideal suspension, are fit to general case. In such a way, we use the concept of ideal suspension (which differs from the real one by the utterly small initial concentration, providing the negligible influence of the triplets and large aggregates) for the convenience of consideration and this procedure does not restrict the validity of expressions for k_{11} and b_{11} obtained on this basis.

The equations system in Equation (32), after neglecting the terms of higher than second order with respect to the small parameter, namely initial concentration of the single particles z_{10}, turns into the much more simple system for our ideal suspension:

$$\mathrm{d}z_1/\mathrm{d}t = z_{10}^2 + 2b_{11}z_2, \tag{42}$$

$$\mathrm{d}z_2/\mathrm{d}t = k_{11}z_{10}^2 - b_{11}z_2. \tag{43}$$

As follows from Equations (42) and (43), the decrease of the concentration of single particles is quadratic with respect to small parameter z_1 and, hence we may to consider that concentration of single particles remains invariable during the coagulation process in our ideal suspension, $z_1 \approx z_{10}$.

In the course of time the concentration of the doublets z_2 will increase due to aggregation process, as reflected by the first term in right-hand side of Equation (42). Therefore, the total rate of doublets disaggregation increases, as reflected by the second term in right-hand side of Equation (42). As a result, after long enough time, exceeding $\tau_{eq} = 1/b_{11}$, the concentration of doublets in ideal suspension has to reach its equilibrium level

$$z_{2eq} = \frac{1}{2}\frac{k_{11}}{b_{11}}z_1^2. \tag{44}$$

On the other hand, the finite values of booth constants, k_{11} and b_{11}, give rise to the possibility of some equilibrium between the processes of aggregation of single particles and disaggregation of doublets. The level of this equilibrium may be expressed on basis of the equilibrium statistical physics [145] through the value of the second virial coefficient B_2, which determines the equilibrium relationship between the concentrations of single particles z_1 and doublets $z_{2\text{eq}}$, in suspension:

$$z_{2\text{eq}} = -z_1^2 B_2. \tag{45}$$

By summarizing Equations (44) and (45) one gets

$$B_2 = -\frac{1}{2}\frac{k_{11}}{b_{11}}, \tag{46}$$

$$B_2 = -2\pi a^3 \int_0^\infty \left[\exp\left(-\frac{V(u)}{k_B T}\right) - 1\right](u + 2)^2 \mathrm{d}u. \tag{47}$$

Taking into account that the interacting particles together with thin solvate layers are modeled as rigid spheres of the radius $a + \delta$ (see Figure 3.1) one rewrites Equation (47) as

$$B_2 = -2\pi a^3 \int_{2\delta/a}^\infty \left[\exp\left(-\frac{V(u)}{k_B T}\right) - 1\right](2 + u)^2 \mathrm{d}u. \tag{48}$$

The lower integration limit is the minimal possible distance between two particles.

By summarizing Equations (46) and (48) one gets

$$\frac{k_{11}}{b_{11}} = 4\pi a^3 \int_{2\delta/a}^\infty \left[\exp\left(-\frac{V(u)}{k_B T}\right) - 1\right](2 + u)^2 \mathrm{d}u. \tag{49}$$

By means of this relation the thickness of the structural layer was calculated, according to the measured velocity constants k_{11} and b_{11} (Table 3.9).

The kinetics of coagulation adds disaggregation of diluted colloid suspensions are booth controlled by the space dependence of the energy of interaction between two colloid particles. Therefore, the *transparency* of the nonremovable thin layer of water molecules near the surface of every particle for van der Waals interactions is assumed.

The finite energy leads to the conclusion that the dielectric properties of the bulk water and the structured layer at the particle surface are closely related. If there were pronounced differences in these properties, they would give rise to infinite bonding energy at the contact of these structured layers [144].

C. Experimental Results: Slow Coagulation

In slow coagulation of hydrosols not all particle collisions lead to coagulation. There are different ways that the double-layer repulsion can influence the coagulation rate. An energy barrier can prevent that every collision or only a fraction of collisions become effective with respect to aggregation. On the other hand, the depth of the primary minimum may be diminished so that particles can escape and this deaggregation can reduce the number of eventually effective Brownian collisions. This was already considered in the chapter on rapid coagulation. Hydrophobic colloids such as metal or silver iodide sols coagulate essentially irreversibly in the primary minimum provided the particle dimensions are small enough such that secondary minimum coagulation does not occur. With more hydrophilic particles such as oxidic particles or latex particles, slow coagulation because of an energy barrier may also occur. However, reversibility becomes more important for the time evolution of the singlets and multiplets of all kinds of particle aggregates. Adsorbed water molecules preventing direct contact of the particles [51–62] or a *rough* surface of the particles, e.g., a hairy surface layer (lattices), diminish the minimum energy in the contact position of aggregated particles.

A deaggregation of doublets or linear aggregates under the condition of slow barrier coagulation has a low probability since the total energy for this process, the sum of $V_{max} + V_{min}$, is rather high in comparison with their kinetic energy.

A relationship between the Fuchs stability ratio W and the electrolyte concentration was derived by Reerink and Overbeek [146] neglecting the reversibility of coagulation.

With several approximations a linear relationship between log W and log c was obtained

$$\log W = -K_1 \log c_{el} + K_2. \tag{50}$$

The stability ratio decreases with the addition of an electrolyte until the electrolyte concentration reaches a critical value above which the energy barrier between approaching particles disappears and W becomes equal to unity (critical coagulation concentration). The slope of the stability curve $d(\log W)/d(\log c)$ is related to the particle radius, Stern potential, and valency of the counterions

$$\frac{d(\log W)}{d(\log c)} = 2.15 \times 10^7 \frac{a}{\nu^2} \tan h^2(0.00973\,\nu\psi_\delta) \tag{51}$$

with a in nm and ψ_δ in mV. ν, the valence of the counterions, according to the relation (50) the Stern potential can be obtained. The relationship between the Hamaker constant A and $d(\log W)/d(\log c)$ is given in the following equation [146]:

$$A = \left[1.37 \times 10^{-36}\left(\frac{d(\log W)}{d(\log c)}\right)^2\right] \times \left[a^2\nu^2 c_{crit}\right]^{1/2} \tag{52}$$

with a in cm and the critical coagulation concentration in $10^{-3}\,M$. According to Equation (51), the stability ratio should be proportional to the particle radius and inversely proportional to the square of the counterion valency. It is also dependent on the Stern potential. Some results of different authors are summarized in Table 3.10 and Table 3.11. From the data in Table 3.11 the Stern potential (Equation (51)) and the Hamaker constant were calculated. The results are summarized in Table 3.11. There is no plausible reason for different values of the Hamaker constant with increasing surface charge density.

The classical DLVO theory [148,149] predicts a pronounced dependence of the stability ratio W for slow coagulation on the radius of the colloidal particles.

There are many experimental data, which indicate that there is no such effect [129,146,150,151]. In particular, the

Table 3.10 Values of the slope $\log w/\log c$ curves for different dispersed systems as a function of particle radius and counterion valency

Sol	$d(\log W)/d(\log c)$	Particle radii (nm)	Electrolyte	Ref.
Silver iodide	−10.6	12.5	KNO_3	[97]
	−5.9	26.0	KNO_3	
	−7.3	100.0	KNO_3	
	−8.0	12.5	$Ba(NO_3)_2$	
	−8.0	26.0	$Ba(NO_3)_2$	
	−8.0	32.5	$Ba(NO_3)_2$	
	−11.1	100.0	$Ba(NO_3)_2$	
	−15.8	26.0	$La(NO_3)_2$	
Selenium	−2.3	70.5	KCl	[108]
	−2.1	84.0	KCl	
	−2.1	94.0	KCl	
	−2.0	102.0	KCl	
	−1.7	91.0	KCl	
	−2.4	91.0	$BaCl_2$	
	−1.4	91.0	$LaCl_3$	
Latex	−2.42	300	$Ba(NO_3)_2$	[109]
	−2.78	515	$Ba(NO_3)_2$	
	−2.71	1213	$Ba(NO_3)_2$	
	−2.70	1840	$Ba(NO_3)_2$	
	−1.59	2115	$Ba(NO_3)_2$	
	−0.9	120	$Ba(NO_3)_2$	[110]
	−0.34	714	$Ba(NO_3)_2$	
	−3.11	30	HNO_3	
	−1.35	51	HNO_3	
	1.15	121	HNO_3	[125]
	−0.98	435	HNO_3	

proportionality of $d(\log W)/d(\log c)$ with the particle size, has not been found.

At first, the theoretical values for W are higher than those found in experiments, and secondly, W is only slightly sensitive on the variation of particle radius. Finally, the critical coagulation concentration is dependent on the particle size.

Table 3.11 Stability analysis for polystyrene latex of the same radius ($a = 9.8$ nm) but different surface charge density [98]

Latex no.	As/m^2	d(log W)/ d(log c)	Critical coagulation concentration MgSO$_4 \times 10^{-3}$ dm	A (10^{-21} J)	ψ_δ (10^{-3} V)
1	10.20	3.18	29.5	4.08	−13.1
2	9.29	2.75	30.0	3.50	−11.9
3	9.10	2.34	34.8	2.76	−11.20
4	6.48	2.20	53.8	2.08	−10.85
5	5.02	1.80	61.0	1.52	−9.75
6	4.67	1.45	93.0	1.07	−8.78
7	1.01	1.07	60.0	0.96	−7.52

To explain these discrepancies there have been several suggestions to refine the theory by including extra terms, such as hydrodynamic interaction between colliding particles, hydrophobic interaction, etc. [23,24,65,152–157]. However, this has not been satisfactory.

The investigations of Shulepov and Frens [158,159] are interesting. They observed that surface roughness, rather than particle size, determines the geometrical factor of the interparticle interaction in the kinetics of slow Brownian co-agulation of electrostatically stabilized colloidal particles. They include surface roughness describing characteristic length scales (the amplitude δ and the wavelength λ of the surface roughness) into the calculation of van der Waals at-traction, double-layer repulsion energies, and their effect on coagulation rate. It is found that surface roughness must strongly affect the relation of the stability ratio W with the curvature of the particle surfaces.

The relationship between characteristic size of the sur-face features and area of the maximal double-layer overlap is embodied in parameter $L = \lambda^2/\delta d$, with d is the particle diameter. The theoretical calculations show that a transition in the character of the stability diagram would be expected at $L = \lambda^2/\delta d = 1$. The experiments explore the differences in sta-bility diagrams for one and the same colloidal material (mono-disperse gold sols, see Table 3.12). Table 3.12 and Figure 3.12

Table 3.12 Particle size and roughness character-istics measured from TEM images [159]

$<d>$ (nm)	d (nm)	δ (nm)	$L = \lambda^2/\delta d$
92	99 ± 3.3	0.62 ± 0.11	0.80 ± 0.14
83	85 ± 3.7	0.62 ± 0.09	0.93 ± 0.12
42	55 ± 1.7	0.62 ± 0.13	1.44 ± 0.21
35	38 ± 1.3	0.70 ± 0.15	1.80 ± 0.29
22	29 ± 1.3	0.76 ± 0.11	2.20 ± 0.28

λ is taken as 7 nm.

demonstrate that a sudden transition in the shape of the $\log W - \log c$ diagrams occurs at $L = \lambda^2/\delta d = 1$. Moreover, a comparison of the experimental stability diagram (Figure 3.12) with the theoretical one (Figure 3.13) shows that it is possible to fit theoretically the experimental data without the need for any adaptable parameter [159].

The experiments on slow coagulation can be analyzed in the same manner as described for rapid coagulation (Equation (32)) by fitting the experimental curves for the time evolution of singlets, doublets, and triplets with calculated curves by

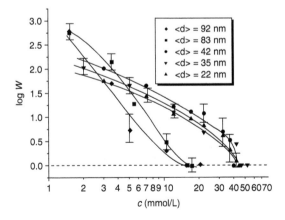

Figure 3.12 Experimental stability diagram for a gold sol with characteristics given in Table 3.12 [159].

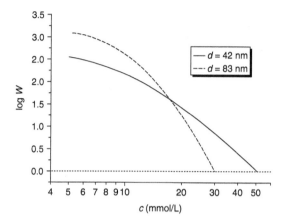

Figure 3.13 Theoretical stability diagram [159].

(Table 3.13) variation of k_{ij} and b_{ij}. Experimental results are shown polystyrene and for polyvinyl acetate particles with the radius $0.3\,\mu m$ different electrolyte concentrations (KNO_3).

In Figure 3.14 and Figure 3.15 experimental results are shown for polyvinyl acetate particles at an initial particle concentration of $z_0 = 1.88 \times 10^8\,cm^{-3}$ and an electrolyte concentration of 10^{-1} and $6 \times 10^{-3}\,M$. (The critical coagulation

Table 3.13 Coagulation and deaggregation constant for polystyrene particles (radius $0.3\,\mu m$) and polyvinyl acetate particles (radius $0.3\,\mu m$) depend on KNO_3 concentration

	Polystyrene		Polyvinyl acetate	
Electrolyte concentration mole	Coagulation constant × $10^{-18}\,m^3$/sec	Deaggregation constant × $10^{-4}\,s^{-1}$	Coagulation constant × $10^{-18}\,m^3$/sec	Deaggregation constant × $10^{-4}\,s^{-1}$
5×10^{-1}	4.9	1.4	6.1	0.6
10^{-1}	–	–	3.0	3.0
5×10^{-2}	2.3	1.4	2.6	3.0
10^{-2}	1.1	1.6	–	–
6×10^{-3}	0.94	1.7	1.5	3.0

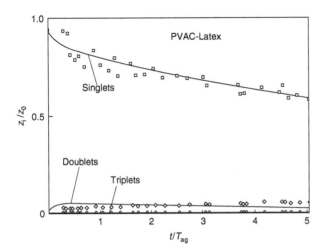

Figure 3.14 Slow coagulation of polyvinyl acetate particles: $z_0 = 1.88 \times 10^8 \, \text{cm}^{-3}$ in $10^{-1} \, \text{KNO}_3$, $t_{\text{max}} = 145 \, \text{min}$, $k_{ij} = 3.0 \times 10^{-18}$ m^3/sec, $b_{11} = 3 \times 10^{-3} \, \text{s}^{-1}$.

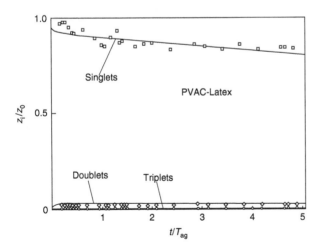

Figure 3.15 Slow coagulation of polyvinyl acetate particles: $z_0 = 1.88 \times 10^8 \, \text{cm}^{-3}$ in $6 \times 10^{-3} M$ KNO_3, $t_{\text{max}} = 145 \, \text{min}$, $k_{ij} = 1.5 \times 10^{-18} \, \text{m}^3/\text{sec}$, $b_{11} = 3 \times 10^{-3} \, \text{s}^{-1}$.

concentration for rapid coagulation equals $5 \times 10^{-1} M$.) For a
definite electrolyte concentration the forward rate constant is
almost independent of the size of the aggregates. The coagulation constant decreases with decreasing electrolyte concentration but much lower than the expected. The deaggregation
constant for doublets and triplets is independent of the electrolyte concentration. From this result one can draw the conclusion that the repulsion forces are not (or not only)
dependent on the double-layer repulsion. Probably steric repulsion forces due to the *hairy layer* around the particles
prevent a deep primary minimum.

Another peculiarity of latex particles may be due to the
possibility of that the charged groups are distributed within
the hairy layer as in polyelectrolytes [160].

From the coagulation constant and the deaggregation
constant one can estimate the height of the energy barrier
and the depth of the primary minimum. Equation (49) can be
applied in the generalized case in accordance with the model
of linear aggregates in which it was assumed that k_{ij} and b_{ij}
are independent of the number of particles within the aggregate (at least in the early stages of coagulation).

By using of the relationship $u = (r - 2a)/a$ and the assumption that δ_{ij} is the distance between an i-mer and j-mer
at which the interaction forces become dominant, one gets

$$\frac{k_{ij}}{b_{ij}} = \int_{2a}^{2a+\delta_{ij}} [\exp(-V/k_{\mathrm{B}}T) - 1] 4\pi r^2 \mathrm{d}r. \tag{53}$$

If one models the interaction energy–distance curve between
two interacting particles by a rectangular potential minimum
and a rectangular maximum of widths δ_1 and δ_2 [147], respectively, Equation (53) can be written as

$$\frac{k_{ij}}{b_{ij}} = 16\pi a^2 \delta_1 \left[\exp\left(-\frac{V_{\min}}{k_{\mathrm{B}}T}\right) - 1\right]. \tag{54}$$

If we assume that only electrostatic forces are responsible for
the repulsion, then the characteristic distance δ_1 can be approximately determined by the thickness of the electrical
double layer (κ^{-1}).

Insofar as V_{max} and V_{min} do not depend much on δ_1 respectively δ_2 and if δ_1 and δ_2 are of the same order of magnitude, one obtains [147]:

$$W = \frac{1}{2}(\kappa a)^{-1} \exp\left(\frac{V_{max}}{k_B T}\right) \qquad (55)$$

and

$$\frac{k_{ij}}{b_{ij}} = 16\pi a^2 \kappa^{-1}\left[\exp -\left(\frac{V_{min}}{k_B T}\right) - 1\right]. \qquad (56)$$

If one now analyzes the experimental results in Table 3.14 with the help of Equations (55) and (56), the height of the barrier and the depth of the energy minimum can be estimated. The result is surprising in that neither the barrier nor the energy minimum is much dependent on the electrolyte concentration. From this result one should come to the conclusion that the repulsion has also a steric component.

In these experiments reversibility was assumed only for the small aggregates; the higher aggregates seemed to be linked together irreversibly. But in some other papers on latex coagulation, reversibility was claimed for all kinds of aggregates resulting in an equilibrium state [109,114,140].

Table 3.14 Height of energy barrier and depth of minimum calculated from kinetic data in comparison with height of barrier calculated from potential

c_{el} (m)	ζ (mV)	$\frac{V_{bar,\zeta}}{k_B T}$	$\frac{V_{bar,kin}}{k_B T}$	$\frac{V_{min,kin}}{k_B T}$
Polystyrene				
6×10^{-3}	95	–	6.3	−7.2
2×10^{-2}	103	1.9×10^3	7.2	−8.8
5×10^{-2}	58	1.2×10^3	6.6	−10.5
1×10^{-1}	53	0.8×10^3	6.7	−11.1
Polyvinyl acetate				
2×10^{-2}	32	–	7.1	−15.2
5×10^{-2}	28	–	7.6	−15.6
1×10^{-1}	26	–	7.2	−15.3

Smith and Thompson [161] using streaming ultramicro-scopy investigated equilibrium coagulation of polystyrene latex particles of the radius 0.8 μm in $10^{-2} M$ KNO$_3$ and for graphite particles of the radius 0.15 μm as a function of the initial particle concentration. From these data the primary minimum energy was calculated (see Table 3.15). The minimum energy decreases with increasing particle concentration. This is the opposite effect that would result from nonlinear aggregates. Cahill et al. [114,140] described further efforts to generate slow aggregation; the success has been limited so far by an inability to generate equilibrium coagulation. They observed a mixture of reversible and irreversible processes. This is in conformity with the result described before.

Thompson and Pryde [162,163] investigated slow coagu-lation of polystyrene particles with a radius $a = 0.5$ μm, stabilized by a monolayer of n-dodecyloctaethylene glycol monoether. The electrolyte concentration was high enough to suppress electrostatic interaction. They measured the influence of the initial particle number and temperature on the rate of coagulation. The minimum energy varied from $-18 k_B T$ at $z_0 = 0.30 \times 10^8$ cm^{-3} to $-22 k_B T$ at $z_0 = 6.9 \times$

Table 3.15 Minimum energy as a function of initial particle concentration for polystyrene ($a = 0.8$ μm) and graphite particles ($a = 0.15$ μm) in $10^{-2} M$ KNO$_3$

Latex		Graphite	
$V_{\min}$		$V_{\min}$	
$z_0 \times 10^7$ cm^{-3}	$k_B T$ units	$z_0 \times 10^7$ cm^{-3}	$k_B T$ units
1.6	−11.8	0.5	−17.2
4.4	−11.9	1.25	−13.7
6.0	−11.6	2.0	−15.0
6.0	−11.4	9.5	−13.4
11.5	−10.7	10.0	−13.0
18.8	−10.6	12.5	−12.6
38.8	−10.3	18.2	−12.3
60.0	−9.4	26.0	−11.9
−	−	32.0	−11.9

$10^8\,\mathrm{cm}^{-3}$. The same latex was investigated at different temperatures. The minimum energy increased from $-13.4\,k_BT$ at 288 K to $14.5\,k_BT$ at 318 K ($z_0 = 2 \times 10^{-8}$ to $3 \times 10^{-8}\,\mathrm{cm}^{-3}$).

Slow coagulation was investigated by Frens [164] with fresh sols of silver iodide, prepared by slowly adding silver nitrate solution with stirring to a solution of potassium iodide. The sol was dialyzed until the conductivity reached 10^{-3} per Ω/m (for about 10–15 days) and then aged for 5 days at 353 K. The stability ratio was determined from the initial slope of the optical extinction–time curve. The initial slopes for rapid coagulation with potassium, barium, and lanthanum counterions varied in the ratio 1.50:1.00:1.2. The authors measured the potential at the stability ratio of 10 for the three electrolytes and diluted the dispersion, thus minimizing the complications with the increasing aggregate size. The results for two sols are summarized in Table 3.16. The effect of the counterion valency can be seen. The Schulze–Hardy rule, may formally express it.

The aggregation of large latex particles that coagulate in the secondary minimum is reversible. This was investigated by Jeffrey and Ottewill [165]. Latex particles of a radius 2.55 and 1.3 μm were measured by simple in situ counting of the number of singlets, doublets, and sometimes triplets as a function of time, using an optical microscope. The experiments were performed in a 50:50 volume-to-volume mixture of D_2O and H_2O to minimize sedimentation. The initial particle concentration was about 10^7–10^8 particles/cm^3. The experimental data indicate that at low electrolyte concentration the singlet-to-doublet ratio initially falls with

Table 3.16 Electrophoresis results at $W = 10$

Electrolyte	Sol a ($a = 16\,\mathrm{nm}$)		Sol b ($a = 33\,\mathrm{nm}$)	
	$c_{10} \times 10^{-3}\,M$	ζ (mV)	$c_{10} \times 10^{-3}\,M$	ζ (mV)
KNO_3	55	$60 + 5$	65	$60 + 5$
$Ba(NO_3)_2$	2.3	$22 + 7$	2.6	$37 + 7$
$La(NO_3)_2$	0.7	$15 + 2$	0.1	$17 + 2$

time but it appears to approach a constant value later. The value of this steady-state ratio decreased with the concentration of salt added and the time taken up to reach steady state increased. The experimental curves were compared with calculated curves, obtained for different models. Figure 3.16 shows curves of the singlet-to-doublet ratio as a function of time for irreversible coagulation (curve 1), for equal deaggregation rate, e.g., $b_{ij} = $ constant (curve 2), and for semireversible coagulation, e.g., doublets and linear aggregates may disaggregate. However, compact aggregates are linked together irreversibly (curve 3). From the experimental data described in Ref. [165] one can make the following statements: the bigger particles show initially some mobility within the flocs, even under the condition of rapid coagulation at $0.1\,M$ sodium chloride, but no dissociation was found. Dispersions of lower electrolyte concentrations 0.067 and $0.05\,M$ showed

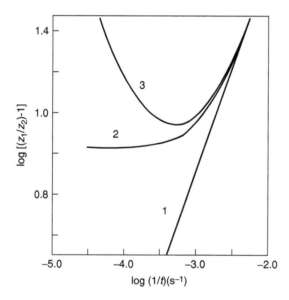

Figure 3.16   Singlet-to-doublet ratio as a function of time. Calculated to (a) model 1, irreversible aggregation; (b) model 2, equal values for all deaggregation constants; (c) model 3, semireversible, deaggregation of doublets and linear aggregates.

marked mobility within flocs. Coagulation of a doublet with a singlet always gave a *linear* triplet, but not usually straight. Observations with the smaller latex were consistent with those of the bigger particles. Doublets formed reversibly. In lower electrolyte concentrations reversible aggregation was obtained.

The Smoluchowski theory of irreversible coagulation predicts a linear decrease of the singlet-to-doublet ratio. This was found in Ref. [165] for rapid coagulation. The value obtained from the integration of Equation (32) was $k_{11} = 11 \times 10^{-18}\,\mathrm{m^3/}$ sec. This value is higher than expected from Smoluchowski theory when the hydrodynamic interaction is taken into consideration. In the same paper [165], slow coagulation was also considered. At low electrolyte concentration the breakup was seen only for a single interparticle bond. The experimental data indicate that at low electrolyte concentration the singlet-to-doublet ratio initially falls with time, but the ratio approaches a constant value later. The value of this steady state decreases with the electrolyte concentration and the time to reach the steady state increases. For the smaller latex the forward rapid rate constant was $5 \times 10^{-18}\,\mathrm{m^3/sec}$. The results for the deaggregation constant as a function of electrolyte concentration for the two latex samples are summarized in Table 3.17. The table summarizes the experimental steady-state single-to-doublet ratios and the best fit to these data according to the model for reversible coagulation.

Table 3.17 Results for deaggregation constants as a function of electrolyte concentration for two latex samples

| Latex a ($a = 2.55\,\mu$m) | | Latex b ($a = 1.30\,\mu$m) | |
| $k_{11} = 11 \times 10^{-18}\,\mathrm{m^3/sec}$ | | $k_{11} = 5 \times 10^{-18}\,\mathrm{m^3/sec}$ | |
C_{NaCl} (mol/dm^3)	b_{11} (s^{-1})	C_{NaCl} (mol/dm^3)	b_{11} (s^{-1})
0.00	0.009	0.00	0.0155
0.019	0.004	0.038	0.00679
0.0305	0.0025	0.067	0.00456
0.00476	0.0019	0.1	0.0028
0.00724	0.0005	0.011	0.000885

It is worth reiterating that Equation (35) is a particularly useful way of describing the aggregation of colloidal dispersions.

REFERENCES

1. Einstein, A., *Ann. Phys.*, 17, 549–560, 1905.
2. O'Neill, M.E., *Proc. Cambridge Philos. Soc.*, 65, 543, 1969.
3. Cooley, M.D.A. and O'Neill, M.E., *Proc. Cambridge Philos. Soc.*, 66, 407, 1969.
4. Einstein, A., *The Theory of the Brownian Movement*, Dover, New York, 1956.
5. Belongia, B.M. and Baygents, J.C., *J. Colloid Interf. Sci.*, 195, 19–31, 1997.
6. Stilbs, B., *Prog. NMR Spectrosc.*, 19, 1, 1987.
7. Dunlop, P.J., Harris, K.R., and Young, D.J., Determination thermodynamic properties, in *Physical Methods of Chemistry*, Rossiter, B.W. and Baetzold, R.C., Eds., Vol. 6, John Wiley & Sons, New York, 1992, p. 236.
8. Zhao, J.J., Bae, S.S., Xie, F, and Granick, S., *Macromolecules*, 34, 3123–3126, 2001.
9. Schumacher, G.A. and van de Ven, Th.G.M., *Faraday Disc. Chem. Soc.*, 83, 75, 1987.
10. Krystev, G.A., Dakova, D.I., and Dimitrov, V.S., *Colloid J.*, 60, 491–498, 1998.
11. Cramer, S.E., Jeschke, G., and Spiess, H.W., *Colloid Polym. Sci.*, 280, 569–573, 2002.
12. Tawari, S.L., Koch, D.L., and Cohen, C., *J. Colloid Interf. Sci.*, 10.1006/jcis.2001, 7646, 2001.
13. Piazza, R. and Degiorgio, V., *J. Phys. Condens. Matter*, 8, 9497, 1996.
14. Koenderink, G.H., Zhang, H., Aarts, D.G.L.A., Lettinga, M., Philipsea, A.P., and Nägele, G., *Faraday Discuss.*, 123, 335–354, 2003.
15. Vadas, E.B., Goldsmith, H.L., and Mason, S.G., *J. Colloid Interf. Sci.*, 43, 630–648, 1973.
16. Vadas, E.B., Cox, R.G., Goldsmith, H.L., and Mason, S.G., *J. Colloid Interf. Sci.*, 57, 308, 1976.
17. Cornell, R.M., Goodwin, J.W., and Ottewill, R.H., *J. Colloid Interf. Sci.*, 71, 254–266, 1979.

18. Reynolds, P.A. and Goodwin, J.W., *Colloid Surf.*, 11, 145–154, 1984.
19. Will, S. and Leipertz, A., *Progr. Colloid Polym. Sci.*, 104, 110, 1997.
20. Pecora, R., *J. Nanoparticle Res.*, 2, 123–131, 2000.
21. von Smoluchowski, M., *Phys. Z.*, 17, 557–599, 1916.
22. Derjaguin, B.V. and Muller, V.M., *Dokl. Akad. Nauk. SSSR* (Engl transl.), 176, 738, 1967.
23. Spielman, L.A., *J. Colloid Interf. Sci.*, 33, 562, 1970.
24. Honig, E.P., Roebersen, G.J., and Wiersema, P.H., *J. Colloid Interf. Sci.*, 36, 97–112, 1971.
25. McGown, D.N.L. and Parfitt, G.D.J., *J. Phys. Chem.*, 71, 449–450, 1988.
26. Zholkowskij, E.K. and Dukhin, S.S., *Kolloidn. Zh.*, 46, 392, 1984.
27. Dukhin, S.S. and Lyklema, J., *Langmuir* 3, 94–98, 1987.
28. Muller, V.M., *Kolloidn. Zh.*, 50, 1119, 1988.
29. von Smoluchowski, M., *Phys. Chem.*, 92, 129–168, 1917.
30. Vicsek, T. and Family, F., *Phys. Rev. Lett.*, 52, 1669, 1984.
31. Meakin, P., Vicsek, T., and Family, F., *Phys. Rev. B*, 31, 564, 1985.
32. Ziff, R.M., Grady, E.D., and Meakin, P., *J. Chem. Phys.*, 82, 5269, 1985.
33. Meakin, P., *Phys. Rev. Lett.*, 51, 1119, 1983.
34. Kolb, M., Botet, R., and Jullien, R., *Phys. Rev. Lett.*, 51, 1123, 1983.
35. Kolb, M., *Phys. Rev. Lett.*, 53, 1653, 1984.
36. Martynov, G.A. and Bakanov, S.P., in *Investigation of Surface Forces*, Isd. Nauk, Moscow, 1961, pp. 220–229.
37. Friedländer, S.K. and Wang, C.S., *J. Colloid Interf. Sci.*, 22, 126–132, 1966.
38. Wang, C.S. and Friedländer, S.K., *J. Colloid Interf. Sci.*, 24, 170–179, 1967.
39. Cohen, E.R. and Vaughan, E.U., *J. Colloid Interf. Sci.*, 35, 612–623, 1971.
40. Lee, U.W., *J. Colloid Interf. Sci.*, 92, 315–325, 1983.
41. Bauer, D., Killmann, E., and Jaeger, W., *Colloid Polym. Sci.*, 276, 698–708, 1998.
42. Zepeda, S., Yeh, Y., and Noy, A., *Langmuir*, 19, 1457–1461, 2003.
43. Yu, X. and Somasundaran, P., *Colloids Surf. A*, 89, 277–2781, 1994.

44. Healy, T.W., *The Colloid Chemistry of Silica*, Bergna, H.E., Ed., Advances in Chemical Series 234, American Chemical Society, 1994, pp.147–159.
45. Kadantseva, I., Gritskova, I.A., Yanul, Y.B., Ferderova, V.A., Donchak, V.A., and Timofeevich, N.T., *Colloid J.*, 61, 60–63, 1999.
46. Molina-Bolivar, J.A., Galisteo-González, F., and Hidalgo-Álvarez, R., *J. Colloid Interf. Sci.*, 206, 518–526, 1998.
47. Zimehl, R. and Priewe, J., *Progr. Colloid Polym. Sci.*, 101, 116–129, 1996.
48. Glazman, Y.M., *Faraday Disc Chem. Soc.*, 42, 255–266, 1966.
49. Killmann, E. and Sapuntzjis, P., *Colloids Surf. A*, 86, 229–238, 1994.
50. Romero-Cano, M.S., Puertas, A.M., and de las Nieves, F.J., *J. Chem. Phys.*, 112, 8654–8659, 2000.
51. Chang, Y.-M. and Chang, P.-K., *Colloids Surf. A*, 211, 67–77, 2002.
52. Higashitani, K., Kondo, M., and Hatade, S., *J. Colloid Interf. Sci.*, 142, 204–213, 1991.
53. Baldwin, J.L. and Dempsey, B.A., *Colloids Surf. A*, 177, 111–122, 2001.
54. Hancer, M., Celik, M.S., and Miller, J.D., *J. Colloid Interf. Sci.*, 235, 150–161, 2001.
55. Eremenko, B.V., Malysheva, M.L., Bezuglaya, T.N., Savitskaya, A.N., Kozlov, I.S., and Bogodist, L.G., *Colloid J.*, 62, 50–55, 2000.
56. Rohrsetzer, S., Pászli, I., and Csempesz, F., *Colloid Polym. Sci.*, 276, 260–266, 1998.
57. Peschel, G. and van Brevern, O., *Prog. Colloid Polym. Sci.*, 84, 405–408, 1991.
58. Ludwig, P. and Peschel, G., *Prog. Colloid Polym. Sci.*, 77, 146–151, 1988.
59. Peschel, G., Belouschek, P., Müller, M.M., Müller, M.R., and König, R., *Colloid Polym. Sci.*, 260, 444–451, 1982.
60. Johnson, G.A., Lecchini, S.M.A., Smith, E.G., Clifford, J, and Pethica, B.A., *Faraday Disc. Chem. Soc.*, 42, 120–133, 1966.
61. Baldwin, J.L. and Dempsey, B.A., *Colloids Surf. A*, 177, 111–122, 2001.
62. Lichtenfeld, H., Shilov, V., and Knapschinsky, L., *Colloids Surf. A*, 142, 155–163, 1989.

63. Sonntag, H. and Strenge, K., *Coagulation Kinetics and Structure Formation*, VEB Deutscher Verlag der Wissenschaften, Berlin, 1987, p. 32.
64. Martynov, G.A. and Muller, V.M., *Surface Forces in Thin Films and Dispersed Systems*, Isd. Nauk, Moscow, 1972, pp. 7–34.
65. Frens, G. and Overbeek, J.Th.G., *J. Colloid Interf. Sci.*, 38, 376–387, 1972.
66. Frens, G., *Faraday Disc. Chem. Soc.*, 65, 146, 1978.
67. Heller, W. and Tabibian, R.M., *J. Colloid Sci.*, 12, 25, 1957.
68. Lichtenbelt, J.W.Th., Ras, H.J.M., and Wiersema, P.H., *J. Colloid Interf. Sci.*, 46, 281, 1974.
69. Melik, D.H. and Fogler, H.S., *J. Colloid Interf. Sci.*, 12, 161, 1983.
70. Robertson, A.A. and Mason, S.G., *Pulp and Paper*, Vol. 55, Hag, Canada, 1954, p. 263.
71. Ditter, W., Eisenlauer, J., and Horn, D., *The Effect of Polymers on Dispersion Properties*, Tadros, Th., Ed., Academic Press, London, 1982.
72. Gregory, J. and Nelson, D., *Colloids Surf. A*, 18, 175, 1986.
73. Yotsumoto, H. and Yoon, R.H., *J. Colloid Interf. Sci.*, 157, 426–433, 1993.
74. Yotsumoto, H. and Yoon, R.H., *J. Colloid Interf. Sci.*, 157, 434–441, 1993.
75. Weiss, A. and Ballauf, M., *Colloid Polym. Sci.*, 278, 1119–1125, 2000.
76. Lips, A. and Willis, E.J., *J. Chem. Soc. Faraday 1*, 67, 2979, 1971.
76a. Lips, A. and Willis, E.J., *J. Chem. Soc. Faraday 1*, 69, 1226, 1973.
77. Wu, C. and Chu, B., Light scattering, in *Handbook of Polymer Science. Modern Methods in Polymer Research and Technology*, Tanaka, T., Grosberg, A., and Doi, M., Eds., Academic Press, Boston, MA, 1998.
78. Richards, R.W., Ed., *Scattering Methods in Polymer Science*, Ellis Horwood Ltd, Great Britain, 1995.
79. Brown, W., Ed., *Dynamic Light Scattering: The Methods and Some Applications*, Clarendon Press, Oxford, 1993.
80. Berne, B.J. and Pecora, R., *Dynamic Light Scattering*, Dover Publications, New York, 2000.
81. Holthoff, H., Egelhaaf, S.U., Borkovec, M., Schurtenberger, P., and Sticher, H., *Langmuir*, 12, 5541–5549, 1996.

82. Holthoff, H., Borkovec, M., and Schurtenberger, P., *Phys. Rev. E*, 56, 6945–6952, 1997.
83. Holthoff, H., Schmitt, A., Fernández-Barbero, A., Borkovec, M., Cabrerizo-Wilchez, M.A., Schurtenberger, P., and Hidalgo-Àlvarez, R., *J. Colloid Interf. Sci.*, 192, 463–470, 1997.
84. van Zanten, J.H. and Elimelech, M., *J. Colloid Interf. Sci.*, 154, 1–7, 1992.
85. Bryant, G., Martin, S., Budi, S., and van Megen, W., *Langmuir*, 19, 616–621, 2003.
86. Virden, J.W. and Berg, J.C., *J. Colloid Interf. Sci.*, 149, 528–535, 1992.
87. Polverari, M. and van de Ven, T.G.M., *Colloids Surf. A*, 86, 209–228, 1994.
88. Killmann, E. and Sapuntzjis, P., *Colloids Surf. A*, 86, 229–238, 1994.
89. Chu, B. and Liu, T., *J. Nanoparticle Res.*, 2, 29–41, 2000.
90. Sedlák, M., *Langmuir*, 15, 4045–4051, 1999.
91. Deblois, R.W. and Bean, C.P., *Rev. Sci. Instr.*, 41, 909, 1970.
92. Steinkamp, A., *Rev. Sci. Instr.*, 55, 1375, 1984.
93. Siedentopf, H. and Zsimondy, R., *Ann. Physik.*, 10, 1, 1903.
94. Zsimondy, R., *Physik Zeitschr.*, XIV, 975, 1913.
95. Zsimondy, R., *Z. Phys. Chem.*, 92, 600, 1918.
96. Derjaguin, B.V. and Vlasenko, G.Y., *Doklady Akad. Nauk*, 63, 155, 1948.
97. Derjaguin, B.V. and Vlasenko, G.J., *J. Colloid Sci.*, 17, 605, 1962.
98. Gucker, F.T., Okonski, Ch.T., Jr., Pickard, H.B., and Pitts, J.N., Jr., *J. Am. Chem. Soc.*, 69, 2422, 1947.
99. Gucker, F.T. and Okonski, Ch.T., Jr., *J. Colloid Sci.*, 4, 541, 1949.
100. Lee, P.K. and La, V.K., *Mer. Rev. Sci. Instr.*, 25, 1004, 1954.
101. Watillon, A. and van Grunderbeek, F., *Bull. Soc. Chim. Belg.*, 63, 115, 1954.
102. Mikirov, A.E., *Izvest. Akad. Nauk SSSR Se. Geofiz*, 4, 512, 1957.
103. Ottewill, R.H. and Wilkins, D.J., *J. Colloid Sci.*, 15, 512, 1960.
104. Derjaguin, B.V., Churakov, V.V., and Vlasenko, Y., *Kolloid Z.*, 2, 234, 1961.
105. Davidson, J.A., Macosko, C.W., and Collins, E.A., *J. Colloid Interf. Sci.*, 25, 381, 1967.
106. Davidson, J.A., Collins, E.A., and Haller, H.S., *J. Polym. Sci., Part C*, 35, 235, 1971.

107. Mullaney, P.F., Van Dilla, M.A., Coulter, J.R., and Dean, P.N., *Rev. Sci. Instrum.*, 40, 1029, 1969.
108. McFadyen, P.F. and Smith, A., *J. Colloid Interf. Sci.*, 45, 573, 1973.
109. Buske, N., Gedan, H., Lichtenfeld, H., Katz, W., and Sonntag, H., *Colloid Polym. Sci.*, 258, 1303, 1980.
110. Cummins, P.G., Smith, A.L., Staples, E.J., and Thompson, L.G., *Solid–Liquid Separation*, Gregory, J., Ed., Ellis Horwood, Chichester, 1984, p. 161.
111. Cahill, J., Cummins, P.G., Staples, E.J., and Thompson, L., *Colloids Surf.*, 18, 189, 1986.
112. Pelssers, E.G.M., Thesis, Wageningen, The Netherlands,1988.
113. Pelssers, E.G.M., Cohen Stuart, M.A., and Fleer, G.J., *J. Colloid Interf. Sci.*, 137, 350–372, 1990.
114. Bowen, M.S., Broide, M.L., and Cohen, R.J., *J. Colloid Interf. Sci.*, 105, 605–617, 1985.
115. Broide, M. and Cohen, R.J., *J. Colloid Interf. Sci.*, 153, 493–508, 1992.
116. Fernández-Barbero, A., Schmitt, A., Cabrerizo-Vilchez, M., and Martinez-Garcia, R., *Physica* A, 230, 53–74, 1996.
117. Holthoff, H., Schmitt, A., Fernández-Barbero, A., Borkovec, M., Cabrerizo-Wilchez, M.A., Schurtenberger, P., and Hidalgo-Àlvarez, R., *J. Colloid Interf. Sci.*, 192, 463–470, 1997.
118. Hershberger, L.W., Callis, J.B., and Christian, G.D., *Anal. Chem.*, 51, 1444, 1979.
119. Lichtenfeld, H., Stechemesser, H., and Möhwald, H., *J. Colloid Interf. Sci.* 276, 97–105, 2004.
120. Hunter, R.J., *Foundations of Colloid Science*, Vol 1, Clarendon Press, Oxford, 1991, pp. 154–157.
121. Gedan, H., Lichtenfeld, H., Sonntag, H., and Krug, H.J., *Colloids Surf.* A, 11, 199, 1984.
122. Zsigmondy, R., *Z. Phys. Chem.*, 93, 600, 1918.
123. Westgren, A. and Reitstötter, J., *Z. Phys. Chem.*, 92, 750, 1918.
124. Tuorilla, A., *Kolloidchem Beihefte*, 22, 193, 1926.
125. Derjaguin, B.V. and Kudravzeva, N.M., *Kolloidny Z.*, 26, 61, 1964.
126. Watillon, A., Romerowski, M., and van Grunderbeek, F., *Bull. Soc. Chim. Belg.*, 68, 450, 1959.
127. Ottewill, R.H. and Rastogi, M.C., *J. Chem. Soc. Trans. Faraday Soc.*, 56, 866, 1960.
128. Penners, N.H.G., Thesis, Agricultural University of Wageningen, The Netherlands, 1985.

129. Penners, N.H.G. and Koopal, L.K., *Colloids Surf. A*, 28, 67, 1987.
130. Lichtenfeld, H., Knapschinsky, L., Sonntag, H., and Shilov, V., *Colloids Surf. A*, 104, 313, 1995.
131. Shilov, V., Lichtenfeld, H., Sonntag, H., *Colloids Surf. A*, 104, 321, 1995.
132. Walz, J.Y., *Adv. Colloid Interf. Sci.*, 74, 119, 1998.
133. Bhattacharjee, S., Ko, Ch.H., and Elimelech, M., *Lagmnuir*, 14, 3365, 1998.
134. Rabinovich, Y.I., Adler, J.J., Ata, A., Singh, R.K., and Mougdil, B.M., *J. Colloid Interf. Sci.*, 232, 10, 17, 2000.
135. Kostoglou, M. and Karabelas, A.J., *J. Colloid Interf. Sci.*, 171, 187, 1995.
136. Highuchi, W.J., Okada, R., Stelter, G.A., and Lemberger, A.P., *J. Pharm. Sci.*, 52, 49, 1963.
137. Matthews, B.A. and Rhodes, C.T., *J. Pharm. Sci.*, 57, 558, 1967.
138. Hatton, W.P., McFadyen, P.F., and Smith, A.L., *J. Chem. Soc. Faraday Trans.* I, 70, 655, 1974.
139. Lips, A. and Duckworth, R.M., *J. Chem. Soc. Faraday Trans.*, 67, 2979, 1971.
140. Cahill, J., Cummins, P.G., Staples, E.J., and Thompson, L., *J. Colloid Interf. Sci.*, 117, 406, 1987.
141. Lichtenfeld, H., Sonntag, H., and Dürr, C., *Colloids Surf.*, 54, 267, 1991.
142. Martynov, G.A. and Muller, V.M., *Surface Forces in Thin Films and Dispersed Systems*, Isd Nauk, Moscow, 1961, p. 220.
143. Muller, V.M., *Colloid J.*, 58, 598–611, 1996.
144. Lichtenfeld, H., Shilov, V., and Knapschinsky, L., *Colloids Surf. A*, 142, 155, 1998.
145. Landau, L.D. and Lifschitz, E.M., *Lehrbuch der Theoretischen Physik*, Bd 5, Statistische Physik, Akademie-Verlag, Berlin, 1966, p. 235.
146. Reerink, H. and Overbeek, J.Th.G., *Disc Faraday Soc.*, 18, 74–84, 1954.
147. Sonntag, H., Shilov, V., Gedan, H., Lichtenfeld, H., and Dürr, C., *Colloids Surf.*, 20, 303–317, 1986.
148. Derjaguin, B.V. and Landau, L.D., *Acta Physicochim. USSR*, 14, 633, 1941.
149. Verwey, E.J.W. and Overbeek, J.Th.G., *Theory of the Stability of Lyophobic Colloids*, Elsevier, Amsterdam, 1948.
150. Ottewill, R.H. and Shaw, J., *Disc Faraday Soc.*, 42, 154, 1966.

151. Watillon, A. and Joseph-Petit, A.M., *Disc Faraday Soc.*, 42, 143, 1966.
152. Derjaguin, B.V., *Theory of Stability of Colloids and Thin Films*, Consultants Bureau, New York, 1989.
153. Dukhin, S.S. and Lyklema, J., *Disc Faraday Soc.*, 90, 261, 1990.
154. Prieve, D.C. and Lin, M.M.J., *J. Colloid Interf. Sci.*, 86, 17, 1982.
155. Kihara, H., Ryde, N., and Matijevic, E., *Colloids Surf.*, 64, 317, 1992.
156. Kihara, H., Ryde, N., and Matijevic, E., *J. Chem. Soc. Faraday Trans.*, 88, 23789, 1992.
157. Overbeek, J.Th.G., *Adv. Colloid Interf. Sci.*, 16, 17, 1982.
158. Shulepov, S.Y. and Frens, G., *J. Colloid Interf. Sci.*, 170, 44, 1995.
159. Shulepov, S.Y. and Frens, G., *J. Colloid Interf. Sci.*, 182, 388, 1996.
160. Spitzer, J.J., Midgley, C.A., Slooten, H.G., and Lok, K.P., *Colloids Surf.*, 39, 273, 1989.
161. Smith, A.L. and Thompson, L., *J. Colloid Interf. Sci.*, 77, 557, 1981.
162. Thompson, L. and Pryde, D., *J. Chem. Soc. Faraday Trans. I*, 77, 2405, 1981.
163. Thompson, D., *J. Chem. Soc. Faraday Trans. I*, 80, 1673, 1984.
164. Frens, G., *Colloids Surf.*, 30, 195, 1988.
165. Jeffrey, G.C. and Ottewill, R.H., *Colloid Polym. Sci.*, 268, 170, 1990.
166. Lichtenfeld, H., Stechemesser, H., and Möhwald, H., *J. Colloid Interf. Sci.* 276, 97–105, 2004.
167. Odriozola, G., Schmitt, A., Moncho-Jordá, A., Callejas-Fernández, J., Martinez-Garcia, R., Leone, R., and Hidalgo-Alvarez, R., *Phys. Rev. E*, 65, 031405, 2002.
168. Odriozola, G., Schmitt, A., Callejas-Fernández, J., Martinez-Garcia, R., Leone, R., and Hidalgo-Alvarez. R., *J. Phys. Chem. B.*, 107, 2180, 2003.

4

Structure Formation in Dispersed Systems

Jonas Addai-Mensah and Clive A. Prestidge
Ian Wark Research Institute, University of South Australia,
Adelaide, Australia

I. INTRODUCTION

Dispersed systems form a major part of our natural and engineering environment and are frequently encountered in many industrial processes (e.g., crystallization, flotation, coagulation, and flocculation) and materials involving solid–fluid systems (aerosol, emulsions, paints, cosmetics, detergents, etc.). The dispersion may comprise colloidal size (largest dimension $<3 \, \mu m$) particulate matter (dispersed phase) such as a solid, liquid, or gas surrounded by a continuous medium (dispersion phase) in a fluid, liquid, or gas. Different classes of dispersions may arise depending upon the selection of the dispersed and dispersion phases. A sol is a dispersion of a solid in a liquid while an aerosol is a dispersion of a liquid or a solid in a gas. Suspensions and pastes are examples of solid sols. Other common dispersions are solid–gas (solid aerosols), liquid–gas (liquid aerosol), liquid–liquid (emulsions) and gas–liquid (foam).

Dispersions may be further classified either as *lyophilic* or *lyophobic* if the dispersed phase is solvent attracting or repelling, respectively. Generally, a *chemical group similarity* with the right geometry must exist between the dispersed particles and dispersion medium for lyophilic sols while none exists for lyophobic sols. In the particular case where the solvent is water and the dispersed particles possess surface groups containing electronegative atoms capable of hydrogen or ionic bonding to water, a hydrophilic interaction is displayed. For instance, when fine metal hydroxide particles (e.g., $Fe(OH)_3$ or $Al(OH)_3$) are dispersed in aqueous medium, the polar OH groups at the surface facilitate hydrogen bonding. On the other hand, where the particles do not possess OH groups in their structure (e.g., graphite or a fluorocarbon such as teflon) or at the solid–water interface (e.g., talc: $Mg_3Si_4O_{10}(OH)_2$), the interaction displayed is hydrophobic.

The state of the dispersion may be defined in terms of a structure developed according to the nature of the operative particle interactions and surface forces. Over molecular dimensions, the origin of the interaction forces may be electrostatic (Coulombic), polarization, or chemical (e.g., covalent). A characteristic feature of colloidal dispersions is that the dispersed particles are too small to be influenced by an external gravitational force and yet too large to be in the molecular domain. Depending upon the interfacial chemistry and prevailing hydrodynamics, a dispersion will experience various types of attractive and repulsive particles interaction forces, reflecting the structure and hence, the colloid stability.

A stable dispersion results where an overall or a net repulsive force dominates. An interparticle force may be considered either as long or short ranged if it acts over distances larger than or equivalent to the characteristic dimensions of the solvent molecules, respectively [1]. Stability may be defined with respect to thermodynamics or kinetics. Thermodynamic stability implies that the system is at its lowest accessible free energy state, displaying time invariant properties. Kinetic stability, on the other hand, requires that the system properties do not change perceptibly within a specified time period, even though it may be thermodynamically metastable or unstable. The main aims of this chapter are to describe how

(i)   the structure of colloidal dispersion may be manipulated through the control of the particle interaction forces (Sections II and IV) and

(ii) the structure may be characterized in terms of particle interactions (Section III).

II. INTERACTION ENERGIES AND FORCES

The stability of a colloidal dispersion is determined by the balance of attractive and repulsive forces present. The theoretical treatment of colloidal stability has been studied extensively since the pioneering work by Deraguin and Landau in 1941 [2] and Verwey and Overbeek in 1948 [3], leading to the classic DLVO theory. Simply, the DLVO theory considers only the attractive van der Waals forces and the electrical double layer repulsive forces between two charged surfaces in a liquid media. In certain cases, however, compliance with

DLVO theory is not observed and additional forces of steric or entropic origins (e.g., hydrophobic or solvation and hydration forces) must be invoked to rationalize the colloidal behavior. Depending upon the dispersion interfacial chemistry and mechanical action used, one may consider a number of additional interaction forces due to polymer-bridging, depletion, (electro)steric, hydrophobic, and hydrodynamic interactions. The total interaction energy or potential (V) between the particles may be considered in terms of the contributions by van der Waals (V_{vdw}), electrical double layer repulsion (V_{edl}), hydration (V_{hyd}), steric (V_{ster}), depletion (V_{dep}), hydrophobic ($V_{hydroph}$), and bridging (V_{bridg}) and hydrodynamic (V_H) forces.

$$V_{tot} = V_{vdw} + V_{edl} + V_{ster} + V_{dep} + V_{hyd} + V_{hydroph} \\ + V_{bridg} + V_H. \tag{1}$$

For a given interaction energy, the corresponding force (F) may be obtained by summing up all the pair forces or as the first derivative of the energy with respect to separation distance r between the nuclei of atoms or center of mass of approximately spherical molecules. The energy required for separating a pair of atoms or molecules at a separation h is given by

$$V = -\int_{h}^{\infty} F \, \mathrm{d}r. \tag{2}$$

Therefore, the net force is

$$F = \frac{\partial V}{\partial h}. \tag{3}$$

Clearly, different combinations of forces may operate in a system at any given time and their judicious manipulation through the primary process variables that influence the interfacial chemistry and hydrodynamics will enable greater control of the structure and stability of the dispersion.

A.   van der Waals Interaction Energies and Forces

The first term in Equation (1) is due to van der Waals interactions, which act between all atoms and molecules that may
be neutral, charged, or dipolar. The forces are usually attractive and arise from atomic and molecular interactions involving correlated, electric dipolar fluctuations when the two
entities (e.g., molecules) are within separations close to atomic
and molecular contact. The forces may be attractive or repulsive. Molecules that are uncharged may possess permanent
or induced electric dipole. The total van der Waals energy
between particles is the sum of three distinct types of interactions [4]:

- Orientational or Keesom interactions: permanent molecular dipole creates an electric field that has the
  effect of orienting other permanent dipoles so that
  they attract each other.
- Induced or Debye interactions: permanent dipole induces a dipole in a polarizable atom, molecule, or medium and the induced dipole is oriented such that it is
  attracted.
- Dispersion or London or electrodynamic force: an instantaneous dipole arising from fluctuations in
  electronic charge distribution, which induces and
  attracts other dipoles in surrounding atoms and
  molecules.

Whether or not the first two types of interactions contribute depends upon the presence or absence of permanent dipoles in the system. Dispersion or London forces make up at
least one third of the total van der Waals forces, their ubiqui-
tous presence playing an important role in interfacial phenomena such as flocculation, coagulation or aggregation,
adhesion, physisorption, and surface tension [1]. Using quantum mechanical perturbation theory to solve Schrødinger's
equation for two hydrogen atoms at large distances, London
[5] showed that the attractive dispersion interaction energy,
V_L, between two identical atoms and molecules whose centers
are separated by a distance r is

$$V_L = -\frac{\beta}{r^6},\qquad(4)$$

where β is the London constant, which is related to the polarizability of the participating atoms or molecules and may be defined as

$$\beta = \frac{3\alpha^2 h\nu}{(4\pi\varepsilon_0)^2},\qquad(5)$$

where α is the polarizability of an atom, ν is the orbiting frequency of the electron, h is Planck's constant (6.63×10^{-34} J sec) and ε_0 is permittivity of free space. It is pertinent to note that both the Keesom and Debye interaction energies contributing to the net van der Waals energy also vary inversely as the sixth power of the separating distance r.

The dispersion interaction energy (Equation (4)) is dependent upon the distance between the two atoms or molecules and may be effective over separations ranging from $\approx 3\,\text{Å}$ (e.g., interatomic spacings) up to $10\,\text{nm}$ or distances smaller than the intrinsic wavelength, λ_L. At larger distances ($>10\,\text{nm}$), the dispersion energy force between particles is considerably diminished due to a retardation effect, decreasing inversely as r^7. For two atoms in contact at $r = 3\,\text{Å}$, $V_L = -4.6\times10^{-21}\,\text{J}$, which is equivalent to the energy due to Brownian motion ($1\,\text{kT}$).

1. Macroscopic Particle Interactions

The dispersion interaction forces between two isolated bodies may be considered as additive. In the presence of other molecules, however, the van der Waals forces are not generally pairwise additive, in contrast to Coulomb and gravitational forces. The van der Waals energy between two molecules in a medium will differ from that calculated for two isolated bodies, as retardation effects which become important at large separation distances, are ignored for the latter. Hamaker [6] used a pairwise additivity approach in deriving the van der Waals interactions between condensed, macroscopic bodies. In this case, the total interaction energy between two bodies,

V_{vdw}, can be calculated by integration of all interaction pairs of molecules on surfaces both bodies as

$$V_{vdw} = - \int_{V_1} dV_1 \int_{V_2} dV_2 \left(\frac{\beta q_1 q_2}{r^6} \right), \tag{6}$$

where V_1 and V_2 are the total volume of bodies 1 and 2 and q_1 and q_2 are the molecular number densities, that is molecules per unit volume, and β is the van der Waals constant.

For two planar bodies 1 and 2 in vacuo integration of Equation (4) for a nonretarded energy per unit area

$$V_{vdw} = - \frac{A_{12}}{12 \pi h^2}, \tag{7}$$

where A is the Hamaker constant defined as

$$A = \beta \pi^2 q_1 q_2. \tag{8}$$

Likewise, the nonretarded V_{vdw} interaction energy for other geometries, such as sphere–sphere, sphere–plate, and cylinder–cylinder can also be calculated on the basis of pairwise additivity [1], as shown below.

Two spheres of different radii a_1 and a_2:

$$V_{vdw} = - \frac{A_{12}}{6h} \frac{a_1 a_2}{(a_1 + a_2)}. \tag{9}$$

Sphere of radius a_1 and a planar surface:

$$V_{vdw} = - \frac{A_{12} a_1}{6h}. \tag{10}$$

Two cylinders of equal lengths L and radii a_1 and a_2:

$$V_{vdw} = - \frac{A_{12} L}{12 \sqrt{2} h^{3/2}} \left(\frac{a_1 a_2}{a_1 + a_2} \right)^{1/2}. \tag{11}$$

Interaction energies between macroscopic bodies are useful; however, the interaction forces are easier to measure. It is also easier to obtain analytical expressions for interaction energies of two flat surfaces (e.g., plate) than curved surfaces (e.g., spheres and cylinders). The interaction energy of two planar surfaces may be related to the corresponding force between two curved surfaces (e.g., spheres) through the Derjaguin approximation [7].

Two spheres of different radii a_1 and a_2:

$$F_{\text{sphere-sphere}} = 2\pi \frac{a_1 a_2}{(a_1 + a_2)} V_{\text{plate-plate}}. \tag{12}$$

A sphere–plate (sphere radius a_1):

$$F_{\text{sphere-plate}} = 2\pi a_1 V_{\text{plate-plate}}. \tag{13}$$

For two cylinders of equal radii (a_1) crossed at right angles:

$$F_{\text{sphere-plate}} = 2\pi a_1 V_{\text{plate-plate}}. \tag{14}$$

The Derjaguin approximation is applicable to any type of force law whether attractive, repulsive, or oscillatory, as long as for spheres a_1 and $a_2 \gg h$, the interparticle separation distance.

2. The Hamaker Constant

The fundamental Hamaker equation intrinsically assumes that the two particles are interacting across a vacuum. In most cases, particle–particle interactions occur in nonvacuum conditions, such as air or liquid, and the effect of the intervening medium must be accounted for. The interaction between like particles is always attractive and the effective Hamaker constant between two particles of identical composition, A_{131}, can be calculated from those of the particle, A_{11}, and the dispersing medium A_{33}:

$$A_{131} \approx \left(\sqrt{A_{11}} - \sqrt{A_{33}} \right)^2. \tag{15}$$

The Hamaker constant for most materials lies in the range 5–$100\,kT$ ($10.6\,kT$ for water at $25°C$ and $44.5\,kT$ for AgBr).

The approach used by Hamaker to calculate dispersion interactions was based on a molecular model and assumed that the total interaction was the sum of the interaction of pairs of atoms, ignoring multibody effects. The above approach while useful turns out to be not entirely accurate when other bodies are present and affect the interactions between the two bodies under consideration.

3.   The Macroscopic Approach — Lifshitz Theory

An alternative approach for calculating dispersion interactions between particles is a macroscopic approach proposed
by Lifshitz in 1956 [8]. Each interacting particle and the
intervening medium is considered as a continuum with a
certain dielectric properties [9], thus avoiding the problem of
pairwise additivity. The forces are calculated using quantum
field theory from the bulk electrical and optical properties.
    The calculation for the Hamaker constant by Lifshitz's
approach is notably different from that in Hamaker's approach and is more complex in the former, requiring the di-
electric constants ε_1 and ε_2 and refractive indices, n_1 and n_2, of
the two macroscopic bodies (1 and 2) and those of the medium
(3), ε_3 and n_3, over a wide range of frequencies. By assuming
that the absorption frequencies of the bodies and medium are
the same, the nonretarded Hamaker constant can be calculated for the interaction between two identical particles:

$$A_{131} \approx \frac{3}{4}kT\left(\frac{\varepsilon_1 - \varepsilon_2}{\varepsilon_1 + \varepsilon_3}\right)^2 + \frac{3h\nu_e}{16\sqrt{2}}\frac{(n_1^2 - n_3^2)^2}{(n_1^2 + n_3^2)^{3/2}}, \tag{16}$$

where ν_e is the main absorption frequency in the UV $\approx 3\times10^{15}$
sec^{-1}. A precise theoretical derivation of Equation (16) for
the Hamaker constant and Lifshitz theory is provided
elsewhere [1].

4.   Retardation of van der Waals Forces

As the distance between two atoms is increased to $r \geq 100$ nm,
the dispersion energy is smaller than predicted by Equation
(3) due to a retardation effect. The reason for this effect is the
finite propagation velocity of the electromagnetic signal between the atoms. The time taken for the electric field of atom 1
to induce an electric field in atom 2 and return to atom 1 may
be greater than the time of a phase shift for the fluctuating
dipole. Therefore, the field returning to atom 1 may find the
direction of the instantaneous dipole less favorable for attractive interactions. The retardation effects on the dispersion
energy between two molecules come into play at much smaller

distances in a medium and become important for macroscopic surfaces dispersed in liquids.

B. Electrical Double Layer Interaction Energies and Forces

As two similarly charged particles dispersed in an electrolyte solution approach one another at some separation distance, h, their electrical double layers overlap and result in an electrostatic repulsion. A particle dispersed in aqueous medium may spontaneously acquire a surface charge that influences nearby ions in the adjacent solution. Counter ions are attracted to the electrically charged surface while similarly charged coions are repelled. The charged interface is made up of an essentially fixed surface charge and a diffuse layer of counter ions, which combine to form the electrical double layer (EDL). When the diffuse double layers around two similar particles overlap, the particles experience repulsion, which prevents them from close approach.

If the particle surface consists of ions which are permanently bound to the surface and groups which are directly ionizable, a surface charge can develop, the extent of which is dependent upon the degree of ionization and the pH of the medium. For metal oxides or hydroxides, the surface may act as either a Brönsted acid or Lewis base due to their amphoteric nature. Upon adsorption of either H^+ or OH^-, a net difference between the surface concentrations of H^+ and OH^- may exist. The pH value at which equal surface concentrations of H^+ and OH^- is achieved, leading to a net surface charge of zero, is termed the point of zero charge (PZC). The H^+ and OH^- ions are referred to as potential determining ions (PDI).

Another important influence on the surface charge of metal hydroxides and oxides, can be the adsorption of specifically adsorbed ions [10]. Specifically adsorbed ions can generally be thought of as any ion whose adsorption at the surface is influenced by forces other than the electrical potential. The type of bond formed between the specifically adsorbed ions and particle surface site species determines whether or not the adsorbed ions associates with the surface as an inner sphere (chemical bond) or outer sphere (ion pair complex) or as part of a diffuse double layer surrounding the surface (physical bond). The reactivity and charge of a surface may be affected by inner sphere complexes, including a nominal shift in the PZC. The effect of these adsorbed ions can be determined by surface characterization experiments.

Microelectrophoretic and electroacoustic techniques provide valuable information regarding the particle surface chemistry in terms of electrical nature of a surface and the impact of pH, electrolyte concentration, potential determining and specifically adsorbing ions. In metal hydroxide or oxide system at low pH, the surfaces are generally positively charged due to protonation and negatively charged at high pH due to deprotonation [10]:

$$M - OH + H^+ \rightarrow MOH_2^+ \tag{17a}$$

$$M - OH + OH^- \rightarrow O^- + H_2O \tag{17b}$$

At some pH, the particle zeta potential is equal to zero and this is the iso-electric point (IEP). In the absence of any specific adsorption of electrolyte ions, the inner layer charge density is zero and the PZC will coincide with IEP [11]. However, if they do not coincide, it is probable that the electrolyte is not indifferent and is specifically adsorbed onto the surface or surface contamination is occurring. Specific adsorption of anions causes a decrease in the IEP, less OH^- is required to balance the decreased positive surface charge with an upward shift in the PZC. Anions favor the adsorption of H^+ and a higher pH is required to restore the H^+/OH^- adsorption balance. If electropositive ions (cations) are specifically adsorbed then the reverse processes occur, that is a decrease in the PZC and an increase in the IEP.

1. The Electrical Double Layer

The basis for modern EDL theory is a model introduced by Stern in 1924 [12] who considered that the double layer is divided into two sections by the Stern plane, where the inner region is the so-called Stern layer – a small space separating the ionic atmosphere around a surface from the charged plane just adjacent to the surface. The thickness of the Stern layer is of the order of 1–5 Å and reflects the finite size of the charged groups and ions associated with the surface.

The Stern model essentially integrates the constant capacitance model [13] with the diffuse electrical double layer treatment of the Gouy–Chapman theory [14,15] to describe a particle surface–electrolyte interface. In the inner region of the EDL, the charge and potential distributions, σ and ψ, are determined mainly by the size of the ions or molecules which participate in specific, non-Coulombic, short-range interactions with each other and with the surface (Figure 4.1). As ions and molecules have a finite size, their approach to a surface is limited to one ion radius, and the distance to which an ion can approach the surface is dependent upon the degree of hydration. The Stern layer is composed of ions

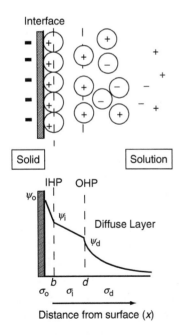

Figure 4.1 Schematic representation of the electrical double layer according to the Gouy–Chapman–Grahame model.

that are specifically adsorbed onto the surface strongly enough to resist displacement by thermal motion, and are therefore regarded as 'fixed' relative to ions in the diffuse double layer. In refining the Stern model, Grahame [16] proposed that the Stern layer can be divided into two planes, the inner Helmholtz plane (IHP) and the outer Helmholtz plane (OHP). The IHP is a plane through the centers of specifically adsorbed ions on the surface, which are desolvated upon adsorption. The distance of closest approach of hydrated ions in the diffuse layer is represented by the OHP, which lies through the center of these ions and is thus the same plane represented by the Stern plane. The inner double layer region can be thought to behave similarly to a capacitor or parallel plate condenser. The potential changes between $0 < x < b$ and $b < x < d$ can be written as

$$\psi_o - \psi_i = \sigma_o \frac{b - 0}{\varepsilon_1}, \tag{18}$$

$$\psi_i - \psi_d = \sigma_d \frac{d - b}{\varepsilon_2}. \tag{19}$$

where ε_1 and ε_2 are permittivities between the surface and the IHP and the OHP, respectively. Generally ε_1 and ε_2 are assumed constant so that the potential in the EDL decays linearly.

The electrokinetic measurements are able to provide an estimate of the diffuse layer potential by determining the potential at the shear plane, the zeta potential (ζ). When a charged particle suspended in an electrolyte solution is subjected to an electric field, the particle and a portion of the EDL layer surrounding the particle will move under its influence. The 'fixed' portion of the EDL, equal to the Stern layer plus an amount of solvent bound to the charged surface, is called as the plane of shear. Thus, the displacement of the plane of shear from the surface is slightly greater than the thickness of the Stern plane, implying that the zeta potential will be slightly less than the surface potential, but this difference is negligible for low potentials.

2. The Thickness of the Electrical Double Layer

The thickness of the double layer $1/\kappa$, where κ is the Debye–Hückel parameter, is dependent upon three parameters – temperature, bulk electrolyte concentration, and permittivity of the solution surrounding the particle. An increase in temperature results in additional thermal motion of the ions dispersed in the ionic atmosphere around the particle, thus increasing the thickness of the diffuse double layer. As the ionic strength is increased, the thickness of the diffuse double layer around the particle is decreased. This phenomenon is referred to as compression of the double layer and is important in studying the colloid stability. The expression for the Debye–Hückel parameter is

$$\kappa^2 = \left(\frac{F^2}{\varepsilon_0 \varepsilon_r RT}\right) I, \tag{20}$$

where F is the Faraday constant, T is the absolute temperature, and R is the universal gas constant. For water at $25°C$, $\varepsilon_r = 78.54$ and κ (nm^{-1}) can be calculated by

$$\kappa = 3.288\sqrt{I}, \tag{21}$$

where I is the ionic strength of the bulk solution defined as

$$I = 1/2 \sum (C_i z_i^2), \tag{22}$$

where C_i and z_i are the concentration (M) and the valency of the ions, respectively. The surface potential of a particle depends not only on the temperature and the composition of the bulk electrolyte solution but also on the charge density of the particle. The surface charge density (σ_0) and potential (ψ_0) can be related as [9]:

$$\sigma_0 = (8n_0 \varepsilon_0 \varepsilon_r kT)^{1/2} \sin h\left(\frac{ze\psi_0}{2kT}\right), \tag{23}$$

where n_0 is the bulk concentration of the ionic species.

When the potential at the surface is small, the Debye–Hückel approximation can be applied and the surface charge is given by

$$\sigma_0 = \varepsilon_0 \varepsilon_r \kappa \psi_0. \tag{24}$$

3. Electrical Double Layer Interactions

In considering the energy of interaction between the double layers of two spherical bodies, Derjaguin [7] proposed

$$V_{\text{EDL}} = \frac{2\pi a_1 a_2}{(a_1 + a_2)} \int_h^\infty F \, dh \tag{25}$$

provided that the double layer thickness was small in comparison with the particle size, $\kappa a \geq 5$. The above expression was derived by treating the interaction energies of the double layers of two spherical particles as a series of contributions from the interactions between infinitesimally small parallel rings, each considered as a flat plate at the same separation, on each sphere as a function of distance.

For the case of two flat plates, the weak overlap approximation leads to

$$V_{\text{EDL}} = \frac{64 n R T \gamma^2}{\kappa} \exp(-\kappa h), \tag{26}$$

where

$$\gamma = \frac{\exp[(ze\psi/\kappa T) - 1]}{\exp[(ze\psi/\kappa T) + 1]}. \tag{27}$$

For a sphere of radius a and a plate at same potential ψ:

$$V_{\text{EDL}} = \varepsilon a \psi^2 \ln[1 + \exp(-\kappa h)]. \tag{28}$$

For two spheres of same radius a, for $\kappa a \gg 1$:

$$V_{\text{EDL}} = \frac{\varepsilon a \psi^2}{2} \ln[1 + \exp(-\kappa h)]. \tag{29}$$

For two spheres of same radius a, for $\kappa a \ll 1$:

$$V_{EDL} = \frac{\varepsilon a^2 \psi^2}{h + 2a} \exp(-\kappa h). \tag{30}$$

The potential ψ, usually the zeta potential, is determined experimentally.

The other situations and models are possible. For instance, Hogg et al. [17] proposed a model (HHF) by considering the Derjaguin approximation and $\kappa a \gg 5$, for small separation distances between spherical particles and small surface potentials. In calculating the interaction energies using the HHF approximation, the Debye–Hückel approximation is assumed, that is the Poisson–Boltzman equation is in its linearized form. The HHF model is

$$V_{EDL} = \frac{\varepsilon \varepsilon_a a_1 a_2 (\psi_{o_1}^2 + \psi_{o_2}^2)}{4(a_1 + a_2)} \left[\frac{2\psi_{o_1} \psi_{o_2}}{(\psi_{o_1} + \psi_{o_2})} \ln \frac{1 + \exp(-\kappa h)}{1 - P \exp(-\kappa h)} \right.$$
$$\left. + \ln(1 - \exp(-2\kappa h)) \right]. \tag{31}$$

For two identical spheres the HHF approximation reduces to

$$V_{EDL} = \varepsilon \varepsilon_a a \psi_o^2 \ln(1 + \exp(-\kappa h)). \tag{32}$$

The HHF model assumes that the system is either at constant potential or constant charge. This approximation is suitable in situations where the surface potential is low. To alleviate the necessity for making calculations on the basis of either constant charge or constant potential, an alternative approximation, called the linear superposition approximation (LSA) has been proposed [18]. This involved making the assumption that the potential at a point, approximately equal in distance from both particle centers, can be calculated by summing the potentials at that point resulting from the two particles treated as isolated systems. Superposition of single sphere potentials yields an approximation for the repulsive forces between spheres with thin double layers and is not dependent upon the particle boundary conditions of constant potential or constant charge. For two identical particles, LSA becomes

$$V_{\text{EDL}} = 4\pi\varepsilon\varepsilon_r \left(\frac{kT}{ze}\right)^2 \frac{a^2}{h+2a} \psi^2 \exp(-\kappa h), \tag{33}$$

where

$$\psi = \frac{ze}{kT}\psi_0.$$

4.   DLVO Interactions of Colloidal Particles

In predicting the colloidal stability of a system, it is necessary to determine the total interaction energy between particles as they approach each other. The DLVO theory considers the total interaction energy as a function of interparticle distance, and is the summation of the energy resulting from the overlap of the electrical double layers and the attractive van der Waals forces. Thus, the colloidal stability can be quantified in terms of the net interaction energy potential experienced over the interparticle distance.

$$V_T = V_A + V_R. \tag{34}$$

For interaction between two spherical particles, the inter-action energy is

$$V_T = \frac{-Aa}{12h} + 4\pi\varepsilon\varepsilon_a \left(\frac{kT}{ze}\right)^2 \frac{a^2}{h+2a} \psi^2 \exp(-\kappa h). \tag{35}$$

A schematic representation of the total potential energy curve and its constituent attractive and repulsive interaction curves between two colloidal particles as a function of the separation distance is shown in Figure 4.2. The shape of the total inter-action curves depicts three important regions relevant to pre-dicting colloidal stability – two local minima and one local maximum. The deepest minimum, which occurs at very small interparticle distance, is termed the primary minimum and theoretically represents irreversible flocculation of par-ticles because of the overwhelming net force of attraction. The secondary minimum is quite shallow, when compared with the

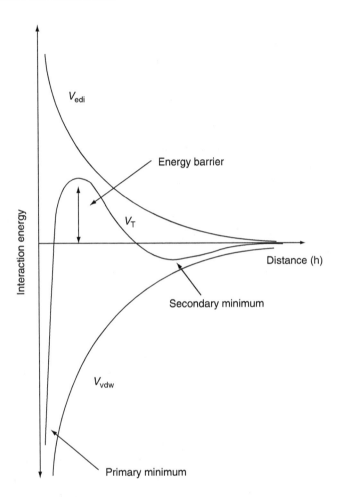

Figure 4.2 A schematic representation of the total interaction energy potential two particles as a function of distance.

primary minimum, and is a consequence of the increasing net attractive force as interparticle distance is increased. Particles, which flocculate or coagulate in the secondary minimum, are more readily dispersed; the relative ease of redispersion is determined by the depth of the minimum in comparison with thermal energy (Brownian motion), kT, of the particles.

The third characteristic feature of the total potential energy curve is a local maximum which is referred to as the potential energy barrier and which is produced by net electrostatic repulsion. The height of this barrier can be directly correlated with colloidal stability and depends upon the magnitude of the zeta potential and the double layer thickness surrounding the particles. If the potential energy barrier is large in terms of kT, then particles are prevented from entering the primary minimum and may flocculate in the secondary minimum or remain dispersed.

The net energy potential and therefore the stability of an aqueous dispersion are influenced by three main parameters: the surface potential, the nature and ionic strength of electrolyte solution, and the Hamaker constant. The surface potential of metal hydroxides or oxides is influenced by PDI, $-H^+$ or OH^-, via dispersion pH adjustment. As the dispersion pH approaches the IEP or the PZC, the V_{edl} is reduced to zero. The valence and the ionic strength of the electrolyte solution are important variables, which determine the thickness of the double layer. Therefore, it is possible to induce coagulation via double layer compression by increasing the ionic strength of the electrolyte. This is the basis of the Schultz–Hardy rule of which coagulation is controlled predominantly by the nature of the added electrolyte and occurs most efficiently by ions of opposite charge and high-charge number. The Hamaker constant is determined by the chemical nature of the particles and the surrounding electrolyte and is therefore difficult to vary, and therefore is considered constant for a specific system. For large values of A, the height of the potential energy barrier is low and depth of the secondary minimum is increased because of the larger attraction between the participating particles. Conversely, small values of A yield a decrease in the attractive forces and therefore an increase in the stability of the system.

C. Non-DLVO Interaction Energies and Forces

Although DLVO theory is an adequate representation of the net potential between two approaching particles, in many systems especially at very small interparticle distances, it fails to account for all the operating interaction forces. In decoupling of the total interaction energy or forces, it is common to calculate and subtract the DLVO interactions, with any residual energy or forces attributed to non-DLVO interactions. These non-DLVO interactions may be monotonically attractive, repulsive, or oscillatory in nature and may be more pronounced and exceed DLVO forces at smaller particle separations [1]. They may include various solvation forces such as attractive hydrophobic forces [19], repulsive, short-range hydration, and solvation forces [20,21], repulsive long-range structural forces [22–24], bridging and steric or polymer effects [1].

Most quantitative knowledge regarding non-DLVO forces has been obtained during the last decade through experimental observations using direct force-measurement techniques (see Section III). Indirect methods used to determine the presence of DLVO and non-DLVO forces include coagulation and rheological studies. By comparing the coagulation behavior of colloidally stable dispersion to the DLVO predicted behavior under various conditions, including pH and high-salt concentrations, it is possible to verify the compliance or noncompliance with DLVO theory. Common documented non-DLVO forces, which may contribute to the overall particles interaction energy, are summarized as below.

1. Hydrophobic Forces

As the name implies, hydrophobic forces are interactions between hydrophobic surfaces and can be long or short range in nature. Long-range forces are measurable at separation distances up to 100 nm, most commonly attractive, and often measurably stronger than predicted van der Waals forces [19], particularly in systems where the Hamaker constant is small. The origin of the monotonically attractive hydrophobic forces has not yet been determined, in spite of their major cause of rapid coagulation of hydrophobic particles in aqueous media [25]. Hydrophobic forces have been measured using surfaces (e.g., silica and mica) chemically hydrophobized with an adsorbed surfactant layer and methylated layer

[26–28]. Israelachvili and Pashley [26] observed that the hydrophobic interactions between two modified mica surfaces decayed exponentially in the range 1–10 nm with a decay length of 1 nm.

Although few studies have yielded accurate and reproducible results, due to the difficulties in working with hydrophobic surfaces and molecules in an aqueous medium, the study of hydrophobic forces is of paramount importance in understanding surface phenomena and behavior of biological membrane structure, protein conformation, and micelle formation [29,30].

2. Hydration Forces

Repulsive, short-range (<10 nm) non-DLVO forces, varying monotonically with separation distance, are often evident in between two hydrophilic surfaces in aqueous media. Referred to as a hydration force, they have been observed in the studies of the swelling of clays and directly measured between mica, silica, and clay platelet surfaces in electrolyte solutions of varying salt concentrations [31–34]. Studies between mica [31,35] and sapphire [36] surfaces in electrolyte solutions, using the SFA indicated that at high salt concentrations and low pH values where surface charge is positive, a repulsive hydration force is detectable. Based on the hydration number of the cation and relative concentrations of anions in present in solution, the hydration force may be associated directly to the energy required to desolvate the surface-bound ions, which may retain a portion of their water of hydration when they bind to the surface [19]. Greater hydration forces of repulsion exist between mica surfaces in NaCl compared with KCl [31].

3. Oscillatory Forces

One main type of solvation force is known as the oscillatory force. As two surfaces in a solution approach a change in the liquid density at the surface occurs. An oscillatory force will result as confined liquid molecules order themselves. The period of this force is approximately equal to the diameter of the molecules and is short ranged – a few molecular diameters [1]. These oscillatory forces have been measured experimentally between mica surfaces in various liquids [37,38]. The origin of the oscillatory forces is not fully known but it is believed to be due to the packing of solvent molecules between smooth surfaces [39,40].

Direct measurement of forces between the basal surfaces of sapphire [1,36] quantified and distinguished repulsive oscillatory forces from hydration and other forces as the surfaces were brought into contact in concentrated NaBr salt solutions at low pH. They were able to distinguish a monotonic, repulsive hydration force with decay lengths ≈ 0.55 nm, an oscillatory force with a periodicity approximately twice the diameter of a water molecule (0.5 nm) and a long-range attractive force with decay lengths of ≈ 12 nm. In addition, it was shown that under specific conditions a repulsive force can be present beyond the thickness of single hydration layer per alumina surface. The hydration forces observed at a low pH regime were associated directly with the binding of anions (Br^-) to the surface [36].

4. Structural Forces

Additional non-DLVO forces may arise especially at high salt concentrations and lead to interparticle repulsion. Structural forces are dependent upon the properties of the intervening solvent including salt type and ionic strength, and the particle surface characteristics (hydrophobicity and hydrophilicity, roughness etc.). In addition, the strength of solvation forces are dependent upon the concentrations of ions present at the interface and how strongly they are bound to the surface; their degree of ion hydration as well as surface hydration and the history of the surface preparation.

Several studies [22–24,41–44] have indicated the reluctance of colloidal size gibbsite crystals to rapidly aggregate and agglomerate during Bayer precipitation from high-ionic strength ($4–6M$) sodium aluminate liquors, contrary to expectation. Certainly, the interactions between slow-growing gibbsite crystals control the agglomeration process and are themselves governed by interparticle forces. At high pH and high electrolyte concentrations ($4-6\,M$, supersaturated caustic aluminate solutions), a strong interparticle repulsion was detected at a particle separation distance of ~40 nm. The repulsive forces, which showed time-dependent attenuation, prevailed at particle approach velocities equivalent to Brownian motion and were attributed to structural forces [12–24].

5. Steric and Electrosteric Forces

The origins of steric or electrosteric repulsion lie in both volume restriction and interpenetration effects, although it is unlikely that either effect would occur in isolation to provide a repulsive force. In many industrial processes where the coagulation of colloidal particles would naturally occur, steric repulsion between particles can be induced by the addition of a polymer, to prevent the approach of the particle cores to a separation where their mutual van der Waals attraction would cause flocculation to occur. Complete particle surface coverage by adsorbed or anchored polymer at high concentration can produce a steric layer that prevents close approach of the particles. The steric layer also acts as a lubricant to reduce the high frictional forces that occur between particles with large attractive interactions. The time-dependent, displaceable, and slow-forming hydrolyzed inorganic layers which lead to repulsive electrosteric forces between mica surfaces in $0.1\,M$ $Cr(NO_3)_3$ electrolyte have been reported [45].

The magnitude of the repulsion resulting from steric forces is dependent upon the surface area of the particle that the polymer occupies and whether the polymer is reversibly or irreversibly attached to the particle's surface.

Adsorbed and nonadsorbing surfactants and polymers are widely used to induce steric stabilization. The principal advantages of steric stabilization over charge stabilization are:

(i)   Provision of stability in nonpolar media where weak electrical effects occur
(ii) Use of higher levels of electrolyte in aqueous media without causing flocculation

(iii) Reduction of electroviscous effects arising from particle charge by the addition of electrolyte without flocculation

(iv) Dispersion stabilization can be achieved at higher particle concentrations.

Graft or block copolymers commonly used as steric stabilizers are designed to have two groups of different functionality, A and B. A is chosen to be insoluble in the dispersion medium and has strong affinity for the particle surface while B is selected to be soluble but have little or no affinity for the particle surface, as shown in Figure 4.3. Other steric interactions, which give rise to short-range repulsion in aqueous dispersions due to bare size of ions present at the particle solution interface at high ion concentrations, have been reported by Franks et al. [46].

6. Hydrodynamic Forces

Mechanical action such as shear or agitation may stabilize colloidal dispersion via hydrodynamic forces, which arise if the intensity is high enough. For a sphere approaching a flat surface, F_H, is dependent on the approach velocity and size of the sphere as well the viscosity of the intervening solution and has been shown by Vinogradova [47] to be

$$F_H = \frac{6\pi\mu a^2 U}{h}, \tag{36}$$

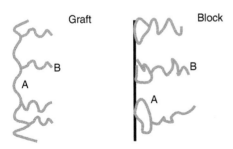

Figure 4.3 Schematic diagram of graft or block copolymers commonly used as steric stabilizers.

where h is the sphere–plate separation distance, U is the particle approach velocity (dh/dt), and μ is the viscosity of solution.

The hydrodynamic force, F_H, between a gibbsite sphere and flat, normalized by sphere radius ($a = R$) measured in concentrated caustic aluminate solutions ([Al] = 2.9 M [NaOH or KOH] = 4.0 M) at 30°C is plotted as a function of particle approach velocity (Figure 4.4). It is demonstrated by the linearity of the plot that Vinogradova's model [47] for predicting the repulsive hydrodynamic interaction force between particles in dispersion is clearly obeyed.

III. CHARACTERIZATION OF STRUCTURES

Both attractive and repulsive interactions between particles in suspension result in the formation of interparticle structure. The structures that form are critical in controlling the physical properties of a suspension and improved knowledge in this area is leading to emerging technologies, e.g., photonic devices from structured concentrated suspensions and electro-

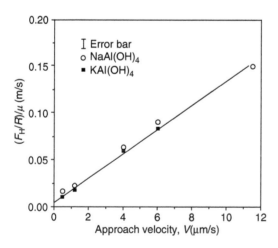

Figure 4.4 $(F_H/R)/\mu$ measured at aged gibbsite (001) crystal face in $NaAl(OH)_4$ and $KAl(OH)_4$ solutions ([Al] = 2.9 M, [NaOH or KOH] = 4.0 M) at 30°C versus particle approach velocity.

rheological fluids. The structures that form in suspension are subtly controlled by the competition between thermal energy and the magnitude of interparticle forces and further modulated by the particle shape, polydispersity, and volume fraction, and external conditions, e.g., mixing and temperature. Suspensions of repulsive spherical particles are fluid-like at low-volume fraction, but form crystalline or glassy structures at high-volume fractions. For nonspherical particles the angular dependency of the pair potential and capillary forces may result in crystalline or liquid crystalline structures. Weakly repulsive or attractive particles may form irreversible particle aggregates and more random structures. The formation and characterization of these aggregated structures is the main focus of the current section. The nature of the structure and the kinetics and thermodynamics of structure formation in suspensions can be characterized both directly and indirectly through a range of experimental techniques. Scattering, optical, and settling techniques are well suited to measurements of dilute suspensions (Section III.A.), whereas rheological approaches (mechanical characterization of structure) are used for more concentrated suspensions (Section III.B). In addition, substantial effort and progress has been made in the measurement of interaction forces on individual particles and in relating these to structure formation (Section III.C).

A. Scattering, Optical, and Settling Approaches

In dilute particulate suspensions an attractive interparticle interaction potential (or even a weak repulsive potential of less than the thermal energy, kT) results in the formation of doublets, triplets, and eventually flocs; this manifests itself as a decrease in particle number and an increase in particle size. The size, density, structure, and fractal dimensions of flocs are dependent on the strength of particle–particle interaction, particle concentration, and mixing conditions. A range of techniques have been used to determine the kinetics and thermodynamics of aggregate formation in dilute solution, e.g., static and dynamic light scattering, turbidity, and particle counting [10]. Aggregate structure is commonly characterized by a range of approaches (see recent review by Bushell et al. [48] for more extensive details), e.g., imaging using optical and electron microscopy; light, x-ray, and neutron scattering, light and x-ray diffraction, settling methods and electrical methods (e.g., dielectric permittivity, electric conductivity, and electroacoustics) as discussed further below.

With respect to the kinetics of aggregation, the von Smoluchowski [49,50] theory describes the Brownian (diffusion limited) coagulation of a dispersion of monodispersed colloidal particles; it assumes that there is no repulsive interaction between interacting particles, that each collision results in coagulation, and that the process is diffusion-controlled. Based on these assumptions, the coagulation of two particles to form a single new aggregate may be described by a second-order process

$$\frac{\mathrm{d}N_t}{\mathrm{d}t} = -k_\mathrm{D}N_t^2, \tag{37}$$

where N_t is the number of primary particles at time t, N_0 is the initial number of primary particles, and k_D is the rate constant of doublet formation. At $t = 0$, $N_t = N_0$ and Equation (37) can be solved:

$$\frac{1}{N_t} - \frac{1}{N_0} = k_\mathrm{D}t. \tag{38}$$

The coagulation rate constant of doublet formation can also be calculated from Brownian hydrodynamics:

$$k_\mathrm{D} = \frac{8k_\mathrm{B}T}{3\eta}. \tag{39}$$

This rate constant is the theoretical maximum coagulation rate and is generally termed the *rapid coagulation rate* constant (k_r). The intrinsic limitation of this rate constant is that it does not account for either surface forces or hydrodynamic interactions, both of which reduce the probability of particle attachment following their collision. For water at 20°C, k_r is theoretically predicted to be $\sim 6 \times 10^{-18}$ m^3/sec, whereas in practice it is in the range 2×10^{-18} to 3×10^{-18} m^3/sec [51]. In the presence of repulsive interactions (e.g., electrostatic or steric), the particles become more colloidally stable and the rate of aggregation is reduced, i.e., it is no longer diffusion-controlled. A slow (or rate limited) coagulation (or floccula-tion) rate constant (k_s) is observed and can be defined:

$$k_s \equiv \frac{k_r}{W}, \tag{40}$$

where the stability ratio (W) is the reciprocal probability of the particles adhering upon collision. W is a useful parameter that is commonly determined from optical or light scattering meas-urements (see below). Theoretically this can also be deter-mined from the area under a plot of the interaction potential against separation ($V(s)$):

$$W = 2 \int_2^\infty \exp\left\{\frac{V(s)}{k_B T}\right\} \frac{ds}{s^2}, \tag{41}$$

where $s = R/a$ and R is the distance between the centers of the approaching spheres of radius a. The stability ratio may also be approximated from the maximum potential energy (V_{max}) in a potential energy versus interparticle separation plot:

$$W \approx \exp\left(\frac{V_{max}}{k_B T}\right). \tag{42}$$

The initial onset of aggregation within dilute dispersions of colloidal particles is classically determined using light scatter-ing or turbidity measurements [10] with more sensitive tech-niques such as x-ray scattering having been used more recently [52]. As particles aggregate in suspension their light scattering properties vary in accordance with the particle size, aggregate size, their refractive index values, and the wavelength of light. For monodispersed dispersions, the light-scattering in-tensity at any angle is proportional to the coagulation rate constant of doublet formation, if the initial particle-number concentration, wavelength, and particle size remain constant [53]. The rate of doublet formation can be expressed as

$$k = \left(\frac{dI}{dt}\right)_{t \to 0} \frac{1}{C_f N_0 I_0}, \tag{43}$$

where I is the scattered light intensity at time t, I_0 is the initial light-scattering intensity, N_0 is the initial particle number concentration, and C_f is a function dependent on particle radius, scattering angle, and the wavelength of the incident light. The stability ratio is then conveniently defined as

$$W = \frac{(k_{\text{D-rapid}})}{(k_{\text{D-slow}})} = \frac{(\mathrm{d}I/\mathrm{d}t)_{t\to0}^{\text{rapid}}}{(\mathrm{d}I/\mathrm{d}t)_{t\to0}^{\text{slow}}}. \tag{44}$$

If I_0, N_0, and C_f are kept constant between experiments, then the stability ratio can be determined from the initial rate change in light-scattering intensity (or turbidity). The absolute rate constant for the formation of doublets (k_D) can then be related to the turbidity as a function of time ($\tau(t)$):

$$\left(\frac{\mathrm{d}\tau}{\mathrm{d}t}\right)_0 = \left(\frac{1}{2}C_2 - C_1\right)k_D N_0^2, \tag{45}$$

where C_1 and C_2 are the scattering cross sections for singlets and doublets, respectively. Various approaches have been used to determine aggregation kinetics from turbidity versus time data, e.g., Maroto and de las Nieves [54] employed an empirical fitting approach using a polynomial function to describe the turbidity versus time data and extracted k_D with good success.

The experimental determination of the stability ratio data as a function of salt is a common approach to characterize the critical coagulation concentration (CCC) of suspensions and to estimate V_{max}. V_{max} values of 5, 15 and $25\,kT$ correspond to stability ratio values of approximately 40, 10^5, and 10^9, respectively [52].

DLVO is reasonably good at predicting CCC values, however the *slow* aggregation region is more difficult to predict due to hydrodynamic interactions, specific ion effects, surface roughness, nonuniform surface charge, and non-DLVO forces. In some cases theoretical calculations of stability ratio plots have been successfully compared with experimental data to gain insight into the range and magnitude of non-DLVO forces [55]. It should be recognized that the turbidity approach is only valid for the initial onset of aggregate formation (i.e., primary particles forming doublets and triplet structures) and for larger aggregates the relationship between particle number, aggregate size, and light scattering behavior is a considerably more complex problem. It is also noted that light scattering or turbidity measurements are also routinely used in a more rudimentary fashion to determine the removal of aggregated particles from a light beam. That is, the decrease in absorbance with time is correlated with the settling of particle aggregates under gravity or a centrifugal field. A relative rate of flocculation can be established from this method, but limited information concerning the floc size or structure. Particle counting using such techniques as Coulter counting and laser obscuration may be used to determine the reduction in particle number during aggregation and when coupled with changes in the particle size distribution measured by laser diffraction or dynamic light scattering give additional insight into the aggregation kinetics.

We now consider cases where aggregation has proceeded for a sufficient time to form structures within suspensions (or flocs) containing in excess of 10 primary particles. The flocs may have an open or compact structure depending on the aggregation mechanism and are adequately described in terms of fractal dimensions (d_f), which can be defined by the number of particles expected within some distance r from the center of an aggregate, i.e., $n = r^{d_f}$. Values of d_f can be determined, or at least approximated, by a variety of techniques, e.g., optical and electron microscopy, light, x-ray, and neutron scattering [48,56,57], and electrical measurements [58]. Solid materials with no free space have a fractal dimension of 3, whereas particulate aggregates are generally in the range 1.4 to 2.5 and strongly influenced by the aggregation mechanism. For volume diffusion-limited aggregation, i.e., where particles have a strongly attractive pair potential and open floc structures are formed, d_f is in the range 1.5 to 1.9. When aggregation is chemical reaction limited, i.e., there is an energy barrier to aggregation that is significantly greater than the thermal energy, particles within a floc can generally move into more closely packed states and the fractal dimension is generally in the range 1.9 to 2.5. In general, salt-induced coagulation results in lower fractal dimensions than for depletion flocculation. Fractal dimensions are controlled by a balance of the mixing conditions and the stability ratio [59]. It is also well established that shear forces and conditions of differential settling lead to flocs with more open structure and lower fractal dimensions. Helsen et al. [58] probed the kinetics of structure formation in graphite suspension after shear from the time-dependent complex dielectric permittivity. In this case diffusion was the rate-controlling factor for structure re-formation.

Computer simulations [48] have been valuable in visualizing aggregate structure and are highly complementary to the experimental methods. The x-rays and neutrons are best suited to characterize structures of less than a micron in size, whereas light probes larger length scales. Small angle light scattering [2,60] has been particularly useful because its detector range is suitable for the size of aggregates generally found in particle separation processes, e.g., dewatering operations. Small angle neutron scattering (SANS) is obviously a less standard technique, but has been used to probe the gelation of silica particles and how the morphology is affected by shear [57].

Direct imagining of flocs is commonly employed either in situ or using cryogenic freezing. However, care must be taken not to disrupt the floc structure during sample preparation. Confocal scanning laser microscopy offers additional potential in this area [61]. For more concentrated suspensions, multiple scattering makes quantitative analysis of particle and floc size and distribution difficult. Multiple scattered light approaches [62] for particle sizing in dense polydisperse colloidal suspensions are also available, but these are not standard. Refractive index matching can be a useful approach and has been used to study structures under flow. Concentrated suspension are also generally opaque, however, it is still possible to use specialized light scattering techniques to investigate their structural behavior, e.g., two-color photon correlation techniques are used to ensure that only single particle motion is measured [63].

Gregory [51,64,65] has developed a flow through optical technique, which determines structural information from the fluctuations in the intensity of light transmitted through a flowing suspension. The mean aggregate size can be determined and when combined with turbidity data the aggregate mass fractal dimension can be estimated. Settling techniques [66] have been utilized to determine the structural compactness of aggregates. Others [48,67] have also developed optical floc density analyzers based on determining the settling velocity and size; these enable the open versus compact nature of flocs to be identified and hence flocculation mechanisms to be identified. Again, these may be used in online monitoring.

Image analysis and settling methods work best for particles that are large and of high contrast, forming structures of low dimensionality; this is complementary to light scattering that is better for small and loose aggregates of low refractive index particles. Imaging methods give specific information on single flocs and hence distributions in floc size and structure, whereas light scattering methods only give an average value, but are much faster and better suited for online analysis.

B. Mechanical Characterization of Structure

Structure in concentrated suspensions can be effectively determined from knowledge of the deformation of the system under the influence of an external stress, i.e., the rheology. The application of rheological techniques for probing structure in concentrated suspensions has developed extensively over the last 20 years, particularly with the advent of highly sensitive stress and strain controlled rheometers that determine rheological parameters over several orders of magnitude of shear rate. Rheological methods are now routinely employed in most industrial and academic laboratories concerned with interparticle interactions and structure formation in concentrated suspensions. Bulk rheology and thermodynamic measurements of model systems have demonstrated the effect of short-range interparticle forces acting on a nanometer scale on the flow properties of suspensions and their processing.

1. Rheological Behavior and
    Rheological Parameters

The shear stress (τ is the external force which acts tangentially per unit area on a deformed body. For viscous liquids (Newtonian liquids) the force per unit area, F/A, is the stress and results in a velocity gradient in the sample (or shear rate), D:

$$\tau = \eta D, \tag{46}$$

where η is the viscosity. For ideal (Hookean) solids the applied stress results in a deformation or strain (γ) and is controlled by the rigidity modulus (G) and governed by

$$\tau = G\gamma. \tag{47}$$

Concentrated suspensions are commonly non-Newtonian and exhibit complex rheological behavior including shear thinning, shear thickening, yielding, and thixotropy (or time-dependent rheology). Shear thickening, or dilatant flow (as commonly observed for wet sand on the beach), is generally a result of the particles within the suspension above a critical volume fraction and particle structuring or ordering occurring on application of shear, i.e., an order–disorder transition takes place at a critical shear. Structured suspensions containing an interparticle network are likely to display shear-thinning behavior, yielding, and thixotropy. In addition, a degree of elastic structure will be present and viscoelasticity will be exhibited. Shear thinning generally arises from the breakdown of aggregates in the shear field. However, it should also be recognized that in some dispersed suspensions a degree of shear thinning may be observed due to Brownian motion dominating the shear flow at low-shear rates, i.e., a random particle arrangement is obtained at low shear. Furthermore, asymmetric particles orientate themselves in the flow lines as the shear rate is increased and also leads to shear thinning.

   Suspensions for which the timescale for structure formation and breakdown is large compared with the rheological measurement time are thixotropic and display time-dependent viscosity behavior. Thixotropic flow is encountered in silica suspensions, clay slurries, and many pharmaceutical and cosmetic formulations [10,68,69]. Antithixotropy is not common in particulate suspensions, but has been reported from the temporary aggregation of attractive particles under the influence of shear [70] or from flocs becoming looser and more open under shear [71]. Thixotropy can be characterized by employing step changes in shear rate or shear stress or from a thixotropic loop [68,69]. The area within a thixotropic loop can be related to the energy delivered to breakdown structure in the suspension. Recently, attention has been paid to understanding the importance of surface forces in controlling thixotropy and reversible cluster formation correlated through short-range lubrication forces [72]. The fact that the reversible shear thickening in concentrated colloidal suspensions is highly sensitive to the surface forces between particles is now utilized to develop model systems that give insight into the interfacial structure. For example, the variation in the onset for shear thickening in the presence of a specific reagent can be utilized to predict adsorbed layer thicknesses with good sensitivity.

In addition to the different flow behaviors observed (see Figure 5), structured suspensions may behave as elastic solids at low-applied stresses. The critical stress below which no flow is observed is termed the yield stress (τ_y). The yielding or plasticity is primarily due to a breakdown in the continuous network that is present within a suspension. A Bingham plastic sample is one where the flow units are not broken down further after yielding and the viscosity is constant, i.e.,

$$\tau = \tau_B + \eta_{pl} D, \tag{48}$$

where τ_B is the Bingham yield stress and η_{pl} the plastic viscosity. In practice few structured suspensions are Bingham but rather they show shear thinning after yielding and are considered pseudoplastic. Numerous rheological models (e.g., the Casson and Herschel Bulkley) have been used to describe the flow behavior of pseudoplastic suspensions and to determine the yield stress values and these have been used to characterize particle interactions in suspension [73]. Many

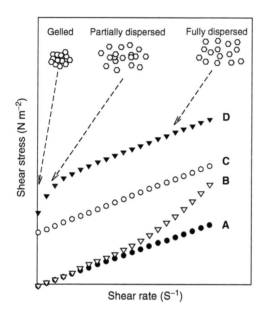

Figure 4.5 A schematic representation of the relationship between shear stress and shear rate for various suspensions: (A) colloidally stable and at relatively low-volume fraction (Newtonian), (B) colloidally stable and at relatively high-volume fraction (dilatant), (C) weakly structured (Bingham), and (D) highly structured (pseudoplastic).

of these flow models have little physical meaning and the parameters extracted are generally empirical with limited fundamental significance. One exception is the Cross model

$$\frac{\eta - \eta_\infty}{\eta_0 - \eta_\infty} = \frac{1}{1 + KD^m},$$ (49)

where η_0 is the zero shear viscosity, η_∞ infinite shear viscosity, and m the power law index, which was developed to describe the shear thinning nature of suspensions in terms of the kinetics of aggregation. This model accounts for the rate constants for the rupture of interparticle chains of flocculated particles by the Brownian motion and shear motion and the rate constant for Brownian rebuilding of floc structure. These rate constants can be determined from a linear reconstruction of the Cross model.

Even though the concept of a true yield stress has been extensively debated and many so-called *yield stress materials*, including suspensions, show flow, or creep below their yield stress; the yield stress is arguably the most widely used parameter for describing the level of structure in concentrated suspensions. It is probably best to consider the yield stress in terms of the abrupt change in the flow behavior to a less resistive state, i.e., a transition from solid-like to liquid-like behavior that can occur over a narrow or wide range of stresses, i.e., *brittle* or *ductile* failure, respectively. Subsequently, there is also much debate over yield stress measurement, e.g., see review by Cheng [74]. If the yield stress is determined from the relationship between the stress and strain in a dynamic state then the yield stress is considered a dynamic (extrapolated or apparent) value. Dynamic yield stresses are influenced by fluid flow through a structured suspension, in addition to the elastic structure, and are highly dependent on the model used for their determination. Furthermore, the onset of nonlinearity caused by slip at the wall of the rheological sensor may also be mistaken as the yielding point. The *vane technique* originally developed by Nguyen and Boger [75,76] is a popular method to determine the yield stress and avoids the problems associated with extrapolated yield stress determination. A bladed vane is slowly rotated in the suspension at a constant low-shear rate and the torque on the vane measured as a function of time. The geometry of the vane tool minimizes slip and the maximum torque is assumed to correspond to the shear failure of the structure within the sample; τ_y is then determined from the maximum torque and the vane dimensions. A yield stress can also be determined from the stress at which the linear viscoelastic region in a small amplitude oscillatory test (see below) is broken down, i.e., the onset of nonlinear behavior. Given the difficulties in yield stress definition and in establishing a technique and model independent value, the yield stress is best considered as a practical engineering parameter. With this in mind, it has been extensively used in the determination of structure in suspensions and correlating structure with interparticle interaction through various models.

The viscoelastic structure within a suspension can be determined either statically, i.e., through stress relaxation at constant strain or *creep* tests at constant stress, or dynamically using an oscillatory measurement. Stress relaxation involves application of a constant strain to the sample. For a viscous sample the stress relaxes to zero, whereas an elastic sample shows no stress relaxation and the stress is related to the relaxation modulus G:

$$G(t) = \tau(t)/\gamma. \tag{50}$$

A linear viscoelastic response is where the stress relaxation is linearly related to the strain. For colloidal suspensions the linear viscoelastic range is typically observed for strains up to 10^{-4} and highly sensitive rheometers are required for measurement in the linear region. For nonlinear viscoelastic behavior

$$G(t, \gamma) = \tau(t, \gamma)/\gamma. \tag{51}$$

In oscillatory measurements a small amplitude sinusoidal stress or strain is applied to the sample and the resultant stress or strain monitored. The phase lag (δ) between stress and strain is determined and the viscous (G'') and elastic (G') dynamic moduli are calculated from $\tan \delta = G''/G'$. The yield stress may be determined from the low-frequency elastic modulus (G'_0) and the corresponding strain (γ_0) data in the linear viscoelastic region

$$\tau_y = G'_0 \gamma_0. \tag{52}$$

The frequency response of G' and G'' gives insight into the relaxation processes within structured suspension and the high-frequency elastic modulus is used to relate to interaction forces [73,77,78]. There is an increasing interest in studies using nonlinear stresses or strain to probe structure, i.e., large amplitude oscillatory measurements [79,80]. Under these conditions the sinusoidal strain wave is deformed and the change in wave shape can be used to gain an understanding of the structure dynamics.

2. Suspension Rheology and Particle Interaction

For suspensions of noninteracting particles Krieger and Dougherty [81] have described an empirical relationship between the relative viscosity (η_r) and particle volume fraction (ϕ):

$$\eta_r = (1 - \phi/\phi_m)^{-[\eta]\phi_m}, \tag{53}$$

where $[\eta]$ is the intrinsic viscosity and ϕ_m the maximum packing fraction. For noninteracting hard spheres ϕ_m can be established and there is agreement with the volume fraction for close packing (0.64 for random close packing and 0.71 for hexagonal close packing). In practice, due to particle anisotropy, polydispersity, electroviscous effects, and particle interaction, ϕ_m deviates from the close packing value. Particles with significant repulsive potentials have an effective particle size and particle volume fraction that are greater than for the hard-sphere case. Deviations from hard-sphere behavior have been used to determine the electrical double layer thickness or steric layer thickness [73].

Suspensions of electrostatically attractive particles (e.g., clays), or where electrostatic repulsive forces are removed by the addition of salt (coagulated suspensions) or flocculated through polymeric bridging or depletion are often termed *gelled* and their rheology is extensively non-Newtonian. Such aggregated or flocculated suspensions generally display a particle size increase and enhanced settling behavior. However, as the volume fraction of particles exceeds a critical point (the gel point) a continuous network is formed and settling is ceased. Rheological measurement is effective at probing the structure within such gelled suspensions [82].

A number of approaches have been made to relate the particle interaction behavior to the rheological parameters and these are dependent on the strength of the structure. For weakly flocculated suspensions the yield stress has been related to the particle interaction energy, e.g., Tadros [73] used the following expression to relate the particle separation energy (E_{sep}) and yield stress of latex particles under depletion flocculation

$$\tau_y = \left[\frac{3\phi_s n E_{\text{sep}}}{8\pi a^2}\right], \tag{54}$$

where n is the number of particle contacts in a flocculated network (8 for random packing and 12 for close packing). For strongly aggregated suspensions, e.g., coagulated, Firth and Hunter [83] and Firth [84] developed the elastic floc model based on DLVO theory, which describes the yield stress in terms of particle interaction and the flow of fluid through the flocculated network of particles; this has been further refined in recent years. Scales et al. [85] showed that for particle suspensions where only van der Waals and electrostatic forces are operative, the yield stress can be rationalized:

$$\tau_y = K_{\text{struc}}\left[\frac{A_H}{12H^2} - \frac{2\pi\varepsilon_0\varepsilon\kappa\zeta^2 e^{-\kappa H}}{1 + e^{-kH}}\right], \tag{55}$$

where A_H is the Hamaker constant, H is the interparticle distance, $\varepsilon = \varepsilon_0 D$ with ε_0 is the permittivity of vacuum and D is the dielectric constant, ζ is the interactive potential well approximated by zeta potential, κ is inverse Debye length, and K_{struc} is a network structural term dependent on the particle size, solid volume fraction, and the mean coordination number. The terms in the brackets correspond to van der Waals and electrostatic forces, respectively. This approach has successfully described the yield stress as a function of pH and salt concentration for many oxide particle systems, e.g., alumina, zirconia, and silica, and has been further extended to account for volume fraction effects [85–87]. The relationship between yield stress, zeta potential, and pH is schematically represented in Figure 4.6.

For suspensions displaying DLVO behavior and where the floc structure is independent of the zeta potential, Equation (55) can be simplified to

$$\tau_y = \tau_{y,\max} - k_y\zeta^2, \tag{56}$$

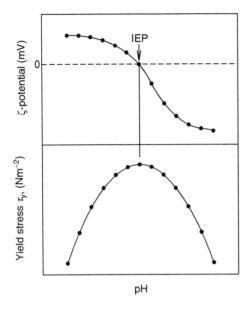

Figure 4.6 A schematic representation of the relationship between zeta potential and the shear yield stresses and pH for a concentrated suspension stabilized by DLVO forces.

where $\tau_{y,max}$ is the maximum value of the yield stress, which occurs at the IEP of the particles and k_y is a constant. Many suspension rheology studies have confirmed that the yield stress scales linearly with the square of the particle zeta potential [22,85–89]. This approach has been extended to account for additional interactions to those from electrostatic and dispersion forces, by the addition of a non-DLVO contribution to the yield stress (τ_y'):

$$\tau_y = \tau_{y,\,max} - k_y\zeta^2 + \tau_y', \tag{57}$$

τ_y' has been shown to be positive in sign and contribute significantly to the overall yield stress for suspensions of sulfide mineral particles [88,89] and surfactant-coated zirconia [86], where hydrophobic attractive forces are in action. In contrast, τ_y' has been shown to be negative for gibbsite suspensions [22] at pH values greater than the IEP. Under similar conditions,

atomic force microscopy (AFM) has revealed a repulsive interaction force between gibbsite particles and this is considered
to be due to either hydration or a steric layer at the particle
surface [22]. In many industrially important processing environments, e.g., gibbsite particles during Bayer processing or
red mud processing, the particle interaction behavior may be
highly dynamic and rheological measurements have been
used to probe the time-dependent structure formation in
such concentrated suspensions [44,90–92].

Recently [72] progress has been made in the application
of high-frequency rheological measurements to probe equilibrium microstructure in suspensions. Torsional resonators
apply frequencies in the range 10–1000 krad/sec and the rheology is determined within timescales shorter than those experienced in Brownian motion, i.e., where the particle diffuses
a distance less than its radius. Direct measurements of electrostatic stabilization forces in concentrated suspension have
been undertaken using rheological methods, which overcome
the concentration limitations of particle electrophoresis or
electroacoustics. The presence of slip at the surface of colloidal
particles and its effect on particle microstructure and suspension rheology is also an active area of investigation.

3. Compressive Flow

The intense interest to improve the dewatering performance
of colloidal suspensions has lead to an increased interest in
the rheological behavior of suspensions under compression
[93,94]; this is highly complementary to the shear rheology.
Upon compression of a suspension a compressive yield stress
$P_y(\phi)$ and hindered settling function, $R(\phi)$ can be defined.
A number of methods have been developed to determine
these parameters, including equilibrium batch settling,
centrifugation and pressure filtration rig testing [94]. The
simplest method for estimating the gel point of a suspension
under compression is through an equilibrium batch-settling
test. The pressure filtration technique is a comprehensive
method involving the application of a uniaxial pressure to a
suspension held in a cylindrical cell. Once the pressure

exceeds the local stress the suspension yields and fluid is discharged, i.e., dewatered.

The compressive yield stress $P_y(\phi)$ gives the network strength of a gelled suspension and has a strong dependence on the solids loading, suggesting a network of interpenetrating fractal aggregates. Shear yielding shows a weaker dependency on ϕ reflecting that the number of interparticle bonds are strained. Ongoing research is focused on gaining further understanding of the interplay between normal and lateral (friction) forces between particles and the compressional suspension rheology.

C. Interaction Forces on Individual Particles

To further our understanding of the structures in colloidal suspensions it is essential to gain direct information concerning the interaction forces between particles. A range of approaches is now available to measure interfacial forces, including total internal reflectance microscopy (TIRM), the surface forces instrument (SFA), laser tweezers, and colloid probe microscopy.

TIRM measures the equilibrium potentials between a particle and a surface from Brownian fluctuation in separation distance using an evanescent wave technique. It has a force resolution as low as 0.01 pN, a distance resolution of ~1 nm and is best suited to energies below $1\,kT$. TIRM has been used to determine weak repulsive forces between a number of different particles and a flat glass plates with various coatings. Particle diffusion [95] and hydrodynamics [96] have been quantitatively probed using time-dependent TIRM investigations.

SFA measures the forces between functionalized mica surfaces over relatively large areas and the separation distance is measured to less than 0.1 nm with interferometric techniques [1]. Though SFA is limited in the materials that can be investigated, it has been highly successful in determining osmotic and elastic forces between adsorbed polymer layers, frictional forces, nanorheological properties, and slip phenomena. Interaction forces between polymer-coated surfaces determined by the SFA have been correlated with rheological parameters for concentrated suspensions of colloidal particles (latex) with equivalent adsorbed layers [74,77,78]. The high-frequency limit of storage modulus (G'_∞) was numerically compared to values predicted from the interparticle force ($F(r)$) data:

$$G'_\infty = -\frac{3n_{co}\phi_{max}}{32\pi r}\left(\frac{\partial F(r)}{\partial r}\right),\qquad(58)$$

where n_{co} is the coordination number and ϕ_{max} the maximum attainable volume fraction. Good qualitative agreement was observed, however, the values calculated from the measured forces generally exceeded the rheological values by an order of magnitude.

The laser tweezers technique optically traps micron-sized particles in solution and can assemble particles into structures; the distance resolution is ~1 nm [52,97]. The force is then determined with a resolution range of 0.1 to 100 pN from the particle displacement from the trap center. Particles can be placed in flow fields and forces measured, or alternatively passed through a complex fluid, hence the technique can be used as a microrheometer.

Colloid probe microscopy using the AFM is currently the most widely used approach for determining the interactions between individual particles. The AFM was developed for imaging surfaces and has an advantage over scanning tunneling microscopy in that it can be used for nonconducting surfaces and in aqueous solutions at high salt concentrations (e.g., biologically relevant solutions). In 1991 Ducker et al. [33] reported that by attaching a sphere to an AFM cantilever the interaction force between the colloid probe and a surface can be determined. This approach was further developed to determine interaction forces between two particles [98]. Particles are attached to the cantilever using micromanipulators and glued with a chemically inert thermotropic resin or polymerizable glue. Smooth spheres are generally preferred for quantitative measurements; however less ideal particle geometries have been used, including fibers, drug particles and mineral particles as well as droplets and bubbles.

Force versus distance measurements are undertaken by driving the particle toward the surface and monitoring cantilever deflection as a function of displacement. Cantilever deflection is determined from the deflection of a laser beam using a photodiode and substrate movement from a piezocrystal scanner. From knowledge of the cantilever's spring constant, its deflection can be converted into an interaction force using Hooke's law. Force curves (see Figure 4.7) between identical particles with significant surface charge in relatively low-salt concentration solution show repulsive forces the range and magnitude of which are described by the surface potential and Debye length. In fact, zeta potentials estimated from AFM derived force versus distance data are in good agreement with that determined for the same colloidal particles using electrophoresis measurements [99].

For the case where an attractive force is introduced between particles (see Figure 4.8c and d), e.g., oppositely charged particles, bridging polymer [100], depletion force due to nonadsorbed polymers [101], a strong hydrophobic attraction [86,88,89,102], then upon approach of the particle

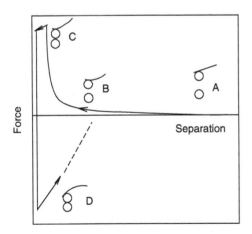

Figure 4.7 A schematic representation of AFM force versus distance curves for particles at different levels of interaction: no interaction (A), repulsion (B), jump into contact (C), and adhesion (D) prior to jumping out of contact.

surfaces, there is a jump into contact and then on retraction there is a minimum force required to separate the particle surfaces, i.e., an adhesion force. It has been recognized that the adhesion force is strongly influenced by the normal force applied to the particles and is also influenced by surface roughness. The adhesion force gives valuable information concerning the strength of particle interaction and correlates with the yield stress of a concentrated suspension of equivalent particles. For hydrophobic ZrO_2 colloids, Leong et al. [86] showed that the yield stress correlated with the hydrophobic and van der Waals forces rather than electrostatic forces, even at pH values away from the IEP. For ZnS particles, Muster et al. [89] showed a positive correlation between the adhesion force and the yield stress, but demonstrated that the surface chemistry is critical for sulfide particles and can be influenced by the surface area to volume ratio. The influence of dissolved gas and surface nanobubble formation is now considered to play a critical role in determining the interaction of hydrophobic particles. The interplay between particle hydrophobicity, solution conditions, the yield stress, and the interaction forces has not been fully explored.

In addition to normal force measurement, AFM is also capable of quantitatively probing lateral or frictional forces between particles and a surface [103–105]. Current research is resolving the role of surface chemistry and surface roughness in controlling interparticle friction and its relationship to the dynamics of floc structure and strength. The integration of experimentally determined interparticle force data into theoretical models that predict the rheology, and therefore the development of a fully predictive structure–property relationship for concentrated suspensions is the challenge for the future.

IV. STRUCTURE FORMATION AND CONTROL THROUGH INTERFACIAL CHEMISTRY

A. Introduction

Destabilization of a colloidal dispersion may be achieved through interfacial chemistry modification, which introduces conditions conducive for a net attraction between particles to dominate. Mineral particle surfaces undergo protonation and become positively charged or deprotonation and become negatively charged, depending upon the pulp pH. At high pH (>9) where most mineral pulp particles are negatively charged, colloidally stable dispersions with little rheological structure (low-yield stress) are formed as a result of the electrostatically repulsive interactions between particles. Destabilization may be achieved through pH modification and the use of chemical additives (coagulants, polymeric flocculant, and surfactant).

Where pH is modified to the IEP, the electrical double layer repulsion may be eliminated or suppressed to allow the particles to approach each other within the range of attractive van der Waals forces for coagulation to occur in the absence of non-DLVO repulsive forces. Alternatively, particle bridging may be achieved through the addition of a high-molecular weight (>10^6 Da) polymeric flocculant (e.g., polyacrylamide). The type of flocculant or coagulant metal ion (salt) used, in conjunction with the dispersion pH and temperature, plays an influential role in species adsorption mechanisms and the modification of particle surface chemistry; impacting upon particle interactions, floc structure, and network formation.

Recourse to the literature shows that in addition to van der Waals forces and polymer bridging or depletion forces, additional non-DLVO attractive interactions may be operative [46]; these include ion–ion correlation forces, charge patch interactions, ion bridging, and precipitation of another phase on the surfaces of the particles. For example, non-DLVO attraction between mica, alumina, and zirconia particles at small separations in the presence of structure making salts (e.g., $LiNO_3$ and $NaIO_3$) that act as coagulants, have been observed [46,106]. Shubin and Kekicheff [106] ascribed the additional interparticle adhesive force to a hydration bridging effect caused by the Li^+ ions sharing their hydration water with the mica surfaces.

In some situations, even under conditions of high ionic strength, where $1/\kappa < 0.1$ nm, coagulation may not readily occur. Repulsive structural forces, which are not due to electrical double layer interactions, have been directly measured between gibbsite surfaces in synthetic supersaturated Bayer liquor to account for the unusual colloidal stability [22–24,44]. Strong, monotonically repulsive, non-DLVO forces (F) were initially observed between a gibbsite plate and a sphere of radius (R) [22–24]. Upon aging in situ, the sphere radius normalized force (F/R), which was long ranged (up to 40 nm), attenuated and disappeared completely with time as shown in Figure 4.8.

The magnitude of the repulsion was ~25% greater for the basal (001) than the prismatic (non-(001)) face of the crystal and ~20% greater in sodium than potassium-based liquors. Likewise, the subsequent decrease in repulsion was also faster for the prismatic face (Figure 4.9). The variation of the interparticle forces due to crystal face and alkaline metal ion type is believed to be associated with the operative inter-facial chemistry.

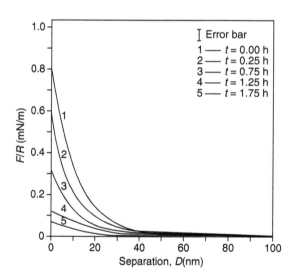

Figure 4.8 *F/R* as a function of separation distance between gibb-site flat (001 face) and sphere in supersaturated Bayer liquor (4.0 *M* NaOH, 2.9 *M* Al(III)) at 30°C as function of time, indicating structural repulsion within the first 2 h. Particle approach velocity = 0.59 μm/sec.

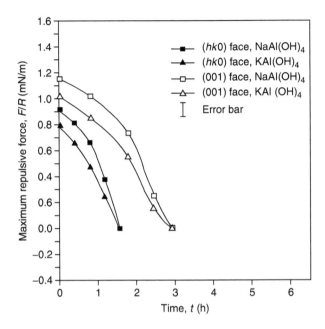

Figure 4.9 The influence of liquor cation and crystal face on the maximum repulsive forces between gibbsite surfaces in synthetic Bayer liquors (4.0 M NaOH or KOH, 2.9 M Al(III)) at 30°C as a function of time.

B. Coagulation

Coagulation of, and structure formation in, stable colloidal dispersions by the addition of metal ions has been widely studied [107–118]. The interparticle repulsive energy barrier may be overcome by a simple increase in ionic strength, facilitating particle aggregation and settling under gravity as a result of electrical double layer compression.

In aging gibbsite–caustic aluminate dispersions, coagulation occurs, albeit after an induction time [23,24,41–44,90,91]. Subsequent to attenuation of the structural repulsion indicated in Figure 4.8 and Figure 4.9, the F/R as a function separation distance curves of aged (>2 h) systems showed characteristically nonmonotonic forces upon approach (Figure 4.10, top part) [23,24]. The shape of the force curves is

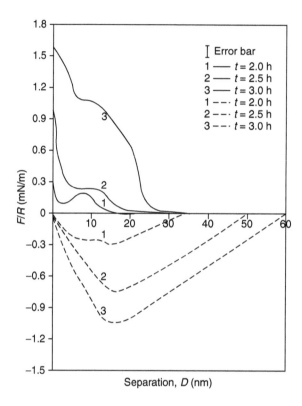

Figure 4.10 F/R as a function of separation distance between gibbsite flat (001 face) and sphere in supersaturated Bayer liquor (4.0 M NaOH, 2.9 M Al(III)) at 30°C as function of time, indicating the presence of a compressible surface structure and adhesive forces after 2 h (full line = approach, dashed line = retraction). Particle approach velocity = 0.59 μm/sec.

indicative of a change in nanomechanical structure as the applied loading force is increased. Thus, a developing structured layer is encountered upon approach of the particles and an adhesive force (separation or pull-off force) observed on retraction (Figure 4.10, bottom part). With time, both the thickness and density of the structured layer and the concomitant adhesive force increase. Interparticle adhesion was stronger and observed earlier in sodium than potassium

based liquors; this is in agreement with rheology data presented below.

Shear stress versus shear rate curves, thixotropic structure and viscoelastic behavior of gibbsite suspensions are
shown in Figure 4.11–Figure 4.14 [24,44,91]. The observed
rheological changes reflect particle aggregation and network
structure formation which occurs as a precursor to cementation as a final step of agglomeration. Increased shear thinning
and pronounced thixotropy (flow curve hysteresis) is evident
upon aging in Figure 4.11. Given that crystal growth and
agglomeration effect are insignificant, it is the change in the
gibbsite particle interaction potential that results in extensive
particle aggregation, i.e., the formation of a gibbsite particle
network with the observed thixotropy.

The area within a flow curve hysteresis loop equates to
the energy per unit volume applied to the suspension per unit
time to breakdown microstructure. Figure 4.12 shows that the
rate and extent of thixotropic structure formation is greater in
sodium than the potassium based liquors, reflecting alkali
metal ion (Na^+ versus K^+) effects on gibbsite particle interactions. The elastic modulus (Figure 4.13) and shear yield
stresses are directly related to particle interaction forces.

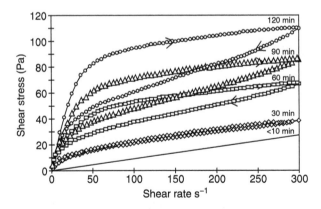

Figure 4.11 Flow curves for a sodium-based Bayer liquor ($4.0\,M$
NaOH, $2.9\,M$ Al(III)) seeded with 20% (w/w) colloidal gibbsite at
65°C.

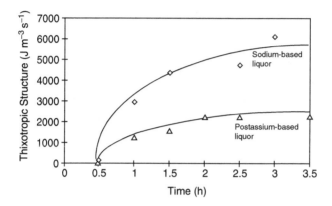

Figure 4.12 Thixotropic structure in aging Bayer liquor (4.0 M NaOH or KOH, 2.9 M Al(III)) seeded with 20% (w/w) colloidal gibbsite at 65°C.

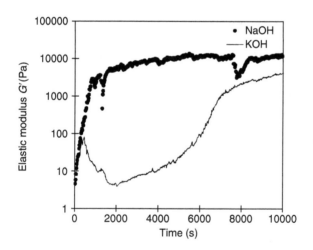

Figure 4.13 Viscoelastic structure development of colloidal gibbsite dispersed at 25% (w/w) in synthetic Bayer liquors (2.9 M Al(III) and 4.0 M NaOH or KOH) at 65°C.

In fresh dispersion of colloidal gibbsite in Bayer liquor, the particle network structure is initially weak and dominated by viscous forces, i.e., the viscous modulus (G'' is much greater than the elastic modulus (G'), consistent with AFM data, which showed that repulsive forces initially dominate. Upon aging there is a dramatic increase in the elastic structure, as exemplified in Figure 4.13. With time, as the adhesive forces develop, G' becomes greater than G'' and the particle network strengthens. The precise form of the viscoelastic structure development is directly related to the interfacial chemistry as particles form aggregates by coagulation and then cement together by deposition of a growth layer. The rate of elastic structure development is considerably faster in sodium compared with potassium based liquors, again confirming the influential role of the cation in controlling the interfacial chemistry and aggregation process.

1. Hydrolyzable Metal Ions

Unlike monovalent metal ions Na^+ and K^+, hydrolyzable, multivalent metal ions such as Mn(II) and Ca(II), if present in solution at minimum concentrations (e.g., 10^{-4}–$10^{-3} M$) and certain pH values (e.g., Mn(II) at pH 7.5 and Ca(II) at pH 10.5), can hydrolyze and specifically adsorb onto the mineral particle surfaces [107–121]. Electrical double layer compression, a reduction and reversal in surface charge and zeta potential, a shift in IEP, and surface nucleation of metal hydrolysis species can take place, leading to particle aggregation, mechanical entrapment, or heterocoagulation [107–132]. Thus, effective coagulation of a stable dispersion at a particular pH may be achieved by choosing a multivalent metal ion, which will hydrolyze and specifically adsorb at that pH. These features are demonstrated by studies [121,125] involving kaolinite clay dispersions and Mn(II) and Ca(II) ions reported below.

In the presence of Mn(II) ions the magnitude of the zeta potential decreased with increasing metal ion concentration, accompanied by a shift in the IEP to a higher pH value (Figure 4.14). The IEP increased with increasing metal ion concentration and, at the highest concentration studied $(0.1 M)$, a marked decrease of zeta potential to values close to zero at >pH 7 occurred. This change in electrokinetic behavior is consistent with the hydrolysis and adsorption model [108–113], suggesting that there is specific adsorption of

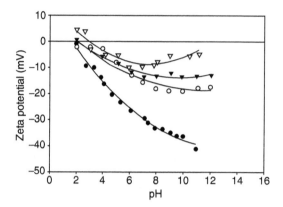

Figure 4.14 Zeta potential of kaolinite particles versus pH at $10^{-2} M$ KNO$_3$, at 8 wt.% solids at Mn(II) ion concentrations: 0 (•), $10^{-3} M$ (○), $10^{-2} M$ (▼), and $10^{-1} M$ (▽).

positively charged Mn(II) hydrolysis products at the kaolin-ite–aqueous solution interface. Addition of Ca(II) ions to the kaolinite dispersions had a substantial effect on zeta potential at the pH >9 (Figure 4.15). The zeta potential reversed from negative to positive values at pH >10 in the presence of

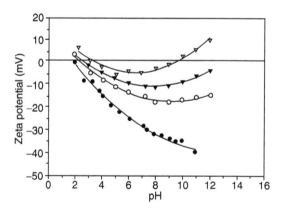

Figure 4.15 Zeta potential of kaolinite particles versus pH at $10^{-2} M$ KNO$_3$ at 8 wt.% solids at Ca(II) ion concentrations: 0 (•), $10^{-3} M$ (○), $10^{-2} M$ (▼), and $10^{-1} M$ (▽).

$10^{-1}M$ Ca(II). The charge reversal at this pH may be explained in terms of specific adsorption of hydrolyzed Ca(II) species (e.g., $Ca(OH)^{+}$) and, possibly, surface nucleation of $Ca(OH)_2$ [107–112]. Similar observations have been reported for Ca(II) ions in barite [114], kaolinite [115], and quartz and hematite [120] dispersions.

Figure 4.16 shows typical flow curves for kaolinite dispersions in the presence of Mn(II) metal ions at pH 7.5. All dispersions showed non-Newtonian, pseudoplastic flow behavior with no hysteresis. Figure 4.17 displays the influence of Mn(II) and Ca(II) ion on the shear yield stress of kaolinite dispersions at pH 7.5. A decrease in yield stress is observed with an increase in metal ion concentration, contrary to expectation. Compression of the electrical double layer and zeta potential reductions accompanied by a primary minimum aggregation may normally be expected at higher ionic strength in the manner of DLVO theory, leading to an increase in yield stress.

This anomalous rheological behavior may be rationalized in terms of heterogeneous charge properties of the basal and edge faces of the kaolinite crystal, which results in three different modes of particle association; edge-to-face, edge-to-edge,

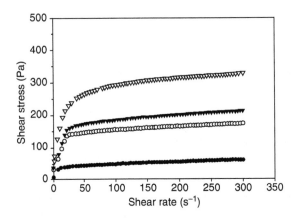

Figure 4.16 Shear stress versus shear rate curves for kaolinite dispersions at 32 wt.% solids as a function of Mn(II) ion concentration: 0 ($\triangledown$), $10^{-3}M$ ($\blacktriangledown$), $10^{-2}M$ ($\circ$), $10^{-1}M$ ($\bullet$).

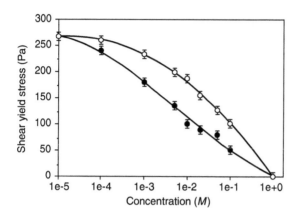

Figure 4.17 Shear yield stress of 32 wt.% solid kaolinite dispersion at pH of 7.5 as a function of Mn(II) (•) and Ca(II) (○) ion
concentration.

and face-to-face. Different combinations of electrical double
layers govern the interaction potentials for the three types of
association. It appears that under high ionic strengths of
the metal ions, the associations led to repulsive rather
than attraction forces, reminiscent of basal face-dominated
interactions.

In another example, the influence of a simple salt
(NH_4Cl) as a coagulant and the specific adsorption of NH_4^+
ions on iron oxide colloid chemistry, particle interactions, and
dewatering behavior are demonstrated in the pH range 2–10
[130]. In the absence NH_4Cl, the zeta potential of the particles
in $10^{-3} M$ KNO_3 plotted as a function of pH (Figure 4.18)
showed a decrease from +63 to −58 mV as the pH increased
from 2 to 10, with an IEP at pH 5.6, as expected [128–134].

The presence of 5×10^{-3} to $5 \times 10^{-1} M$ NH_4Cl in solution
had a profound effect on both the zeta potential and yield
stress, impacting on the dewatering behavior of the dispersions. The data exhibited in Figure 4.18 showed that at
5×10^{-3} and $5 \times 10^{-2} M$ NH_4Cl, the zeta potential decreased
with increasing pH as expected, the magnitude of which de-
creased dramatically with increasing NH_4Cl concentration.
This was accompanied by a shift of the IEP from pH 5.6 to

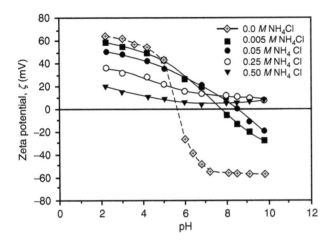

Figure 4.18 The zeta potential of iron oxide particles (a 20 wt.% solids dispersion) as a function of pH at different NH_4Cl concentrations, 22°C and $10^{-3} M$ KNO_3 background electrolyte.

7.8 and 8.5, as the NH_4Cl concentration increased to 5×10^{-3} and $5 \times 10^{-2} M$, respectively. This indicates that specific adsorption of electropositive ions (NH_4^+) occurred onto negatively charged particles present above the initial IEP (pH 5.6). A further increase in NH_4Cl concentration up to $5 \times 10^{-1} M$ led to a significant decrease in zeta potential and complete charge reversal at pH >8. As clearly shown in Figure 4.18, at 2.5×10^{-1} and $5 \times 10^{-1} M$ NH_4Cl, the zeta potential remained practically constant (<7 mV) above pH 8.

The lack of an IEP or the reversal of charge displayed by the particles at both higher NH_4Cl concentration and high pH (>9) confirms that the NH_4^+ ions specifically adsorbed onto the iron oxide particle surfaces. Similar specific adsorption behavior for hydrolyzable Fe(III) and Co(III) ions and the accompanying charge reversal of hematite particles at pH >7 have been reported [131,132].

Both the yield stress and settling rate of 20 wt.% solid dispersions in the absence of NH_4Cl were low at pH values lower or higher than the IEP where the zeta potential and hence, interparticle electrostatic repulsion was the greatest

(Figure 4.19). The maximum yield stress and settling rate occurred near the IEP where the zeta potential was close to zero and the total interaction potential was controlled by attractive van der Waals forces.

The yield stresses of 60 wt.% solid dispersions measured as a function of NH_4Cl concentration at different pH values indicated that both NH_4Cl concentration and solid loading had a marked effect on the particle interactions (Figure 4.20). The maximum in the yield stress occurred for all NH_4Cl concentrations at pH 8 where the zeta potentials of the particles were zero or close to zero. The significantly enhanced yield stress is attributed to both van der Waals attractive forces and an additional interparticle "cementation" effect resulting from specific NH_4^+ ion adsorption, particularly at high NH_4^+ ion concentration [46].

C. Structure Formation Due to Polymer Action (Flocculation)

Polymeric flocculant mediated mineral dispersion destabilization and structure modification are directly linked to polymer–particle interactions, floc structure characteristics, and

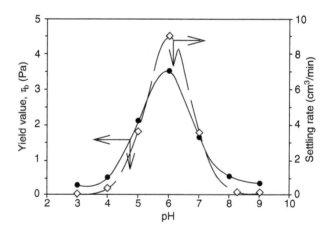

Figure 4.19 The variation of shear yield stress and settling rate with pH of iron oxide dispersions at 20 wt.% solid, 22°C and $10^{-3} M$ KNO_3 background electrolyte.

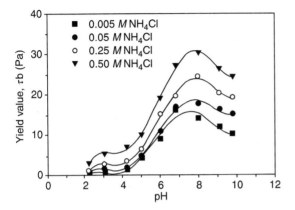

Figure 4.20 Shear yield stress of 60 wt.% solid iron oxide dispersion as a function of pH at different NH_4Cl concentrations, $22°C$ and $10^{-3} M$ KNO_3 background electrolyte.

sediment dewatering behavior [150]. The key factors controlling effective flocculation induced by interparticle bridging by polymer segments include flocculant characteristics such as molecular weight, charge density, functionality and dosage, mineral particle zeta potential, particle size and surface area and solid concentration, presence of simple and hydrolysable metal ions, pH, temperature, and the degree of shear and compression (see Figure 4.21).

From a theoretical basis, Mackenzie [135] and Healy [136] proposed that the mechanisms which play important

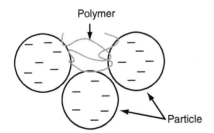

Figure 4.21 Schematic illustration of bridging flocculation between negatively charged particles and a nonionic polymer.

roles in polymer adsorption onto mineral particles are chemical, electrostatic, hydrogen bonding, van der Waals interactions, and hydrophobic interactions. With respect to the overall driving force for polymer adsorption, the standard free energy of adsorption (ΔG^o_{ads}) of a polymer at the mineral particle–solution interface may be represented as the sum of the standard free energies of the individual contributions

$$\Delta G^o_{ads} = \Delta G^o_{chem} + \Delta G^o_{elec} + \Delta G^o_{h\text{-}b} + \Delta G^o_{vdw} + \Delta G^o_{h\text{-}phob},$$

(59)

where the subscripts are chem = chemical, elect = electrostatic, h-b = hydrogen bonding, vdw = van der Waals interactions, and h-phob = hydrophobic interactions.

For a particular mineral particle–polymer pair, the mineral particle surface–solution chemistry and polymer functionality, molecular structure, and conformation will determine the sign and the magnitude, and hence the importance, of the various terms in Equation (59). Flocculation may occur by a number of mechanisms including polymer bridging, charge compensation or neutralization, polymer–particle surface complex formation, and depletion flocculation, or by a combination of these mechanisms [64].

1. Bridging Flocculation

Bridging flocculation occurs as a result of adsorption of individual polymer chains onto several particles simultaneously, forming molecular bridges between the adjoining particles in the floc. This is facilitated at a low polymer surface coverage on the particles. For nonionic polymers such as polyethylene oxide (PEO), polymer adsorption onto particle surfaces is achieved via hydrogen bonding, as is the case for flocculating negatively charged particles with anionic flocculants.

2. Charge Neutralization

On the other hand, charge neutralization becomes a major mechanism for polyelectrolytes where significant particle surface sites with charge opposite to that of ionic polymer functional groups is present (Figure 4.22).

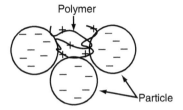

Polymer

Particle

Figure 4.22 Schematic illustration of a charge neutralization flocculation mechanism between negatively charged particles and a cationic polymer.

3. Electrostatic Patch Mechanism

This occurs when particles have a fairly low density of immobile surface charges and the adsorbing polymer has a high charge density (Figure 4.23). In this instance, it is not physically possible for each charged group on the polymer chain to be adjacent to a charged surface site and regions of excess charge develop even though the particle as a whole may be electrically neutral. Particles having this *mosaic* type of charge distribution may interact in such a way that positive and negative *patches* come into contact, giving quite strong attachment [64].

4. Depletion Flocculation

A less common mechanism is depletion flocculation. It occurs in situations where free nonadsorbing polymer molecules

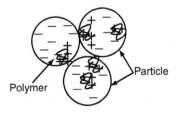

Particle

Polymer

Figure 4.23 Schematic illustration of electrostatic patch flocculation mechanism between negatively charged particles and a cationic polymer.

present in the dispersion are excluded from the space between two approaching particles due to their large size. The resulting osmotic pressure forces the particles to flocculate [137]. The most important characteristics are the molecular weight and the charge density of the polymer.

Mechanical actions such as shear or compression are also well documented to exert a significant influence on the overall efficiency of flocculation for colloidal dispersions [138–149]. In sedimentation processes, as the particles attain higher concentrations in the compression zone deep down within thickener or clarification equipment, a point is reached where further settling is restrained by hydrodynamic forces and the mechanical support from layers of particles below. A self-supporting particle network structure is invariably formed at a certain volume fraction (gel point).

The influence of depth of pulp and external mechanical shear imparted to the pulp by raking and pickets action are very important in the ultimate consolidation of the dispersion to a higher solid loading. Excessive shear may not only deform or fragment the polymer chains, but also cause floc rupture. The flow field induced by shear or compression may influence the adsorbed polymer conformation and layer thickness, improving the particle bridging and packing efficiency and, therefore, the pulp dewaterability. If more than sufficient drag is exerted on the adsorbed polymer chains, desorption from the particle surface may occur.

The use of shear-sensitive floc forming polymers (e.g., high-molecular weight PEO) has been of considerable interest in recent years due to their ability to produce thickened tails of solid loadings significantly greater than those achieved with conventional high-molecular weight polymers (e.g., homo- and co-polyacrylamides [PAM]) under moderate shear conditions [138–149].

High-molecular weight polymers, which produce strong dispersion structures and shear-sensitive flocs at certain dosages, are characteristically flexible, linear chain of atoms connected by chemical bonds (backbone), and simple pendant of atoms. They possess a large number of degrees of freedom and the ability to change conformation or shape by rotation about bonds (single) bonds in the backbone. Thermal motion is strong compared with the energy barriers associated with backbone rotation. This means that for a high-molecular weight polymer with long chains, a wide spectrum of conformations will be available. The presence of bulky pendant groups (e.g., Figure 4.24) may significantly limit conformation changes, making some desired conformation unfavorable.

Prior to polymeric flocculant adsorption onto mineral particles, its conformation in the aqueous phase, as determined by solvent (water)–polymer segment interactions, plays a crucial role in the flocculation performance. As discussed in Fleer et al. [139], the polymer–solvent interaction may be described by the Flory–Huggins parameter χ defined as the energy change associated with the transfer of segment from *pure* polymer to *pure* solvent phase. For a good solvent, the Flory–Huggins parameter $0 < \chi$, solvent and polymer segments have the same polarity and polarizability. Strong polymer–solvent interactions occur and lead to polymer chain expansion or swelling, a condition conducive to improved interparticle bridging. At low polymer concentration, polymer segments and chains repel each other; molecules are reluctant to interpenetrate, although as the polymer concentration increases, the chains start to overlap.

For the case where $0 < \chi < 0.5$, solvent-dependent net attraction between polymer segments occurs, whereas at $\chi = 0.5$, the θ-point, the effective polymer–polymer segment repulsion vanishes. Ideally, the polymer chains start to

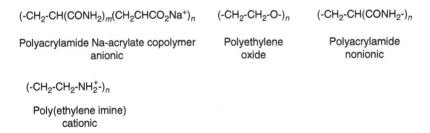

Figure 4.24 begins with chemical structures:

$(\text{-CH}_2\text{-CH(CONH}_2)_m(\text{CH}_2\text{CHCO}_2\text{Na}^+)_n$

Polyacrylamide Na-acrylate copolymer
anionic

$(\text{-CH}_2\text{-CH}_2\text{-O-})_n$

Polyethylene
oxide

$(\text{-CH}_2\text{-CH(CONH}_2\text{-})_n$

Polyacrylamide
nonionic

$(\text{-CH}_2\text{-CH}_2\text{-NH}_2^+\text{-})_n$

Poly(ethylene imine)
cationic

Figure 4.24 Examples of typical polymers used in flocculating colloidally stable mineral dispersions.

interact between themselves. For the case $\chi = 0.5$, poor solvency results in polymer–polymer interactions more preferred to polymer–solvent interactions. This invariably leads to a more coiled or contracted polymer chain conformation, a condition which is detrimental to interparticle bridging. The Flory–Huggins parameter χ is dependent on temperature, polymer concentration, pH, metal ion type, and concentration.

D. Flocculation

The type of polymer used in flocculation plays a pivotal role in particle interactions, network structure, and strength [123–127,138–149]. The effect of nonionic PEO (molecular weight $\sim 2.5 \times 10^6$ Da) and anionic PAM (molecular weight $\sim 2.7 \times 10^6$ Da) flocculants on the shear yield stress and corresponding energy of separation (E_{sep}) (floc network strength estimated from Equation (54)) is shown in Figure 4.25 and Figure 4.26, for colloidal kaolinite and smectite clay mineral dispersions, respectively [149].

An increase in shear yield stress and energy of separation with increasing flocculant concentration is clearly evident for

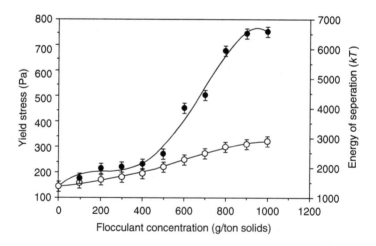

Figure 4.25 Shear yield stress and energy of separation of 40 wt.% kaolinite dispersion as a function of PEO ($\bullet$) and anionic PAM ($\circ$) concentration at pH 7.5 and $10^{-3} M$ KNO$_3$.

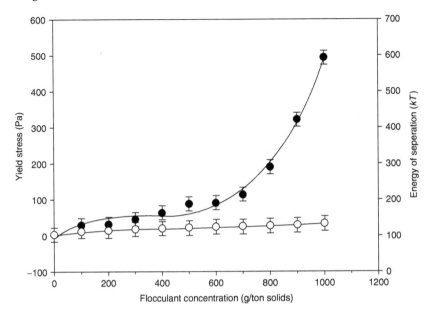

Figure 4.26 Shear yield stress and energy of separation of 20 wt.% smectite dispersion as a function of PEO (●) and PAM (○) concentration and in $10^{-2} M$ Ca(II) ions at pH 7.5 and $10^{-3} M$ KNO$_3$.

both clay dispersions. The yield stresses show that a strong polymer structure type dependency prevailed under flocculation, flocs, and their network structures, which are much stronger for PEO than PAM. For instance, the increase in yield stress is dramatic above 500 g PEO per ton of solid, but no such behavior is displayed by PAM. This concentration can be seen as a critical PEO concentration above which significant particle interactions and network structure formation occur. The estimated separation energies as a function of flocculant concentration were similar to those of the shear yield stress, increasing from 1500 to 7000 kT for PEO and 1300 to 2000 kT for PAM as the polymeric concentration increased (Figure 4.25). Likewise for smectite dispersions, E_{sep} increased from 100 to 600 kT for PEO and 100 to 120 kT for PAM as the polymer concentration increased (Figure 4.26).

Similarly, a strong polymer structure type dependency was shown in the settling rates of the flocculated dispersions, particularly at flocculant concentrations above 500 g per ton of solid as exemplified in Figure 4.27 and Figure 4.28 for both clay dispersions. For the PAM-based dispersions, the initial settling rate increased gradually with increasing concentration up to 1000 g per ton of solid and finally reached a plateau, while for PEO concentrations greater than 600 g per ton solids, a strong increase in settling rates with flocculant concentration was observed.

1. Synergistic Effect of Coagulants (Metal Ions)
    and Flocculants

The use hydrolyzable metal ions, Mn(II) and Ca(II), in conjunction with nonionic PEO and or anionic PAM flocculants to modify the particle interactions (shear yield stress) and

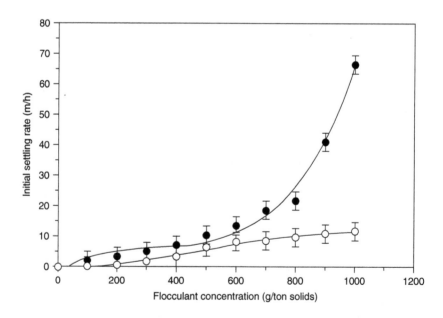

Figure 4.27   Initial settling rates of 8 wt.% solid kaolinite dispersions at pH 7.5 and $10^{-3}M$ KNO_3 as a function of PEO (•) and anionic PAM (○) concentration.

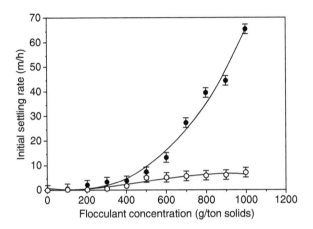

Figure 4.28 Initial settling rates of 8 wt.% solid smectite dispersions as a function of PEO (•) and PAM (○) concentration at pH 7.5, in $10^{-3} M$ KNO$_3$.

enhance the dewatering behavior of negatively charged kaolinite and Na-exchanged, swelling smectite dispersions are shown in Figure 4.29–Figure 4.32 [149].

Typical shear yield stresses of 8 wt.% solid smectite dispersions measured as a function of Mn(II) and Ca(II) ion concentration at pH 7.5 without flocculant are displayed in Figure 4.29, while those for flocculated 32 wt.% solid dispersions with and without Mn(II) and Ca(II) ions concentration are exhibited in Figure 4.30. In the absence of Mn(II) and Ca(II) ions, the yield stress was ~58 Pa for the 8 wt.% dispersion. A decrease in the yield stress is observed with an increase in metal ion concentration, an observation, which is contrary to the DLVO theory. An increase in yield stress is normally expected at higher ionic strength due to the compression of the electrical double layer and concomitant particle zeta potential reduction, leading to a primary minimum aggregation. The adsorption of metal ions appears to have screened the attractive edge (+) and basal (−) face interactions and as a result, a net repulsive basal face associations-dominated at higher ionic strength. This appears to be facilitated by the fact that the interlayer sodium ions were

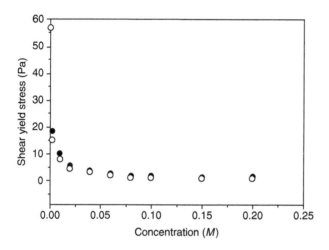

Figure 4.29 Shear yield stress of 8 wt.% solids smectite dispersions at pH of 7.5 as a function of Mn(II) (●) and Ca(II) (○) ion concentration.

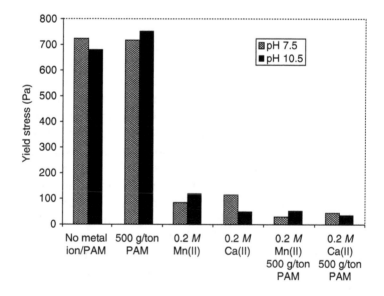

Figure 4.30 Shear yield stress of 32 wt.% solids smectite dispersions as a function of flocculant and metal ion concentrations at pH 7.5 and 10.5.

replaced by the divalent Mn(II)/Ca(II) ions, suppressing the swelling of smectite at higher ionic strength. Consequently, a network structure breakdown resulted and reflected a lowering of the yield stress.

Figure 4.30 indicates a significant increase of particle interactions occurring in the Na-exchanged smectite dispersion at high-solids loading (32 wt.%). The addition of both metal ions and flocculant led to relatively lower yield stresses, an outcome, which is conducive to enhanced dispersion dewaterability (Figure 4.31 and Figure 4.32). The interparticle bridging by polymer that did not reflect an increased shear yield stress may be due to the specific modification of basal and edge face interaction forces that led to the predominance of basal face–face type of platelet aggregation or flocculation.

The settling behavior of 8 wt.% solid kaolinite and Na-exchanged smectite clay dispersions at pH 7.5 and 10.5 and 500 g PEO per ton solids are shown in Figure 4.31 and Figure

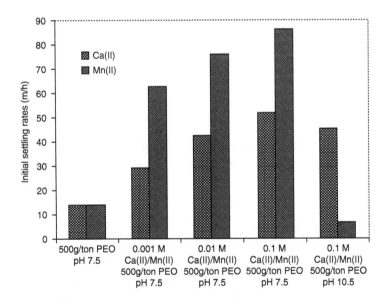

Figure 4.31 The initial settling rate of 8 wt.% kaolinite dispersion at pH 7.5 and 10.5 in the presence of Mn(II) and Ca(II) ions and PEO.

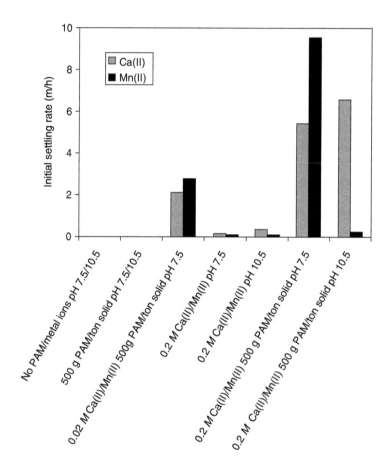

Figure 4.32 The initial settling rate of 8 wt.% smectite dispersions at pH 7.5 and 10.5 in the presence of Mn(II) and Ca(II) ions and 500 g PAM per ton of solid.

4.32 [149]. Both Mn(II) and Ca(II) ions have a significant effect on the settling rates. The addition of $10^{-3}\,M$ Mn(II) or Ca(II) ions prior to the addition of 500 g PEO per ton of solids dramatically increased the settling rates to $\sim$60 m/h for Mn(II) and 28 m/h for Ca(II) ions. Strong pH dependency and metal ion hydrolysis effects are clearly indicated, as higher settling rates were obtained with Ca(II) than Mn(II) ions at pH 10.5 and Mn(II) than Ca(II) ions at pH 7.5, consistent with of the interfacial chemistry modification as predicted by the James and Healy pH-mediated metal hydrolysis model.

2. Influence of pH and Temperature

The influence exerted by dispersion pH and temperature on floc structure and behavior is demonstrated in Figure 4.33 (pH) and Figure 4.34–Figure 4.37 (temperature) for flocculated kaolinite dispersions [149]. In the presence of PAM, the settling rates increased as the dispersion pH increased up to an optimum pH of 6–7, and then gradually decreased as the pH was further increased. This observation may be attributed to the fact that at pH <5 (pK_a ~5) the PAM is nonionic and has a more coiled conformation which is not effective for bridging, as the conformation-directing sodium acrylate species are not ionized or dissociated. Upon increasing pH up to ~6.5, the settling rate increased to a maximum. At this pH, it appears

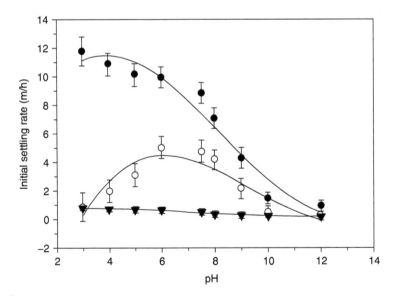

Figure 4.33 Initial settling rates of 8 wt.% solid kaolinite dispersions in $10^{-3} M$ KNO$_3$ and flocculated at 500 g PEO (●), PAM (○) per ton of solid and with no flocculant (▼) as a function of pH.

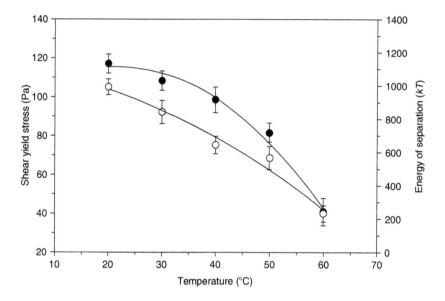

Figure 4.34 Shear yield stress and energy of separation of 28 wt.% kaolinite dispersions flocculated at 500 g per ton of solids PEO (●) and PAM (○) concentration as a function of temperature at pH 7.5 and $10^{-3} M$ KNO$_3$.

that anionicity increases and the anionic PAM acquires an extended conformation, which enables it to interact effectively with more kaolinite surface sites for improved flocculation efficiency. As pH is further increased beyond 7, poor flocculation and dewatering resulted, presumably due to the fact that the polymer stretches out further as a result.

3. Effect of pH on Dewatering Behavior

For nonionic PEO, which does not exhibit pH-dependent hydrolysis or charging behavior in the pH range 3–12, no significant conformation changes are expected. However, a decrease in the electropositive (Bronsted acid sites) at the kaolinite particle surface with increasing dispersion pH occurred and, significantly impacted upon the flocculation and dewatering performance of PEO. Hence, as the pH was increased, the settling rate decreased monotonically.

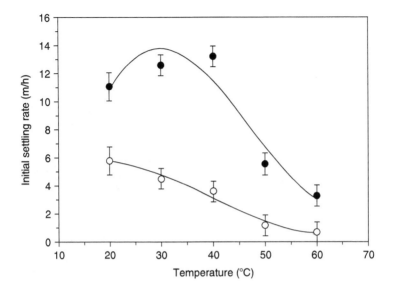

Figure 4.35 Initial settling rates of 8 wt.% solid kaolinite dispersions flocculated at 500 g PEO (●) and PAM (○) per ton of solid as a function of temperature at pH 7.5 and $10^{-3} M$ KNO$_3$.

4. Effect of Temperature on Dispersion Behavior

The profound effect that temperature may have on the flocculated particle network structure, shear yield stress, and settling rates is exemplified by colloidal kaolinite dispersions flocculated with PEO and PAM in the temperature range 20–60°C (Figure 4.34–Figure 4.37). A marked decrease in shear yield stress and energy of separation with increasing temperature, reflecting a polymer structure type effect on the particle interactions, is observed. E_{sep} decreased from 1200 to $260\,kT$ for PEO and from 1000 to $200\,kT$ for PAM as the temperature increased. The lower yield stress observed at higher temperatures ($>40°C$) is attributed to polymer conformation changes and reduced hydrogen bonding that occurred and led to weakened polymer-mediated interparticle bridging forces.

A slight increase in flocculation performance of PEO and hence settling rates is observed as the dispersion temperature

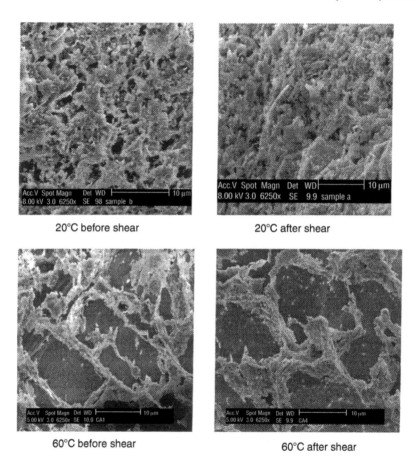

20°C before shear   20°C after shear

60°C before shear   60°C after shear

Figure 4.36 Scanning electron micrographs of kaolinite dispersions flocculated at 500 g per ton of solid PEO and pH 7.5 and $10^{-3}\,M$ KNO$_3$ before shear and after shear, at different temperatures.

was increased up to 40°C. However, beyond 40°C the settling rates decreased significantly. Upon flocculation with PAM, the initial settling rates decreased monotonically with temperature in the range from 20 to 60°C. Furthermore, PEO resulted in higher settling rates than did PAM and this is attributed to differences in the polymer structure-related characteristics. The settling behavior suggests that an optimum temperature exists for enhanced flocculation with PEO, shown to be

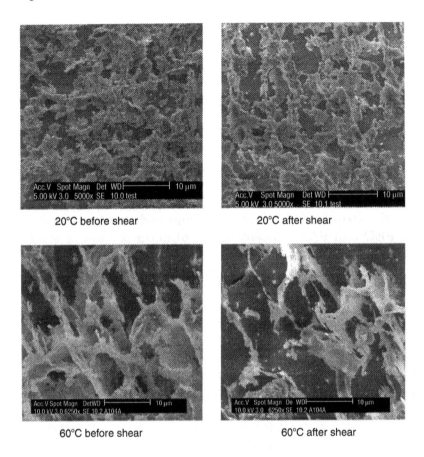

20°C before shear

20°C after shear

60°C before shear

60°C after shear

Figure 4.37 Scanning electron micrographs of kaolinite dispersions flocculated at 500 g per ton of solid PAM and pH 7.5 and $10^{-3} M$ KNO$_3$ before shear and after shear at different temperatures.

around 40°C. At dispersion pH of 7.5 and temperatures from 20 to 60°C, the zeta potentials of the kaolinite particles were in the range, -33 to -43 mV, implying that the ionic PAM and nonionic PEO polymer chains adsorbed onto the particles via hydrogen bonding. Notably, hydrogen bonding can be quite strong and involves interaction energies in the range 10–40 kJ/mol in contrast with $\sim$1 kJ/mol associated with van der Waals bonds [1].

The reduced polymer bridging performance, reflecting decreased settling rates with increasing temperature for both flocculants may be explained in terms of expanded to contracted conformation changes in tandem with weakened or reduced hydrogen bonding. Both the viscosity and hydrodynamic diameter of a 0.1 wt.% PEO and PAM polymer solutions, indicative of polymer–solvent H-bonding interaction and polymer conformation, decreased slightly upon increasing temperature in the range from 20 to 40°C, but sharply upon further increase to 60°C.

Scanning electron microscopy (SEM) micrographs revealing the structure of flocculated kaolinite dispersions produced at PEO and PAM concentrations of 500 g per ton of solid at 20°C and 60°C, with and without moderate shear at 250 rpm by a two-blade, 40 mm impeller, are shown in Figure 4.36 and Figure 4.37, respectively [149]. A more compacted structure of PEO flocculated slurries produced at 20°C rather than at 60°C is clearly evident, and is consistent with the settling behavior at these temperatures. The flocs at 20°C were denser aggregates and less prone to breakage upon shear, while at 60°C, the floc structure was more fragile and open with large voids of trapped supernatant. For the PAM-flocculated slurries, the floc structures were more open and less compact in comparison with PEO at all temperatures. As the temperature was increased, the floc structures became more open, with larger voids, which were easily destroyed upon shear.

REFERENCES

1. J.N. Israelachvili, *Intermolecular and Surface Forces*, 3rd ed., Academic Press, London, UK, 1994.
2. B.V. Deraguin and L.D. Landau, *Acta. Phys. Chim. URSS*, 14, 633, 1941.
3. E.J.W. Verwey and J.Th.G. Overbeek, *Theory of the Stability of Lyophobic Colloids*, Elsevier, Amsterdam, 1948.
4. P.C. Hiemenz, *Principles of Colloid and Surface Chemistry*, 2nd ed., Marcel Dekker, New York, 1986.
5. F. London, *Z. Phys.*, 63, 245, 1930.
6. Hamaker, Physica, 4, 1058, 1937.

7. B.V. Derjaguin, *Kolloid Z.*, 69, 155, 1934.
8. E.M. Lifshitz, *Sov. Phys. JETP*, 2, 73, 1956.
9. D.J. Shaw, *Introduction to Colloid and Surface Chemistry*, 3rd ed., Butterworth & Co.,London, 1980.
10. R.J. Hunter, *Introduction to Modern Colloid Science*, Oxford University Press, New York, 1993.
11. J.S. Noh and J.A. Schwarz. *J. Colloid Interf. Sci.*, 130, 30, 1989.
12. O. Stern, *Z. Electrochem.*, 30, 508, 1924.
13. H. von Helmholtz, *Ann. Phys. Wiedemann*, 7, 337, 1879.
14. G. Gouy, *J. Phys. Chem.*, 9, 457, 1910.
15. D.L. Chapman, *Philos. Mag.*, 25, 475, 1913.
16. D.C. Grahame, *Chem. Rev.*, 41, 441, 1947.
17. R. Hogg, T.W. Healy, and D.W. Fuerstenau, *Trans. Faraday Soc.*, 62, 1638, 1966.
18. G.M. Bell, S. Levine, and L.N. McCartney, *J. Colloid Interf. Sci.*, 33, 335, 1970.
19. R.G. Horn, *J. Am. Ceram. Soc.*, 73, 1117, 1990.
20. W.A. Ducker, Z. Xu, D.R. Clarke, and J.N. Israelachvili, *J. Am. Ceram. Soc.*, 2, 437, 1994.
21. M.L. Fielden, R.A. Hayes, and J. Raltson, *Langmuir*, 12, 3721, 1996.
22. J. Addai-Mensah, J. Dawe, R.A. Hayes, C.A. Prestidge, and J. Ralston, *J. Colloid Interf. Sci.*, 203, 115, 1998.
23. J. Addai-Mensah and J. Ralston, *J. Colloid Interf. Sci.*, 215, 124, 1999.
24. J. Addai-Mensah, C.A. Prestidge, and J. Ralston, *Miner. Eng.*, 12, 655, 1999.
25. Z. Xu and R.-H. Yoon, *J. Colloid Interf. Sci.*, 134, 427, 1990.
26. J.N. Israelachvili and R.M. Pashley, *Nature*, 306, 249, 1982.
27. P.M. Claesson, P. Herder, P. Stenius, J.C. Eriksson, and R.M. Pashley, *J. Colloid Interf. Sci.*, 109, 31, 1986.
28. J.L. Parker, D.L. Cho, and P.M. Claesson, *J. Phys. Chem.*, 93, 6121, 1989.
29. C. Tanford, *The Hydrophobic Effect*, Wiley, New York, 1980.
30. N. Israelachvili, *Acc. Chem. Res.*, 20, 415, 1987.
31. R.M. Pashley, *J. Colloid Interf. Sci.*, 83, 531, 1981.
32. P.M. Claesson, R.G. Horn, and R.M. Pashley, *J. Colloid Interf. Sci.*, 100, 250, 1984.
33. W.A. Ducker, T.J. Senden, and R.M. Pashley, *Nature*, 353, 239, 1991.
34. P.M. Claesson, R. Kjellander, P. Stenius, and H.K. Christenson, *J. Chem. Soc. Faraday Trans.*, 1(82), 2735, 1986.

35. R.M. Pashley and J.N. Israelachvili, *J. Colloid Interf. Sci.*, 97, 446, 1984.
36. A. Grabbe and R.G. Horn, *J. Colloid Interf. Sci.*, 157(2), 375, 1995.
37. R. Kjellandler, S. Marcelja, R.M. Pashley, and J.P. Quirk, *J. Chem. Phys.*, 92, 4399, 1990.
38. P.M. McGuiggan and R.M. Pashley, *J. Colloid Interf. Sci.*, 124(2), 560, 1988.
39. R.G. Horn and J.N. Israelachvili, *Chem. Phys. Lett.*, 71(2), 192, 1980.
40. D.M. LeNeveau, R.P. Rand, and V.A. Parsegian, *Nature*, 259, 601, 1976.
41. C. Misra, The Precipitation of Bayer Aluminium Hydroxide, Ph.D. dissertation, University of Queensland, Brisbane, Australia, 1970.
42. G.C. Low, Ph.D. dissertation, University of Queensland, Brisbane, Australia, 1975.
43. D. Ilievski, Ph.D. dissertation, University of Queensland, Brisbane, Australia, 1991.
44. C.A. Prestidge, I. Ametov, and J. Addai-Mensah, *Colloids Surf. A*, 157, 137, 1999.
45. R.M. Pashley, *J. Colloid Interf. Sci.*, 102, 23, 1984.
46. G.V. Franks, S.B. Johnson, P.J. Scales, D.V. Boger, and T.W. Healy, *Langmuir*, 15, 4411, 1999.
47. O.I. Vinogradova, *Langmuir*, 11, 2213, 1995.
48. G.C. Bushell, Y.D. Tan, D. Woodfield, J. Raper, and R. Amal, *Adv. Colloid Interf. Sci.*, 95, 1, 2002.
49. M. von Smoluchowski, *Phys. Z.*, 17, 557, 1916.
50. M. von Smoluchowski, *Z. Phys. Chem.*, 92, 129, 1917.
51. J. Gregory, *Crit. Rev. Environ. Control*, 19. 185, 1989.
52. J. Wagner, W. Hartl, and R. Hempelmann, *Langmuir*, 16, 4080, 2000.
53. H. Holthoff, S.U. Egelhaaf, M. Borkovec, P. Schurtenberger, and H. Sticher, *Langmuir*, 12, 5541, 1996.
54. J.A. Maroto and F.J. de las Nieves, *Colloids Surf. A*, 132, 153, 1998.
55. D.R. Snoswell, J. Duan, D. Fornasiero, and J. Ralston, *J. Phys. Chem. B*, 107, 2986, 2003.
56. D.W. Schaefer, J.E. Martin, P. Wiltzius, and D.S. Cannell, *Phys. Rev. Lett.*, 52, 2371, 1984.
57. C.D. Muzny, G.C. Straty, and H.J.M. Hanley, *Phys. Rev. E, Sci. instr.*, 40, R675, 1994.

58. J.A. Helsen, R. Govaerts, G. Schoukens, J. De Graeuwe, and J. Mewis, *J. Phys. E, Sci. Instr.*, 11, 139, 1978.
59. R. Amal, J.R. Coury, J.A. Raper, W.P. Walsh, and T.D. Waite, *Colloids Surf. A*, 46, 1, 1990.
60. J.L. Burns, Y.-D. Yan, G.J. Jameson, and S.R. Biggs, *Langmuir*, 13, 6413, 1997.
61. L.G.B. Bremmer, B.H. Bijsterbosch, P. Walstra, and T. Van Vliet, *Adv. Colloid Interf. Sci.*, 46, 117, 1993.
62. Z. Sun, C.D. Tomlin, and E.M. Sevick-Muraca, *Langmuir*, 17, 6142, 2001.
63. P.N. Segre, W. van Megan, P.N. Pusey, K. Schatzel, and W. Peters, *J. Modern Opt.*, 42, 1929, 1995.
64. J. Gregory, *J. Colloid Interf. Sci.*, 105, 357, 1985.
65. J. Gregory, *Colloids Surf. A*, 31, 231, 1988.
66. C. Allain, M. Cloitre, and F. Parisse, *J. Colloid Interf. Sci.*, 178, 411, 1996.
67. J. Farrow and L. Warren, in *Coagulation and Flocculation — Theory and Applications*, B. Dobias, Ed., Marcel Dekker, New York, 1993, p. 391.
68. H.A. Barnes, J.F. Hutton, and K. Walters, *An Introduction to Rheology*, Elsevier, Amsterdam, 1989.
69. J. Mewis, *J. Non-Netwonian Fluid Mech.*, 6, 1, 1979.
70. C. Chang and P.A. Smith, *Rheol. Acta*, 35, 382, 1996.
71. Kanai, *Rheol. Acta.*, 34, 303–310, 1995.
72. N.J. Wagner and J.W. Bender, The role of nanoscale forces in colloid dispersion rheology, *Mater. Res. Soc. Bull.*, February, 100, 2004.
73. Th.F. Tadros, *Adv. Colloid Interf. Sci.*, 68, 97, 1996.
74. D.C.-H. Cheng, *Rheol. Acta*, 25, 542, 1986.
75. Q.D. Nguyen and D.V. Boger, *J. Rheol.*, 29(3), 335, 1985.
76. Q.D. Nguyen and D.V. Boger, *J. Rheol.*, 27, 321, 1983.
77. Th.F. Tadros, *Langmuir*, 6, 28, 1990.
78. Th.F. Tadros, W. Liang, B. Costello, and P.F. Luckham, *Colloids Surf. A*, 79, 105, 1993.
79. J.A. Yosick and A.J. Giacomin, *J. Non-Newtonian Fluid Mech.*, 66, 193, 1996.
80. M. Wilhelm, *Macromol. Mater Eng.*, 287, 38, 2002.
81. I.M. Krieger and T.J. Dougherty, *Trans. Soc. Rheol.*, 3, 137, 1959.
82. R.G. Larson, *The Structure and Rheology of Complex Fluids*, Oxford University Press, New York, 1999.
83. B.A. Firth and R.J. Hunter, *J. Colloid Interf. Sci.*, 57, 248, 1976.

84. B.A. Firth, *J. Colloid Interf. Sci.*, 57, 257, 1976.
85. P.J. Scales, S.B. Harbour, and T.W. Healy, *AIChE J.*, 44, 538, 1998.
86. Y.K. Leong, D.V. Boger, P.J. Scales, T.W. Healy, and R. Buscall, *J. Chem. Soc. Chem. Commun.*, 639, 1993.
87. Y.K. Leong, P.J. Scales, T.W. Healy, and D.V. Boger, *Chem. Soc. Faraday Trans.*, 89, 2473, 1993.
88. C.A. Prestidge, *Colloids Surf. A*, 126, 75, 1997.
89. T. Muster, G. Toikka, R.A. Hayes, C.A. Prestidge, and J. Ralston, *Colloids Surf. A*, 106, 203, 1996.
90. J. Addai-Mensah, A. Gerson, C.A. Prestidge, I. Ametov, and R. Ralston, *Light Metals*, 159, 1998.
91. C.A. Prestidge and I. Ametov, *J. Cryst. Growth*, 209, 924, 2000.
92. Q.D. Nguyen and D.V. Boger, *Rheol. Acta*, 24, 427, 1985.
93. G.M. Channel, K.T. Miller, and C.F. Zukoski, *AIChE J.*, 46, 72, 2000.
94. R.G. De Kretser, P.J. Scales, and D.V. Boger, *Compressive Rheology: An Overview*, Rheology Reviews 2003, D.M. Binding and K. Walters, Eds., British Society of Rheology, 2003, p. 125.
95. M. Bevan and D. Prieve, *Langmuir*, 15, 7925, 1999.
96. D. Prieve, *Adv. Colloid Interf. Sci.*, 82, 95, 1999.
97. E.M. Furst and A.P. Gast, *Phys. Rev. E, Sci. Instr.*, 61, 6732, 2000.
98. P.G. Hartley, I. Larson, and P.J. Scales, *Langmuir*, 13, 2207, 1997.
99. I. Larson, C.J. Drummond, D.Y.C. Chan, and F. Griesser, *Langmuir*, 13, 2109, 1997.
100. S.R. Biggs, *Langmuir*, 11, 156, 1995.
101. A. Milling and S.R. Biggs, *J. Colloid Interf. Sci.*, 170, 604, 1995.
102. T.H. Muster and C.A. Prestidge, *J. Pharm. Sci.*, 91, 1432, 2002.
103. G. Toikka, R.A. Hayes, and J. Ralston, *J. Adhesion Sci. Technol.*, 11, 1479, 1997.
104. A. Feiler, I. Larson, P. Jenkins, and P. Attard, *Langmuir*, 16, 10269, 2000.
105. G. Bogdanovic, F. Tiberg, and M.W. Rutland, *Langmuir*, 17, 5911, 2001.
106. V.E. Shubin and P. Kekicheff, *J. Colloid Interf. Sci.*, 155, 108, 1993.
107. R.O. James and T.W. Healy, *The Adsorption of Metal Ions on Colloidal Minerals*, University of Melbourne, Australia, 1968.

108. R.O. James and T.W. Healy, *J. Colloid Interf. Sci.*, 40, 53, 1972.
109. R.O. James and T.W. Healy, *J. Colloid Interf. Sci.*, 40, 52. 1972.
110. R.O. James and T.W. Healy, *J. Colloid Interf. Sci.*, 40, 65, 1972.
111. G.R. Wiese and T.W. Healy, *J. Colloid Interf. Sci.*, 51, 434, 1975.
112. R.O. James, P. Stiglich, and T.W. Healy, *J. Colloid Interf. Sci.*, 59, 142, 1975.
113. I. Ece, *J. Inclusion Phenomena Macrocyclic Chem.*, 33, 155, 1999.
114. Z. Sadowski and R.W. Smith, *Miner. Metall. Process.*, 114, 1987.
115. G. Atesok, P. Somasundaran, and L.J. Morgan, *Colloids Surf.*, 32, 127, 1988.
116. R. de Kretser, Ph.D. thesis, University of Melbourne, Australia, 1995.
117. I. Larson and R.J. Pugh, *J. Colloid Interf. Sci.*, 208, 399, 1998.
118. J. Duan and J. Gregory, *Adv. Colloid Interf. Sci.*, 100, 475, 2003.
119. R. Hogg, *Int. J. Miner. Process.*, 58, 223, 2000.
120. V.R. Manukonda and I. Iwasaki, *Miner. Metall. Process.*, 217, 1987.
121. P. Mpofu, J. Addai-Mensah, and J. Ralston, *J. Colloid Interf. Sci.*, 261, 349, 2003.
122. K. Bremmell and J. Addai-Mensah, *J. Colloid Interf. Sci.*, in press.
123. P. Mpofu, J. Addai-Mensah, and J. Ralston, *J. Colloid Interf. Sci.*, 271, 145, 2004.
124. P. Mpofu, J. Addai-Mensah, and J. Ralston, *Int. J. Miner. Process.*, 71, 247, 2003.
125. P. Mpofu, J. Addai-Mensah, and J. Ralston, Proceedings of the XXII Mineral Processing Congress, Cape Town, South Africa, 29 September–3 October, 2003.
126. P. Mpofu, J. Addai-Mensah, and J. Ralston, *Miner. Eng.*, 17, 41–423, 2004.
127. P. Mpofu, J. Addai-Mensah, and J. Ralston, *Int. J. Miner. Process.*, in press.
128. T. Hiemstra, J.C.M. De Witt, and W.H. Van Riemsdijk, *J. Colloid Interf. Sci.*, 133, 105, 1989.
129. G.A. Parks, *Chem. Rev.*, 65, 177, 1965.
130. J. Addai-Mensah and J. Ralston, *J. Hydrometall.*, 74, 221–231, 2004.
131. H. Hirosue, N. Yamada, E. Abe, and H. Tateyama, *Powder Technol.*, 54, 27, 1988.

132. E. Matijevic, R.J. Kuo, and H.P. Kolny, *J. Colloid Interf. Sci.*, 80(1), 509, 1981.
133. D.J. Glenister and T.M. Abbot, Dewatering Practice and Technology Symposium, Brisbane, Australia, 105, 1989.
134. V. Hackley and M. Andersonm, *Langmuir*, 5, 191, 1989.
135. Mackenzie, in D. Malhotra and W.F. Riggs, Eds., *Chemical Reagents in the Mineral Processing Industry*, Society of Minerals Engineers, Colorado, USA, 1989.
136. T.W. Healy, *J. Macromol. Sci. Chem. A*, 8, 603, 1974.
137. R.J. Hunter, *Zeta Potential in Colloid Science: Principles and Applications*, Academic Press, London, 1981.
138. J.B. Farrow, R.R.M. Johnston, K. Simic, and J.D. Swift, *Chem. Eng. J.*, 80, 141, 2000.
139. G.J. Fleer, M. Cohen Stuart, J.M.H.M. Scheutjens, T. Cosgrove, and B. Vincent, *Polymers at Interfaces*, Chapman & Hall, London, 1993.
140. R. Hogg, *Colloids Surf. A*, 146, 253–263, 1999.
141. E. Koksal, R. Ramachandran, P. Somasundaran, and P. Maltesh, *Powder Technol.*, 62, 253, 1990.
142. R.A. Pradip, A. Kulkarni, S. Gundiah, and B.M. Moudgil, *J. Colloid Interf. Sci.*, 32, 259, 1991.
143. B.J. Scheiner and G.M. Wilemon, *Flocculation Biotechnol. Separation Syst.*, 4, 175, 1987.
144. P. Somasundaran and B.M. Moudgil, *Reagents in Mineral Technology*, Marcel Dekker, New York, 1988.
145. D.A. Stanely and B.J. Scheiner, US Bureau of Mines Report number 19021, 1986.
146. D.A. Stanely, B.J. Scheiner, and S.W. Webb, *Miner. Metall. Process.*, 577, 1984.
147. A. Sworska, J.S. Laskowski, and G. Cymerman, *Int. J. Miner. Process.*, 60, 153, 2000.
148. M.L. Taylor, G.E. Morris, P.G. Self, R. St. C. Smart, *J. Colloid Interf. Sci.*, 250, 28, 2002.
149. P. Mpofu, Ph.D. dissertation, University of South Australia, Adelaide, Australia, 2004.
150. R.S. Farinato, S.-Y. Huang, and P. Hawkins, *Polyelectrolyte-Assisted Dewatering in Colloid–Polymer Interactions: Fundamentals to Practice*, R.S. Farinato and P.L. Dubin, Eds. John Wiley and Sons, New York, 1999, p. 3.

5

Ion Adsorption on Homogeneous and Heterogeneous Surfaces

Luuk K. Koopal

Laboratory of Physical Chemistry and Colloid Science, Wageningen University,
Wageningen, The Netherlands

I. INTRODUCTION

For the behavior of colloidal particles in solution the surface charge is an important factor for many applications. The sign and magnitude of the surface charge are directly related to colloid stability (see e.g., [1]) and they profoundly influence the adsorption characteristics of particles. Examples of such behavior are numerous. Ions adsorbed at the surface will change the particle charge, the surface, and Stern layer potential and hence the electrostatic particle–particle interaction. In the case of adsorption of organic ions structural effects may also affect the particle–particle interaction [2].

   In this chapter, the nature and origin of surface charges will be discussed, followed by a treatment of ion adsorption on charged surfaces and the formation of the electrical double layer. According to their behavior, we shall classify the various ions into three groups: charge-determining ions, specifically adsorbing ions, and indifferent ions. In the case of specifically adsorbing ions a distinction will be made between inorganic and organic ions. With the inorganic ions a further distinction will be made between monovalent ions, which in general adsorb only weakly and multivalent ions that may adsorb strongly. For the organic ions the amphiphilic character may lead to an additional lateral interaction between the adsorbed molecules, even when such interaction is hardly noticed in bulk solution. A special section is devoted to surface heterogeneity. The overall emphasis of the chapter will be on the physical and mechanistic description of ion adsorption. For a comparison of modeling and experimental results emphasis is given to adsorption behavior on mineral surfaces and metal oxides in particular.

A.   Surface Charge Determining Ions

In general, the surface charge itself is the result of ion adsorption and different mechanisms of surface charge formation can be recognized. In the case of sparingly soluble ionic solids dispersed in an aqueous electrolyte solution equilibrium exists between the ions making up the surface of the crystals and these same ions in solution. The solubility product of the material determines the concentrations of these constituent ions. The (Galvani) potential of the solid will be fully determined by thermodynamic adsorption equilibrium in accordance with a Nernst equation. These ions are therefore referred to as potential determining. The potential-determining ions for a given solid are usually apparent from its chemical composition. The best-studied example of this behavior is colloidal AgI [3].

For metal oxides H^+ and OH^- are considered as potential-determining ions of the second kind. The dissociation product of water determines the relation between the concentrations of these ions in solution. With metal oxide particles the H^+ and OH^- ions are not constituent ions, but they react with the surface metals to form surface groups and, in principle, in thermodynamic equilibrium the Nernst equation does not apply. In order to avoid confusion with respect to Nernstian behavior, it is more appropriate to call the potential-determining ions of the first and second kind *charge-determining ions*.

A similar mechanism of surface charge formation occurs with solids that contain ionizable groups at their surface. This situation occurs for polymer lattices and many natural particles. Also in this case H^+ and OH^- are generally considered as the charge-determining ions.

The presence of charge-determining ions requires that the primary surface charge density, σ_s (C/m^2), is defined as

$$\sigma_s = F(\Gamma_{P^+} - \Gamma_{N^-}), \tag{1}$$

where F is the Faraday constant, P^+ and N^- are the charge-determining ions, for instance, Ag^+ and I^- in the case of AgI, or H^+, and OH^- in the case of metal hydroxides or oxides, lattices, and natural particles. The quantities Γ_{P+} and Γ_{N-} represent the adsorption of P^+ and N^- in mol/m^2, respectively.

The adsorption of surface charge-determining ions is followed by a positive adsorption of counterions and a negative adsorption or exclusion of coions in the immediate vicinity of the surface, in order to satisfy the requirements of electroneutrality. The charged interface plus the solution region in which the surface charge is counterbalanced is called the *electric double layer*. The charges in solution are mobile, contrary to those on the surface, which are generally assumed to be localized. Due to the equilibrium between thermal and electrostatic forces a *diffuse* layer is formed.

The special mechanism through which particles may be charged is that of isomorphic substitution. The crystal lattice of the solid may contain a net positive or negative charge arising from interior defects or lattice substitution. This net charge is compensated by an equivalent ionic charge at the surface. In contact with water the compensation ions dissociate to form the counterions of the double layer. This type of double layer formation is important in describing the behavior of certain ion exchanging minerals, such as zeolites and clay minerals [4].

B. Specifically Adsorbing Ions

Ions, other than the charge-determining ions, may also have a specific (i.e., noncoulombic) affinity for the surface. Such ions may be strongly adsorbed by, for instance, the formation of a surface complex. In other cases the adsorption may be aided by hydrophobic interactions, hydrogen bond formation, or by London-dispersion forces. The latter interactions are particularly important for organic ions. We will refer to all these ions as *specifically adsorbing ions* in order to distinguish them from charge-determining ions. It is often assumed that the charge of these specifically adsorbing ions is located at some distance from the surface, whereas the charge of the charge-determining ions is assumed to coincide with the surface plane. For complex systems the difference between specifically adsorbing ions and charge-determining ions is not always clear and is ultimately determined by definition. Specific adsorption, which leads to complex formation with a surface group, is often called *surface complexation.*

C.   Indifferent Ions and Indifferent Electrolytes

The last group of ions is those that are not charge determining and not specifically adsorbing, they solely exhibit coulomb interactions with the surface. These ions are termed *indifferent ions.* For sake of convenience it is frequently assumed that most simple monovalent ions are indifferent. Consequently, 1:1 electrolytes can be considered as indifferent electrolytes, specifically when the electrolyte concentration is not too high. The presence of an indifferent 1:1 electrolyte in solution often defines the background electrolyte concentration for the adsorption of charge determining and specifically adsorbing ions.

II.   THE ELECTRICAL DOUBLE LAYER

A.   Gouy–Chapman Model

Before discussing the surface charge itself, the solution side of the double layer will be considered. The formation of an electrical double layer is easily imagined in the case of a nonpermeable rigid charged surface in contact with an aqueous electrolyte solution. The ions in the solution respond to the field force near the interface by taking up new positions until a distribution is reached that represents equilibrium between thermal and electrostatic forces. The analysis of the ion distribution, while it resembles the Debye–Hückel [5] theory of strong electrolytes, developed some 10 years later, was carried out by Gouy [6] and independently by Chapman [7]. To improve the Gouy–Chapman (GC) model, Stern [8] proposed to divide the double layer in two parts: (1) adjacent to the surface a layer of thickness d_1 of the order of molecular dimensions, called the Stern layer, and (2) a diffuse layer of point charges. This approach is called the Stern–Gouy–Chapman (SGC) model. The SGC model will be treated in Section II.B.

The analysis of the diffuse double layer according to the model of Gouy and Chapman involves several assumptions. The charge on the solid surface is uniformly distributed, the ions making up the diffuse layer are point charges, the solvent is a structureless medium whose only influence is through its dielectric constant and the surface is planar.

A basic requirement of the space charge in the double layer is that Poisson's equation be satisfied:

$$\frac{d^2\psi}{dx^2} = -\frac{\rho(z)}{\varepsilon_0\varepsilon_r}, \tag{2}$$

where ψ is the double layer potential, ε_0 the permittivity of the vacuum, ε_r the relative dielectric permittivity of the medium and $\rho(z)$ the net space charge per unit volume at distance z from the plane where the diffuse layer starts. A Boltzmann relation gives the distribution of ions in the potential field

$$n_i(z) = n_i(\infty)\exp(-\nu_i F\psi(z)/RT), \tag{3}$$

where $n_i(z)$ is the concentration and ν_i the valence (including sign) of ions of kind i at distance z from the surface where the potential is $\psi(z)$, $n_i(\infty)$ is the bulk concentration of i, and R and T have their usual meanings.

The space charge density ρ in the diffuse layer is the net sum of the positive and negative ion concentrations:

$$\rho(z) = \sum_i \nu_i F n_i(z). \tag{4}$$

Equations (2)–(4) give the differential equation

$$\frac{d^2\psi}{dz^2} = -\frac{F}{\varepsilon_0\varepsilon_r}\sum_i \nu_i n_i(\infty)\exp\left(\frac{\nu_i F\psi(z)}{RT}\right) \tag{5}$$

that describes the variation of ψ with location in the diffuse double layer. Equation (5) is known as the Poisson–Boltzmann (PB) equation for flat double layers. For a symmetrical indifferent electrolyte with valence ν and concentration n, Equation (5) integrates to give

$$\left(\frac{d\psi}{dz}\right) = -\left(\frac{8\,RTn}{\varepsilon_0\varepsilon_r}\right)^{1/2} \sinh\left(\frac{\nu F\psi(z)}{2\,RT}\right). \tag{6}$$

For asymmetrical electrolytes Equation (6) is approximately correct if ν is the valence of the ion of charge opposite to that of the surface and n its bulk concentration. A second integration gives the potential distribution

$$\tanh\left\{\frac{\nu F\psi(z)}{4\,RT}\right\} = \tanh\left(\frac{\nu F\psi_d}{4\,RT}\right)\exp\left(-\kappa z\right), \tag{7}$$

where ψ_d is the potential at the onset of the diffuse layer. This potential equals the surface potential, ψ_s, if the onset of the diffuse layer coincides with the surface layer. The function κ^{-1} is the Debye-length or the *double layer thickness*. The inverse Debye length is defined by the ionic strength

$$\kappa = \left(\frac{F^2 \sum_i n_i \nu_i^2}{\varepsilon_0\varepsilon_r RT}\right)^{1/2}. \tag{8}$$

For small potentials, in the so-called Debye–Hückel (DH) limit, Equation (7) reduces to

$$\psi(z) = \psi_d \exp\left(-\kappa z\right). \tag{9}$$

At distance $z = \kappa^{-1}$, $\psi(z = \kappa^{-1}) = \psi_d/e$, illustrating why κ^{-1} is also called double layer thickness. In the DH limit electrostatic interactions can be treated by the linearized form of the PB equation.

When the surface charge is compensated by ions in the diffuse layer only, the electroneutrality of the system requires that

$$\sigma_s = -\int_0^\infty \rho\,dz = -\sigma_d \tag{10}$$

i.e., the surface charge is equal but opposite in sign to the diffuse layer charge, σ_d. Using Equation (2) we can integrate the above expression

$$-\sigma_d = \varepsilon_0\varepsilon_r \int_0^\infty \left(\frac{d^2\psi}{dx^2}\right) dz = -\varepsilon_0\varepsilon_r \left(\frac{d\psi}{dz}\right)_{z=0}. \tag{11}$$

Insertion of Equation (6) in (11) gives

$$-\sigma_d = (8\,RT\varepsilon_0\varepsilon_r n)^{1/2} \sinh\left(\frac{\nu F\psi_d}{2\,RT}\right). \tag{12}$$

Equation (12) can also be used to find ψ_d if $-\sigma_d$ is known:

$$\psi_d = \frac{2\,RT}{\nu F} \operatorname{arc\,sin} h\left\{\frac{-\sigma_d}{(8\,RT\varepsilon_0\varepsilon_r n)^{1/2}}\right\}. \tag{13}$$

For small values of ψ_d Equation (12) reduces to the DH limit:

$$-\sigma_d = \varepsilon_0\varepsilon_r\kappa\psi_d. \tag{14}$$

Equation (14) also shows the physical meaning of κ^{-1}, namely as the plate distance of a condenser with a medium with relative permittivity ε_r in-between the plates.

For surfaces for which ψ_d is fixed Equation (7) describes how ψ decreases from its initial value ψ_d with increasing distance from the surface at any given concentration of indifferent electrolyte. However, for surfaces of constant charge and $\sigma_s = -\sigma_d$, ψ_d will depend on electrolyte concentration (Equation (13)); once ψ_d is determined, Equation (7) will apply.

The potential distribution according to Equation (7) is shown in Figure 5.1(a). It can be seen that ψ decreases progressively with increasing distance. The values of ψ at $\kappa z = 0$ are identified as ψ_d. For any combination of the electrolyte concentration c_s and z_s, κ can be found from Equation (8). Also shown are the results obtained with Equation (9), the low potential approximation. The charge-potential relationship as given by Equation (12) is illustrated in Figure 1(b) for a monovalent indifferent electrolyte. For a given value of ψ_d the surface charge ($\sigma_s = -\sigma_d$) is proportional to $n^{1/2}$, i.e., to the square root of the background electrolyte concentration.

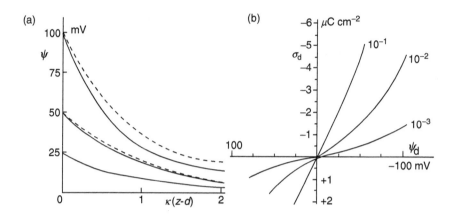

Figure 5.1 (a) Potential decay in a flat diffuse double layer at $T = 25°C$. Full curves, Equation (7); dashed curves, Equation (9). (b) Surface charge as a function of the surface potential, Equation (12). The concentration of monovalent electrolyte is indicated.

The GC theory also allows us to calculate the adsorption in the diffuse layer, i.e., the surface excess amounts of anions and cations that compensate the surface charge:

$$\Gamma_{\pm,\mathrm{d}} = \int_0^\infty \{n_i(z) - n_i(\infty)\}\,dz. \tag{15}$$

Solution of this integral [9] gives for the cations:

$$\Gamma_{+,\mathrm{d}} = (2\,RT\varepsilon_0\varepsilon_{\mathrm{r}}n/\nu^2F^2)^{0.5}[\exp(-\nu F\psi_\mathrm{d}/2\,RT) - 1] \tag{16}$$

and for the anions

$$\Gamma_{-,\mathrm{d}} = -(2\,RT\varepsilon_0\varepsilon_{\mathrm{r}}n/\nu^2F^2)^{0.5}[\exp(\nu F\psi_\mathrm{d}/2\,RT) - 1] \tag{17}$$

In spite of its shortcomings, the Gouy–Chapman analysis of the double layer has been useful in developing an understanding of colloid stability and ion adsorption. The error caused by the neglect of the finite size of the ions can largely be corrected by separating the double layer in a Stern layer and a diffuse layer.

The presented equations all apply to flat double layers and symmetrical electrolytes. For nonsymmetrical electrolytes the treatment is essentially the same but mathematically more complicated. Nir and Bentz [10,11] give solutions for such electrolytes and electrolyte mixtures. General numerical solutions to the Poisson–Boltzmann equation for arbitrary geometry have been presented by James and Williams [12] and Gilson and Honig [13]. Several authors review analytical solutions for spheres and cylinders [14–17].

B. Stern–Gouy–Chapman Model

In the SGC theory the double layer is divided into a Stern layer, adjacent to the surface with a thickness d_1 and a diffuse (GC) layer of point charges. The diffuse layer starts at the Stern plane at distance d_1 from the surface. In the simplest case the Stern layer is free of charges. Even in this case the presence of a Stern layer has considerable consequences for the potential distribution. Without charge in the Stern layer the potential drops linearly from the surface potential ψ_s to the potential at the Stern plane, ψ_d (Figure 5.2).

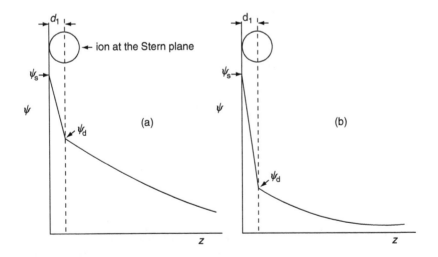

Figure 5.2   Potential decay for the SGC model (a) in the absence of specific adsorption; (b) with counterions specifically adsorbed in the Stern layer.

Often ψ_d is considerably lower than ψ_s. This is of particular importance for the colloidal stability because the potential ψ_d of the Stern layer is the leading potential in the expressions for the electrostatic interaction rather than the surface potential [14]. The potential drop $(\psi_s - \psi_d)$ is related to the Stern layer capacitance C_1:

$$\psi_s - \psi_d = \sigma_s/C_1, \tag{18}$$

where C_1 is defined as

$$C_1 = \varepsilon_0 \varepsilon_1/d_1 \tag{19}$$

with ε_1 the relative permittivity of the medium in the Stern layer. The value of ε_1 is affected by the structure of the solvent near the surface, especially by the degrees of freedom of the dipoles. As compared with the bulk solution, this freedom can be substantially reduced by the presence of the surface. For metal hydroxide or oxide and aqueous solution interfaces C_1 has a high value, and ε_1 turns out to be of the same order of magnitude as ε_r of the bulk solution. However, for interfaces such as the AgI-aqueous and Hg-aqueous solution interface, ε_1 is much smaller than ε_r in bulk solution and C_1 is low [3]. In principle ε_1 is also a function of the charge density, but in practice ε_1 and C_1 are often considered as constants. Once C_1 is known, ψ_s is obtained by combining Equations (18) and (13).

$$\psi_s = \frac{\sigma_s}{C_1} + \frac{2RT}{\nu F} \, \text{arc sin h} \left\{ \frac{-\sigma_d}{(8RT\varepsilon_0\varepsilon_r n)^{1/2}} \right\}. \tag{20}$$

van Riemsdijk and coworkers [18,19] and Yeh and Berkowitz [20] have presented explanations for the magnitude of C_1 near different types of surfaces. Hiemstra and van Riemsdijk [18] assume that for the primary hydration sheet of the surface ions dielectric saturation is reached, so that ε_r is about 6 [21], and that in the secondary hydration sheet ε_r is close to that of bulk water ($\varepsilon_r = 78$). The assumption of dielectric saturation of the first hydration sheet also ensures that the surface charge has little effect on ε_1. For hydrated AgI the distance of closest approach of a (hydrated) counterion to the surface is of the

order of two water diameters. Considering each of these layers
as a plate condenser with a plate distance of about 0.3 nm (one
water molecule) the capacitances are roughly $0.2 \, F/m^2$ for the
first and $1.7 \, F/m^2$ for the second layer. The total Stern layer
capacitance is dominated by the smallest value and about
$0.2 \, F/m^2$. For metal hydroxide or oxide surfaces the primary
hydration sheet of the metal ion is formed by the (protonated)
surface oxygens. The closest distance of approach of a hydrated counterion to the surface oxygens is only the water
diameter of the secondary hydration sheet. The capacitance
of the Stern layer is therefore much higher than for AgI or Hg
and about $1.7 \, F/m^2$. In a detailed study of TiO_2 van Riemsdijk
and coworkers [19] come to the conclusion that for metal
oxides two Stern layer capacitances occur in practice. For
well-crystallized oxides a relatively low Stern layer capacitance of $0.9 \pm 0.1 \, F/m^2$ applies and for less well-structured
oxides a relatively high value of $1.6 \pm 0.1 \, F/m^2$. Capacitances
considerably higher than $1.7 \, F/m^2$ can only be reached if the
surface groups protrude a small distance (of the order of a
water diameter) in the solution, as may be the case for non-
porous silica. Also for charged groups on polymer latex
particles the value of C_1 depends on the distance the surface
groups protrude in the solution. For *hairy* surfaces a high
capacitance value is to be expected.

The Stern layer concept can be used with or without
specific adsorption. In the absence of specific adsorption the
first consequence of the SGC model as compared to the GC
model is that the distinction between charge-determining ions
and other ions becomes more pronounced: ψ_s is the leading
potential for the charge-determining ions, ψ_{2d} that of the
indifferent ions. In the absence of specific adsorption $-\sigma_d$ in
Equation (20) can be replaced by σ_s.

In the presence of specific adsorption, with a charge
density equal to σ_1, one has to decide at which plane these
ions adsorb. Two simple choices are: (1) the specifically
adsorbed ions are located at the Stern plane and (2) these
ions are placed at the surface plane. For both options the
SGC model can be used. In the SGC model with specifically
adsorbing ions located at the Stern plane the expression for ψ_s

as given by Equation (20) is valid. This choice is appropriate for ions that form outer-sphere complexes with the surface sites or for ions that have no affinity for the proton sites. The diffuse layer charge can be calculated if both σ_s and σ_1 are known. Electroneutrality requires that

$$\sigma_1 + \sigma_s + \sigma_d = 0. \tag{21}$$

Alternatively σ_d can be obtained from Equation (12) when ψ_d can be derived from experimental data, for example, by equating ψ_d to the electrokinetic potential ζ [22]. For low values of the ionic strength this is a reasonable approximation.

When the specifically adsorbing ions are placed at the surface plane instead of at the Stern plane (strong affinity and perfect screening of the surface charge), Equation (20) has to be modified to

$$\psi_s = \frac{\sigma_s + \sigma_1}{C_1} + \frac{2RT}{\nu F} \text{ arc sin h}\left\{\frac{-\sigma_d}{(8RT\varepsilon_0\varepsilon_r n)^{1/2}}\right\}. \tag{22}$$

For counterions that are forming inner-sphere complexes with the surface groups the charge can also be placed partly at the surface plane and partly at the Stern plane. This *charge distribution model* will be considered in detail in Section IV.D. Also in this situation only one Stern layer capacitance is required for a basic description of the characteristics of the double layer.

The idea of specific adsorption of ions, i, at the Stern plane has been originally suggested by Stern [8]. Stern implicitly assumed that the sites for specific adsorption were independent of the sites for the charge-determining ions (no site competition) and proposed the following modification of the Langmuir equation to describe the adsorption

$$\frac{\theta_i}{1 - \theta_i} = \frac{c_i}{55.5} \exp\left(\frac{-\Phi_i}{RT} - \frac{\nu_i F \psi_d}{RT}\right), \tag{23}$$

where θ_i is the fraction of surface sites covered with specifically adsorbing ions i, $c_i/55.5$ is the mole fraction of i, Φ_i the *specific* Gibbs energy of adsorption, and $\nu_i F \psi_d$ the coulombic contribution to the adsorption energy. The use of ψ_d implies that the centers of charge of i are at the Stern plane. Later

Equation (23) is mainly applied for specific adsorption of organic ions. We will return to the Stern equation in Section V on the adsorption of organic ions. For adsorption of inorganic ions site competition with protons is nowadays the most common assumption and this leads to a more complicated expression than Equation (23). In Section IV this type of specific adsorption of inorganic ions will be considered.

In the next sections a refinement (Section III.C) and some critical comments on the SGC and related models are mentioned together with some alternative treatments (Section III.D). Apart from that discussion we will adhere to the SGC model. This model incorporates a fair amount of detail and accounts for both short-range (Stern layer) and long-range (PB diffuse layer) electrostatic interactions. Furthermore, experience has shown that the SGC model works very well in practice.

C. Triple Layer Models

For specifically adsorbing counterions that are forming inner-sphere complexes with the surface groups often a distinction is made between the surface plane, the plane where the centers of charge of the partially dehydrated specifically adsorbed ions reside, the inner Helmholtz plane, and the Stern plane at distance d_1, also called the outer Helmholtz plane. The double layer model composed of an inner and outer Helmholtz layer plus a diffuse layer is mostly called the *triple layer* (TL) model. The potential drop over the inner and the outer Helmholtz layers can be described by an inner and an outer Helmholtz layer capacitance, each having its own relative dielectric permittivity and thickness. Although for ideally flat surfaces such as the mercury–solution interface it is physically elegant to make a distinction between an inner and an outer Helmholtz plane [9,10], it is of less relevance for ordinary solid surfaces. In practice, the extent of solid surface irregularities makes it often impossible to obtain unique values for the capacitances of both Helmholtz layers. In Section III.D.3. a brief discussion is given of the use of the TL model for metal oxide surfaces.

D.   Alternative Treatments of the Double Layer

For a discussion of the limitations of the GC theory various reviews can be consulted [9,23–27]. Reeves [9] gives a general review. Torrie and Valleau [23] and Gunnarsson and coworkers [24,25] have discussed the use of the Poisson–Boltzmann equation by comparing the results with Monte Carlo simulations. In general it can be remarked that for monovalent counter ions the Poisson–Boltzmann equation works well, for higher valences the neglect of ion–ion correlation effects make the results of the Poisson–Boltzmann approach somewhat questionable. Reviews on statistical methods to treat the electrostatic interactions have been presented by, for example, Carnie and Torrie [26] and Attard [27].

Also the SGC-type models have been criticized (see e.g., Ref. [28]), but as stated above, experience has shown that models that employ a division of the double layer into a diffuse (long-range) and a nondiffuse (short-range) part work very satisfactorily and keep the theoretical treatments relatively simple. Illustrative examples of the improvements that result after incorporation of a basic Stern layer in double layer models can be found in the paper by Larson and Attard [29]. Also in his review Attard [27] gives an important place to the Stern layer.

The use of modern liquid theory is briefly summarized by Larson and Attard [29], who discuss the issue of Nernstian surface potentials. Instead of the mean-field PB theory they apply the hypernetted chain (HNC) integral equation to account for ion size and ion correlation in the diffuse layer. For the total double layer the HNC diffuse layer is combined with a Stern layer (SHNC model).

Wingrave [30], Abbas et al. [31], and M. Gunnarsson et al. [32] have revived the discussion on ionic activity coefficients. Wingrave's analysis is based on the linearized Poisson–Boltzmann equation and leads for the solution phase to the DH model for the ionic activity coefficient in which the size of the ions is taken into account. The analysis for the interface is an extension of model, it follows the method of Onsager and Samaras [33] for the description of the surface tension of electrolyte solutions (see also Ref. [34]), and the *primitive interfacial model* of Attard [27]. In this latter model the charges of the surface ions and the specifically adsorbed ions are all placed in the Stern layer. The resulting expressions of the activity coefficients for the charge-determining ions and the specifically adsorbing ions are very similar to the DH expression for the ion activity in solution with the valence of the ion replaced by the effective charge per site. Due to the fact that the method is based on the linearized Poisson–Boltzmann equation the results will only hold for surfaces with a low charge density. No numerical comparison is made with the classical double layer models.

Abbas et al. [31] give a corrected DH model (CDH model) for the solution phase for charged particles with a size ranging from 0 to ∞ based on the generalized van der Waals theory [35]. The latter theory has shown its merits in the description of adsorption and surface tension phenomena in simple fluids. For the repulsion between screening ions the CDH model takes excluded volume and so-called hole corrections of electrostatic interactions into account. For 1:1 salt solutions the CDH and DH theories start to deviate around $0.4\,M$; for 2:2 electrolytes the deviation becomes apparent around $0.1\,M$. The CDH results for simple salts agree well with MC calculations. The advantage of the CDH theory is that it works well and is still not too complicated to be further extended to flat surfaces and surface complexation.

The extension to surfaces (surface charging or SC model) has been made by Gunnarsson et al. [32], who combined the CDH theory with a site-binding model (see also Section III). In the SC model a charge-free Stern layer is introduced for further regulation of the screening. The model reflects in many ways the more classical treatments that combine a site-binding model with a solution of the mean-field PB equation; the main difference is the more advanced treatment of the diffuse double layer with ion pair correlations and excluded volume effects. However, the model is still restricted to low surface charge densities. A more sophisticated model for higher charges recently appeared [36]. Calculated results of the SC model are compared with experimental results for a goethite surface and with results obtained with the SGC model. The conclusion is that incorporation of a Stern layer in the SC model is crucial for a good result at low values of the surface charge. For low ionic strength and surface charges larger than $0.05\,C/m^2$ it is not possible to get good predictions with the SC model. The SGC model gives good results for all electrolyte concentrations.

In conclusion we may say that especially the studies of Gunnarsson and coworkers [31,32] and Larson and Attard [29] open new perspectives for a modern description of the diffuse double layer and ion adsorption. Yet, except for the SHNC model [29], the classical models are still superior for the description of experimental results at higher charge densities. For interesting but complex situations, like metal ion or oxyanion binding to mineral surfaces from a solution with a constant background electrolyte concentration, the SGC and related models will still hold their position for quite some time. Most of the classical treatments can be considered as an equilibrium to which the mass action law is applied. The activity of the ions at the Stern plane or the surface plane is calculated semiempirically by multiplying the ion concentration with the mean DH activity coefficient for the ions in solution and a Boltzmann factor containing the relevant (smeared-out) electrostatic potential(s). Hence, the major *ingredients* are involved, but sophistication is lacking. Fundamental errors due to neglecting ion pair correlations will not easily obstruct the situation because fitted intrinsic ion affinities and a fitted Stern layer capacity can easily 'absorb' such errors (see also Ref. [37]).

E. Nonrigid Particles

The SGC model is developed for solid particles. It can be applied to organic colloids as long as these are rigid and nonpermeable. The treatment of permeable and swelling particles is outside the scope of this chapter and only some literature references will be given. A simple way to describe permeable particles with fixed charges is the Donnan model [38–41]. In this model the particle charge is restrained in (about) the particle volume by a hypothetical membrane permeable to small ions only. The membrane separates two phases: the bulk solution and the Donnan phase. The Donnan phase volume or Donnan volume is the solution volume within the membrane. The Donnan volume contains the primary charge that is fixed at the solid matrix and the counter charge. With a sufficiently high fixed charge density the local double layers overlap strongly and result in approximately one smeared-out electrostatic potential ψ_D inside the entire Donnan phase. Similarly as in the GC theory ψ_D is defined with respect to the bulk solution. At the surface of the membrane the electrostatic potential drops to zero. Some analytical expressions for the relation between the Donnan potential and the fixed charge density for both the simple Donnan model and a Stern–Donnan model with two potentials inside the Donnan volume can be found in Ref. [38].

More sophisticated models for permeable and semipermeable charged particles that serve specific purposes can be found in the literature. Tanford [42] gives, in Chapters 7 and 8 of his book on macromolecules, a basic description of semipermeable spherical particles and linear polyelectrolytes. The derived analytical expressions apply to low potentials (DH limit). Another treatment of semipermeable spheres is that by Stigter and Dill [43] who also include swelling. Modern treatments of flexible polyelectrolytes in solution including coil expansion have been presented by Odijk and Houwaart [44], Fixman and Skolnick [45], and by Stigter [46]. Stigter [47] has given a brief review of the various elements of the theory by applying it to the expansion of DNA in salt solutions.

III. SURFACE IONIZATION

A. Adsorption of Primary Charge-Determining Ions: Outline

In the previous section the presence of a surface charge has been presumed and the solution side of the double layer was analyzed. In this section the adsorption of charge-determining ions will be discussed. The adsorption of charge-determining ions has received considerable attention in literature. An integrated approach has been presented by Healy and White [48], who formulated the statistical thermodynamic backgrounds for adsorption of charge-determining ions. Essentially their treatment extends the Langmuir model for adsorption by considering not only *intrinsic* or *chemical* interactions with the surface, but also electrostatic interactions. For the latter interactions the mean-field PB equation is used. For a discussion of more sophisticated models for the electrostatic interaction the reader is referred to Section II.D.

The adsorption of charge-determining ions can be seen as the inverse of ionization of discrete surface groups. Briefly one therefore speaks of site binding or surface ionization. Healy and White [48] mainly considered the surface ionization in the presence of an indifferent electrolyte. Other authors, see for example, the review by James and Parks [49], considered surface ionization in combination with *surface complexation*: a complex is formed between a charged surface site and a specifically adsorbed inorganic ion. Surface complexation in combination with surface ionization leads to *competition* for the surface sites and to an effective screening of the surface charge. It enhances the surface ionization.

It is also possible to consider specific adsorption without site competition. This type of specific adsorption may occur for organic ions and large inorganic ions. In this case the adsorption of the charge-determining ions is only affected by the electrostatic screening mechanism and not by site competition.

In the following we will discuss first adsorption isotherm equations for charge-determining ions in the absence of specific adsorption. In Section IV isotherm equations for surface charge formation in the presence of specific adsorption and surface complexation will be presented, and in Section V adsorption of organic ions on charged surfaces. In our treatment protons and hydroxyl ions are considered to be the set of charge-determining ions, for other charge-determining ions the same line of reasoning can be followed provided the point of zero charge and the point of zero potential coincide. Note that this is, for instance, not the case for the AgI–electrolyte interface [3].

B. Monofunctional Surfaces

The ionization of a homogeneous surface with one type of surface group, S_jH, in equilibrium with an aqueous solution containing an indifferent electrolyte and an acid or base, can be described as

$$S_j^- + H^+ \rightleftharpoons S_jH; \quad K_{j,H}^* \tag{24}$$

with the equilibrium condition

$$K_{j,H}^* = \frac{\{S_jH\}}{\{S_j^-\}\{H^+\}}, \tag{25}$$

where $\{\ \}$ denote surface site activities (mol/m^2) or solution activities (mol/dm^3) and $K_{j,H}^*$ is the equilibrium affinity for proton association at site j. The equilibrium affinity $K_{j,H}^*$ is directly related to the Gibbs energy of adsorption which contains a *chemical* or *specific* contribution, $\Delta_a g_{j,H}^0$, and a mean-field electrostatic contribution, $F\psi_s$:

$$K_{j,H}^* = \frac{1}{55.5} \exp\left\{\frac{-\Delta_a g_{j,H}^0}{RT} - \frac{F\psi_d}{RT}\right\}. \tag{26}$$

The units of $K_{j,H}^*$ are l/mol, and the factor 55.5 is the molar concentration of water. By introducing the *intrinsic affinity'* constant, $K_{j,H} = (1/55.5) \exp\{\Delta_a g_{j,H}^0/RT\}$, Equation (26) can be simplified to

$$K_{j,H}^* = K_{j,H} \exp\left(-F\psi_s/RT\right), \tag{27}$$

where $K_{j,H}$ is a measure of the affinity of a proton for site j at the pristine point of zero charge (pristine PZC) where $\psi_s = 0$. To some approximation discreteness of charge effects can be included in $\Delta_a g_{j,H}^0$ [37,48]. In making the above separation all generic electrostatic effects are incorporated in the coulombic term. Specific contributions accrue in either the intrinsic affinity (when such contributions are independent of ψ_s), or the surface activity coefficients.

Assuming ideal (perfect) behavior in the surface layer or mutual cancellation of the (specific) surface activity coefficients, the ratio $\{S_jH\}/\{S_j^-\}$ in Equation (25) can be replaced by $\theta_{j,H}/\theta_{j,R}$, where $\theta_{j,H}$ is the fraction of sites j associated with a proton: $\{S_jH\}/N_s$ and $\theta_{j,R}$ that of the reference sites: $\{S_j^-\}/N_s$, hence $\theta_{j,R} = 1 - \theta_{j,H}$. N_s is the density of surface sites, in this case

$$N_s = \{S_jH\} + \{S_j^-\}. \tag{28}$$

Substitution of Equation (27) in Equation (25) and using site fractions gives

$$\frac{\theta_{j,H}}{\theta_{j,R}} = K_{j,H}\{H^+\} \exp\left(-\frac{F\psi_s}{RT}\right) \tag{29}$$

or in a simplified notation

$$\frac{\theta_{j,H}}{\theta_{j,R}} = \frac{\theta_{j,H}}{1 - \theta_{j,H}} = K_{j,H}\{H_s\}, \tag{30}$$

where $\{H_s\}$ is defined as

$$\{H_s\} = \{H^+\} \exp\left(F\psi_s/RT\right). \tag{31}$$

The proton activity, $\{H^+\}$, in mol/dm^3 is directly related to the pH or can be found from the proton concentration, $[H^+]$, and the mean ion activity coefficient, $\gamma_\pm$, in bulk solution The latter is obtained from tabulated values [50] or calculated with the Debye–Hückel theory or one of its extensions [34,50]. The activity $\{H_s\}$ can be seen as the proton activity at the solution side of the surface plane if equilibrium (24) is treated as mass action equilibrium. Equation (30) is the proton adsorption isotherm, it is equivalent to the Langmuir equation, but $\{H_s\}$ is not an experimentally assessable quantity.

The surface charge density due to the dissociation of the acid surface groups

$$\sigma_s = -F\{S_j^-\} \tag{32a}$$

can easily be obtained from Equation (30):

$$\sigma_s = -FN_s\left(1 - \theta_{j,H}\right) = -FN_s\left[1 + K_{j,H}(H_s)\right]^{-1}. \tag{32b}$$

Thus, provided $\{H_s\}$, N_s, and $K_{j,H}$ are known σ_s can be calculated. In order to obtain $\{H_s\}$, ψ_s should be known, but it is difficult, if not impossible, to measure ψ_s directly. At best one can measure *changes* in ψ_s as a function of the solution concentration of charge-determining ions using the ISFET technique [51–53]. Therefore, in most cases ψ_s has to be calculated by using one of the models for the double layer described in Section II. Sometimes the electrokinetic or ζ-potential [22] is substituted for ψ_s, but this is a rather poor approximation. In general ζ is not too different from ψ_d, and ψ_d is often considerably smaller than ψ_s (Figure 5.2).

For the basic SGC model ψ_s can be obtained from Equation (20) and, in the absence of specific adsorption, $-\sigma_d = \sigma_s$. Combination of Equations (20) and (29) leads to an implicit equation from which $(1 - \theta_{j,H})$ or σ_s can be calculated. For low potentials where ψ_d and σ_d are related by Equation (14) we find

$$\sigma_s = -FN_s \left[1 + K_{j,H}\{H^+\} \exp\left\{ -\frac{F\sigma_s}{RT}\left(\frac{1}{C_1} + \frac{1}{\varepsilon_0\varepsilon_1\kappa} \right) \right\} \right]^{-1}.$$

$$(33)$$

Equation (33) shows that even in this simple case σ_s cannot be written explicitly and that σ_s is a function of N_s, C_1 and κ (or the salt concentration). Some results, expressed as $(1 - \theta_{j,H})$ versus pH, calculated with the full equation for σ_s are shown in Figure 5.3. In Figure 5.3(a) and (b) the screening effect of the indifferent electrolyte is shown for two different values of N_s, in Figure 3(c) the effect of C_1 on the charging behavior is illustrated. For low Stern layer capacitances the screening of the surface charge by electrolyte is very poor which makes it difficult to charge the surface. Note that if we plot $(1 - \theta_{j,H})$ versus pH_s, only one curve is obtained independent of N_s, κ, or C_1 (see Equation (30)).

Equation (29) can also be seen as the expression for the surface potential. By taking the logarithm one finds

$$\psi_s = \frac{2.3\,RT}{F}\left[(pK_{j,H} - pH) - \log\frac{\theta_{j,H}}{1 - \theta_{j,H}} \right].$$

$$(34)$$

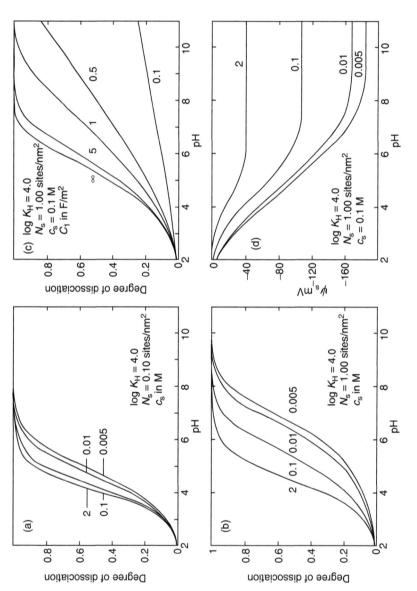

Figure 5.3 Surface ionization as a function of pH: (a) and (b) effect of the salt concentration c

Some results are shown in Figure 5.3(d). By realizing that $\theta_{j,H}$ is related to σ_s, it follows that the surface potential is not just a function of $(pK_{j,H} - pH)$ but also of the surface charge. Only at 50% dissociation is the potential given by the difference $(pK_{j,H} - pH)$. As a consequence the potential is highly non-Nernstian. This is also clear from Figure 3(d).

In a practical situation, when we have measured a series of $\sigma_s(pH)$ curves at different concentrations of indifferent electrolyte, we may replot the curves as $\sigma_s(pH_s)$ curves through calculation of ψ_s by Equation (20) for a series of C_1 values [54b]. The double layer characteristics are adequately described with the SGC theory when, for a specific C_1 value, the $\sigma_s(pH_s)$ curves merge into a master curve. When the observed master curve does not reflect the Langmuir isotherm, the surface is heterogeneous [54,55] (see also Section VI). A prerequisite for this procedure is that the background electrolyte is indeed indifferent (no specific adsorption).

Equation (32b) in combination with Equation (20) can be used to describe, for example, the charging of polymer latex with carboxylic surface groups. As the carboxylic groups tend to protrude in the solution a high Stern layer capacitance is required for the description. James et al. [56] have modeled the properties of such latex by combining a site-binding model including surface complexation with the TL model. The authors did not discuss the possibility of describing their system with the much simpler SGC model in the absence of specific adsorption.

Hiemstra et al. [18,57–60] have described the surface charge data of silica, presented by Bolt [61], with Equation (32b) in combination with the SGC model (Equation (20)). A theoretical justification was given for the $\log K_H$ ($= 7.5$) value used. The fitted value for the Stern layer capacitance depends on the estimate of N_s and the presence of cation pair formation with the surface sites. For $N_s = 8$ sites/nm^2 the capacitance C_1 ranges between 2.5 and 2.9 F/m^2. This high value can be understood if it is assumed that the reactive groups protrude from the surface [62]. In combination with the DLVO theory this model description can also be used to explain the pH-variation of the interaction force at contact between a silica sphere and the 101 face of quartz [60,63]. It should be noted here that the surface charge and surface potential of silica and quartz behave quite differently from most other metal oxides [38,59,64] (see also Section III.D.1).

For surfaces with positive groups an analogous treatment can be followed. For surfaces containing more than one type of surface group Equation (30) can be seen as a *local* isotherm representing the adsorption on one type of site. We return to this in the section on heterogeneous surfaces.

C. Zwitterionic surfaces

A zwitterionic surface is a surface with two *independent* types of surface groups, one positive and the other negative. This type of surface is encountered with polymer lattices [65,66], or can be seen as an idealized representation of the surface of many natural organic particles [42]. When the two types of groups are mixed in a regular or random way, so that the same (smeared-out) electrostatic potential applies to both types of groups, the classical double layer models can be used to describe the $\psi_s(\sigma_s)$ relation [67]. For the site-binding part of the modeling two reactions have to be considered:

$$S_1^- + H^+ \rightleftharpoons S_1H, \ K_{1,H} \tag{35}$$

and

$$S_2 + H^+ \rightleftharpoons S_2H^+, \ K_{2,H} \tag{36}$$

The local isotherm equations can be derived similarly as in Section III.B. The surface charge density σ_s is obtained as

$$\sigma_s = F\{-N_{1,s}(1 - \theta_{1,H}) + N_{2,s}\theta_{2,H}\}, \tag{37}$$

where $N_{j,s}$ is the density of groups j. An important characteristic of this type of surface is the point of zero net charge, i.e., the pH for which

$$1 - \theta_{1,H} = \frac{N_{2,s}}{N_{1,s}}\theta_{2,H}. \tag{38}$$

Harding and Healy [68] have applied a site-binding model, based on Equations (35) and (36) in combination with the GC and the TL models, to describe $\sigma_s(pH)$ curves of amphoteric polystyrene lattices. The description based on the GC model (i.e., SGC with $C_1 = \infty$) fitted the data very well. The very high Stern layer capacitance can be understood by the fact that the surface groups protrude in the solution. An extension of the model was used to describe metal ion adsorption curves.

D. Mineral Oxide Surfaces

1. 2-pK, 1-pK, and MUSIC Model

The description of the surface charge development of amphoteric metal (hydr)oxides is more complicated than that of the simple mono- or bifunctional surfaces and to some extent controversial. The first model attempts assume that the surface contains only one type of surface group that can be protonated. In the oldest approach the reference site can adsorb *two* protons by two consecutive reactions:

$$SO^- + H^+ \rightleftharpoons SOH; \quad K_{H,1} \tag{39a}$$

and

$$SOH + H^+ \rightleftharpoons SOH_2^+; \quad K_{H,2} \tag{39b}$$

each reaction is characterized by its own pK_H value, the SO^-, SOH, and SOH_2^+ groups represent three surface species of the same basic site [49,69–75]. We will call this approach the "2-pK" model. The main shortcoming of the 2-pK model is that it can be shown theoretically (see below) that the difference between the two pK_H values, ΔpK_H, should be about $14\,pH$ units. In practice, however, ΔpK_H is used as a fitting parameter and for $\Delta pK_H \ll 14$ the obtained constants have no physical significance. Therefore, the use of the 2-pK models should be discouraged.

An alternative description has been suggested and is mainly used by van Riemsdijk and coworkers [76–79] and

Gibb and Koopal [80]. This model is based on only one protonation reaction and the adsorption of *one* proton:

$$SOH^{1/2-} + H^+ \rightleftharpoons SOH_2^{1/2+}; \quad K_H \tag{40}$$

Equation (40) characterizes the amphoteric behavior by only two surface species, $SOH^{1/2-}$ and $SOH_2^{1/2+}$, and one pK_H value. We will call this approach the "1-pK" model. The main shortcoming of this 1-pK model is that the formal charge of the reference site (1/2) is fixed.

To illustrate the reasoning behind the two models, the surface of an idealized uncharged metal hydroxide or oxide is shown schematically in Figure 5.4. The metal ions are assumed to be in hexa-coordination with six $-OH(H)$ groups as in most metal oxides.

In Figure 5.4(a) the 1-pK model is illustrated. In this model each *oxygen group* at the surface is seen as a surface site that may associate with either one or two protons. According to Pauling's bond valence concept the charge of the metal ion is equally distributed over its surrounding oxygen groups and its degree of neutralization can be expressed per bond according to the principle of local charge neutralization [81,82]. This leads to reduced bond valences and fractional charge numbers for the surface groups. The metal ion (3+) in Figure 4(a) is assumed to have six bond valences of 1/2+

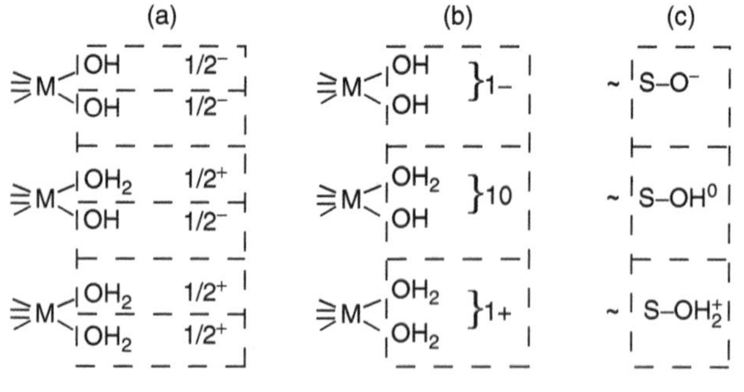

Figure 5.4 Schematic representation of an idealized metal oxide surface: (a) 1-pK model, (b) and (c) 2-pK models.

each in coordination with oxygens. The surface oxygen $(2-)$ is coordinated with the metal ion and one or two protons. The formal charge of the $-OH$ surface group is thus $(+1/2 - 2 + 1 =) -1/2$, that of the $-OH_2$ group $(+1/2 - 2 + 2) = +1/2$. The idealization of Equation (40) is that (a) not all surface groups on metal hydroxides or oxides have the coordination as shown in Figure 5.4(a) and (b) the charge distribution applies to a symmetrical surrounding of the central ion.

In Figure 5.4(b) the 2-pK version of the same metal hydroxide or oxide surface is presented. In the 2-pK model each *metal ion* in the surface layer is seen as a surface site, so that two $-OH(H)$ groups are combined to one site. This has the formal advantage that no fractional charges are present, but the disadvantage is that two proton binding-sites are combined in one site. Figure 5.4(c) can be interpreted as a simplified notation of Figure 5.4(b).

Alternatively, Figure 5.4(c) can be considered as one *oxygen* group in such coordination that the group charge equals -1 for the unprotonated species, 0 for the first protonation step, and $+1$ for the second one. Within the concept of Pauling's bond valence this representation is appropriate for silica where the charge of the Si^{4+} ion is distributed over four oxygens. For Si^{4+} in coordination with one surface oxygen the proton adsorption reactions are then properly described by Equations (39a) and (39b). If it is assumed that only protonation reaction (39a) can occur in the normal pH range, the description of the surface charge of silica as mentioned in Section III.B. will follow. The assumption that the second protonation reaction will not occur is approved within the MUSIC model [58] (see below).

It should be noted that the special position of silica (with protonation reaction (39a) instead of (40)) also explains why the charging of silica is quite different from that of the metal oxides that follow reaction (40). The main reason is that the pristine PZC for silica is reached asymptotically (all groups in the SOH^0 form or $\theta_H = 1$), whereas the pristine PZC of the other oxides is reached when the surface is covered with equal amount of $SOH^{1/2-}$ and $SOH_2^{1/2+}$, or $\theta_H = 0.5$. This difference has consequences for the shape of the charge–pH curves and the dependence of the surface potential on the pH. The difference between silica and the other metal oxides is explained in more detail in Refs. [38,59,64]. By using a 2-pK model with a too small ΔpK value Larson and Attard [29] have reached a wrong conclusion with respect to the Nernstian behavior of the surface potential of silica (see also Section IV.D.5).

Equations (39a) to (40) are simplifications of reality, not only because of the shortcomings mentioned, but also because they apply to an idealized oxide surface with only one type of surface group. From spectroscopy it is well known that several types of oxygen-containing groups may occur on a metal hydroxide or oxide surface [62,83,84]. In order to avoid these shortcomings Hiemstra and coworkers [18,57,58,85,86] have presented a new, detailed, description of surface ionization of metal hydroxides or oxides called *MUSIC* or multisite complexation.

We will first treat the original MUSIC method [18,57,58] "MUSIC-1." Just like the 1-pK method, MUSIC-1 is based on Pauling's [82] concept of bond valence, but on top of that it is taken into account that on metal oxide surfaces more than one type of surface group is present. From crystallographic point of view, in general, three different types of surface oxygen groups can be found: singly, doubly, and triply coordinated with metal cations of the solid. Each of these types of groups may adsorb, in principle, one or two protons according to the following reactions:

$$-\text{Me}_n - \text{O}^{n\nu-2} + \text{H}^+ \rightleftharpoons -\text{Me}_n - \text{OH}^{n\nu-1}; \quad K_{1,\text{H}} \qquad (41)$$

and

$$-\text{Me}_n - \text{OH}^{n\nu-1} + \text{H}^+ \rightleftharpoons -\text{Me}_n - \text{OH}_2^{n\nu}; \quad K_{2,\text{H}} \qquad (42)$$

where ν is the bond valence and n the number of metal ions that are bonded (coordinated) to the oxygen group. Equations (39a) to (40) are special cases of these reactions. The first advantage of MUSIC-1 over the 1-pK and 2-pK models is that it conforms much better with crystallographic data and as such physically speaking, it is more realistic than the simplified models. The fact that MUSIC is fairly realistic has two further advantages. The first advantage is that it is possible to predict, a priori, the intrinsic proton affinity constants $K_{1,H}$ and $K_{2,H}$ for a series of $-O(H)(H)$ groups present on metal hydroxide or oxide surfaces on the basis of the behavior of hydroxylated cations in solution, using the Gibbsite surface as reference [57]. It turns out that the pK_H values for the common metal hydroxides or oxides are about 4-pK units different from that of the oxo–hydroxo complexes of the comparable cations in solution. An exception is made for the pK_H value of silica that is very similar to the pK_H value of SiOH groups in solution. The second advantage is that, by using the predicted proton affinity constants, it could be derived that in the normal pH window not all of the proton reactions that are, in principle, possible will take place [58]. This is due to the fact that the ΔpK_H value of two consecutive protonation steps on the same oxygen [see Equations (41) and (42)] is very large (about 14 units). It means that in practice only one protonation step per type of reactive group is important. Moreover, only sites that have their pK_H values in the normal pH range will contribute to changes in the surface charge density. Calculations show that several groups are neutral in the ordinary pH range. Therefore, these sites will not contribute to the surface charge of a particular crystal plane on which they are present. The fact that the ΔpK_H value of two consecutive protonation steps is very large is on different grounds also predicted by Borkovec et al. [87,88].

Depending on the crystal plane composition, the same metal hydroxide or oxide may exhibit quite different charging characteristics. For a prediction of the charging behavior of an oxide the crystal plane areas and the surface groups on each crystal plane must be known. When the composition of the surface groups and their $pK_{j,H}$ values are known the pristine PZC can be predicted.

For the calculation of the surface charge as a function of the pH in a simple 1:1 electrolyte solution the MUSIC-1 model can be combined with the SGC model. The first assumption to be made is that the different types of groups lead to one smeared-out surface potential. Second, an assumption has to be made with respect to the role of the monovalent salt ions.

The simplest situation is that the 1:1 electrolyte is indifferent and that the salt ions adsorb in the diffuse layer only; one step more advanced is to assume that these ions also adsorb specifically at the Stern plane in the form of outer-sphere complexes. The MUSIC-1 SGC model has been applied to a range of metal oxides like gibbsite, goethite, and silica [18,57,58], rutile or anatase [58,89,90], cerium oxide [91], and magnetite [92].

MUSIC-1 has received criticism from two sides. Bleam [93] has indicated a weakness in the calculation of the proton affinity constants. He argued that the surroundings of the metal centers in the surface are more important than the distance between the metal ion and the adsorbed proton as used in Ref. [57]. Especially the number of surrounding oxygens and hydrogen bonding at the solution side should be taken into account. Schwertmann [94] has pointed out that the main crystal plane of goethite, an important mineral in soil science, is the 110 face and not the 100, 010, and 001 faces that were used for the predictions [58]. At the 110 face different surface groups will be proton active than on the 100, 010, and 001 face. The criticism has prompted Hiemstra and co-workers [85,86] to refine the MUSIC model.

The refined model, which we will call MUSIC-2, uses the actual bond valence instead of the Pauling bond valence. The actual bond valence is strongly related to the bond length [95], and this means that an asymmetry in the surrounding of the metal centers will lead to an unequal distribution of the charge over the ligands. The shorter the distance, the higher the charge attributed to that bond. This aspect is of particular importance for goethite. Also hydrogen bonding is explicitly taken into account in MUSIC-2. The refined approach shows that the undersaturation of the oxygen charge is the basis for the variation in proton affinity. This undersaturation depends on the number of metal ions coordinated to the oxygen, the metal ion–oxygen distance, and the presence of hydrogen bonds.

With MUSIC-2 new predictions are made of the proton affinity constants and of the pristine PZC values of the most important metal oxides. It is concluded that the main conclusions with respect to the predicted pK_H values in MUSIC-1 remain valid, but that the underlying details are different. MUSIC-2 also explains why the proton affinity constant for silica (with SiOH groups protruding in solution) is close to that of the Si–oxo–hydroxo complexes in solution, whereas the proton affinity constants of metal oxides such as rutile and anatase, hematite, goethite, and gibbsite (with *closely packed* oxygens) are about 4-pK units different.

The insights gained with MUSIC-2 are used for the description of proton adsorption to silica and quartz [60,85], goethite [85,86,96,97], gibbsite and bayerite [85,98], rutile and anatase [19,85], lepidocrocite [86], and hematite [60,86]. For these descriptions the MUSIC model was combined with the SGC model for the double layer and assuming that the charging of the different sites leads to only one smeared-out surface potential. The MUSIC approach has also consequences for the description of surface complexation (specific adsorption and inner-sphere complexes); this will be discussed in Section IV.

When little information is present about the structure of an oxide surface, or when a simple approach is preferred, the best solution is to neglect the site heterogeneity and to describe the surface protonation with a *generalized 1-pK model* with a proton binding reaction similar to that presented in Equation (40) but with an open choice for the formal charges. Although in principle not correct, the generalized 1-pK-SGC model may serve as the physically most realistic simplified model to describe the charging of metal hydroxides or oxides. In view of the large interest in metal oxides, and the fact that often no detailed knowledge is present, the availability of a simplified but reasonably realistic model is very welcome next to the advanced MUSIC model.

In Section III.D.2 the generalized 1-pK model will be worked out in combination with the SGC double layer model. For a given metal oxide the approximate values of the proton affinity constant and site density to be used in the generalized 1-pK model can be derived from the insights gained with the MUSIC model. In view of the difficulties involved in finding a unique set of model parameters with especially the 2-pK-TL model [37,99–102] some a priori knowledge about two parameters is a tremendous step forward towards a less arbitrary way of modeling the metal oxide charging and complexation behavior. Lützenkirchen [103] has shown that even without prior knowledge of some of the parameters the 1-pK SGC model should be considered as the best choice model with respect to goodness of fit and uniqueness of the estimated parameters. Because of their importance in the older literature the 2-pK models for surface charging will be briefly discussed in Section III.D.3. although it is recommended not to use this model anymore.

In Section IV we will discuss specific adsorption. In many cases even simple 1:1 electrolytes may not be considered as fully indifferent electrolytes. In Section IV.C, it will be shown how the generalized 1-pK SGC model can be extended to incorporate outer-sphere surface complexation of monovalent ions and model predictions of experimental results will be discussed. In Section IV.D, specific adsorption of multivalent ions will be discussed, including the extension of the generalized 1-pK model for inner-sphere surface complexation. The last part of Section IV is concerned with a brief discussion of the 2-pK model and surface complexation and with some remarks on overviews that have appeared.

2.   Generalized 1-pK SGC Model

In this section, a general description is presented of surface charging according to Equations (41) or (42). The result is a *local* isotherm equation for a specific type of group on a heterogeneous metal hydroxide or oxide surface. The model can also be used as the best option for a simplified homogeneous treatment of the overall charging behavior of metal hydroxides or oxides.

Equations (40)–(42) can be written in a simplified notation as

$$S_j^p + H^+ \rightleftharpoons S_j H^{p+1}; \quad K_{j,\mathrm{H}}, \tag{43}$$

where p ($-2/3$, $-1/2$, $-1/3$, -1) denotes the bond valence (including sign) of surface group j, S_j^p is considered as the reference site R. The proton adsorption isotherm corresponding to

Equation (43) is equivalent to Equation (29) or (30) with $\theta_{j,\mathrm{H}} = \{\mathrm{S}_j\,\mathrm{H}^{p+1}\}/N_\mathrm{s}$ and $\theta_{j,R} = \{\mathrm{S}_j^p\}/N_\mathrm{s}$ with

$$N_\mathrm{s} = \left\{\mathrm{S}_j^p\right\} + \left\{\mathrm{S}_j\mathrm{H}^{p+1}\right\}. \tag{44}$$

The surface charge density σ_s is given by

$$\sigma_\mathrm{s} = FN_\mathrm{s}\{p\theta_{j,R} + (p+1)\theta_{j,\mathrm{H}}\} = FN_\mathrm{s}(\theta_{J,\mathrm{H}} + p). \tag{45}$$

Substitution of Equation (30) in Equation (45) results in

$$\sigma_\mathrm{s} = pFN_\mathrm{s}\left[\frac{1 + ((p+1)/p)K_{j,\mathrm{H}}\{\mathrm{H_s}\}}{1 + K_{j,\mathrm{H}}\{\mathrm{H_s}\}}\right]. \tag{46}$$

As before $\{\mathrm{H_s}\}$ can be calculated by combining Equations (31) and (20) with $\sigma_\mathrm{s} = -\sigma_\mathrm{d}$, if the basic SGC double layer model is used. In general σ_s is a function of N_s, C_1, and κ. Some results for $p = -1/2$ are shown in Figure 5.5, where σ_s is plotted versus pH. The curves at different ionic strengths intersect each other at the pristine point of (net) zero charge (pristine PZC). At this point the (smeared-out) surface potential ψ_s is zero, so that κ does not affect the proton adsorption.

In the pristine PZC $\theta_{j,\mathrm{H}} = -p$ and $\psi_\mathrm{s} = 0$, thus $\{\mathrm{H_s}\} = \{\mathrm{H}_0^*\}$ the proton activity at the pristine (*) PZC. The relationship between pH_0^* and $\log K_{j,\mathrm{H}}$ is found by substitution of these values in Equation (30):

$$\mathrm{pH}_0^* = \log K_{j,\mathrm{H}} - \log\left(\frac{-p}{p+1}\right). \tag{47}$$

For $p = -1/2$, $\mathrm{pH}_0^* = \log K_{j,\mathrm{H}}$; for $p = -2/3$, $\mathrm{pH}_0^* = \log K_{j,\mathrm{H}} - 0.3$, and for $p = -1/3$, $\mathrm{pH}_0^* = \log K_{j,\mathrm{H}} + 0.3$. For $p = -1$ the PZC is reached asymptotically at low pH.

A long-standing issue concerning metal hydroxide or oxide surfaces is the behavior of the surface potential versus the pH [29,37,38,49,64,104–106]. By combining Equations (29) and (47) we find

$$\psi_\mathrm{s} = \frac{2.3\,RT}{F}\left[(\mathrm{pH}_0^* - \mathrm{pH}) - \log\frac{(p+1)\theta_{j,H}}{(-p)\theta_{j,R}}\right]. \tag{48}$$

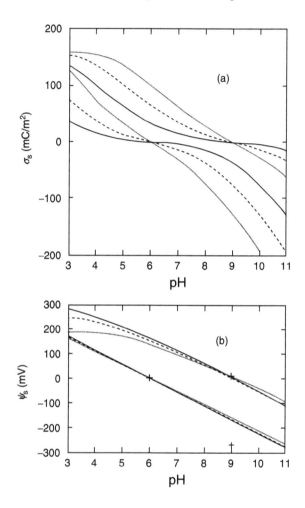

Figure 5.5   Surface charge and surface potential according to the generalized 1-pK SGC model for $p = -1/2$, $\log K_H = 6$ or 9, $C_1 = 2\,\mathrm{F/m^2}$. The effect of c_s (-10^{-3}, $---\,10^{-2}\ldots10^{-1}\,M$) on the $\sigma_s(\mathrm{pH})$ curves and $\psi_s(\mathrm{pH})$ is shown in (a) and (b), respectively, for two values of N_s ($N_s = 8$ sites/nm^2 for $\mathrm{pH}_0^* = 6$ and $N_s = 2$ sites/nm^2 for $\mathrm{pH}_0^* = 9$).

According to Equation (48) at the pristine PZC ($\mathrm{d}\psi_s/\mathrm{dpH}$) is Nernstian. Away from the PZC the second term on the RHS also contributes and $\psi_s(\mathrm{pH})$ is, in principle, non-Nernstian.

The degree of nonideality depends mainly on N_s; this can be shown more clearly by writing Equation (48) as

$$\psi_s = \frac{2.3\,RT}{F}\left[(pH_0^* - pH) - \log\frac{(p+1)\{(-p)+(\sigma_s/FN_s)\}}{-p\{(p+1)-(\sigma_s/FN_s)\}}\right].$$
(49)

For large values of N_s, σ_s/FN_s is small, even relatively far away from the pristine PZC, so that the second term on the RHS of Equation (49) is small. In that case ψ_s is pseudo-Nernstian. For low values of C_1, σ_s remains relatively small; this may also contribute to pseudo-Nernstian behavior. In Figure 5.5(b) some ψ_s(pH) curves are shown as a function of N_s and c_s for oxide surfaces with $p = 1/2$ and $C_1 = 2\,F/m^2$.

In the literature some experimental results regarding the behavior of ψ_s(pH) have been reported for metal (hydr)oxides. An iron oxide electrode (hematite, baked on platinum) studied by Penners et al. [107] showed a Nernstian response in KNO_3 solutions up to about 3 pH units away from the PZC. Also a ruthenium oxide electrode behaved in a pseudo-Nernstian manner [108,109]. Van Hal et al. [53] using the ISFET technique show that Al_2O_3 and Ta_2O_5 behaved pseudo-Nernstian. The slope of the ψ_s(pH) curve for Al_2O_3 is $-54\,mV$ that of Ta_2O_5 $-58\,mV$. Silica is non-Nernstian, especially around the PZC. The slope of the ψ_s(pH) curve in the pH range 6–12 is about $-45\,mV$ for SiO_2. This corresponds with our view that silica is different from the other metal oxides. The model of Larson and Attard [29] is in contradiction with this result (see also Section IV.D.4). Van Hal [53] also compared the experimental results with model results based on the 1-pK SGC ($p = -1/2$ and $pK_H = pH_0^*$ for Al_2O_3 and Ta_2O_5) and the 2-pK SGC model. Both models could describe the results well.

A comparison of the 1-pK SGC model (for a given value of p) with experimental results is relatively simple if the pristine PZC of the oxide is known. In that case C_1 and N_s are the only adjustable parameters. For N_s preferably a *theoretical* estimate based on other information, such as crystal plane structures, should be used. C_1 can be obtained by fitting Equation (46) to the experimental data using a least squares method.

An example of the fit which can be obtained by using the 1-pK SGC model ($p = -1/2$) to describe experimental results is shown in Figure 5.6 for the rutile–KNO$_3$ solution interface [57]. The fit is obtained using a theoretical estimate for N_s (12.2 sites/nm^2), equating pK_H to pH$_0^*$ ($= 5.8$) and adjusting C_1. For C_1 a value of 1.33 F/m^2 is found. In view of recent insights, the theoretical estimate of N_s used is too high, because the Ti$_3$O^0 surface groups are *inactive* in the charging

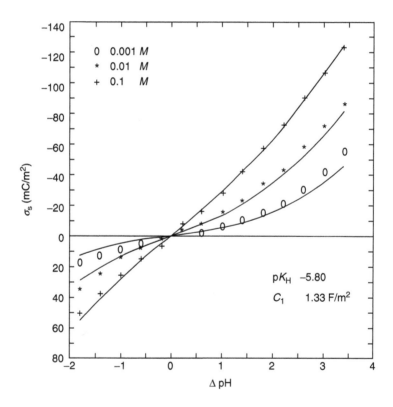

Figure 5.6   Surface charge as a function of pH for the rutile/KNO$_3$ solution system. *Experimental* points are obtained from Yates [62], the curves are calculated with the 1-pK SGC model with $p = -1/2$, log $K_H = 5.8$, $N_s = 12.2$ sites/nm^2 (1.9 C/m^2), and $C_1 = 1.33$ F/m^2 (from W.H. van Riemsdijk, G.H. Bolt, L.K. Koopal, and J. Blaakmeer, *J. Colloid Interf. Sci.*, 109, 219, 1986. With permission.).

process. A better estimate is 8 sites/nm^2 [58] that counts the TiOH$^{1/3-}$ and Ti$_2$O$^{2/3-}$ groups only. However, the result is not very sensitive to the actual value of N_s as long as σ_s is small as compared with FN_s. The value of the Stern layer capacitance is somewhat lower than the theoretical estimate of 1.7 F/m^2. This is largely due to the fact that an average pK_H is used instead of the two values suggested by Hiemstra et al. [18,57].

The above way to describe the charging of a metal oxide surface is the simplest option. The first step towards more sophistication is to take specific adsorption of the background electrolyte into account. We return to this in Section IV where also a further comparison of model predictions and experimental results will be presented.

3. Brief Discussion of the 2-pK Models

The combination of the classical 2-pK model with a double layer model gives the least realistic description of the charging of metal hydroxide or oxide surfaces because the two pK_H values have no physical significance at all (to have significance they should be about 14 pK units apart). The model therefore merely serves as a mathematical description of the charging behavior and the advice is *not to use this model* any more. Yet, in the past the 2-pK model has served an important role, therefore a brief discussion will be given of the various 2-pK models. Two older reviews (dating back to around 1980) that consider the most important literature on the 2-pK model are those of Westall and Hohl [110] and James and Park [49].

Schindler and Stumm and coworkers [e.g., 70,71,111–113] mainly used the constant capacitance model in combination with a graphical method to obtain the pK_H values. The graphical determination of the pK_H values starts with the assumption that the ΔpK value is large [37], however, the obtained ΔpK values are still much smaller than what is physically realistic. The method is simple, but has the disadvantage that the pK_H values are just fitting parameters and that for each salt concentration a new capacitance value is required. Lützenkirchen [114] has shown that when a numerical

routine is used to estimate the parameters substantial difficulties may arise in the effort to obtain unique parameters.

Huang and coworkers [115–117] and Dzombak and Morel
[118] have made tabulations based on the 2-pK GC model. The
ΔpK value is used as a fitting parameter (without physical
significance) and the model is very simple.

The most popular is the 2-pK TL model [e.g.,
62,72,74,75,100–102,118–127]. Although the 2-pK-TL model
can describe most σ_s(pH) results very well, it is impossible to
obtain a unique set of parameters on the basis of σ_s(pH) and
ζ(pH) curves [37,99,101]. Hence, for the same results several
descriptions based on the same way of modeling can be found.
This makes the disadvantage that the parameters have no
physical meaning even worse. Lützenkirchen et al. [100]
have shown that by applying a number of physical constraints
this problem can be avoided. However, in view of the overall
weakness of the 2-pK model this is only practical and no
principle advantage.

Furthermore, it should be noted that in many of the cited
studies an a priori value of 0.2 F/m^2 is assumed for the outer
Helmholtz layer capacitance. This value is derived from studies of the AgI and Hg electrolyte solution interface [62,74,128].
However, as explained in Section II.B, for the mineral oxides a
high Stern layer capacitance and thus a high outer Helmholtz
layer capacitance is realistic. A low capacitance value prevents effective screening of the charge. Therefore, for the
highly charged oxides, a strong specific adsorption in combination with a high inner Helmholtz layer capacitance is required to describe the experimental results with the TL
model and $C_2 = 0.2$ F/m^2. Consequently the specific adsorption
is overestimated and a conflicting picture emerges for the
values of the inner and outer Helmholtz layer capacitances.

Bowden et al. [73] and Janssen and Stein [129] have used
different kinds of multilayer models in combination with the
2-pK model. Parameter evaluation with such models is faced
with similar problems as those that occur for the 2-pK-TL
model. The values obtained for the parameters depend on
(implicit) a priori assumptions about their magnitude or on
the initial input values in the iteration procedure.

IV. SPECIFIC ADSORPTION

A. General Aspects

In Section III it has been assumed that as a first-order approximation specific adsorption of the background electrolyte can be neglected. For most surfaces in contact with an ordinary background electrolyte composed of monovalent ions, this is a reasonable approximation as long as the electrolyte concentration and the surface charge are not too high. However, for intermediate and high concentrations of 1:1 electrolytes or with divalent counterions specific adsorption is a common phenomenon. Direct studies of the adsorption of monovalent ions have been made using ion radiolabeling techniques [120,122,130]. According to Sprycha [122] strong specific adsorption of both anions and cations may occur around the PZC in (about) equal amounts. For multivalent ions both radio-tracer e.g. [32] and various in situ spectroscopic techniques like EXAFS [e.g., 131–134], NMR, and ESR [e.g., 135–137] or IR [e.g., 138–143] have been used.

In the literature regarding the adsorption of ions onto metal hydroxides or oxides it is often assumed that adsorption occurs through *surface complexation*. Proof for this assumption is based on spectroscopic studies. Surface complexation is a particular form of specific adsorption in which the adsorbate ion forms a complex with the ionizable surface group. A surface site can then either be in its basic reference state, be occupied with a proton (charge-determining ion), with a complexing cation, or with a proton plus a complexing anion. Surface complexation therefore corresponds to *site competition*. Complexation will be treated in more detail in Sections IV.C (monovalent ions) and IV.D (multivalent ions).

In the presence of specific adsorption of background electrolyte the often observed *common intersection point* (CIP) of a series of $\sigma_s(\text{pH})$ curves measured at different values of the ionic strength can no longer be identified with the pristine PZC [144,145], unless the specific adsorption of anions and cations is fully symmetrical around the PZC. In general, the CIP differs for different electrolytes and the PZC shifts with the ionic strength and is dependent on the type of ions. Moreover, the PZC is no longer equal to the isoelectric point (IEP). At the IEP the electrokinetic charge, that is the charge within the plane of shear [22], is zero. A difference between IEP and PZC is a very clear indication of specific adsorption.

Electrokinetic measurements are also very useful to obtain a qualitative or semiquantitative insight into specific adsorption. With metal oxides, for example, measurements can be made as a function of the pH and the electrolyte concentration or the type of electrolyte. Some typical results are presented in Figure 5.7 showing the behavior of rutile particles as a function of pH in $Ca(NO_3)_2$ (Figure 5.7(a)) and in a series of different nitrates at constant nitrate salt concentration (Figure 5.7(b)) [146]. In $Ca(NO_3)_2$ the Ca^{2+} ions adsorb specifically at sufficiently high pH. Raising the pH increases the adsorption with the result that the mobility (and thus ζ) decreases and even reverses sign. This trend becomes more pronounced if the Ca^{2+} concentration increases. At $1.67 \times 10^{-3}\,M$ the mobility remains positive over the entire pH range. Figure 5.7(b) shows the behavior in different nitrates; it is easily seen that the affinity sequence for specific adsorption on rutile can be written as $Ba \gg Sr > Ca > Mg$. In $NaNO_3$, which contains no specifically adsorbing ions, the behavior is quite simple: to the left where $\sigma_s > 0$, also $\zeta > 0$ and to the right both are negative. For $Ba(NO_3)_2$ cation adsorption is extremely strong, the electrokinetic charge of the particles remains positive for all pH values, even when $\sigma_s < 0$. Many other examples of the use of electrokinetic measurements to study specific adsorption can be found in the literature [e.g., 147–151].

In general the adsorption of multivalent ions is found to be pH dependent. For a given initial cation concentration a rapid increase in uptake of the multivalent cations usually occurs over a narrow pH range [113,147,152–154]. The affinity of a certain surface for different divalent cations can be studied by comparing the cation uptake as a function of pH at a given total cation concentration [see e.g., 155–158]. A typical example is shown in Figure 5.8. With oxyanions in general a gradual decrease in adsorption with increasing pH is found [see e.g., 159–161].

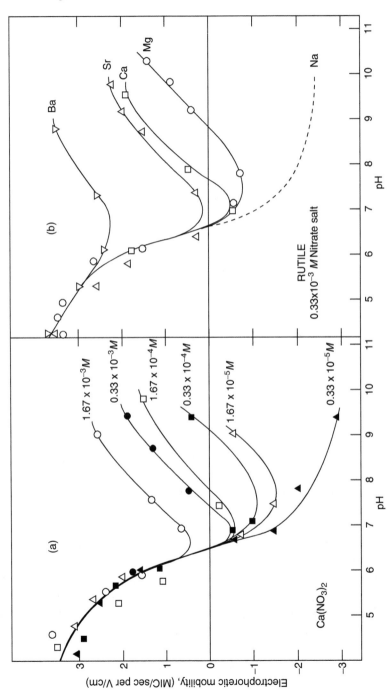

Figure 5.7 Electrophoretic mobility versus pH curves for rutile in the presence of (a) different concentrations of Ca(NO₃)₂, and (b) different alkaline-earth cations at $0.33 \times 10^{-3} M$ concentration (from D.W. Fuerstenau, D. Manmotian, and Raghavan, in *Adsorption from Aqueous Solutions*, P.H. Tewari, Ed., Plenum Press,

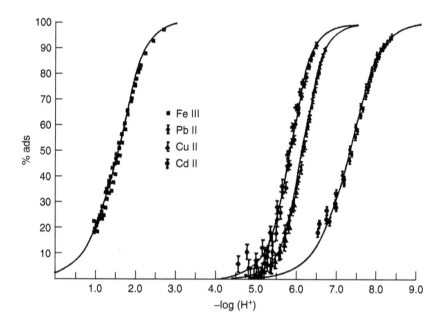

Figure 5.8 Adsorption of metal ions on amorphous silica as a function of pH (from P.W. Schindler, B. Fürst, R. Dick, and R. Wolf, *J. Colloid Interf. Sci.*, 55, 469, 1976. With permission).

An older review on specific adsorption of cations on oxides and phyllosilicate clays has been given by Kinniburgh and Jackson [162]. A general discussion of the adsorption of both cations and oxyanions with emphasis to soil systems can be found in Chapters 4 and 5 of *The Surface Chemistry of Soils* by Sposito [154]. The book *Adsorption of Metals on Geomedia* [163] contains many contributions and references regarding adsorption of metal ions and radioisotope ions on mineral and clay surfaces. Jenne [164] discusses data analysis; he presents a wide range of different models used to describe the adsorption including surface complexation based on the 2-pK model and reviews factors that control the adsorption. Payne et al. [165] and Pabalan and coworkers [166,167] use a 2-pK surface complexation type of modeling to describe the sorption of Uranium[VI] or Neptunium[V] on model minerals. Crawford et al. [168] review the literature on heavy metal ion adsorption to hydrous metal oxides. Baumgarten [169] gives a review of cation and anion adsorption on aluminum hydroxides or oxides. A brief overview of more recent work on oxyanion adsorption on minerals is presented in the introduction of the paper of Geelhoed et al. [170].

B. Specific Adsorption on Independent Sites

With specific adsorption, in general, the sites for the adsorbate ions may be independent of the sites for charge-determining ions. Considerable size differences between the charge-determining ions and the specifically adsorbing ions can be a good reason to assume independent sites for both adsorbates. The Stern adsorption isotherm, Equation (23), describes specific adsorption without site competition in an ideal (perfect) monolayer. Although the sites are independent, surface ionization and specific adsorption still affect each other through the electric field. The adsorption of charge-determining ions is governed by ψ_s, that of the specifically adsorbing ions by ψ_d. The relationship between the two monolayer adsorptions and the diffuse layer adsorption is the electroneutrality equation. The advantage of taking independent sites for the specifically adsorbing ions is that the precise nature of the ion–surface complex does not have to be specified.

In modern literature the specific adsorption of inorganic ions onto metal oxide surfaces is, in general, treated on the basis of surface complexation and site competition. However, in the older literature several studies exist that consider the sites for the charge-determining ions and those for the inorganic ions as independent. Bowden et al. [73,171], Barrow et al. [172], and Fokkink et al. [151,173] have analyzed ion adsorption on metal hydroxides or oxides under the assumption that the specifically adsorbing ions are not in competition with the charge-determining ions. Bowden and Barrow use a (multicomponent) Stern equation in combination with a *multilayer* double layer model to describe specific adsorption of both anions and cations (including metal ions). A discussion of their way of modeling in relation to other adsorption models has been given by Barrow [174]. Hingston [175] has presented a review on anion adsorption with particular attention to the type of modeling of Bowden and Barrow.

Fokkink et al. [151,173] have studied metal ion adsorption on oxides, in particular hematite and rutile. In the first paper [151] they concentrate on the co-adsorption ratio, r_{OH}, that is the number of OH^- ions co-adsorbing with each specifically adsorbing metal ion at a given pH. The most important message of this paper is that it shows that r_{OH} can be explained on a purely electrostatic basis, i.e., without assuming an exchange reaction. The double layer model used is the SGC theory combined with the Nernst equation to calculate the surface potential.

In the second paper [173] the temperature dependence of cadmium adsorption on rutile and hematite is studied as a function of the surface charge density. The metal ion adsorption is described by the Frumkin [176] equation (see Section V, Equation 81), which extends Equation (23) with a nonelectrostatic lateral interaction term:

$$\frac{\theta_M}{1 - \theta_M} = c_M K_M \exp\left(\chi\theta_M - \frac{v_M F \psi_d}{RT}\right) \qquad (50)$$

Again the surface potential ψ_s is calculated using the Nernst equation and ψ_d with the SGC model. For the introduction of χ no specific motivation has been given. The cadmium adsorption isotherms have been fitted with $N_{s,M}$, K_M, and χ as fitting parameters. $N_{s,M}$ and K_M are (approximately) independent of the initial surface charge and salt concentration. However, χ is not constant, mostly repulsive, and strongly dependent on the chosen value of $N_{s,M}$. The authors refrain from an interpretation of χ. In their conclusions they state that the active groups in the oxide–solution interface behave very similarly to bulk OH^- ions with respect to both H^+ and Cd^{2+} binding and that their results form "a direct indication for the occurrence of 'surface complex formation' in the case of cadmium binding on oxides." The question arises, why the authors did not treat the adsorption as surface complexation. van Riemsdijk et al. [78] had shown before that cadmium adsorption on

hematite and amorphous iron oxide could be described very
well with a surface complexation model without the additional
parameter χ.

C. Surface Ionization and Complexation of Monovalent Ions

In this section surface complexation of monovalent ions with
metal oxides will be discussed. The starting point is the generalized 1-pK model for proton adsorption on a homogeneous
or pseudo-homogeneous oxide surface based on the equilibrium Equation (43) with S_j^p as reference site and Equations
(30) and (31) for $\theta_{j,\mathrm{H}}$ and $\{H_s\}$, respectively. Except for the
protonation reaction two additional equilibriums have to be
considered to describe surface complexation. With, say, KNO_3
as electrolyte, the complexation equation for the cation is

$$S_j^p + K^+ \rightleftharpoons S_j^p K^+; \quad K_j(K^+), \tag{51}$$

where p can take the values -1, $-2/3$, $-1/2$, $-1/3$. The equilibrium condition is (compare Equation (30))

$$K_j(K^+) = \frac{\{S_j^p K^+\}}{\{S_j^p\}\{K_d^+\}} \tag{52}$$

and

$$\{K_d^+\} = \{K^+\}\exp(-F\psi_d/RT). \tag{53}$$

In Equation (53) ψ_d enters instead of ψ_s because it is assumed
that the complexing ions are located at the Stern plane. For the
complexation of the anions we also take S_j^p as reference site:

$$S_j^p + H^+ + NO_3^- \rightleftharpoons S_j H^{p+1} NO_3^-; \quad K_j'(NO_3^-) \tag{54}$$

with equilibrium condition

$$K_j'(NO_3^-) = \frac{\{S_j^p H^{p+1} NO_3^-\}}{\{S_j^p\}\{H_s^+\}\{NO_{3,d}^-\}} \tag{55}$$

and

$$\{NO_{3,d}^+\} = \{NO_3^-\} \exp(+F\psi_d/RT). \tag{56}$$

The total site density N_s equals in this case

$$N_s = \{S_j^p\} + \{S_j^p K^+\} + \{S_j H^{p+1}\} + \{S_j H^{p+1} NO_3^-\}. \tag{57}$$

The surface charge density σ_s, obtained by potentiometric titration of the surface groups with either H^+ or OH^-, includes all groups that have reacted with a proton

$$\begin{aligned}
\sigma_s = F[p\{S_j^p\} + p\{S_j^p K^+\} + (p+1)\{S_j H^{p+1}\} \\
+ (p+1)\{S_j H^{p+1} NO_3^-\}].
\end{aligned} \tag{58}$$

The charge *neutralized* by complexation, σ_1, is

$$\sigma_1 = F[\{S_j^p K^+\} - \{S_j H^{p+1} NO_3^-\}] \tag{59}$$

and the remaining part of the surface charge density is compensated in the diffuse layer

$$\begin{aligned}
\sigma_d = -F[p\{S_j^p\} + (p+1)\{S_j^p K^+\} + (p+1)\{S_j H^{p+1}\} \\
+ p\{S_j H^{p+1} NO_3^-\}].
\end{aligned} \tag{60}$$

Equations (58) to (60) are in accordance with the electroneutrality equation (21). The adsorptions of K^+ and NO_3^- in the *diffuse* layer follow from Equations (16) and (17), respectively. For the calculation of ψ_d and ψ_s the SGC model can be chosen using, respectively, Equations (20) and (13).

The effect of complexation on the $\psi_s(pH)$ relation can also be studied. From Equation (48) with the relevant expressions for $\theta_{j,H}$ and $\theta_{j,R}$ we find

$$\frac{F\psi_s}{2.3RT} = (pH_0^* - pH) - \log\left[\frac{(p+1)(N_s^c - \{S_j^p\})}{-p(N_s^c - \{S_j H^{p+1}\})}\right] \tag{61}$$

with $N_s^c = N_s - (\{S_j^p K^+\} + \{S_j H^{p+1} NO_3^-\})$ The factor $(p+1)/(-p)$ in the logarithmic term takes into account the asymmetry of the charges. According to Equation (61) $N_s^c (< N_s)$ is important rather than N_s. However $\{S_j^p\}$ and $\{S_jHp^{+1}\}$ are also smaller than without complexation. Therefore, for high values of N_s, the second term on the RHS will again be small. It can be concluded that as long as N_s is high the Nernst equation is a reasonable approximation. Equation (61) also shows that the pristine PZC, pH_0^*, is the reference point for the surface potential. When the PZC (pH_0) at a given salt concentration is used instead of pH_0^* the surface potential is overestimated for specific adsorption of cations and underestimated for anions. For simple monovalent ions $(pH_0^* - pH_0)$ is of the order of 0.1 pH units for a tenfold increase in salt concentration. For strongly adsorbing ions this difference may become much larger.

Important indicators for ion complexation are the shifts of the PZC and the IEP with increasing salt concentration. According to Equation (58) in the PZC ($\sigma_s = 0$):

$$p\{S_j^p\} + p\{S_j^p K^+\} + (p+1)\{S_jH^{p+1}\} + (p+1)\{S_jH^{p+1}NO_3^-\} = 0.$$

$$(62)$$

From Equation (62) it follows that for preferential adsorption of, say K^+, at the PZC $(p+1)\{S_jH_p^{+1}\}$ and $-p\{S_j^p\}$ are no longer equal. Moreover, the densities $\{S_j^p K^+\}$ and $\{S_j H^{p+1}NO_3^-\}$ will depend on the KNO_3 concentration, so that the PZC depends also on $c(KNO_3)$. A quantitative measure of the shift of pH_0, the pH of the PZC, follows from Equations (30), (47), (59), and (62):

$$pH_0 = pH_0^*$$

$$- \log\left[\frac{\{S_jH^{p+1}\}}{\{S_jH^{p+1}\} + \{S_jH^{p+1}NO_3^-\} + (p/(p+1))\{S_j^p K^+\}}\right],$$

$$(63)$$

where pH_0^* is the pristine PZC (see Equation (47)). Only when the total charge of cation complexes plus anion complexes at the PZC is zero one finds $pH_0 = pH_0^*$. For a nonzero charge balance a shift in the PZC will be observed. When the charge of the cation complexes outweighs that of the anion complexes, i.e., $p\{S_j^p K^+\}/(p+1)+\{S_j H^{p+1}NO_3^-\}<0$, pH_0 is smaller than pH_0^*, or the PZC shifts to the left.

The condition for the IEP ($\sigma_d = 0$) is

$$p\{S_j^p\} + (p+1)\{S_j^p K^+\} + (p+1)\{S_j H^{p+1}\} + p\{S_j H^{p+1} NO_3^-\} = 0.$$

(64)

The stronger the cation complexation is, the smaller is $p\{S_j^p\} + (p+1)\{S_j H^{p+1}\} + p\{S_j H^{p+1} NO_3^-\}$ and the higher the pH has to be to match Equation (64). The pH(IEP) thus shifts to higher pH values with increasing $c(KNO_3)$.

The various surface group densities, the PZC, the IEP and the surface potentials can be calculated iteratively for known values of the affinity constants, C_1, N_s, and p. An elegant algorithm for the calculation has been proposed by Westall and coworkers [177,178]. Westall's calculation scheme can also be used for more complicated electrolyte solutions. Based on this type of approach several computer codes have been developed (e.g., MINEQL [179] and ECOSAT [180]) that can be used to calculate the adsorptions or to fit the model (FITEQL [181]) to an experimental data set.

van Riemsdijk et al. [58] have successfully used the 1-pK SGC model with $p = -1/2$; log $K_H = -8.4$; $N_s = 8$ sites/nm^2 and $C_1 = 1.72 \, F/m^2$ to describe the results obtained by Breeuwsma and Lyklema [182] for the hematite and KNO_3 solution interface. The log K_H value has been equated to pH$_0^*$. The fitted Stern layer capacitance corresponds well with the theoretical estimate. Outer-sphere complexation is neglected. This type of modeling corresponds well with the MUSIC approach. According to Hiemstra et al. [58], ordinary hematite can be approximated as a variable charge surface in which the charge development is dominated by protonation of $FeOH^{1/2-}$ groups. Comparable results can be obtained with the hematite data of Atkinson et al. [69].

Charge pH curves of hematite have also been described by Gibb and Koopal [80], Schudel et al. [183], and Gunnarsson et al. [184,185] using the 1-pK SGC model. In all cases good fits could be obtained for $p = 1/2$ and log $K_H \approx$ pH$_{PZC}$ provided outer-sphere complexation of the salt ions was also incorporated. N_s values ranged from 5 to 8 sites/nm^2 and C_1 from 0.8 to 1.5 F/m^2. Gibb found asymmetrical pair formation and a dif-

ference between the PZC and the IEP, Schudel and Gunnarsson both used symmetrical pair formation around the PZC.

In Figure 5.9 the results of Gibb and Koopal [80] obtained
for hematite and rutile in KNO_3 solutions are shown.

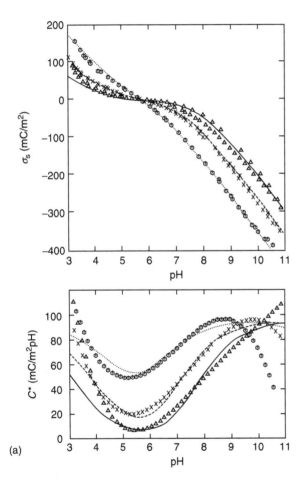

(a)

Figure 5.9 (a) Surface charge σ_s and operational differential capacitance C^* of a rutile suspension (a) and hematite suspension (b) in
KNO_3 solution as a function of the pH. Symbols: experimental
points ($\Delta = 0.002\,M$, $x = 0.01\,M$, $0 = 0.1\,M$), curves calculated with
the 1-pK SGC model. The BET areas and the model parameters are
shown in Table 5.1 (from A.W.M. Gibb and L.K. Koopal, *J. Colloid
Interf. Sci.*, 134, 122, 1990. With permission).

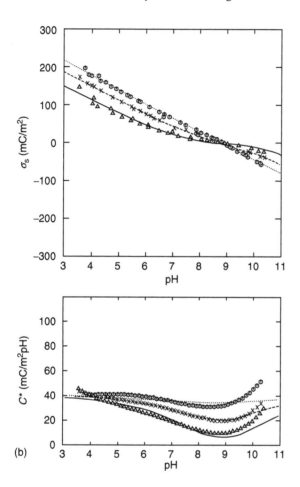

Figure 5.9 (*continued*)

Figure 5.9(a) shows the measured and calculated $\sigma_s(\text{pH})$ and $C^*(\text{pH})$ results for rutile, Figure 5.9(b) those for hematite. For both oxides the calculations are based on the generalized 1-pK SGC model with $p = -1/2$. The capacitance C^* is the slope of the $\sigma_s(\text{pH})$ curve at a given ionic strength. The values of N_s are based on theoretical estimates [58] and log K_H is set equal to pH_0^*. Optimization of the parameters C_1, log $K(\text{K}^+)$, and log $K'(\text{NO}_3^-)$ was performed by a least squares method. A way to estimate the initial values has been discussed in the original paper [80]. The results are summarized in Table 5.1. In both cases the fit of the σ_s(pH) curves is slightly better than when surface complexation is neglected. The capacitance curves are however quite sensitive for the magnitudes of log K(K+) and log K'(NO$_3^-$). Use of the capacitance curves is therefore essential in obtaining physically realistic values of the model parameters.

The value obtained for C_1 (rutile) is higher than that for rutile prepared by Yates (see Figure 5.7) and slightly higher than the theoretical value. This points to a poorly structured surface [19]. The low value of C_1 (hematite) point towards a highly structured surface [19].

Experimental proof of surface complexation is presented by the values of the PZC and the IEP as measured for both oxides at different KNO$_3$ concentrations. The results are shown in Table 5.2. At 0.002 M it has been assumed that surface complexation can be neglected, so that the pristine PZC is given by the measured IEP at this salt concentration. The other PZCs were determined from a series of subsequent *short range* titration curves.

It is also possible to calculate the components of charge: σ_s, σ_d, σ_{K^+}, and $\sigma_{NO_3^-}$ using the parameters presented in Table 5.1. Results are reproduced in Figure 5.10 for the lowest (0.002 M) and the highest (0.1 M) electrolyte concentrations. At low ionic strength surface complexation is very small

Table 5.1 Surface chemical characterization of rutile (TiO$_2$) and hematite ($\alpha-$Fe$_2$O$_3$) according to the 1-pK SGC model[a]

(M)	A (BET) (m^2/g)	N_s (sites/nm^2)	C_1 (F/m^2)	log K_H	log K(K$^+$)	log K(NO$_3^-$)
TiO$_2$	44.5	8	2.2	5.9	−0.2	4.8
αFe$_2$O$_3$	32.3	8	0.8	9.1	−0.1	8.3

Source: From A.W.M. Gibb and L.K. Koopal, *J. Colloid Interf. Sci.*, 134, 122, 1990. With permission.
   BET area is also indicated

Table 5.2   Experimental IEP and PZC values of rutile and hematite

	0.002		0.01		0.1	
c (M)	IEP	PZC	IEP	PZC	IEP	PZC
TiO_2	5.9	5.9	6.1	5.8	6.4	5.7
αFe_2O_3	9.1	9.1	9.3	9.0	9.7	8.9

Source: From A.W.M. Gibb and L.K. Koopal, *J. Colloid Interf. Sci.*, 134, 122, 1990. With permission.

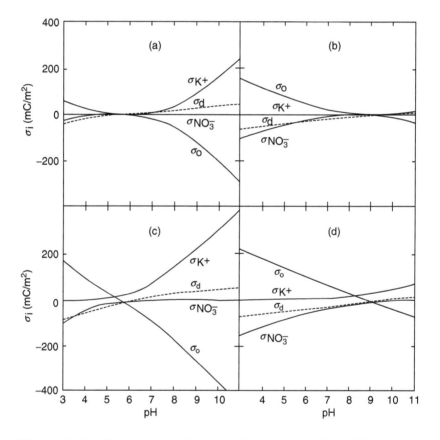

Figure 5.10   Components of charge of rutile (a and c) and hematite (b and d) in $0.002\,M$ (a and b) and $0.1\,M$ (c and d) KNO_3 as calculated with the 1-pK SGC model. For parameters: see, Table 5.1 (from A.W.M. Gibb and L.K. Koopal, *J. Colloid Interf. Sci.*, 134, 122, 1990. With permission).

around the PZC, but considerable at large $\Delta pH(=pH_0^* - pH)$ values. At high ionic strength surface complexation at the PZC is small but clearly present, it increases strongly away from the PZC. The diffuse layer charge remains relatively low ($<80 \, mC/m^{-2}$) for all conditions. Note that Equations (17) and (19) in the paper of Gibb and Koopal [80] are incorrect (compare Equations (59) and (60) given above); however, for the calculations the correct equations have been used.

Recently, several sets of charge pH curves of TiO_2 (rutile, anatase, and Degussa P25) in various 1:1 electrolyte solutions have been described by Bourikas et al. [19] using the MUSIC model with two independent protonation reactions ($p = -1/3$; $N_s = 6$ sites/nm^2 and $p = 2/3$; $N_s = 6$ sites/nm^2), but assuming that the two pK_H values were the same. Good fits could be obtained by fitting the Stern layer capacitance and the ion pair formation constants and taking the pK_H value close to the PZC. For the capacitance of the Stern layer two groups of values were found. For the well crystallized samples about $0.9 \, F/m^2$ and for the less well structured surfaces about $1.6 \, F/m^2$. The binding constants for the anions follow the sequence $Cl > NO_3 > ClO_4 > I$, that of the cations $Cs < K < Na < Li$.

Of the iron oxides not only hematite is investigated, also goethite and amorphous iron oxide have been considered by Hiemstra et al. For amorphous iron oxide (data Benjamin and Leckie [186]) with a PZC of 7.9 a reasonable result is obtained by using the 1-pK SGC model (with $p = -1/2$, log $K_H = 7.9$, $Ns = 8$ sites/nm^2, $C_1 = 1.54 \, F/m^2$), but some specific adsorption of K^+ and NO_3^- is required [77]. A specific aspect of amorphous iron oxides is that the particles are so small that at low ionic strength the SGC model may overestimate the electrostatic interactions.

The charge versus pH curves of goethite with a PZC of 9.3 can also be described with the 1-pK SGC model (with $p = -1/2$, log $K_H = 9.3$, $N_s = 6$ sites/nm^2, $C_1 = 1.1 \, F/m^2$) [187]. Rietra et al. [97] consider the charging of goethite in different 1:1 electrolytes using the model with symmetrical pair formation around the PZC (see Section IV.C).

Also the minerals gibbsite and bayerite have been considered by Hiemstra et al. [57,58,79,85,98]. Both can be modeled with the 1-pK SGC model ($p = -1/2$ and $\log K_H \approx pH_{PZC}$) taking into account the protonation of the singly coordinated surface groups only. The site densities are about 8 sites/nm^2. The Stern layer capacitance values range between 0.9 and 1.5 F/m^2, just as for TiO$_2$. For gibbsite it is important to compare the surface charge density on the basis of the edge area with that on the basis of the total (BET) area. The obtained charge densities at pH values below the PZC should be in the range of 0–250 mC/m^2. When the entire BET area is reactive mean site densities of 4 sites/nm^2 are used in the modeling. For a good result it is also necessary to take some specific adsorption of the salt ions at the Stern plane (outer-sphere complexes) into account.

D.   Complexation of Multivalent Ions

1.   General Aspects

In practice, most systems do not only contain monovalent ions, but also multivalent ions and these ions adsorb strongly on most surfaces. The adsorption mechanism is probably similar to that of ion–ion complexation or association in bulk solution [e.g., 188,189], but geometric factors will also be important.

The model description of complexation of multivalent ions with surface groups is faced with considerable complexity. The first complication arises because of the speciation in bulk solution: multivalent anions may react with protons, and many multivalent cations hydrolyze at high pH values. A similar situation might occur at the surface: for one ionic species different surface complexes are formed at different pH values or different surface complexes in equilibrium with each other exist at a given pH. Second, when multivalent ions react with the surface sites for the charge-determining ions the stoichiometry of the binding reaction is important. However, due to complexation the electric field near the surface changes and this change will also affect the adsorption of charge-determining ions. The net effect is an exchange ratio (or co-adsorption ratio) of charge-determining and multivalent ions that is only partly due to the stoichiometry of the reaction. The exchange ratio is thus not an immediate clue to the stoichiometry; see for more detail Rietra et al. [190]. A third complication is the surface heterogeneity. The results for proton adsorption on metal hydroxides or oxides illustrate the type of difficulties that have to be faced with heterogeneous surfaces. For multivalent ions that may form bidentate complexes, the situation is even more complicated.

All these complications make a unique description of surface complexation of multivalent ions extremely difficult, if not impossible. A priori assumptions regarding the type of adsorbing species and the type of surface complexes formed cannot be avoided and different opinions will exist as to the *most reasonable* assumptions. Pragmatism will be a good guide. Spectroscopic information on existing surface complexes can help to decide what the most probable structure of a surface complex is and to reduce the possible number of adsorbed species.

A drastic and simple way out is to describe adsorption as if the sites for surface complexation are entirely independent of the surface groups for the charge-determining ions [73,151,171–174]. This option has been discussed in Section IV.B. A disadvantage of this approach is that it is not very realistic for small inorganic adsorbate ions for which complexation is highly probable or even proven by spectroscopy. For large ions that cover several surface groups, so that only a part of the surface groups can participate in the binding, the approximation of independent sites seems more realistic.

With surface complexation models the adsorbate-surface group reaction has to be specified. In favorable cases where sufficient knowledge is available with respect to the adsorbents and the type of complexes formed, the MUSIC model can be used. The MUSIC model is very well suited to take spectroscopic information into account. When the bond valence principle of Pauling is also used to model the complexation, it is often appropriate to distribute the charge of the complexing ions over the Stern plane and the surface plane

[191]. This version of the MUSIC model is called CD-MUSIC. Similarly as for the protons it should, in principle, be possible to predict the complexation constants. At the moment no such treatment exists and the complexation constants involved have to be treated as adjustable parameters. van Riemsdijk et al. [78,96,170,192–198] have published a series of studies on multivalent ion complexation in which the CD-MUSIC model has been used.

In many cases only little information will be available on both the surface structure and the type of complexes present and the CD-MUSIC model cannot be used directly. In that case the most realistic approximation for the metal oxide surfaces is to consider the surface as pseudo-homogeneous and to describe the protonation with the generalized 1-pK SGC model. For the extension to surface complexation the complexation reactions have to be specified, similarly as for monovalent ion complexation. Also in this simplified version of the CD-MUSIC model it is possible to distribute the charge of the complexing ions over the surface plane and the Stern plane. The CD-MUSIC model and its simplified version based on the generalized 1-pK model will be discussed in Section IV.D.2.

By far the most studies on multivalent ion complexation have the 2-pK model as a starting point. As explained above, for metal hydroxides or oxides this simplification is physically less satisfactory than the 1-pK approach, nevertheless a brief summary will be given of this work in Section IV.D.3.

Fitting of the models to experimental results should preferably be based on adsorption data covering different values of multivalent ion concentration, pH, and (background) salt concentration. In order to avoid over-interpretation the number of complexation reactions should be kept to a minimum.

2.   Complexation and the MUSIC model:
     CD-MUSIC

On the way to an improved model for surface complexation with metal oxides van Riemsdijk et al. [78] have first studied the adsorption of cadmium ions on hematite and amorphous iron oxide. In this study surface heterogeneity has been considered explicitly by assuming a continuous distribution of proton and metal ion affinities. The conclusion was reached that the two oxides could be approximated as pseudo-homogeneous with respect to the surface groups present. Several models could describe the adsorption of cadmium, but preference is given to modeling based on the 1-pK SGC theory. To incorporate cadmium complexation the 1-pK model based on Equation (43) with $p = -1/2$ is extended by allowing the cadmium species to adsorb at the Stern plane. Adsorption at the surface plane gave poor results. Solution speciation was considered on the basis of the known hydrolysis reactions. The cadmium adsorption as a function of pH and cadmium concentration could be described with only one additional parameter, the cadmium complexation constant. Calculations indicated that the amount of cadmium adsorbed as $SOH^{1/2-}Cd^{2+}$ was negligible compared to $SOH^{1/2-}CdOH^+$, although the predominant species in solution was Cd^{2+} (the authors did not consider bidentate complexation). The results as observed for hematite are shown in Figure 5.11, the symbols are the experimental results at four pH values, the curves the theoretical prediction. The quality of fit for amorphous iron oxide (not shown) is very similar. Moreover pK(CdOH) is almost the same for hematite (-6.41) and amorphous iron oxide (-6.97).

In the same study it has been shown that electrostatic effects can explain nonstoichiometric proton–metal in exchange ratios. For the surface complexation a monodentate complex was assumed, yet the theoretical proton–metal ion exchange ratio ranged from 1.6 to 1.9 in accordance with the experimental observations. As stated before, Fokking et al. [173] showed in a later study that such ratios also might occur if the metal ion adsorption sites are entirely independent of the proton adsorption sites, provided the Stern plane and the surface plane are very close. These authors assumed that all metal ions adsorbed as Me^{2+} species and that hydroxo complexes could be neglected. However, this assumption is not crucial. In the case of adsorption of hydroxo complexes one proton is released by the formation of the complex, the remainder upon adsorption.

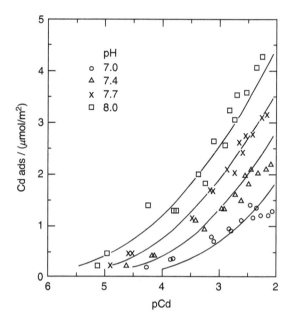

Figure 5.11 Cadmium adsorption on hematite at different pH values ($0.1 M KNO_3$; $20°C$). Symbols: experimental points; curves: pseudo homogeneous 1-pK SGC model with $p = -1/2$, log $K_H = 8.4$, $N_s = 8$ sites/nm^2, $C_1 = 1.72$ F/m^2, log $K_{CdOH} = 6.41$ (from W.H. van Riemsdijk, J.C.M. De Wit, L.K. Koopal, and G.H. Bolt, *J. Colloid Interf. Sci.*, 116, 511, 1987. With permission).

The second study of surface complexation closely related to the MUSIC model is concerned with bicarbonate adsorption on amorphous iron oxide [192]. Amorphous iron oxide can be regarded as water-rich iron oxide in which the coordination of the surface oxygens with iron is not precisely known. van Riemsdijk and Hiemstra assumed that the relevant protonation reaction for the description of the surface charge is that of the singly coordinated surface hydroxyl:

$$FeOH^{-1/2} + H^+ \rightleftharpoons FeOH_2^{1/2+}; \quad \log K_{2.1} = 10.7. \quad (65)$$

The adsorption reaction of the bicarbonate ion is also assumed to take place with the singly coordinated group only. The complexation reaction is described using the reduced valence principle of Pauling [82]:

$$FeOH_2^{1/2+} + HCO_3^{1-} \rightleftharpoons FeO_3^{1-} - COOH^{2/6-} + H_2O; \quad K(HCO_3^-).$$
(66)

The carbon center ($z = 4+$) distributes its positive charge over three surrounding ligands, attributing to each ligand on average a valence $v = + 4/3$. The Fe center attributes $+1/2$ to its surrounding ligands. This implies that the overall charge of the surface oxygen will be $(1/2) + (4/3) - 2 = -(1/6)$ and that of the oxygen plus the hydroxyl removed from the surface $-(2/6)$. For the location of both charge fractions the surface plane (net charge $-1/6$) and the Stern plane ($-2/6$) have been chosen. The separation of the charges of the surface complex over the surface layer and the Stern layer is essential for a satisfactory description with one additional constant only. The results could not be fitted on the basis of the reaction

$$FeOH_2^{1/2+} + HCO_3^{-1} \rightleftharpoons Fe^{1/2+} - HCO_3^{-1} + H_2O.$$
(67)

Based on Equations (65) and (66) and assuming a site density $N_s = 8$ sites/nm^2 and $C_1 = 2.23 \, F/m^2$ the adsorption data of CO_2 on amorphous iron oxide as a function of pH, measured by Zachara et al. [199], could be well described with log $K(HCO_3^-) = 4.4$. With this log $K(HCO_3^-)$ value the influence of CO_2 on the apparent PZC of goethites could also be explained.

   Equation (66) shows the charge distribution approach with complexation and it deserves some further attention. The use of the reduced valence principle of Pauling to describe oxyanions such as HCO_3^-, SO_4^-, HPO_4^{2-}, PO_4^{3-} etc. agrees with the MUSIC model and is less crude than the assumption that all charge is located in the center of the ion. Moreover, once an assumption is made with respect to the number of ion-surface ligands, the charge balance over the surface layer and the Stern layer follows from a simple count without the introduction of adjustable parameters. The fraction of charge that is brought to the surface plane reduces the surface charge, that is to say it gives complete screening, whereas the screening efficiency of the charge at the Stern plane depends on the

Stern layer capacitance. The overall result is a better screening than when the entire charge is placed at the Stern layer. This result is similar to the introduction of an inner Helmholtz plane, which also leads to an improved screening of the surface charge. However, in the latter case at least one extra adjustable parameter is required.

It is also possible to check the effect of the number of ion-surface ligands: more the ligands, better the screening. This is in fact illustrated by comparing the results based on Equations (66) and (67). Equation (66) represents a one-ligand binding with good screening and Equation (67) an outer-sphere complex with poor screening. Modeling based on Equation (67) could not explain the results.

An important aspect for the ligancy of an ion–surface complex is the stereochemistry of the surface compared to that of the ion. The more closely the molecular arrangement of the ion fits the stereochemistry of the surface oxygens, the more likely is a multidentate bond. For example, the HCO_3^- ions have a triangular structure with close packed oxygens. The iron oxides have an octahedral coordination and the distance between the surface oxygens is larger than in HCO_3^- due to the size of the Fe^{3+} ion. These differences reduce the probability of forming tri- or bidentate surface complexes. A review of the molecular structure of anion surface complexes is given by Parfitt [200], specific attention is paid to solid soil components.

So far the discussion has been restricted to anions. For cations a similar approach can be followed. Especially in the case of transition metal ions the coordinating solvent molecules become in effect coordinating ligands to the metal ion. For these ions the reaction with the surface groups can therefore also be seen as a ligand exchange. The number of ligands involved in the metal ion–solvent coordination can be estimated from the calculated primary coordination number, or from measured primary hydration numbers. According to Pauling [82] the primary coordination number or ligancy can be calculated on the basis of the cation/anion radii ratio. For most of the di- and trivalent ions a primary coordination with six or eight water molecules is found, except for the very small cations. Primary hydration numbers are difficult to measure and the numbers found for the metal ions range from about 6 to 14, depending on the method of investigation [201,202]. The high values are not so surprising because in an octahedral coordination eight or more water molecules are only some 15% further away from the central cation than the six primary coordination molecules. According to Burgess [201] the NMR peak area method measures only the hydration molecules in the first sheet and this method indicates a value of about six for most metal cations. This value agrees well with the primary coordination number. For surface complexation studies a ligancy of six for ordinary (alkaline earth) metal ions seems therefore appropriate.

By way of illustration surface complexation of Cd^{2+} with $FeOH^{1/2-}$ groups will be considered somewhat more closely. The complexation reactions for this system can be written as

$$FeOH^{1/2-} + H_2O - Cd - (OH_2)_5^{2+} \rightleftharpoons FeOH^{1/6-}$$
$$- Cd - (OH_2)_5^{10/6+} + H_2O \tag{68}$$

for a monodentate complex, or

$$2(FeOH^{1/2-}) + (H_2O)_2 - Cd - (OH_2)_4^{2+}$$
$$\rightleftharpoons (FeOH)_2^{2/6-} - Cd - (OH_2)_4^{8/6+} + 2H_2O \tag{69}$$

for a bidentate complex. These reactions will occur at relatively low pH. At high pH a subsequent deprotonation of the surface complex may occur. Taking the monodentate reaction as an example

$$FeOH^{1/6-} - Cd - (OH_2)_5^{10/6} \rightleftharpoons FeOH^{1/6-}$$
$$- Cd - (OH_2)_4(OH)^{4/6+} + H^+. \tag{70}$$

This reaction can be compared with the solution reaction

$$HOH^{2/6+} - Cd - (OH_2)_5^{10/6+} \rightleftharpoons HOH^{2/6+}$$
$$- Cd - (OH_2)_4(OH)^{4/6+} + H^+. \tag{71}$$

The (intrinsic) deprotonation constant of the surface complex will be considerably lower than that in the bulk solution; the negative $FeOH^{1/6-}$ group (Equation (70)) inhibits deprotonation whereas the positive $HOH^{2/6+}$ group (Equation (71)) promotes deprotonation. According to the acidity rules of Pauling and taking into account the charge difference only, the deprotonation constant of Equation (70) is probably about $10^{-2.5}$ times that of Equation (71). Similarly, deprotonation of the bidentate complex can be compared with deprotonation of $Cd(OH)^+$ in solution. Deprotonation of the bidentate complex will, however, only occur in strongly alkaline solutions. A further difference with deprotonation in bulk solution is that deprotonation of a surface complex is governed by the proton activity at the Stern plane and not by $\{H^+\}$. For a negatively charged surface, which promotes cation complexation, $\{H_d^+\} < \{H^+\}$ and deprotonation of the surface complex is inhibited as compared to the situation at the PZC. For super-equivalent complexation (charge reversal due to complexation) $\{H_d^+\} < \{H^+\}$ and deprotonation of the complex is favored.

From reactions (68) and (70) one may infer that adsorption of Cd^{2+} followed by deprotonation is essentially the same as adsorption of $Cd(OH)^+$

$$FeOH^{1/2-} + Cd(OH_2)_5OH^{1+} \rightleftharpoons FeOH^{1/6-}$$
$$- Cd - (OH_2)_4(OH)^{4/6+} + H_2O. \tag{72}$$

Hence, the surface complexation constants of Cd^{2+} and $Cd(OH)^+$ are related by the deprotonation constant of the surface complex.

In the present example the surface is treated as pseudo-homogeneous. In general this is a simplification of reality. From a physical point of view it is a drawback that one has to make such simplifications, but in practice they are often unavoidable. The advantage of the above way of modeling is that, apart from the complexation constants, a further specificity is introduced by giving each type of complex its own *screening ability* without introducing adjustable parameters. Venema et al. [193,194] have considered Cd adsorption on goethite by considering more than one surface group (see below).

In calculations with the reduced valence type of modeling one has to take into account that the charge of a surface complex is located partly at the surface plane and partly at the Stern plane. Binding of an ion X^z with n_x ligands to the surface groups brings a charge fraction $n_x v_x z_x$ to the surface plane (v_x being the reduced valence of each ligand) and a fraction $(1 - n_x v_x)z_x$ to the Stern plane. That is to say, the total specifically adsorbed charge of X, $\sigma_{1,x}$, has to be separated into two contributions: $n_x v_x \sigma_{x,1}$ for the surface plane and $(1 - n_x v_x)\sigma_{1,x}$ for the Stern plane. The total charge at the surface plane, σ_0, is thus

$$\sigma_0 = \sigma_s + \sum_x n_x v_x \sigma_{1,x}, \tag{73}$$

where σ_s is due to charge-determining ions and the summation covers the specific adsorption. The charge at the Stern plane becomes

$$\sigma_1 = \sum_x (1 - n_x v_x)\sigma_{1,x}. \tag{74}$$

In order to calculate ψ_s, Equation (20) can be used, however σ_s should be replaced by σ_0.

The equation for the calculation of the activity of X^z near the surface, $\{X_{sd}^z\}$, becomes

$$\{X_{sd}^z\} = \{X^z\} \exp\left\{ \frac{n_x v_x z_x F(\psi_s - \psi_d)}{RT} + \frac{z_x F \psi_d}{RT} \right\}, \tag{75}$$

where the subscript sd indicates that the charge of X is separated over the surface plane and the Stern plane. Up to the Stern plane all ions are treated as point charges and $z_x F \psi_d$ is the work required to bring the ions from the bulk solution to the Stern layer. From here a fraction $n_x v_x z_x$ of the charge is brought to the surface, this requires an extra amount of work $n_x v_x z_x F(\psi_s - \psi_d)$.

van Riemsdijk and coworkers have used metal ion [192–194,198] and oxyanion binding [170,190,194–198] on goethite to illustrate the CD-MUSIC model in detail. For studies in which not only a 1:1 electrolyte is present, but also a divalent cation or an oxyanion, it may be necessary to divide the Stern layer in an inner and an outer Helmholtz layer. In this case the capacitance of the entire Stern layer should be determined in the presence of a 1:1 electrolyte (surface sites form outer-sphere complexes with ions located at the Stern plane). With metal ions or oxyanions present inner-sphere complexes are formed and the locus of charge is the inner Helmholtz plane *and* the surface plane (charge distribution). The values of the outer and inner Helmholtz layer capacitances are both adjustable parameters with the condition that the two capacitors in series should have the same overall capacitance as the Stern layer (see Ref. [187]).

In accordance with MUSIC-2 in which the actual bond valence is introduced, the charge distribution over the surface plane and the inner Helmholtz plane should be fine-tuned by fitting. This has been done in the work of Venema [194] on cadmium and phosphate adsorption on goethite. The results for phosphate show that the charge distribution is close to what is expected on the basis of the simple Pauling bond valence. Hiemstra and van Riemsdijk [195] and Geelhoed et al. [170] have modeled phosphate adsorption on the basis that only the singly coordinated surface groups are active with a site density that is the weighted average of the two crystal faces present. Venema et al. [194] have used the same approach but the crystal faces are considered separately. The binding of the phosphate with the singly coordinated surface groups can be monodentate (including one proton to charge the reference site positive) or bidentate (with coadsorption of two or three protons).

The results for cadmium [193,194] show that the singly coordinated groups at the 110 face are most important. Cadmium may form mono- and bidentate complexes with these groups. The groups at the 021 edge may form a tridentate surface complex. At low cadmium loading the cadmium is mainly bound to the edges. At higher cadmium loading the 110 face becomes important. The monodentate binding at the 110 face turns out to be important with the cadmium adsorption in the presence of phosphate. Phosphate enhances the cadmium adsorption, because the adsorbed phosphate ions decrease the positive charge near the surface. In a recent study the interaction between phosphate and calcium has been investigated [198]. Also in this case it is possible to predict the adsorption behavior with the CD-MUSIC model with parameters derived from single-ion measurements. The fact that the model can explain the competition effects is highly relevant for the prediction of the adsorption behavior in soils.

Oxyanion adsorption to goethite has been studied in detail by Rietra et al. [97,190,196–198]. For the modeling of the results it has been assumed that two types of surface groups that have the same intrinsic proton affinity constant are active. Furthermore, microscopic information derived form spectroscopy is used in combination with the bond valence concept of Pauling to estimate the interfacial charge distribution of the adsorbed anion–surface site complexes. In Ref. [190] the adsorption of nine different anions has been studied with special attention to the proton–ion adsorption stoichiometry.

For the different ions studied Figure 5.12 gives a schematic representation of the surface coordination derived form spectroscopy and the allocation of the charge of the ion over the surface plane and the Stern plane according to the bond valence concept. It is shown that the structure of the adsorbed complex is a major factor ruling the pH dependence of the adsorption and that the overall proton–ion stoichiometry is almost exclusively determined by the interfacial charge distribution of the complex. It follows that the proton–ion adsorption stoichiometry is determined primarily by the electrostatic interaction of the ion with the surface site. A similar stoichiometry can only be explained by a similar ion–surface structure, and conversely, if the stoichiometry is different the adsorbed complexes will be different. Hence, when no spectroscopic data are available, the proton–ion stoichiometry can be used to estimate the charge distribution and to predict which complexes can be identified by spectroscopy.

In Ref. [196] sulfate adsorption on goethite is studied in detail over a large range of sulfate concentrations, pH, values

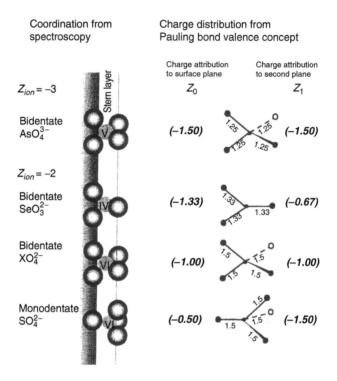

Figure 5.12   Schematic representation of the surface coordination derived form spectroscopy and the allocation of the charge of the ion over the surface plane and the Stern plane according to the bond valence concept for different oxyanions adsorbed to goethite (from R.P.J.J. Rietra, T. Hiemstra, and W.H. van Riemsdijk, *Geochim. Cosmochim. Acta.*, 63, 3009, 1999. With permission).

and electrolyte concentrations. All the data can be modeled with one adsorbed species if it is assumed that the charge of adsorbed sulfate is spatially distributed as predicted by the proton–ion adsorption stoichiometry over the surface plane and the Stern plane. This result also compares well with spectroscopic data. The exact charge distribution is used as adjustable parameter (actual bond valence) and differs slightly from the ideal Pauling charge distribution. It is also shown that modeling based on the 2-pK TL model leads to less good results. In Ref. [197] selenate adsorption is studied and compared with sulfate adsorption. The adsorption behavior of both ions is very similar. Spectroscopic results for selenate suggest outer-sphere complexes at high pH and a monodentate inner-sphere complex at low pH. Selenate adsorption can be modeled using the surface complexes found from spectroscopic data.

3. Surface Complexation and the 2-pK Models

For the description of multivalent ion adsorption on metal oxides with the 2-pK models the same *schools* exist as for proton adsorption. Schindler and Stumm and coworkers [111–113,148,157,161,204] prefer to use the 2-pK constant capacitance model for both cation and anion adsorption. With metal ion adsorption the proton–metal ion exchange ratio is explained by a combination of metal–surface reactions, including different metal species and mono- and bidentate surface complexes. A clear review of this work has been given by Schindler [148]. Golberg and Sposito [205,206], Golberg and Glaubig [207,208], and Sposito et al. [209] also use the constant capacitance model to describe adsorption of various anions on mineral surfaces, including clay minerals. Lützenkirchen [114] has discussed the numerical parameter optimization for the 2-pK constant capacitance model.

Leckie [123] and coworkers model cation [75,210] and anion [211,212] adsorption on iron oxides with the 2-pK TL model. Davis and Leckie [75,211] place the specifically adsorbing ions at the inner Helmholtz plane, whereas Hayes and Leckie [210] and Hayes et al. [212] make a distinction between strongly and weakly adsorbing ions. The strongly adsorbing ions are assumed to form inner-sphere complexes with their charge center located at the surface plane, the weakly adsorbing ions form outer-sphere complexes with their charge center at the inner Helmholtz plane. The distinction between weakly and strongly adsorbing ions is based on the ionic strength dependency of the adsorption. The adsorption of strongly adsorbing ions (Cd^{2+}, Pb^{2+}, Ni^{2+}, SeO_3^{2-}) shows a weak dependence on ionic strength, whereas that of weakly adsorbing ions (Mg^{2+}, Ca^{2+}, Ba^{2+}, SeO_4^{2-}) shows a strong dependence.

Katz and Hayes [102,213] use cobalt adsorption data to illustrate how the 2-pK TL model can be used at moderate and high surface coverage. Specific attention is paid to optimal fitting of the model parameters. In relation to parameter optimization of the 2-pK-TL model Lützenkirchen et al. [100] discuss specific constraints that can be used to obtain the binding constants of monovalent ions.

An application of the 2-pK model that deserves special attention is the work of Finch and coworkers [214] on the modeling of proton and ion adsorption to sulfide minerals. Instead of the classical 2-pK-TL model with a fixed outer layer capacitance they adjust the outer layer capacitance. Illustrations are presented for sphalerite in NaCl plus some ferrous ions. The presence of ferrous ions may activate the flotation of sphalerite. The model is also applied to boehmite in KNO_3.

Huang et al. [117] use the 2-pK GC model to describe heavy metal ion adsorption on various hydrous oxides and activated carbons. To obtain the protonation constants a graphical method is used. Several species of the same multivalent ion are allowed to adsorb and for divalent species mono- and bidentate complexes may form.

Dzombak and Morel [118] also use the 2-pK GC model to describe ion adsorption onto amorphous iron oxides. They apply a numerical parameter optimization. Their motivation for the use of the 2-pK GC model to describe the protonation reactions is its simplicity. Also for the adsorption of multivalent ions the simplest options for complexation have been chosen to describe the results. The authors make a distinction between strong and weak adsorption sites and consider in some cases also surface precipitation (see also [215]). Apart from the modeling results a compilation and critical evaluation is given of all ion adsorption studies on amorphous iron oxide. Due to its simplicity and the presence of the database this work is often cited and followed by others.

4.   Recent Overviews

Inspection of recent articles, reviews, or book chapters shows that the MUSIC concept is still (partially) unknown, or its implications are insufficiently realized or not accepted. In 1996, a book on inorganic sorbents appeared that contained various contributions discussing the metal oxide–solution interface. The chapter written by Koopal [38] gives a similar view as expressed in Section III.D.1 and III.D.2, and specific attention is given to the fundamental difference between silica and gibbsite type surfaces. A similar view is expressed in a brief review in Ref. [64]. In the chapter by Rudzinski et al. [124] the 2-pK model is used in combination with a generic surface heterogeneity to give an alternative interpretation of the difference between PZC and IEP. The authors cite the MUSIC approach but see no harm in expanding the 2-pK model. More recently Rudzinski et al. [125] discuss adsorption at heterogeneous solids in general and pay specific attention to the metal oxide interface in Section II. In this case also heats of adsorption are considered as additional source of information. Tertykh and Yanishpolskii [216] discuss ion adsorption on silica as complex formation without reference to the classical articles in this field. Kallay et al. [217] devote their chapter to a description and interrelation between existing theoretical models (mainly 2-pK and generalized 1-pK models) considering data on surface charge densities, electrokinetics, PZC and IEP and calorimetry.

Kallay et al. [218] have edited a book in which Kallay contributes a chapter on the enthalpy of surface charging and on the interpretation of interfacial equilibriums on the basis of adsorption and electrokinetic data [219]. Experimental results can be explained with both the 2-pK and the (generalized) 1-pK model. In general it may be concluded from Kallay's contribution that fitting the models to experimental result does not allow a firm conclusion regarding preference for a given model. The main chapter on charged solid–liquid interfaces in Kallay's book is written by Kosmulski et al. [126]. First these authors discuss the various models (including 1-pK, 2-pK, and MUSIC) by comparing the experimental charge density of hematite with results obtained with the different models. They do not give a ranking of the models, but in Sections IV and V of their chapter the 2-pK model is almost exclusively used. In this overview the authors also present the graphical way to obtain the intrinsic affinities. Koopal et al. [37] have shown that this method implicitly assumes that ΔpK_H is large. Obtained values for ΔpK_H are however not as large as those predicted by MUSIC, therefore the resulting values have no physical meaning.

Contescu and Schwartz [220] use the general concepts of acidity and basicity to discuss the origin of the acid–base properties of solid surfaces covering ideally polarizable surfaces, oxides, clays, zeolites, activated carbon, and pseudo-liquid materials. In the section on oxides the 2-pK, 1-pK, and MUSIC type of modeling is discussed. The protonation constants predicted with MUSIC for several oxides compare well with the apparent proton binding constants derived by converting potentiometric proton titrations in proton affinity distributions.

In 2001, a book with the title *Oxide Surfaces* has appeared. In this book the equilibrium adsorption at the solution–metal oxide interface is treated by Wingrave [127], who is also the editor of the book. In this treatment only a 2-pK model type of approach is presented together with its extensions to surface complexation with both anions and cations, hence, little news in this respect. The interesting part is that Wingrave also provides a discussion on the ionic activity coefficients near the interface [30] (see Section II.C). No detailed comparison is made between the mean-field PB model (in the form of the SGC model, for example) and Wingrave's new model.

Larson and Attard [29] have recently discussed the surface charging of silver iodide and several mineral oxides. They use the 2-pK model in combination with a double layer model derived from a combination of a HNC model for the diffuse layer in combination with a Stern layer (SHNC model; see also Section II.D). The authors seem unaware of the 1-pK and MUSIC models. Specific attention is given to the issue of the Nernstian behavior of the surface potential. The calculations of the 2-pK SHCN model are compared with experimental results and with results based on the GC and SGC models in which the surface potential is calculated with the Nernst equation. The general conclusion derived from this work is that around the PZC the surface potential of AgI and the oxides is pseudo-Nernstian even for silica. We have to disagree with the conclusion for silica and also for AgI some critical comments can be made.

The scepsis with respect to the conclusion for silica has two reasons: (1) The experimental results of silica on which the conclusion is based [221] show a PZC at about pH = 3; at this low pH the potentiometric titration technique is not reliable. Therefore, the points in the pH range from 3.5 to 2 are highly inaccurate and the sudden change to positive charges is an artifact. (2) The 2-pK model with a ΔpK value less than 10–14 units is a very poor model for this surface [38,59,64] and it leads to the erroneous conclusion that the surface potential of silica is pseudo-Nernstian near its PZC. For a more appropriate analysis of silica Refs. [38,59,64] should be consulted. For the other mineral surfaces the same conclusion is reached in Section III.D.3 using the generalized 1-pK SGC model. It should be noted that with the 1-pK SGC model pseudo-Nernstian behavior is more pronounced than with the 2-pK SGC model, because the first model is based on two surface species and the second model on three and this has implications for the (near-)constancy of the surface configurational entropy.

With respect to AgI Larson and Attard have assumed that the point of zero surface potential, corresponds with the PZC of AgI. However, this is not the case. Studies with adsorption of uncharged molecules have shown a shift in the PZC of AgI of as much as 190 mV and this is explained with the presence of oriented water dipoles [3,222]. A more appropriate model for the charging of AgI (taking into account the dipole potential near the surface) has been presented by de Keizer and Lyklema [223].

Another recent treatment of mineral surfaces is that by Gunnarsson et al. [31,32] (see also Section II.C). In this work the generalized 1-pK model is combined with a new double layer model. The new double layer model is based on the generalized van der Waals theory, which allows incorporation of the volume of the ions and ion–ion correlations. For bulk solution this leads to new expressions for the ion activity (corrected DH) [31]. For the surface the CDH model is combined with the presence of a Stern layer (1-pK SCDH model) [32]. Calculated results are compared with experimental results on goethite in $NaClO_4$ and with calculations based on the 1-pK SGC model. The 1-pK SCDH model gives good results for low surfaces charges or high ionic strength. The 1-pK SGC model gives the best representation of the experimental results over the entire salt concentration and surface charge range.

V. ADSORPTION OF ORGANIC IONS

A. Introduction

Adsorption of organic ions will in general change the interfacial region more strongly than the adsorption of inorganic ions because organic ions are amphiphilic in nature. They are composed of a strongly hydrophilic ionic part and an organic part that is apolar or weakly polar. As a consequence of the adsorption both the overall surface charge and the wettability may change. This makes adsorption of organic electrolytes particularly important in relation to colloidal stability. Their effect on the colloidal stability depends on the mode of adsorption, which may be a function of the concentration and the structure of the organic ion.

The nonpolar part of the molecule contributes to the adsorption by, for instance, π-electron interaction or hydrophobic attraction. In the latter case unfavorable contacts between water molecules and the apolar part of the organic molecule exchange for favorable apolar–apolar contacts. The decrease in unfavorable water–apolar contacts can be achieved by both surface–adsorbate contacts and specific lateral adsorbate–adsorbate contacts. These noncoulombic lateral interactions that may occur with organic ions are a complication that is not encountered with inorganic ions.

When particle and organic ion are oppositely charged, the ionic group may adsorb through electrostatic (coulombic) and specific (noncoulombic) attractions, similarly to an inorganic ion. However, as soon as sufficient material is adsorbed the specific lateral attraction may start to play an important role and this may even lead to strong super-equivalent adsorption. This is for instance the case with surfactant adsorption. A further complication is that the ionic group is not necessarily fully dissociated so that some of the molecules may adsorb as polar but uncharged species.

Other complications that are generally neglected with inorganic ions but that may be important for organic ions are the effects of size, shape, and flexibility of the molecules. In order to accommodate an organic ion on the surface, it may be necessary to replace several solvent molecules instead of only one, as has been assumed with inorganic ions. The shape and flexibility of organic molecules may range from compact and rigid to short flexible chains. Compact and rigid molecules may be restricted in their orientation if they become adsorbed, chain type molecules may adapt their conformation. Mostly these complications are neglected in the descriptions of the adsorption.

Another often-used simplification is that it is assumed that the sites for adsorption of the organic ions are independent of that of the protons and the other inorganic ions, i.e., there is no site competition. This does not imply that the surface charge has to be independent of the adsorption of the organic component. As shown by Fokkink et al. [151] electrostatic interactions may lead to a proton–ion exchange ratio or surface charge adjustment without any site competition. In some studies lateral interactions are neglected but site competition is taken into account. This is often done for adsorption of organic ions on mineral oxides.

A further difference with the inorganic ions is that the adsorption of organic ions will change the electrostatic capacitance of the Stern layer due to a decrease in the relative permittivity or to an outward displacement of the Stern plane. Both will lead to a lowering of the Stern layer capacitance and this may lead to a lesser screening of the surface charge so that charging of the surface may become more difficult. This effect will not be treated in further detail, the general impact of the effect of the Stern layer capacitance on the surface charge can be obtained from the equations given in Section II.A and B.

In the present section emphasis will be given to the effect of the lateral interaction on the form of the adsorption isotherm equation and initially site competition will be neglected. Moreover, the structure of the organic ions will be neglected. This means that the equations that are presented are only suited to describe simple organic ions and that the models are only a primitive approximation for the adsorption surfactants. In Section V.D. some attention will be given to adsorption of (multivalent) organic ions to mineral surfaces and site competition.

B. Strong Organic Electrolytes

The easiest way to incorporate lateral interactions in the adsorption model is by assuming regular behavior (a mean-field approximation) in both the solution and the monolayer. Hildebrand et al. [224] introduced the regular mixture as one in which (1) the components mix without volume change and (2) the entropy of mixing is given by that corresponding to an ideal mixture. This implies that the molecules should be similar in size and that the arrangement of the molecules is completely random. As a consequence of the random mixing assumption, the number of nearest neighbor contacts between a molecule i and j is simply proportional to the product of their mole fractions. Taking into account nearest neighbor contacts only and restricting the adsorption to a monolayer, the adsorption equilibrium for a binary system becomes [225–228]:

$$\frac{\theta_1}{1-\theta_1} = \frac{x_1}{1-x_1} K_{12} \exp 2\chi_{12}(\langle\theta_1\rangle - x_1), \qquad (76)$$

where θ_1 is the fraction of the organic component in the monolayer, $<\theta_1>$ a weighted fraction, x_1 the mole fraction 1 in solution, K_{12} the adsorption constant, which is a measure of the preference of the surface for the organic component (1) with respect to the solvent 2, and χ_{ij} the so-called Flory–Huggins interaction parameter. The expression for χ_{ij} is

$$x_{ij} = \frac{N_A Z}{RT} \{\varepsilon_{ij} - 0.5(\varepsilon_{ii} + \varepsilon_{ij})\} \qquad (77)$$

with N_A is the Avogrado's number, Z is the number of nearest neighbors to a central molecule and ε_{ii}, ε_{jj}, and ε_{ij} the pairwise interaction potentials, which should be seen as (local) Gibbs energies rather than as enthalpies (see Flory [229]). Note that according to its definition $\chi_{ii} = \chi_{jj} = 0$. When $\chi_{ij} = 0$ for all i and j, the mixture behaves ideally or perfect [230]. Since ε_{ij}, ε_{ii}, and ε_{jj} are in general of negative sign, positive values of χ_{ij} result if like contacts are preferred over unlike contacts and negative values of χ_{ij} indicate that unlike contacts are favored.

The site fraction $<\theta_1>$ in the monolayer accounts for the fact that adsorbed molecules are in contact with neighbors in the monolayer, in the solid surface and in the bulk solution. The mole fractions of the organic molecule in the solid phase $(=0)$ and in the bulk solution $(=x_1)$ are different from that in the monolayer $(=\theta_1)$. The fraction $<\theta_1>$ is therefore defined as

$$< \theta >= \lambda_1 0 + \lambda_0 \theta_1 + \lambda_1 x_1, \qquad (78)$$

where λ_0 is the fraction of nearest neighbors in the monolayer and λ_1 that with the bulk solution or the surface, so that $\lambda_0 + 2\lambda_1 = 1$.

It can be shown [231] that in a regular binary mixture, an incipient instability occurs for $\chi_{12} = 2$ at $x_1 = 0.5$. For $\chi_{12} > 2$ regular mixtures are unstable and at sufficiently high mole fractions a phase separation in two stable phases occurs in the bulk solution. For this reason Equation (76) can only be used up to the mole fraction for which phase separation occurs. In the monolayer incipient instability occurs for $\chi_{12} = 4$ at $\theta_1 = 0.5$ and for $\chi_{12} > 4$ a two-dimensional 2D phase separation or *2D condensation* occurs. At the condensation step a gaseous monolayer is in equilibrium with a liquid like monolayer.

For adsorption from dilute solution (i.e., $K_{12} \gg 1$ and $\theta_1 \gg x_1$), a situation frequently occurring in practice, Equation (76) can be simplified to the Frumkin–Fowler–Guggenheim (or FFG) [176,232] equation:

$$\frac{\theta_1}{1 - \theta_1} = c_1 K_1 \exp(\chi \theta_1), \qquad (79)$$

where $K_1 = K_{12}/55.5$ (l/mol) and $\chi = 2\lambda_0\chi_{12} \approx \chi_{12}$. For adsorption of organic molecules from dilute aqueous solutions Equation (79) should be seen as an improved version of the Langmuir equation. The latter equation results for $\chi = 0$. Due to the presence of the lateral interaction term Equation (79) is a much more versatile equation than the Langmuir equation.

A discussion of the regular behavior model in relation to its application to experimental results has been given by, for instance, Prausnitz et al. [231]. They indicated that Equations (76) and (79) can also be used for spherical or cubic organic molecules that are larger than the solvent molecules. For chain-type molecules, such as surfactants, the two equations are approximations that do not take into account the possible changes in chain conformations and the use of volume fractions (rather than mole fractions) is appropriate. In practice however, the FFG equation (79) is also frequently used to get (qualitative and quantitative) insight into surfactant adsorption.

A set of typical adsorption isotherms according to Equation (79) is shown in Figure 5.13 for different values of χ. Adsorption due to association of molecules in the monolayer increases with increasing χ and for high values of χ surface condensation occurs. With respect to the values of χ, it should be noted that χ considerably larger than 2 will occur only for solute molecules that are larger than the solvent molecules. For negative values of χ repulsion occurs and the adsorption isotherm remains lower than the Langmuir ($\chi = 0$) isotherm.

So far uncharged organic solutes have been considered. However, Equation (79) can easily be adapted to describe the adsorption of organic ions by replacing the concentration of organic solute in bulk solution by $c_{a,1}$ [48]:

$$c_{a,1} = c_1 \exp(-\nu_1 F\psi_a/RT). \tag{80}$$

In Equation (80) ψ_a is the potential at the center of charge of the adsorbed organic ion and ν_1 the valence (sign included) of the ion. The magnitude of ψ_a depends on (1) the charge due to the adsorbed ion itself, (2) the surface charge of the adsorbent, (3) the background electrolyte concentration, and (4) the position of the center of charge of the adsorbed ions relative to that of the surface. Substitution of Equation (80) into Equation (79) gives

$$\frac{\theta_1}{1 - \theta_1} = K_1 c_1 \exp\left(\chi\theta_1 - \frac{\nu_1 F \psi_a}{RT}\right) \tag{81}$$

Equation (81) can be seen as a simple equation for the adsorption of organic ions from dilute solutions. The sites for adsorption of the organic ion are assumed to be entirely independent of those for the charge-determining ions. For $\chi = 0$, Equation (81) reduces to the Stern equation (23) and when also $\theta_1 \ll 1$ the so-called Stern–Graham equation results. These equations are frequently used in the older literature for surfactant adsorption [e.g., 233–235].

For small or flat adsorbing organic ions ψ_a can be approximated reasonably well by the Stern layer potential ψ_d or by the electrokinetic or ζ potential. The ζ potentials of the adsorbent particles can be measured as a function of the adsorption or the concentration of the organic ion. Once the ζ potential is known $c_{a,1}$ can be calculated. To check whether or not the substitution of ζ for ψ_a is correct, adsorption isotherms of the organic ion measured at different salt concentrations can be replotted as a function of $c_{a,1}$. The different isotherms will merge into one master curve if the substitution is valid. The master curve thus obtained can be compared with isotherms such as those shown in Figure 5.13 to establish K_1 and χ. In practice, adsorption measurements in combination with ζ potential measurements have mostly been used for a semiquantitative interpretation of the adsorption behavior of organic ions on the basis of Equation (81) with $\psi_a = \zeta$ (see, for example, Ref. [235]). Quantitative use of Equation (81) after substitution of ζ for ψ_a is rare. This may be due to the fact that it is difficult to obtain accurate values of ζ.

If no experimental approximation of ψ_a is available Equation (81) can be worked out by using a simple double layer representation such as the Gouy–Chapman model (Equation (13) with $\sigma_d = -(\sigma_s + \sigma_1)$) or the Debye–Hückel (Equation (14)) model. Examples of this type of modeling have been worked out for the adsorption of *p*-nitrophenol on carbons

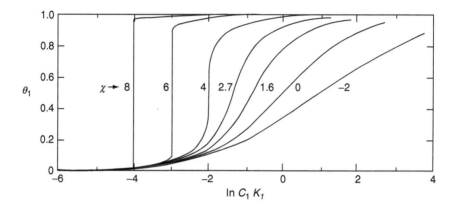

Figure 5.13   Adsorption isotherms according to the FFG equation (79), for a series of χ values (indicated with the curves).

[236,237] and for *tetra*-alkylammonium and bipyridinium ions on AgI [238].

Koopal and Keltjens [236] have used the Debye–Hückel theory to work out Equation (81) in somewhat more detail. The advantage of the Debye–Hückel approximation is that charge and potential are directly proportional, with a proportionality constant related to the square root of the ionic strength (see Equations (14) and (8)). As a consequence the resulting organic ion adsorption equation is relatively simple, but it illustrates the basic principles very well. By further assuming that (1) the organic ions adsorb at the Stern plane and (2) the surface charge of the adsorbent adapts to the adsorption of the organic ion in a linear manner, Equation (81) can be rewritten as:

$$\frac{\theta_1}{1 - \theta_1} = c_1 K_1 \exp\left(-\frac{\nu_1 F \sigma_s^u}{\varepsilon_0 \varepsilon_w \kappa RT}\right)$$
$$\exp\left[\left(\chi - \frac{(\sigma_s^c - \sigma_s^u + \sigma_{1,\max})\nu_1 F}{\varepsilon_0 \varepsilon_w \kappa RT}\right)\theta_1\right], \tag{82}$$

where σ_s^u and σ_s^c are surface (proton) charge densities sign included for $\theta_1 = 0$ (u) and $\theta_1 = 1$ (c), respectively, and

$\sigma_{1,\max} = \nu_1 F \Gamma_{1,\max}$ where $\Gamma_{1,\max}$ is the maximum adsorption density of the organic component. For variable charge surfaces σ_s^u and σ_s^c depend on both the concentration of charge-determining ions (mostly pH) and the ionic strength. However, at constant pH and ionic strength σ_s^u, σ_s^c, and κ are all constants and Equation (82) simplifies to a special form of the FFG equation (see Figure 5.13). For constant charge surfaces, i.e., the surface charge density is independent of the adsorption of the organic ion, Equation (82) can be simplified using $\sigma_s = \sigma_s^u = \sigma_s^c$ independent of θ_1 and the ionic strength.

A very similar treatment based on a linear dependence of ψ_a on θ_1 leads to an equation expressing θ_1 as a function of ψ_a^u and ψ_a^c [236, 238]. The relation obtained is slightly more general than Equation (82), because it is not limited by the Debye–Hückel approximation. In the Debye–Hückel approximation both treatments are equivalent [236].

Although Equation (82) is based on the Debye–Hückel approximation ($\psi_a < 50\,\mathrm{mV}$), it clearly shows how intrinsic and electrostatic interactions affect the adsorption. Hydrophobic and other intrinsic interactions between surface and adsorbate and laterally between the adsorbate molecules, are incorporated through K_1 and χ, respectively, the surface charge (pH, c_s) effects are taken into account by σ_s^u and σ_s^c and κ incorporates the effect of the (square root of) the ionic strength. Equation (82) also shows that the electrostatic interaction can be separated into two components: one depending on the initial state of the surface and the valence of the organic ion (electrostatic surface–adsorbate interaction), the other depending on the extent of adsorption of the organic component (lateral electrostatic interaction). The electrostatic surface–adsorbate interaction can be either attractive (ν_1 and σ_s have opposite signs) or repulsive (similar signs for ν_1 and σ_s). The lateral electrostatic interaction is always repulsive.

An interesting aspect of Equation (82) is that it shows that for $\chi > 0$ a Langmuir type isotherm may still result if the lateral chemical attraction is just balanced by the lateral electrostatic repulsion. This balance can be attained by adjusting the salt concentration and the pH ($\sigma_s^c - \sigma_s^u$ will in general depend on both).

In the case that surface and adsorbate are oppositely charged, an increase in salt concentration results according to Equation (82) in a decreasing attraction with the surface and a decrease in the lateral repulsion. In other words, at high salt concentrations the adsorption starts at higher concentrations, but the slope of the isotherm is steeper than at low-salt concentrations. For strongly hydrophobic ions (large molecules with a high χ) isotherms measured at different salt concentrations may intersect. This phenomenon has been studied by De Keizer et al. [145]. Figure 5.14 shows results obtained for tetrabutylammonium ions adsorbed on silver iodide in the presence of KNO_3. The adsorption isotherms pass through a common intersection point or CIP (see Figure 5.14a). At $10^{-3}\,M$ KNO_3 this CIP corresponds with the IEP of the particle–adsorbate system. At $10^{-1}\,M$ KNO_3 the IEP and CIP are slightly different due to specific adsorption of K^+ (see Figure 5.14b). The explanation for the observed behavior is that the CIP corresponds with compensation of the surface charge by the adsorbate, i.e., with ψ_a equal to zero. If it is assumed that $\psi_a = 0$ corresponds with $\zeta = 0$, the CIP corresponds with the IEP. The position of the CIP is a function of the surface charge: the higher the surface charge, the more organic ions have to be adsorbed before the charge compensation point (CIP) is reached. For constant charge surfaces the CIP can thus be used to calculate the surface charge. For variably charged surfaces the CIP indicates the actual primary surface charge in the presence of the adsorbed organic ions. By comparing the surface charge of the bare surface at the given pH and salt concentration with the surface charge derived from the CIP an estimate can be obtained of the surface charge adjustment due to the adsorption of the organic ion. The surface charge adjustment due to adsorption of the organic ion can also be found from the shift in the equilibrium pH if the charge of the organic ion is independent of the pH and the proton is the charge-determining ion.

CIPs have also been observed with isotherms of ionic surfactants measured at different salt concentrations and constant pH [239,240].

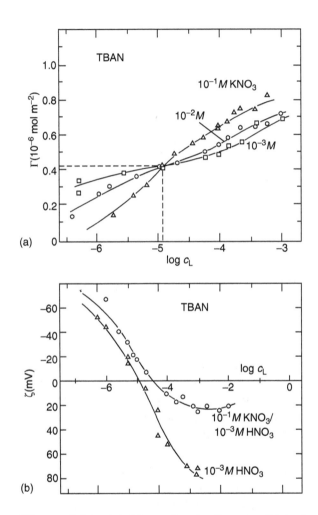

Figure 5.14 (a) Adsorption isotherms of tetrabutylammonium nitrate (TBAN) on silver iodide at different concentrations of KNO_3 and pAg $\approx$11.5. (b) Zeta-potential of AgI as a function of the TBAN concentration for two KNO_3/HNO_3 concentrations at pAg = 11.6 (from A. De Keizer, M.R. Böhmer, T. Mehrian, and L.K. Koopal, *Colloids Surf.*, 51, 339, 1990. With permission).

When the maximum adsorption is known or when it can be estimated using geometrical considerations, it is instructive to investigate the adsorption of organic ions in more detail by plotting the results as

$$\ln \frac{\theta_1}{1 - \theta_1} \frac{1}{c_1} \text{ versus } \theta_1.$$

The intercept of the curve at $\theta_1 = 0$ is an indication of the overall affinity of the organic ion for the surface. The slope of the curve indicates the magnitude of the overall lateral interaction. A positive slope indicates attraction, a negative slope repulsion. If the plot is linear the FFG equation can be applied. If the slope equals zero the lateral attraction and repulsion compensate each other and the Langmuir equation can be used. If a curved line results the FFG equation cannot be used, curvature is a strong indication that the Debye–Hückel approximation for the relationship between charge and potential should be replaced by, for example, the GC approximation.

C. Weak Organic Electrolytes

In practice many organic ions are not fully ionized and the adsorption behavior of weak organic ions is more complicated than that of strong organic electrolytes. The complication is that two species of the organic component are present: the undissociated and the dissociated organic compound. Now there are several options. (a) Only the charged component adsorbs. This may be the case when the charged group forms an inner-sphere complex with a mineral surface site, or for a dissociated aromatic organic component adsorbing at an aromatic surface (π-electron donor–acceptor interaction). (b) Only the uncharged molecules adsorb. This may be the case when the molecules interact with their apolar part with the surface and the charged species are repelled from the surface region (due to the surface charge). (c) Both the charged and the uncharged components adsorb. In situation (a), the model described in Section V.B. can be used. However, for mineral surfaces it seems realistic to consider explicitly proton adsorption and site competition (see Section V.D.) Situation (b) is a limiting case of situation (c) for highly charged surfaces with a charge sign opposite to that of the organic ion. The simplest way to describe situation (c) is to assume that (1) the chemical interactions (both with the surface and laterally) and (2) the maximum adsorption, $\Gamma_{1,\text{max}}$, are equal for both species. In other words, the difference between the two species is due to the electrostatic interactions. The summation of the adsorptions of the uncharged and the charged component then leads to the following equation:

$$\frac{\theta_1}{1-\theta_1} = c_1(1-\alpha_1)K_1\exp(\chi\theta_1) + c_1\alpha_1K_1\exp\left(\chi\theta_1 - \frac{\nu_1 F\psi_a}{RT}\right),$$

(83)

where α_1 is the degree of ionization in the bulk solution. The other symbols have the same meaning as in Equation (81). Equation (83) can be used in combination with, e.g., the Debye–Hückel or the Gouy–Chapman model. If ν_1 and ψ_a have the same sign and when the degree of dissociation is small, it is a reasonable approximation to neglect the second contribution on the RHS. This approximation becomes better for larger absolute values of ψ_a. This leads to situation (b). Equation (83) has been used to describe organic acid adsorption to carbon [241] and graphite [242] surfaces. The work of Haderlein et al. [243,244] on the adsorption of nitrophenols on clays shows that both the degree of dissociation and the counterion of the clay are important. However, the authors do not present a quantitative evaluation of the charge effects.

In principle it is of course also possible to consider the adsorption of two organic ions that compete for the same surface sites. This situation is, for instance, important with adsorption on activated carbons. However, these surfaces are heterogeneous and heterogeneity is certainly as important as a detailed description of lateral and electrostatic interactions. In the section on surface heterogeneity (VI) adsorption competition will be briefly discussed.

D. Adsorption of Organic Ions and Site Competition

This situation is considered in literature for organic ion adsorption to mineral surfaces. The description of the adsorption is analogous to the description of the complexation of protons and mono- or multivalent ions to mineral surfaces. The situation is fully equivalent if both the intrinsic or specific lateral interaction in the adsorbed layer and the effect of the adsorption on the Stern layer capacitance are neglected, as is generally done. In the case of outer-sphere complexation the center of charge of the organic ion is located at the Stern plane. For inner-sphere complexation the charge can be placed on either the inner Helmholtz layer or the charge distribution model can be used with part of the charge at the surface plane and part at the Stern plane.

The situation becomes more complicated when the organic ions have more than one functional group. Several studies have appeared in literature (e.g., [245–258]) in which the adsorption of organic acids with two and more carboxylic acid groups to mineral surfaces are studied. In this situation multiple possibilities are open for surface complexation. The first systematic studies are those of Stumm et al. [245–247]. In order to simplify the situation Bowers and Huang [248,249] and Kallay and Matijevic [250] both show that for given conditions a kind of pseudo-Langmuir equation can describe the results. The affinity becomes a coefficient that depends on pH and sometimes the maximum amount adsorbed. A somewhat similar strategy has been followed by Sigg and coworkers [251,252] to determine the number of sites involved in the reaction and the affinity coefficient at a given pH. Due to the large number of possible modes of complexation the modeling becomes increasingly complicated. A possible option is to take only the most prominent surface species into account to describe the adsorption. This has been done by, for example, Dzombak and coworkers [253–255] and by Boily et al. [256]. Filius et al. [257,258] argue that this is a feasible option as long as a relatively small number of components are involved. However, for adsorption in a multicomponent mixture these authors recommend a full speciation treatment. To reduce the possible number of surface species of the organic component spectroscopic data are used. In a first paper the CD-MUSIC approach is followed [257], in the second paper [258] a heterogeneous complexation model is used that calculates the mean mode of the adsorbed ion, which is defined by an overall affinity, charge distribution, and reaction stoichiometry. This new way of modeling is applied to the data of Boily et al. [256] on benzenecarboxylic acids adsorbed on goethite. Most of the more recent work on these well-defined organic acids is done to get a better feeling for the way in which the adsorption of natural organic matter (fulvic or humic acids) can be described.

VI. SURFACE HETEROGENEITY

A. Introduction

In addition to the description of ion adsorption on homogeneous interfaces it is necessary to describe the effect of surface heterogeneities, because most surfaces will be heterogeneous. Heterogeneity may be due to the presence of different crystal planes, different surface groups on one crystal plane, plane imperfections, surface impurities, etc. Surface heterogeneity has already been mentioned with regard to adsorption on metal hydroxide or oxide surfaces. With respect to modeling adsorption on heterogeneous surfaces one must strive towards a satisfactory, but not excessive, degree of detail otherwise too many adjustable parameters become involved.

The basis concepts concerning the adsorption on heterogeneous surfaces were originally developed with reference to gas adsorption. Langmuir [259] was one of the first who addressed the problem. Since then a considerable amount of work on heterogeneity has appeared. Monographs on the subject have been presented by Jaroniec and Madey [260] and Rudzinski and Everett [261]. General reviews are those by Jaroniec [262], Jaroniec and Bräuer [263], and House [264]. In the present section, emphasis is given to ion adsorption from aqueous electrolyte solution, a topic that has received attention only relatively recently (e.g., [38,54,55,64,67,77,78,80, 124,125,265–268]).

The present section consists of three parts. The first part is concerned with general aspects of the extension of the existing models for monocomponent ion adsorption on homogeneous surfaces to heterogeneous surfaces. The second part is concerned with multicomponent ion adsorption, i.e., surface charging and complexation. The third part discusses the possibilities of deriving the actual heterogeneity from experimental adsorption isotherms.

B. Ion Adsorption in a Monolayer

1. Generalized Adsorption Equation

For ion adsorption on homogeneous surfaces there is only one site type S_j and the adsorption of an ion 1, from a 1:1 background electrolyte solution, into a monolayer adjacent to the surface can be described with an equation equivalent to Equation (30):

$$\theta_{j,1}^* = \frac{K_{j,1} c_{1,\delta}^*}{1 + K_{j,1} c_{1,\delta}^*},\tag{84}$$

where $\theta_{j,1}^*$ is the fraction of sites S_j covered by 1, $K_{j,1}$ is the intrinsic affinity constant of ion 1 for site type j, $c_{1,\delta}^*$ is the 'effective' concentration of ion 1 at the plane of adsorption, δ indicates the plane of adsorption. The asterisk is added to indicate that $c_{1,\delta}^*$ is an 'effective' concentration. Expressions for $c_{1,\delta}^*$ (and $\theta_{j,1}^*$) depend on the type of heterogeneity and specific lateral interactions, as will be shown below. For instance, Equation (84) applies to adsorption of organic ions if the specific lateral interaction is included in the effective concentration $c_{1,\delta}^*$.

For proton adsorption on metal oxides Equation (84) applies when the generalized 1-pK model (Section III.D.2) is used to represent the proton association. In the case of the 2-pK model (Section III.D.1) the situation is more complicated (see Ref. [77] for a discussion). However, use of the 2-pK model is strongly discouraged for other reasons as well (see Section III.D.1).

For a heterogeneous surface there are several types of surface sites and Equation. (84) should be applied to each site type S_j present. The isotherm for the homogeneous groups of sites as expressed by Equation (84) is mostly called the *local* isotherm. The *overall* isotherm can simply be described as the sum of the local contributions provided that the effect of lateral interactions is accounted for in the local isotherm. For a discrete distribution of intrinsic affinity constants the total surface coverage, $\theta_{t,1}$, is thus

$$\theta_{t,1} = \sum_j f_j \theta_{j,1}^* \tag{85}$$

where f_j is the fraction of sites of type j and $\theta_{j,1}^*$ is the local coverage of sites j with ion 1. For a discussion of f_j see van Riemsdijk et al. [265,266]. For a continuous distribution of affinity constants one may calculate the overall fraction of the sites covered with ion 1 as

$$\theta_{t,1} = \int_\Delta \theta^*(K_1,\ c_{1,\delta}^*) f(\log K_1) \mathrm{d}\log K_1, \tag{86}$$

where $f(\log K_1)$ is the distribution function of $\log K_1$, Δ is the relevant range of integration and $\theta^*(K_1, c_{1,\delta}^*)$ is the local isotherm function (Equation (84)).

Equations (85) and (86) are generalized adsorption isotherm equations. The generalized adsorption equation applies to organic and inorganic ion adsorptions provided the sites for adsorption are specific for that ion. Equations (85) and (86) are not suited for surface complexation, because this is a form of multicomponent adsorption with different ions competing for the same sites. It will be clear that also for heterogeneous surfaces the description of specific adsorption on independent sites is far simpler than that of surface complexation. In order to work out (85) or (86) a further distinction has to be made regarding the type of heterogeneity.

In general, two extreme cases of heterogeneity are considered: *patchwise* heterogeneity with noninteracting patches, and *random* or *regular* heterogeneity. This distinction is important as soon as lateral interactions come into play. For a random and a regular distribution of the different site types (a) the specific lateral interactions will depend on the *total* coverage of the surface and the lateral interaction parameter and (b) the electrostatic interactions can be calculated on the basis of the average, smeared-out, electrostatic potential of the *entire* plane of adsorption. For a patchwise heterogeneous surface the equal energy sites are present in patches, each patch covering a macroscopic fraction of the total surface area. In the case of noninteracting patches it is assumed that the lateral interactions are restricted to the patch, that is to say (a) the specific lateral interactions are proportional to the *local* coverage and the lateral interaction parameter and (b) each patch develops its own, smeared-out, *patch* potential that is not influenced by the potential of other patches. Especially in the case of long-range interactions ideal patchwise heterogeneity will only occur for large patches, such as well-defined large crystal planes. In the case of ion adsorption the characteristic dimensions of the patches should be at least several times the Debye length in order to call it a patch. When the patches are smaller than the Debye length the random approximation can be used. Hence, with ion adsorption the complication occurs that the distinction between patchwise or random heterogeneity depends on the ionic strength. A more extensive discussion on random and patchwise heterogeneity with regard to the electrostatic effects has been given by Koopal and van Riemsdijk [67]. With organic ion adsorption both short-range specific lateral interactions and long-range electrostatic interactions may occur simultaneously. In this case the effect of the long-range electrostatic interactions will dominate and hence also here the characteristic dimension of the patches should be larger than the Debye length in order to consider the surface as patchwise heterogeneous.

For random heterogeneity, where the potential in the plane of adsorption is assumed to be uniform over the entire surface, $c_{1,\delta}^{*}$ appearing in Equation (84) should be written as

$$c_{i,\,\delta}^{*} = c_1 \exp\left(\chi \theta_{t,\,1}\right) \exp\left(-\,\nu_1 F \psi_{\delta,\,t}/RT\right), \tag{87}$$

where the first Boltzmann factor accounts for the specific lateral interactions with χ the lateral interaction parameter and the second for the electrostatic interactions with $\psi_{\delta,t}$ is the potential of the entire adsorption plane. The potential $\psi_{\delta,t}$ can be calculated with the SGC model (or an alternative double layer model). There is only one set of double layer parameters (σ_s and C_1) required for the entire surface. This situation will be worked out in more detail in Section VI.B.2. The first exponent on the RHS of (87) cancels when specific lateral interactions are absent ($\chi = 0$).

For patchwise heterogeneity with noninteracting patches lateral interactions are restricted to the patch, each patch has its own, smeared-out, potential and $c_{1,\delta}^*$ becomes

$$c_{i,\delta}^* = c_1 \exp(\chi\theta_{p,1}) \exp(-\nu_1 F\psi_{\delta,p}/RT), \tag{88}$$

where $\psi_{\delta,p}$ is the potential of the plane of adsorption for patch p. As before, the potentials $\psi_{\delta,p}$ can be calculated with the help of the SGC model. It will be clear that constructing a predictive model for patchwise systems requires rather detailed information of not only the fractions of sites, but also of the double layer parameters of each patch ($\sigma_{s,p}$, $C_{1,p}$). Adsorption on patchwise heterogeneous surfaces will de considered further in Section VI.B.3. In the absence of specific lateral interactions Equation (88) can be simplified using $\chi = 0$.

It is also possible that a patch itself is randomly or regularly heterogeneous. This situation may occur with metal hydroxides with exposure of different crystal planes, each occupied with different types of surface groups. This situation is best treated as discrete heterogeneous both with respect to the situation per crystal plane and the summation of the plane charges to obtain the overall surface charge. Only when all intrinsic affinities, site fractions per patch, and the double layer parameters for each patch are known it is possible to make a predictive model. In the MUSIC type modeling [57,58,85,86] the situation is mostly simplified by assuming that the different crystal planes (patches) all have the same electrostatic potential.

2. Random Heterogeneity

For a random heterogeneous surface, with a uniform electrostatic potential over the whole surface, the description of ion

binding is still relatively simple. The only additional characteristic as compared to adsorption on a homogeneous surface
is the distribution function of intrinsic affinity constants.
Once this distribution function is known, the overall isotherm
can easily be calculated using Equation (85) or (86) combined
with Equation (87).

For certain specific distribution functions it is possible to
solve the integral equation (86) analytically [77,78,267], as
has been done for uncharged surfaces [262,263]. The equations thus obtained are Freundlich-type equations [266–268].
In the case of ion adsorption and the so-called *Sips-distribu-
tion* [269,270] Stern–Langmuir–Freundlich (SLF) type equations result. The name emphasizes that the local isotherm is
of the Langmuir type (with the effective concentration $c_{i,\delta}^*$),
that the adsorption free energy also contains an electrostatic
contribution as indicated by Stern and that the surface is
heterogeneous (Freundlich) The most common SLF equation
reads

$$\theta_{t,1} = \frac{\left(\tilde{K}_1 c_{i,\delta}^*\right)^m}{1 + \left(\tilde{K}_1 c_{i,\delta}^*\right)^m}, \tag{89}$$

where $\tilde{K}_1$ is the weighted average K_1-value corresponding
with the maximum in the distribution function of the $\log K_1$
values pertaining to Equation (86), and m is a parameter
corresponding to the width of the distribution function. Note
that for $m = 1$ and $\chi = 0$ (see Equation (87)) the homogeneous
Stern–Langmuir equation (23), (30), or (84) results and for
$m = 1$ and any value of χ the Stern–FFG equation (81). The
distribution function corresponding to Equation (89) has been
obtained using a Stieltjes transform [262,264,269,270]. In
Figure 5.15 a series of distribution functions corresponding
to Equation (89) is shown, together with the isotherms for
$\chi = 0$. The distributions are symmetrical around $\log \tilde{K}_1$ and
nearly Gaussian. For $m = 1$ a Dirac-delta function results. The
isotherms are plotted as a function of $\log \tilde{K}_1 c_{1,\delta}$ to normalize
the results. The slope of the isotherms decreases with increas-
ing heterogeneity. For $m = 1$ the Langmuir isotherm results.

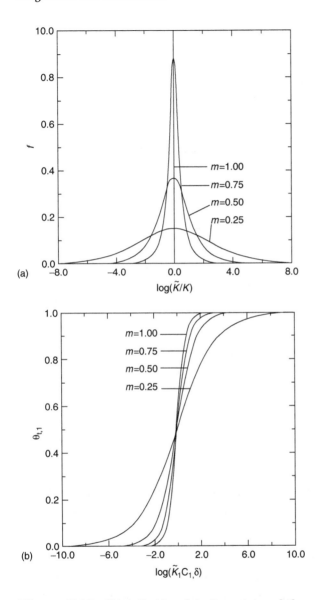

Figure 5.15 Distribution functions (a) and the corresponding SLF adsorption isotherms (b), see Equation (89), for different values of m (from W.H. van Riemsdijk, G.H. Bolt, and L.K. Koopal, in *Interactions at the Soil Colloid–Soil Solution Interface*, G.H. Bolt, M.F. de Boodt, and M.H.B. Hayes, Eds., Kluwer Academic Publishers (M. Nijhoff), Dordrecht, 1990, chapter 3. With permission).

By using $c_{1,\delta}$ instead of c_1 the shape of the isotherm attains the characteristic features of the heterogeneity only, electrostatic effects are incorporated in $c_{1,\delta}$ (see Equation (87)).

A major advantage of the use of Equation (89) over a discrete approach to the heterogeneity is that only two parameters $(\tilde{K}_1, m)$ are required to describe the distribution function. For a discrete approach two parameters $(f_j, K_{j,1})$ are required for each site type. Provided the maximum value of $\Gamma_{t,1}$ is known $(\theta_{t,1} = \Gamma_{t,1}/\Gamma_{t,1,\mathrm{max}})$, the parameters $\tilde{K}_1$ and m can be easily obtained from the linearized version of Equation (89):

$$\log \frac{\theta_{t,1}}{1 - \theta_{t.1}} = \log (\tilde{K}_1)^m + m \log c_{1,\delta}^*. \tag{90}$$

In order to apply Equation (90) to experimental results $c_{1,\delta}^*$ and therefore χ and ψ_δ are required. Normally χ is an adjustable parameter (for simple inorganic ions $\chi = 0$) and ψ_δ has to be calculated using a double layer model. In the case of proton adsorption ψ_s $(= \psi_\delta)$ can be calculated with the SGC model (parameters C_1 and the salt concentration c_s) provided the surface charge density is known. For other ions the adsorption plane corresponds in the simplest case to the Stern plane, i.e., $\psi_\delta = \psi_d$, and the GC model (parameter c_s) suffices for the calculation of ψ_d. In this situation both the surface charge density and the charge density due to adsorption of ion i must be known.

For proton adsorption the adequacy of the applied double layer model can be checked when adsorption data are available at different values of the ionic strength. In the absence of specific adsorption of the background electrolyte all isotherms should merge into a master curve when the adsorption is plotted as a function of pH_s [54,55]. For other inorganic ions adsorption data at various pH values and salt concentrations should be available. In this case the adsorption as a function of $\log c_{1,d}$ should give a master curve.

In order to illustrate the effect of the electrostatic interactions on the adsorption a series of $\sigma_s(pH)$ curves for a heterogeneous acid surface is shown in Figure 5.16. The heterogeneity is the same as in Figure 5.15, i.e., based on

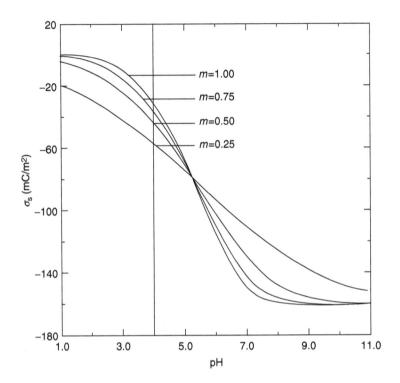

Figure 5.16 Series of $\sigma_s(\text{pH})$ curves based on the SLF equation (89) for an acid colloid and different values of m. The electrostatic interactions are calculated with the SGC model. Parameter values: $\log \tilde{K}_H = 4$, $N_s = 1$ site/nm^2, $C_1 = \infty$ and $c_s = 0.1\,M$ (from W.H. van Riemsdijk, G.H. Bolt, and L.K. Koopal, in *Interactions at the Soil Colloid–Soil Solution Interface*, G.H. Bolt, M.F. de Boodt, and M.H.B. Hayes, Eds., Kluwer Academic Publishers (M. Nijhoff), Dordrecht, 1990, chapter 3. With permission).

Equation (89) and a given set of m values. For the calculation of the electrostatic interaction the SGC model is used. Apart from m the adopted parameter values are $\log \tilde{K}_H = 4$, $N_s = 1$ site/nm^2, $C_1 = \infty$ and $c_s = 0.1\,M$. The curves of Figure 5.16 can be transformed to those in Figure 5.15(b) by replotting data as $1 - (\sigma_s/\sigma_{s,\text{max}})$ versus $\log\log \tilde{K}_H \{H_s\}$.

In Figure 5.16 the pH for which the degree of dissociation reaches 0.5 (i.e. $\sigma_s = -0.5N_sF$) is considerably larger than log

$\tilde{K}_{\mathrm{H}}$. Due to the electrostatic interactions the surface group dissociation is inhibited. The point for which $\sigma_{\mathrm{s}} = -0.5N_{\mathrm{s}}F$ corresponds to the common intersection point of the isotherms. At this point $(\tilde{K}_{\mathrm{H}}\{H_{\mathrm{s}}^{+}\})^{m} = 1$ for all values of m. Since the value of the potential ψ_{s} at a given salt concentration depends solely on σ_{s}, it is fully determined by $\theta_{\mathrm{t,H}}$. Accordingly, the surface potential at $\theta_{\mathrm{t,H}} = 0.5$ is independent of m and thus also the pH is the same, leading to the observed CIP. The difference between $\mathrm{pH}_{\mathrm{cip}}$ and $\log \tilde{K}_{\mathrm{H}}$ is a measure of the electrostatic interactions. Replotting the data as $\sigma_{\mathrm{s}}(\mathrm{pH}_{\mathrm{s}})$ will lead to a CIP characterized by $\mathrm{pH}_{\mathrm{s}} = \log \tilde{K}_{\mathrm{H}}$.

When little is known about the electrostatic effects Equation (89) can be written as

$$\theta_{t,1} = \frac{\Gamma_1}{\Gamma_{1,\,\mathrm{max}}} = \frac{(K_{1,\,\mathrm{app}}c_1)^m}{1 + (K_{1,\,\mathrm{app}}c_1)^m}, \tag{91}$$

where in general, $K_{1,\mathrm{app}} = \tilde{K}_1 \exp(\chi\theta_{t,1}) \exp(-\nu_1 F\psi_\delta/RT)$ is an apparent affinity coefficient characterizing the overall selectivity of ion 1 for the surface, Γ_1 is the adsorption density of ion 1, $\Gamma_{1,\mathrm{max}}$ its saturation value. Even when $\chi = 0$, ψ_δ is a function of c_1 or $\theta_{1,t}$ and $K_{1,\mathrm{app}}$ is not a constant. Often the maximum adsorption $\Gamma_{1,\mathrm{max}}$ is also used as fitting parameter. Equation (91) is a rather flexible three-parameter equation that may serve to describe many adsorption isotherms. When only a limited range of c_1 values is covered Equation (91) reduces to the classical Freundlich equation.

A series of equations comparable to Equation (91) has been tested by Kinniburgh et al. [271] to describe the pH and concentration dependence of metal ion adsorption on a ferrihydrite. As local isotherm a Langmuir equation has been used in which the concentration (activity) was replaced by an activity ratio: $\{M\}/\{H\}^a$, the power a being the average number of protons released upon adsorption of M. The advantage of considering the adsorption as an exchange reaction is that the variation of the electrostatic potential with cation surface coverage is relatively small, so that the electrostatic contributions to the adsorption free energy are, roughly speaking, constant. Furthermore, a suppression of the electrostatic effects was achieved by using a high $(1M)$ concentration of indifferent electrolyte. The remaining electrostatic interactions are probably accounted for through a. All Freundlich-type equations could fit the data reasonably well. A good fit could also be obtained by using a summation of two discrete Langmuir (exchange) isotherms. With the Langmuir–Freundlich equations a very narrow distribution $(m = 0.8)$ was found for both Ca^{2+} and Zn^{2+}. The average values of a were 0.88 (Ca^{2+}) and 1.64 (Zn^{2+}), indicating different adsorption mechanisms for both ions. Benjamin and Leckie [186] found that Cd^{2+} adsorption by ferrihydrite obeyed a simplified SLF isotherm $(\theta_{t,1} \ll 1)$ with $K_{1,app}$ a function of pH, but m a constant.

Dzombak et al. [272,273] discuss some adsorption models, including Equation (91), to describe metal binding on humate materials, but they have not tested Equation (91) for real adsorption data. In a review on the sorption of trace metals by humic materials Sposito [274] also discusses Equation (91) together with a series of other simple adsorption and a metal–proton exchange models.

In the case of proton adsorption Equation (91) can be transformed to the well known (generalized) Henderson–Hasselbalch (HH) equation [275]:

$$pH = pK_H^{\#} + \frac{1}{m} \log\left(\frac{\alpha}{1 - \alpha}\right), \tag{92}$$

where $K_H^{\#} = 1/K_{H,app}$ is an apparent dissociation coefficient and $\alpha = (1 - \theta_{t,H})$ the degree of dissociation of the surface groups. Similarly as $K_{H,app}$, $K_H^{\#}$ is only constant when the electrostatic interactions are independent of α. This is only the case if the proton adsorption occurs through an exchange reaction at constant potential. Equation (92) is mainly used to describe the surface group dissociation of homo-polyelectrolytes. In this case both $K_H^{\#}$ and m are related to the electrostatic interactions. Fitting the equation to experimental results indeed leads to $pK_H^{\#}$ and m values that are dependent on the type and concentration of background electrolyte [275–277].

Freundlich-type equations are also used frequently for the adsorption of organic components and ions to activated carbon [278]. When Equation (91) is used for organic ions the apparent heterogeneity may not only incorporate electrostatic interactions, but also lateral 'chemical' (specific) effects $(\chi \neq 0)$.

3. Patchwise Heterogeneity

With patchwise heterogeneity rather detailed information on the surface properties is required in order to be able to give a predictive description of the adsorption. For each patch the area, site density, affinity constant(s), and Stern layer capacitance have to be known. The extension of the homogeneous models to heterogeneous surfaces composed of noninteracting patches is simply achieved by calculation of $\sigma_{s,p}(\text{pH})$ for each patch using the homogeneous model (with known parameters) and applying the additivity rule.

As a model for patchwise heterogeneity Gibb and Koopal [80] have studied proton adsorption on mixtures of hematite and rutile. Both hematite and rutile could be treated as pseudo-homogeneous and could be described with the 1-pK generalized SGC model ($p = -1/2$; $\text{p}K_H = \text{p}H_0^*$) (see Section IV.C). The surface charge–pH curves of the mixtures could be predicted by weighed addition of the $\sigma_s(\text{pH})$ curves of rutile and hematite at each salt concentration, i.e.,

$$\sigma_{s,t}(\text{pH}) = f_{\text{TiO}_2}\sigma_{s,\text{TiO}_2}(\text{pH}) + f_{\text{Fe}_2\text{O}_3}\sigma_{s,\text{Fe}_2\text{O}_3}(\text{pH}), \qquad (93)$$

where f_i $(i = \text{TiO}_2, \text{Fe}_2\text{O}_3)$ is the area fraction $(N_{s,\text{TiO}_2} = N_{s,\text{Fe}_2\text{O}_3})$. The additivity indicates that lateral electrostatic interaction between the "patches" could be neglected. For the total differential capacitance $(d\sigma_s/d\text{pH})$ a similar additivity rule could be applied.

By treating the hematite–rutile mixtures as patchwise heterogeneous surface, the $\sigma_s(\text{pH})$ curves of the mixtures could be calculated on the basis of the known values of $\text{p}K_{H,p}$, $N_{s,p}$, and $C_{1,p}$ for each patch. The theoretical predictions did correspond very well with the experimental results. The observed and predicted values of the overall PZC also corresponded well. For patchwise heterogeneous surfaces the pH of the PZC is not only a function of the surface composition, but also of the ionic strength. The PZC can no longer be identified with the (near) common intersection point of a series of $\sigma_s(\text{pH})$ curves at different salt concentrations [80]. This point had been stressed before by Koopal and van Riemsdijk [67]. Although identification of the common intersection point with the PZC is principally wrong, the error in doing so is small if the Stern layer capacitances of both patches are similar. In literature on the PZC of heterogeneous oxides, see the brief review by Kuo and Yen [279], the assumption has been made, without further discussion, that the PZC can be identified with the common intersection point.

Hiemstra et al. [57,58,85,86] in their MUSIC model also consider patchwise heterogeneity. The different crystal faces are assumed to develop their own surface potential, without patch–patch interactions. In principle, each crystal plane (patch) has its own surface and Stern layer potentials, whether it contains one or more types of surface groups. The best example of patchwise heterogeneity is gibbsite [57,58]. On the crystalline gibbsite as studied by Hiemstra two types of crystal faces (edge and planar) could be distinguished. The OH groups at the planar face are all doubly coordinated, whereas at the edge face two types of OH groups are found, singly and doubly coordinated. The site densities of these groups follow from crystallographic and spectroscopic information. Up to pH ≈ 10 the $\sigma_s(\text{pH})$ curve of gibbsite is determined by the singly coordinated groups of the edge face, above pH $= 10$ the doubly coordinated OH groups of the planar face start to contribute. The doubly coordinated groups of the edge face remain "inactive" due to the negative potential caused by the dissociation of the singly coordinated OH groups on the edges. By applying a discrete patchwise model for the different crystal planes the $\sigma_s(\text{pH})$ curves of gibbsite can be explained very well. A similar reasoning can be applied to goethite [58]. A comparison of $\sigma_s(\text{pH})$ curves for goethites that differ strongly in BET surface area revealed that the surface group structure on goethites with a small BET area is different from that on goethites with a large BET area (especially due to imperfections at the 100 face).

C. Multicomponent Adsorption

1. Generalized Multicomponent Adsorption Equation

So far we have discussed the situation where only one species adsorbs on the available surface sites. If different ions or more than one ion may adsorb on the same site, as is the case with surface complexation, the local isotherm Equation (84) and the overall adsorption equation as given by Equation (86) no longer hold. For the local adsorption the competitive or multicomponent form of Equation (30) should be used. In Equation (30) the multicomponent character is "hidden" in $\theta_{j,R}$. When i species compete for the reference site S_j, $\theta_{j,R}$ equals $1 - \sum_i \theta_{j,i}$ and the general expression for the competitive Langmuir equation becomes

$$\theta_{j,i}^* = \frac{K_{j,i}c_{i,\delta}^*}{1 + \sum_i K_{j,i}c_{i,\delta}^*}, \tag{94a}$$

where $\theta_{j,i}^*$ is the local isotherm of ion i, $c_{i,\delta}^*$ is used to describe the effective concentration of i at plane δ. The asterisk is included to remember that the expression for $c_{i,\delta}^*$ depends on the lateral interactions and therefore on the nature of the heterogeneity (see Equations (87) and (88)). For the specific situation of complexation of an ion B (located at the Stern plane) with a surface site of type j, on which also a proton (located at the surface plane) can adsorb, the local isotherm equation thus becomes

$$\theta_{j,B}^* = \frac{K_{j,B}\{B_d\}^*}{1 + K_{j,H}\{H_s\}^* + K_{j,B}\{B_d\}^*}. \tag{94b}$$

The expression for the overall complexation of B, $\theta_{t,B}$, with all the different site types is now a double integral equation:

$$\theta_{t,B} = \int_\Delta \int \theta_{j,B}^* f(\log K_B, \log K_H)d\log K_B\, d\log K_H, \tag{95}$$

where $\theta_{j,B}^*$ is given by Equation (94b). For patchwise heterogeneous surfaces $\{B_d\}^* = \{B_d\}_p$ and $\{H_s\}^* = \{H_s\}_p$, whereas for random heterogeneous surfaces $\{B_d\}^* = \{B_d\}_t = \{B_d\}$ and $\{H_s\}^* = \{H_s\}_t = \{H_s\}$, i.e., both are independent of the site type.

The intrinsic affinity distribution $f(\log K_B, \log K_H)$ is a two-dimensional function.

At present there seems little prospect of using Equation (95) for the analysis of ion complexation on heterogeneous surfaces; with its two-dimensional distribution function, Equation (95) is too complicated for a routine analysis. Equation (95) therefore has to be simplified by making a priori assumptions with respect to the heterogeneity and the nature of the adsorption complex. Two extreme situations will be considered:

(1) the distribution of the intrinsic affinity of B is fully independent of that of H,
(2) the distribution of the intrinsic affinity of B is fully coupled to that of H.

2. Fully Independent Intrinsic Affinities

When the distribution of intrinsic affinities of B is assumed to be fully independent of that of the protons, it is logical to further assume that also the sites for B and H can be considered as independent. In that case the situation is the same as that treated in the former section on monocomponent adsorption.

The model can be tested by plotting the adsorption of B as a function of $\{B_d\}$, the activity of B at the Stern plane. Curves at different background electrolyte concentrations (constant pH) or at different pH values (constant background electrolyte concentration) should merge into a master curve when plotted as adsorption versus pB_d. If not, either the double layer model used is incorrect or the sites are not independent. If the double layer model works well for proton adsorption in the absence of B, the assumption of independent sites is most probably incorrect and the option of fully dependent sites should be investigated. De Wit et al. [55,280] give a more extensive discussion and some examples with specific reference to humic materials.

3. Fully Coupled Intrinsic Affinities and Random Heterogeneity

For fully coupled or correlated intrinsic affinity distributions the shape of the distribution of $\log K_B$ is identical to that of $\log$

K_H, the only difference is the position of the $\log K_B$ distribution on the $\log K$ axis, i.e., $\log K_{j,B} = \log K_B^0 + \log K_{j,H}$, or in general:

$$\log K_{j,i} = \log K_i^o + \log K_{j,\text{ref}} \tag{96}$$

for each site type j and each species i and 'ref' indicating the reference component. Restricting ourselves to random surfaces only, the local isotherm equation (94b) can be written in the case of adsorption of B and H as

$$\theta_{j,\text{B}} = \frac{K_{j,\text{H}}K_B^o\{B_d\}}{1 + K_{j,\text{H}}\{H_s\} + K_{j,\text{H}}K_B^o\{B_d\}}, \tag{97}$$

where K_B^o, $\{H_s\}$, and $\{B_d\}$ are independent of the site type and H is the 'ref' component. The overall adsorption equation (95) reduces to

$$\theta_{t,\text{B}} = \int_{\Delta H} \theta_B(K_H K_B^o, \{B_d\}, \{H_s\})f(\log K_H)\mathrm{d}\log K_H, \tag{98}$$

which is comparable to the monocomponent overall adsorption equation. For a fully coupled system for which $f(\log K_H)$ is known, the adsorption of B can now be described with only one additional parameter: K_B^o. This parameter determines the position of $f(\log K_B)$ on the $\log K$ axis. For a more detailed analysis of this situation, see De Wit et al. [55,280].

For a few specific distribution functions Equation (98), with Equation (97) as a local isotherm, can be expressed in an analytical form. For example, the two-component equivalent of Equation (89) reads [78,267]):

$$\theta_{t,\text{B}} = \frac{\tilde{K}_B\{B_d\}}{[1 + (\tilde{K}_B\{B_d\} + \tilde{K}_H\{H_s\})^m](\tilde{K}_B\{B_d\} + \tilde{K}_H\{H_s\})^{1-m}}. \tag{99a}$$

Equation (99a) can be written in a general notation that can be used for all kind of ions as:

$$\theta_{t,i} = \frac{\tilde{K}_i c_{i,\delta}^*}{\sum_i \tilde{K}_i c_{i,\delta}^*} \times \frac{\left(\sum_i \tilde{K}_i c_{i,\delta}^*\right)^m}{1 + \left(\sum_i \tilde{K}_i c_{i,\delta}^*\right)^m}, \tag{99b}$$

where $\tilde{c}_{i,\delta}^*$ is the effective concentration (including lateral interactions) of species i at the adsorption plane δ. The first term on the RHS of (99b) is the fraction of reference sites covered with i and the second term is the total coverage of the reference sites. Equation (99) is a multicomponent Langmuir–Freundlich type equation. For the calculation of $\theta_{t,i}$ the procedure of Westall [177] can be extended to heterogeneous surfaces.

In Ref. [78] some calculated examples are given of multi-solute adsorption according to Equation (99a). Note that at very low concentrations of B ($\tilde{K}_B\{B_d\} = 1$) a linear isotherm may result, provided $K_H\{H_s\}$ is constant. Similarly as for Equation (89) Equation (99a) can also be used to describe multicomponent ion exchange by replacing $\{B_d\}$ by an activity ratio. Dzombak et al. [272,273] have erroneously suggested that Equation (89) could not be extended to multicomponent adsorption.

The multicomponent Langmuir–Freundlich type equations are often used for adsorption on activated carbons (see e.g., Ref. [278]) and for ion adsorption to soil particles (e.g., Ref. [286]).

4. Fully Coupled Intrinsic Affinities, Variable Stoichiometry, and Random Heterogeneity

So far it has been assumed that all the species adsorb with the same stoichiometry. Recently a new model has been developed that describes multicomponent ion adsorption on random heterogeneous surfaces and that allows a different stoichiometry for different components [281,282]. In this case the local isotherm equation should be a competitive Hill-type equation rather than the competitive Langmuir-type equation (94). The Hill-type equation for competitive binding of species i ($i = 1, \ldots, j$) on a group of equal energy sites, j can be written as:

$$\theta_{j,i} = \frac{(K_{j,i}c_{i,\delta}^*)^{n_i}}{1 + \sum_i (K_{j,i}c_{i,\delta}^*)^{n_i}}. \tag{100}$$

Hence, for each component i the coverage of the reference sites is not only characterized by the intrinsic affinity for the reference site, $K_{j,i}$, as with the competitive Langmuir-type equation (94), but also by the stoichiometry factor, n_i. For $n_i = 1$ the competitive L-type equation results. The stoichiometry factor means that each reference site becomes occupied with n_i molecules of i. Therefore, for the calculation of the bound amount, Q_i, n_i is also required:

$$Q_i = n_i \theta_{,j,i} Q_{\max} \qquad (101)$$

where $Q_{\max}$ is the density of reference sites. Equation (101) shows that the binding maximum is different for the different components.

The extension of Equation (100) to the overall binding of species i in the competitive situation on a heterogeneous substrate follows by using Equation (100) as a local isotherm in the integral adsorption equation. In order to arrive at an analytical expression one has to assume as before (1) random heterogeneity and (2) full correlation between the affinity distributions for the different ions. For the affinity distribution the "Sips"-distribution [269,270] is used. The solution of the thus obtained integral binding equation is the nonideal competitive adsorption or NICA equation for the overall binding of species i in the competitive situation [281,282]:

$$\theta_{t,i} = \frac{(\tilde{K}_i c_{i,\delta}^*)^{n_i}}{\sum_i (\tilde{K}_i c_{i,\delta}^*)^{n_i}} \times \frac{\left\{ \sum_i (\tilde{K}_i c_{i,\delta}^*)^{n_i} \right\}^p}{1 + \left\{ \sum_i (\tilde{K}_i c_{i,\delta}^*)^{n_i} \right\}^p}, \qquad (102)$$

where $\theta_{t,i}$ is the fraction of all of the sites occupied by species i, $\tilde{K}_i$ is the median value of the affinity distribution for species i and p is the width of the affinity distribution. For $n_i = 1$ for all i the multicomponent Langmuir–Freundlich equation (99b) results. The total bound amount of component i, $Q_{i,t}$, is now given by

$$Q_{t,i} = \theta_{t,i} n_i Q_{t,\max}, \qquad (103)$$

where $Q_{t,\max}$ is the overall density of the reference sites. Equation (103) reflects again that n_i incorporates the stoichiometry.

Note that for monocomponent adsorption Equation (102) reduces to (89) with $m = np$, this shows that in this case it is not possible to distinguish between stoichiometry and hetero-geneity (with monocomponent adsorption it is common to assume that the reference site automatically has a 1:1 stoichiometry with the adsorbing ion). Only in the case of multi-component adsorption the distinction between stoichiometry and heterogeneity becomes important.

In order to access $Q_{t,max}$ in practice, one has to select a particular component and determine its binding capacity. Most likely the proton will be used to determine the total binding capacity, $Q_{t,max} = Q_{max,H}$, because this ion can be studied without ion competition. Moreover, proton-binding data allow us to extract the site heterogeneity (see Section VI.D) [283–285]. The bound amount of component i should now be scaled by n_i/n_H rather than by n_i alone:

$$Q_{t,i} = \theta_{t,i}(n_i/n_H)Q_{max,H}. \tag{104}$$

Substituting Equation (103) for $\theta_{t,i}$ into Equation (104) defines the basic NICA model in terms of bound amount as a function of the concentrations of the species present. The NICA equation for $\chi = 0$ has mainly been used for the description of ion binding to humic substances [e.g., 281,282,287–293], but this restriction is not essential. In general the NICA equation can be used for competitive adsorption of interacting molecules or ions on heterogeneous surfaces when the different adsorbates do not have the same maximum adsorption.

In general, a bimodal affinity distribution (accounting for carboxylic and phenolic type groups) is required for a good quality of fit when the NICA equation is applied to ion adsorption on natural organic matter. This requires a summation of two NICA equations. Furthermore, an electrical double layer model has to be assumed to convert the charge of the humic acid molecules into the electrostatic potential and this involves simplifying assumptions related to the type of particles and their size, shape, and conformation. In general two types of models are used. The first model is based on the assumption that the organic colloid is impenetrable for ions [e.g., 54,287]. In this case the charge of the particles is smeared-out over the surface of (mostly) spherical particles. The surface potential can be calculated from the surface charge density using an approximate double layer model for hard spheres [287]. The other model considers the colloid as a Donnan phase [287–290]. The charge due to the dissociation of the functional groups leads in this case to a smeared-out Donnan potential which is the same everywhere inside the volume of the particle and zero outside the particle boundary [38–41].

For both models an estimate of the size of the particles is required in order to be able to calculate the specific surface area or the Donnan volume. De Wit et al. [54] have used the surface charge model with a fixed particle radius. The order of magnitude of the particle radius was derived from the assumed geometry of the particles and the density of the humics. Benedetti et al. [288] and others [e.g., 289,290] have used the Donnan model with Donnan volumes depending on the ionic strength. The Donnan volumes used have been larger than the physical volume of the humic acid molecules. Koopal et al. [287] have discussed both the surface charge models and the Donnan models. An advantage of using the Donnan model of Benedeth [288,289] is its ease of calculation.

D. Determination of the Distribution Function

1. Methods to Obtain the Distribution Function

In order to apply the general adsorption equation for monocomponent ion adsorption the intrinsic affinity distribution has to be known. In principle $f(\log K_1)$ can be obtained from adsorption data, if the local isotherm equation is known. Solution of Equation (86) with respect to $f(\log K_1)$ when both $\theta_{t,1}$ and $\theta_{j,1}$ are known is however faced with mathematical instabilities [294]. In view of this two main routes have been followed to solve Equation (86) for $f(\log K_1)$.

The first group of methods is concerned with numerical methods that consider the instability problems explicitly. Examples of these methods are singular value decomposition, CAESAR [295–297], and regularization [297–305]. In these methods the only constraint is that the distribution function is a somewhat smooth function. About the overall shape of the distribution function (e.g., the number of peaks) no assumptions have to be made. Moreover, for the local isotherm function several options are open and both patchwise and random heterogeneity can be considered. Disadvantages of these methods are that a complicated numerical algorithm is used and that information is required over the entire domain of $\theta_{t,1}$. So far these methods have been mainly used in connection with carefully measured (submonolayer) gas adsorption isotherms [263].

The second group of methods uses specific approximations of the local isotherm equation, so that the integral equation can be inverted and an analytical expression results for $f(\log K_1)$ [283–285,306]. A disadvantage of these methods is that always some smoothing of the distribution function results. The effect depends on the quality of the approximation of the local isotherm function. The advantage is that, in principle, a solution for $f(\log K_1)$ is found without much effort. Moreover, when only a part of the overall isotherm is available the distribution function can still be obtained in this range. In practice, where only a *window* of data are available this is an important advantage over the regularization methods. Nederlof and coworkers [283,284] have called this group the Local Isotherm Approximation (LIA) methods. A weakness of the LIA methods is that the calculated distribution functions are rather sensitive to experimental errors in $\theta_{t,1}$. In order to overcome this problem a sophisticated smoothing of the isotherm data is required [284,306,307]. In the Section VI.D.2 the LIA methods will be discussed in somewhat more detail.

A third option to obtain the distribution function is to fit one of the SLF equations (e.g., Equation (89)) or the NICA equation (102) for the overall isotherm to the adsorption data. With the obtained parameters ($\tilde{K}_1$ and m) the distribution function can be constructed (compare Figure 5.15). This method works well when relatively little data are available or when the data are not very accurate. Disadvantages are that a priori choices are made with respect to the shape of the distribution function and that the heterogeneity is assumed to be random. In general it is therefore to be recommended first to use a LIA method to see what kind of distribution function is to be expected. For distributions with more than one maximum a combination of SLF or NICA equations should be used.

With all methods mentioned the *intrinsic* affinity distribution is obtained only when $c_{1,\delta}^*$ is used in the local isotherm instead of c_1. This requires that $\theta_{t,1}$ (or Γ_1) is replotted as a function of log $c_{1,\delta}^*$. For simple inorganic ions it is safe to assume $\chi = 0$ and isotherms measured as a function of the indifferent electrolyte concentration should merge into a master curve, when plotted as $\theta_{t,1}$ ($-$ log $c_{1,\delta}^*$), if the double layer model used is correct [54,55,286]. The master curve thus obtained can be analyzed to obtain the distribution of the intrinsic affinities, $f(\log K_1)$.

When the distribution analysis is directly applied to the experimental isotherms ($\theta_{t,1}$ versus log c_1) instead of to the master curve, apparent affinity distributions are obtained, with $K_{1,\mathrm{app}} = K_1 \exp(\chi\theta_1^*) \exp(-\nu_1 F\psi_\delta^*/RT)$. In this case adsorption on an intrinsically homogeneous surface will still lead to a distribution of affinities because both the specific lateral interaction and ψ_δ^* are dependent on θ_1^*, the local or entire surface coverage. Although the advantage is that one does not have to rely on a double layer model, the disadvantage is that $f(\log K_{1,\mathrm{app}})$ is much less informative with respect to (chemical) heterogeneity than $f(\log K_1)$. Some examples of a comparison of apparent and intrinsic affinity distributions as calculated from proton adsorption isotherms ($\chi = 0$) on heterogeneous colloids have been given by De Wit et al. [55,280].

2. LIA Methods

In order to illustrate the LIA methods the Langmuir type equation (84) is used as local isotherm, and it is assumed that $c_{1,\delta}^*$ can be obtained without difficulties. That is to say, the experimental adsorption isotherms when plotted as $\theta_{t,1}$ versus log $c_{1,\delta}^*$ merge into a master curve. A simple approximation of Equation (84) is the condensation approximation (CA) [283–285,307–310]. The CA replaces Equation (84) by a step function

$$\theta_{j,1}(K_1, c_{1,\delta}^*) = 0 \quad \text{for } K_1 c_{1,\delta}^* < (K_1 c_{1,\delta}^*)_C, \tag{105a}$$

$$\theta_{j,1}(K_1, c_{1,\delta}^*) = 1 \quad \text{for } K_1 c_{1,\delta}^* \geq (K_1 c_{1,\delta}^*)_C, \tag{105b}$$

where $(K_1^*, c_{1,\delta})_C$ corresponds to the intersection point between the true local isotherm and the CA isotherm at $\theta_{j,1} = 0.5$. According to Equation (84) at this point $K_1^* c_{1,\delta} = 1$ or $K_{1,C} = 1/(c_{1,\delta}^*)_C$. Inserting Equation (105) in Equation (86) gives

$$\theta_{t,1} = \int_{\log K_{1,C}}^{\infty} f_{CA}(\log K_1)\, d\log K_1, \tag{106}$$

where the index CA denotes that f_{CA} is an approximation of the distribution function due to the use of the CA as local isotherm. Differentiation of equation (106) with respect to $\log c_{1,\delta}^*$ leads to

$$f_{CA}(\log K_1) = \frac{d\,\theta_{t,1}}{d\log c_{1,\delta}^*} \tag{107a}$$

and

$$\log K_{1,C} = -\log c_{1,\delta}^*. \tag{107b}$$

The conversion of $\log c_{1,\delta}^*$ to $\log K_1$ follows from their relation at the critical point ($\theta_{j,1} = 0.5$). The CA distribution function, $f_{CA}(\log K_1)$, is thus simply found by taking the first derivative of the adsorption isotherm with respect to $(\log c_{1,\delta}^*)$ followed by substitution $-\log c_{1,\delta}^* = \log K_1$. The CA method shows very nicely how, to a first approximation, the isotherm and the distribution function are related.

In order to illustrate the CA method the CA distribution functions corresponding to the SLF equation for $m = 0.9$ and $m = 0.4$ are calculated (see for the isotherms Figure 5.20a). The results are shown in Figure 5.17 together with the true distribution functions. As expected the distribution function of the nearly homogeneous isotherm ($m = 0.9$) is very poorly recovered, nevertheless the peak of the distribution function is located at the correct position. For the heterogeneous surface ($m = 0.4$) the CA distribution is somewhat flatter than the

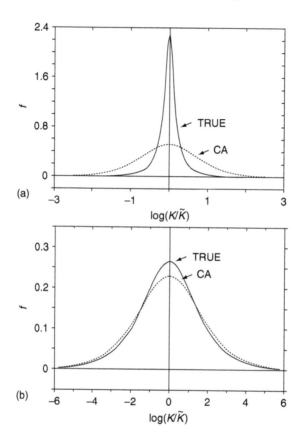

Figure 5.17   The "CA" approximation of two distribution functions corresponding to the SLF equation for (a) $m = 0.9$ and (b) $m = 0.4$ (from M.M. Nederhof, W.H. van Riemsdijk, and L.K. Koopal, *J. Colloid Interf. Sci.*, 135, 410, 1990. With permission).

true distribution, but in general a good representation is achieved. For very wide distribution functions (e.g., $m = 0.2$) $f_{CA}(\log K)$ is nearly identical to $f(\log K)$.

When one is satisfied with the distribution of apparent affinities, the CA method can be applied directly to the $\sigma_s(\text{pH})$ curves. Figure 5.18 gives the CA distributions of the apparent affinity constant as derived from the $\sigma_s(\text{pH})$ curves shown in Figure 5.16. The distribution functions obtained can be compared with those in Figure 5.15(a). Due to the electrostatic

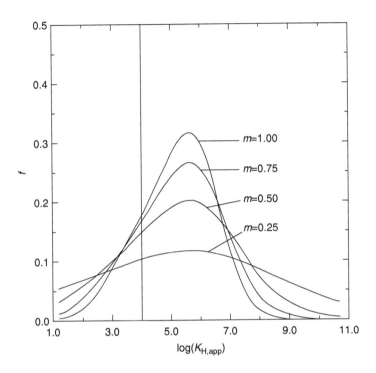

Figure 5.18 The apparent affinity distributions as obtained with the CA method for the isotherms shown in Figure 5.16 (from W.H. van Riemsdijk, G.H. Bolt, and L.K. Koopal, in *Interactions at the Soil Colloid–Soil Solution Interface*, G.H. Bolt, M.F. de Boodt, and M.H.B. Hayes, Eds., Kluwer Academic Publishers (M. Nijhoff), Dordrecht, 1990, chapter 3. With permission).

interactions the apparent affinity distribution is rather wide, even for a homogeneous surface. The CA distribution is therefore a good representation of the true apparent distribution function even for $m = 1$. Another consequence of the electrostatic interactions is that the peak position of the distribution function is shifted to log $K_{H,app}$ values higher than log $\tilde{K}_H$ and the distributions are no longer symmetrical.

Clearly the CA isotherm is a poor approximation of the local isotherm. A considerably better result is obtained if an approximation is used which follows Equation (84) more closely. A series of possibilities have been worked out by

Nederlof et al. [283–285]. A good result can be obtained with the LOGA function, which replaces the local isotherm in the following way:

$$\theta_{j,1}(K_1, c_{1,\delta}^*) = \alpha(K_1 c_{1,\delta}^*)^\beta \quad \text{for } K_1 c_{1,\delta}^* < (K_1 c_{1,\delta}^*)_C, \qquad (108a)$$

$$\theta_{j,1}(K_1, c_{1,\delta}^*) = 1 - \alpha(K_1 c_{1,\delta}^*)^{-\beta} \quad \text{for } K_1 c_{1,\delta}^* \geq (K_1 c_{1,\delta}^*)_C, \qquad (108b)$$

where α and β are adjustable parameters. Note that Equation (108), when plotted as $\theta_{j,1}$ (log K_1, $c_{1,\delta}^*$), is symmetrical in $\theta_{j,1} = 0.5$ just as Equation (84). This symmetry point is used to calculate the critical value $(K_1 c_{1,\delta}^*)_C$. The distribution function corresponding to Equation (108) for $\alpha = 0.5$ is

$$f_{\text{LOGA}}(\log K_1) = \frac{\mathrm{d}\,\theta_{t,1}}{\mathrm{d}\log c_{1,\delta}^*} - \frac{0.189}{\beta^2} \frac{\mathrm{d}^3\theta_{t,1}}{(\mathrm{d}\log c_{1,\delta}^*)^3} \qquad (109a)$$

with

$$\log K_{1,C} = -\log c_{1,\delta}^*. \qquad (109b)$$

The conversion of log $c_{1,\delta}^*$ to log K_1 follows from the value of $(K_1 c_{1,\delta}^*)_C$ at $\theta = 0.5$. For β several suggestions have been made [283] and some of the solutions are in good agreement with previous work [311–315]. In Figure 5.19 the LOGA isotherm is compared with the true local isotherm for three values of β: 0.7 (Figure 5.19a), 0.79 and 1 (Figure 5.19b). The LOGA isotherm for $\beta = 0.7$ is indicated as LOGA1, that of $\beta = 0.79$ as RJ and that of $\beta = 1$ as AS2.

The notation RJ stems from the fact that for $\beta = 0.79$ the distribution function corresponds to an approximate expression for $f(\log K_1)$ derived by Rudzinski and coworkers [311,312]. The notation AS2 indicates that this approximation leads to an expression for $f(\log K_1)$, which is very similar to that obtained with the so-called second-order affinity spectrum method [313–315].

In order to show the capabilities of the LOGA method, three test isotherms, shown in Figure 5.20 with their corresponding distribution functions, have been analyzed for

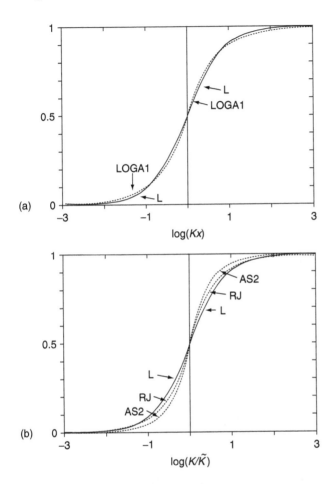

Figure 5.19 The Langmuir isotherm (L) and its "LOGA" approximation (Equation (109)) for three values of β ($\log Kx = \log K_1^* c_{1,\delta}$) (a) $\beta = 0.7$ (LOGA 1), (b) $\beta = 0.79$ (RJ), and $\beta = 1$ (AS2 with $\log a \to 0$) (from M.M. Nederhof, W.H. van Riemsdijk, and L.K. Koopal, *J. Colloid Interf. Sci.*, 135, 410, 1990. With permission).

heterogeneity. The results are shown in Figure 5.21(a) for $\beta = 0.70$ and in Figure 21(b) for $\beta = 0.79$ and $\beta = 1.0$. The results are much better than for the CA method, but the LOGA distributions are still smoother than the true distributions if sharp peaks are present.

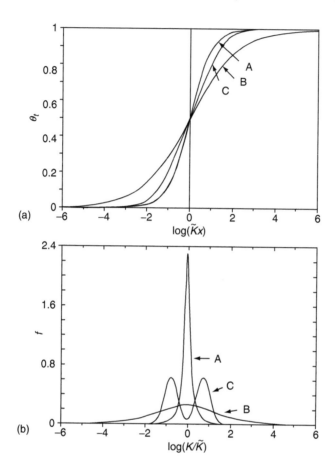

Figure 5.20 (a) Overall adsorption θ_t as a function of $\log \tilde{K}x = \log \tilde{K}_1 c_{1,\delta}^*$ for the three distribution functions shown in (b) and the Langmuir equation as local isotherm (from L.K. Koopal, M.M. Nederhof, and W.H. van Riemsdijk, *Prog. Colloid Polym. Sci.*, 81, 1990. With permission).

So far only highly accurate adsorption isotherms have been analyzed, in practice however, data may have a considerable error. An error of 5% in the θ_t data fully ruins the distribution function; see Nederlof et al. [283,284,306,307]. In order to overcome this problem a presmoothing of the isotherm data with a sophisticated smoothing technique is

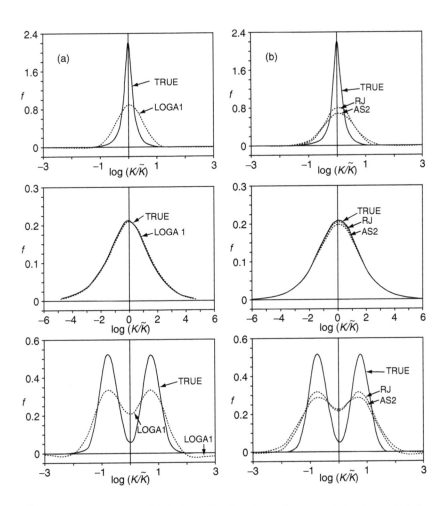

Figure 5.21 The LOGA distribution functions and the true distribution functions A, B, and C as shown in Figure 5.20. (a) The LOGA1 ($\beta = 0.7$) distribution functions, (b) the RJ ($\beta = 0.79$) and AS2 (log a $\rightarrow$ 0) ($\beta = 1.0$) distribution functions (from M.M. Nederhof, W.H. van Riemsdijk, and L.K. Koopal, *J. Colloid Interf. Sci.*, 135, 410, 1990; L.K. Koopal, M.M. Nederhof, and W.H. van Riemsdijk, *Prog. Colloid Polym. Sci.*, 81, 1990. With permission).

required. A good result is obtained with a smoothing spline procedure [316,317] that provides continuous derivatives up to the fourth derivative. The optimal solution is obtained when a smoothing is achieved which corresponds to the experimental error:

$$\sum_{k=1}^{n} \frac{\left((\theta_{t,k} - \theta_{t,k}(\text{sp}))/\theta_{t,k}(\text{sp})\right)}{s_k^2} = n, \qquad (110)$$

where $\theta_{t,k}$ (sp) is the calculated value of $\theta_{t,k}$ after smoothing, s_k is the relative error in $\theta_{t,k}$, and n is the number of points. The value of the smoothing parameter is adjusted until Equation (110) is satisfied. A requirement for this procedure is that s_k is known. If no knowledge about s_k is available the smoothing spline method can be used in combination with cross validation [302,306,317–319]. Nederlof et al. [319] use a generalized cross validation method in combination with one or more physical constraints.

Some results obtained with the LOGA method after smoothing of the data are shown in Figure 5.22 for a 5% (a) and a 0.1% (b) error level in θ_t. The true distribution functions are the same as those in Figure 5.20(b) (A,B,C). The applied smoothing is adequate; no wild oscillations appear. However, the larger the error is, the smoother the distribution function becomes. In the case of a large error, detail in the distribution function cannot be recovered, e.g., the two peaks in the bi-gaussian distribution function merge into one smooth peak. Figure 5.22(a) clearly indicates that with an error level of 5% the simplest solution is to apply the SLF equation (89) or the NICA equation (102). For application of the LOGA method to humic materials see Refs. [55,285,319].

Using the CA or the LOGA method to obtain the intrinsic affinity distribution relies on sufficient knowledge to obtain $c_{1,\delta}^*$, hence, both the electrostatic and the specific lateral interactions have to be known. Moreover, the LOGA method as presented above can only be used for surfaces that are randomly heterogeneous. In this case $\theta_{t,1}$ is known and both the electrostatic and specific lateral interactions can be calculated for a given double layer model and χ parameter. For inorganic

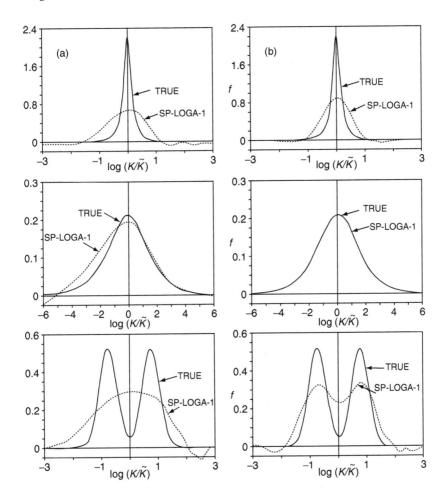

Figure 5.22 Comparison of the LOGA-1 distributions (indicated as SP-LOGA-1) obtained by using Equation (110) and the true distributions for two error levels in θ_t. (a) 5% random error, (b) 0.1% error (accurate results) (from L.K. Koopal, M.M. Nederhof, and W.H. van Riemsdijk, *Prog. Colloid Polym. Sci.*, 81, 1990. With permission).

ions it is safe to assume $\chi = 0$ and only the electrostatic interactions have to be evaluated to obtain $c_{1,\delta}^*$.

For patchwise surfaces with Equation (84) as local isotherm the situation is complex: the lateral interactions and potentials are determined by $\theta_{j,1}$ instead of $\theta_{t,1}$ and calculation of $c_{1,\delta}^*$ requires detailed knowledge of the patches. Even if this information is available, the situation is complicated because the LOGA isotherm, Equation (109), has to be adjusted to the local isotherm equation with specific values of α, β, and $(c_{1,\delta}^* K_1)_C$, depending on the electrostatic model and value of the lateral interaction parameter. Some results for uncharged adsorbates have been presented by Koopal et al. [284].

For strong specific lateral attraction ($\chi \gg 4.5$) the local isotherm is considerably steeper than the Langmuir isotherm. This makes the simple CA method an attractive option.

REFERENCES

1 T.W. Healy, D. Chan, and L.R. White, *Pure Appl. Chem.*, 52, 1207, 1980.

2 J.N. Israelachvilli, *Chemica Scripta*, 25, 7, 1985.

3 B.H. Bijsterbosch and J. Lyklema, *Adv. Colloid Interf. Sci.*, 9, 147, 1978.

4 H. Van Olphen, *Clay Colloid Chemistry*, 2nd ed., Interscience, New York, 1977.

5 P. Debye and E. Hückel, *Physik Z.*, 24, 185, 1923.

6 G. Gouy, *J. Phys.*, 9 (4):457, 1910; *Ann. Phys.*, 7(9), 129, 1917.

7 D.D. Chapman, *Philos. Mag.*, 25(6), 475, 1913.

8 O. Stern, *Z. Elektrochem.*, 30, 508, 1924.

9 R. Reeves, *Comprehensive Treatise of Electrochemistry*, Vol 1. *The Double Layer*, JO'M. Bockris, B.E. Conway, and E. Yeager, Eds., Plenum Press, New York, 1980, chapter 3.

10 S. Nir and J. Bentz, *J. Colloid Interf. Sci.*, 65, 399, 1978.

11 J. Bentz, *J. Colloid Interf. Sci.*, 90, 127, 1982.

12 A.E. James and D.J.A. Williams, *J. Colloid Interf. Sci.*, 107, 44, 1985.

13 M.K. Gilson and B.H. Honig, *Protein. Struct. Func.. Genet.*, 4, 7, 1988.

14 R.J. Hunter, *Foundations of Colloid Science*, Clarendon Press, Oxford, 1987.

15 J. Lyklema, *Fundamentals of Interface and Colloid Science*, Vol. II, Academic Press, London, 1995, chapter 3.

16 H. Ohshima, *Electrical Phenomena at Interfaces*, H. Ohshima and K. Furusawa, Eds., Surfactant Science Series 76, 2nd ed., Marcel Dekker, New York, 1998, chap. 1.

17 H. Ohshima, T.W. Healy, and L.R. White, *J. Colloid Interf. Sci.*, 90, 17, 1982.

18 T. Hiemstra and W.H. van Riemsdijk, *Colloids Surf.*, 59, 7, 1991.

19 K. Bourikas, T. Hiemstra, and W.H. van Riemsdijk, *Langmuir*, 17, 749, 2001.

20 I.C. Yeh and M.L. Berkowitz, *J. Phys. Chem.*, 110, 7935, 1999.

21 J.O'M. Bockris and A.K.N. Reddy, *Modern Electrochemistry*, Vol. 1, Plenum Press, New York, 1976, chapter 2.

22 R.J. Hunter, Zeta potential, in *Colloid Science, Principles and Applications*, Academic Press, London, 1981.

23 G.M. Torrie and J.P. Valleau, *Chem. Phys. Lett.*, 65, 343, 1979.

24 G. Gunnarsson, B. Jönsson, and H. Wennerström, *J. Phys. Chem.*, 84, 3114, 1980.

25 P. Linse, G. Gunnarsson, and B. Jönsson, *J. Phys. Chem.*, 86, 413, 1982.

26 S.L. Carnie and G.M. Torrie, *Adv. Chem. Phys.*, 56, 141, 1984.

27 P. Attard, *Adv. Chem. Phys.*, 92, 1, 1996.

28 I.L. Cooper and J.A. Harrison, *Electrochim. Acta*, 22, 519, 1977.

29 I. Larson and P. Attard, *J. Colloid Interf. Sci.*, 227, 152, 2000.

30 J.A. Wingrave, in *Oxide Surfaces*, J.A. Wingrave, Ed.,, Surfactant Science Series, 103, Marcel Dekker, New York, 2001, chapter 4.

31 Z. Abbas, M. Gunnarsson, E. Ahlberg, and S. Nordholm, *J. Colloid Interf. Sci.*, 243,11, 2001.

32 M. Gunnarsson, Z. Abbas, E. Ahlberg, E. Gobom, and S. Nordholm, *J. Colloid Interf. Sci.*, 249, 52, 2002.

33 L. Onsager and N.N.T. Samaras, *J. Phys. Chem.* 2, 528, 1934.

34 H.S. Harned and B.B. Owen, *The Physical Chemistry of Electrolyte Solutions*, Reinhold Publishing, New York, 1958.

35 S. Nordholm and A.D.J. Haymet, *Aust. J. Chem.*, 33, 2013, 1980.

36 M. Gunnarsson, Z. Abbas, E. Ahlberg, and S. Nordholm. *J. Colloid Interf. Sci.*, 274, 563, 2004.

37 L.K. Koopal, W.H. van Riemsdijk, and M.G. Roffey, *J. Colloid Interf. Sci.*, 118, 117, 1987.

38 L.K. Koopal, in *Adsorption on New and Modified Inorganic Sorbents*, A. Dabrowski and V.A. Tertykh. Eds., Studies in Surface Science and Catalysis, 99, Elsevier, Amsterdam, 1996, chapter 3.5.

39 J.Th.G. Overbeek, *J. Colloid Sci.*, 8, 593, 1953.
40 J.T. Davies and E.K. Rideal, *Interfacial Phenomena*, Academic Press, New York, 1961, chapter 2.
41 G. Bolt, *Soil Chemistry: B, Physico-chemical Models*, G.H. Bolt, Ed., 2nd ed., Elsevier, Amsterdam, 1982, chapter 3.
42 C. Tanford, *Physical Chemistry of Macromolecules*, 5th ed., John Wiley & Sons, New York, 1967.
43 D. Stigter and K.A. Dill, *Biochemistry*, 29, 1262, 1990.
44 T. Odijk and A.C. Houwaart, *J. Polym. Sci. Polym. Phys. Ed.*, 16, 627, 1978.
45 M. Fixman and J. Skolnick, *Macromolecules*, 11 863, 1978.
46 D. Stigter, *Macromolecules*, 15, 635, 1982; 18, 619, 1985.
47 D. Stigter, *Cell Biophys.*, 11, 139, 1987.
48 T.W. Healy and L.R. White, *Adv. Colloid Interf. Sci.*, 9, 303, 1978.
49 R.D. James, G.A. Parks, in *Surface and Colloid Science*, Vol. 12, E. Matijevic. Ed., Plenum Press, New York, 1982, chapter 2.
50 R. Parsons, *Handbook of Electrochemical Constants*, Butterworths, London, 1959.
51 L, Bousse, *J. Chem. Phys.*, 76,5128, 1982.
52 L. Bousse, N.F. De Rooi, and P. Bergveld, *Surf. Sci.*, 135, 479, 1983.
53 R.E.G. Van Hal, J.T.C. Eijkel, and P. Bergveld, *Adv. Colloid Interf. Sci.*, 69, 31, 1996.
54 (a) J.C.M. De Wit, W.H. van Riemsdijk, and L.K. Koopal, *Environ. Sci. Technol.*, 27, 2005, 1993; (b) J.C.M. De Wit, W.H. van Riemsdijk, and L.K. Koopal, in *Heavy Metals in the Hydrological Cycle*, M. Astruc and J.N. Lester, Eds., Selper, London, 1988, p. 386.
55 J.C.M. De Wit, W.H. van Riemsdijk, M.M. Nederhof, D.G. Kinniburgh, and L.K. Koopal, *Anal. Chim. Acta.*, 232, 189, 1990.
56 R.O. James, J.A. Davis, and J.O. Leckie, *J. Colloid Interf. Sci.*, 65, 331, 197.
57 T. Hiemstra, W.H. van Riemsdijk, and G.H. Bolt, *J. Colloid Interf. Sci.*, 133, 91, 1989.
58 T. Hiemstra, J.C.M. De Wit, and W.H. van Riemsdijk, *J. Colloid Interf. Sci.*, 133 105, 1989.
59 W.H. van Riemsdijk and T. Hiemstra, in *Mineral–Water Interfacial Reactions, Kinetics and Mechanisms*, D.L. Sparks and T.J. Grundl, Eds., ACS Symposium Series 715, American Chemical Society, Washington, DC, 1996, chapter 5.
60 T. Hiemstra and W.H. van Riemsdijk, *Langmuir*, 15, 8045, 1999.

61 G.H. Bolt, *J. Phys. Chem.*, 61, 1166, 1957.
62 D.E. Yates, The Structure of the Oxide/Aqueous Electrolyte Interface, Ph.D. thesis, University of Melbourne, Melbourne, 1975.
63 C.M. Eggelston and G. Jordan, *Geochim. Cosmochim. Acta.*, 62, 1919, 1998.
64 L.K. Koopal, *Electrochim. Acta.*, 41, 2293, 1996.
65 A. Homola and R.O. James, *J. Colloid Interf. Sci.*, 59, 123, 1977.
66 I.H. Harding and T.W. Healy, *J. Colloid Interf. Sci.*, 89, 185, 1982.
67 L.K. Koopal and W.H. van Riemsdijk, *J. Colloid Interf. Sci.*, 128, 188, 1989.
68 I.H. Harding and T.W. Healy, *J. Colloid Interf. Sci.*, 107, 382, 1985.
69 I.R. Atkinson, A.M. Posner, and J.P. Quirk, *J. Phys. Chem.*, 21, 550, 1967.
70 P.W. Schindler and H.R. Kamber, *Helv. Chim. Acta.*, 53, 1781, 1968.
71 W. Stumm, C.P. Huang, and S.R. Jenkins, *Croat. Chem. Acta.*, 42, 223, 1970.
72 D.E. Yates, S. Levine, and T.W. Healy, *J. Chem. Soc. Faraday Trans.*, Part I, 70, 1807, 1974.
73 J.W. Bowden, A.M. Posner, and J.P. Quirk, *Aust. J. Soil. Res.*, 15, 121, 1977.
74 J.A. Davis, R.O. James, and J.O. Leckie, *J. Colloid Interf. Sci.*, 63, 480, 1978.
75 J.A. Davis and J.O. Leckie, *J. Colloid Interf. Sci.*, 67, 90, 1978.
76 G.H. Bolt and W.H. van Riemsdijk, in *Soil Chemistry, B. Physico-chemical Models*, 2nd ed., G.H. Bolt, Ed., Elsevier, Amsterdam, 1982, chapter 13.
77 W.H. van Riemsdijk, G.H. Bolt, L.K. Koopal, and J. Blaakmeer, *J. Colloid Interf. Sci.*, 109, 219, 1986.
78 W.H. van Riemsdijk, J.C.M. De Wit, L.K. Koopal, and G.H. Bolt, *J. Colloid Interf. Sci.*, 116, 511, 1987.
79 T. Hiemstra, W.H. van Riemsdijk, and M.G.M. Bruggenwert, *Neth. J. Agric. Sci.*, 35, 281, 1987.
80 A.W.M. Gibb and L.K. Koopal, *J. Colloid Interf. Sci.*, 134, 122, 1990.
81 L. Pauling, *J. Am. Chem. Soc.*, 51, 1010, 1929.
82 L. Pauling, *The Nature of the Electrostatic Bond*, 3rd ed., Cornell University Press, Ithaca, NY, 1967, chapter 13-6.

83 P. Jones and J.A. Hockey, *Trans. Faraday Soc.*, 67, 2679, 1971.
84 L.R. Parfitt, R.J. Atkinson, and R.S.C. Smart, *Soil Sci. Soc. Am. J.*, 39, 837, 1975.
85 T. Hiemstra, P. Venema, and W.H. van Riemsdijk, *J. Colloid Interf. Sci.*, 184, 680, 1996.
86 P. Venema, T. Hiemstra, P.G. Weidler, and W.H. van Riemsdijk, *J. Colloid Interf. Sci.*, 198, 282, 1998.
87 M. Borkovec and G.J.M. Koper, *Langmuir*, 13, 2608, 1997.
88 M. Borkovec, B. Jönsson, and G.J.M. Koper, in *Surface and Colloid Science*, E. Matijevíc, Ed., Vol. 16, Kluwer Academic/Plenum Press, 2001, p. 19.
89 C.E. Giacomelli, M.J. Avena, and C.P. DePauli, *Langmuir*, 11, 3483, 1995.
90 R. Roderigues, M.A. Blesa, and A.E. Regazzzoni, *J. Colloid Interf. Sci.*, 177, 122, 1996.
91 M. Nabavi, O. Spalla, and B. Cabane, *J. Colloid Interf. Sci.*, 160, 459, 1993.
92 L. Vayssieres, Ph.D. thesis, Université Pierre et Marie Curie, URA-CNRS 1466, Paris, 13, 1996.
93 W.F. Bleam, *J. Colloid Interf. Sci.*, 159, 312, 1996.
94 U. Schwertmann and R.M. Cornell, *Iron Oxides in the Laboratory. Preparation and Characterization*, VCH Verlag, Weinheim, Germany, 1991, chapter 5.
95 I.D. Brown, *Chem. Soc. Rev.*, 7, 359, 1978.
96 T. Hiemstra and W.H. van Riemsdijk, *J. Colloid Interf. Sci.*, 225, 94, 2000.
97 R.J.J. Rietra, T. Hiemstra, and W.H. van Riemsdijk, *J. Colloid Interf. Sci.*, 229, 199, 2000.
98 T. Hiemstra, H. Yong, and W.H. van Riemsdijk, *Langmuir*, 15, 5942, 1999.
99 K.F. Hayes, G. Redden, W. Ela, and J.O. Leckie, *J. Colloid Interf. Sci.*, 142, 448, 1991.
100 J. Lützenkirchen, P. Magnico, and P. Behra, *J. Colloid Interf. Sci.*, 170, 326, 1995.
101 L. Righetto, G. Azimonti, T. Missana, and G. Bodoglio, *Colloids Surf. A*, 95, 141, 1995.
102 L.E. Katz and K.F. Hayes, *J. Colloid Interf. Sci.*, 170, 477, 1995.
103 J. Lützenkirchen, *Environ. Sci. Technol.*, 32, 3149,1998.
104 J.M. Kleyn and J. Lyklema, *J. Colloid Interf. Sci.*, 120, 511, 1987.
105 S. Levine and A.L. Smith, *Disc. Faraday Soc.*, 52, 290 1971.
106 Y.G. Berubé and P.L. De Bruyn, *J. Colloid Interf. Sci.*, 27, 305, 1968.

107 N.H.G. Penners, L.K. Koopal, and J. Lyklema, *Colloids Surf.*, 21, 457, 1986.

108 S. Ardizonne, P. Siviglia, and S. Trasatti, *J. Electroanal. Chem.*, 122, 395, 1981.

109 P. Siviglia, A. Daghetti, and S. Trasatti, *Colloids Surf.*, 7, 15, 1983.

110 J. Westall and H. Hohl, *Adv. Colloid Interf. Sci.*, 12, 265, 1980.

111 P.W. Schindler, in *Metal Ions in Biological Systems*, H. Sigel, Ed., Vol. 18: Circulation of Metal Ions in the Environment, Marcel Dekker, New York, 1984, chapter 7.

112 H. Hohl and W. Stumm, *J. Colloid Interf. Sci.*, 55, 281, 1976.

113 P.W. Schindler and W. Stumm, in *Aquatic Surface Chemistry*, W. Stumm. Ed., John Wiley & Sons, New York, 1987, chapter 4.

114 J. Lützenkirchen, *J. Colloid Interf. Sci.*, 210, 384, 1999.

115 C.P. Huang and W. Stumm, *J. Colloid Interf. Sci.*, 43, 409, 1973.

116 C.P. Huang, in *Adsorption of Inorganics at Solid/Liquid Interfaces*, M.A. Anderson and A.J. Rubin, Eds., Ann Arbor Science, Ann Arbor, MI, 1981, chapter 5.

117 C.P. Huang, Y.S. Hsieh, S.W. Park, A.R. Bowers, and H.A. Elliot, in *Metals Speciation, Separation, and Recovery*, J.W. Patterson and R. Passino, Eds., Lewis Publishers, Chelsea, MI, 1987, p. 437.

118 D.A. Dzombak and F.M. Morel, *Surface Complexation Modeling Hydrous Ferric Oxide*, John Wiley & Sons, New York, 1990.

119 D. Chan, J.W. Perram, L.R. White, and T.W. Healy, *J. Chem. Soc. Faraday Trans.* I, 71, 1046, 1975.

120 W. Smit and C.L.M. Holten, *J. Colloid Interf. Sci.*, 78, 1, 1980.

121 L. Bousse, N.F. De Rooij, and P. Bergveld, *Surf. Sci.*, 135, 479, 1983.

122 R. Sprycha, *J. Colloid Interf. Sci.*, 102, 173, 1984.

123 J.O. Leckie, in *Metal Speciation: Theory, Analysis and Applications*, J.R. Kramer and H.E. Allen, Eds., Lewis Publishers, Chelsea, MI, 1988, chapter 2.

124 W. Rudzinski, R. Charmas, and T. Borowiecki, in *Adsorption on New and Modified Inorganic Sorbents*, A. Dabrowski and V.A. Tertykh, Eds., Studies in Surface Science and Catalysis 99, Elsevier, Amsterdam, 1996, chapter 2.2.

125 W. Rudzinski, J. Narkiewicz-Michalek, R. Charmas, M. Drach, W. Piasecki, and J. Zajac, in *Interfacial Dynamics*, N. Kallay, Ed., Surfactant Science Series 88, Marcel Dekker, New York, 2000, chapter 4.

126 M. Kosmulski, R. Sprycha, and J. Szczypa, in *Interfacial Dynamics*, N. Kallay, Ed., Surfactant Science Series 88, Marcel Dekker, New York, 2000, chapter 4.

127 J.A. Wingrave, in *Oxide Surfaces*, J.A. Wingrave, Ed., Surfactant Science Series 103, Marcel Dekker, New York, 2001, chapter 1.

128 J. Lyklema, *Kolloid ZZ Polymere*, 175, 129, 1961.

129 M.J.G. Janssen and H.N. Stein, *J. Colloid Interf. Sci.*, 111, 112, 1986.

130 A. Foissy, A. M'Pandou, J.M. Lamarche, and N. Jaffrezic-Renault, *Colloids Surf.*, 5, 63, 1982.

131 K.F. Hayes, A.L. Roe, G.E. Brown, K.O. Hodgson, J.O. Leckie, and G.A. Parks, *Science*, 238, 783, 1987.

132 G.E. Brown, in *Mineral–Water Interface Geochemistry*, M.F. Hochella and A.F. White, Eds., Reviews in Mineralogy, MAS, 23, 309, 1990.

133 L. Spadini, A. Manceau, P.W. Schindler, and L. Charlet, *J. Colloid Interf. Sci.*, 168, 73, 1994.

134 T.C. Waite, J.A. Davis, T.E. Payne, G.A. Waychunas, and N. Xu, *Geochim. Cosmochim. Acta.*, 58, 5465, 1994.

135 W.F. Bleam, *Adv. Agron.*, 46, 91, 1991.

136 H. Motschi, *Colloids Surf.*, 9, 333, 1984.

137 H. Motschi, in *Aquatic Surface Chemistry*, W. Stumm, Ed., John Wiley & Sons, New York, 1987, chapter 5.

138 M.B. McBride, *Advances in Soil Science*, Vol. 10, Springer-Verlag, New York, 1989, p. 1.

139 M.I. Tejedor-Tejedor and M.A. Anderson, *Langmuir*, 6, 602, 1990.

140 M.V. Biber and W. Stumm, *Environ. Sci. Technol.*, 28, 763, 1994.

141 X. Sun and H. Doner, *Soil Sci.*, 161, 865, 1996.

142 S. Hug, *J. Colloid Interf. Sci.*, 188, 415, 1997.

143 S. Fendorf, M.J. Eick, P. Grossl, and D.L. Sparks, *Environ. Sci. Technol.*, 31, 315, 1997.

144 J. Lyklema, *J. Colloid Interf. Sci.*, 99, 109, 1984.

145 A. De Keizer, M.R. Böhmer, T. Mehrian, and L.K. Koopal, *Colloids Surf.*, 51, 339, 1990.

146 D.W. Fuerstenau, D. Manmotian, and Raghavan, in *Adsorption from Aqueous Solutions*, P.H. Tewari, Ed., Plenum Press, New York, 1981, p. 93.

147 R.O. James and T.W. Healy, *J. Colloid Interf. Sci.*, 40, 53, 1972.

148 P.W. Schindler, in *Adsorption of Inorganics at Solid/Liquid Interfaces*, M.A. Anderson and A.J. Rubin, Eds., Ann Arbor Science, Ann Arbor, MI, 1981, chapter 7.

149 D.D. Hansmann and M.A. Anderson, *Environ. Sci. Technol.*, 19, 544, 1985.
150 M.C. Fuerstenau and K.E. Han, in *Reagents in Mineral Technology*, P. Somasundaran and B.M. Moudgil, Eds., Surfactant Science Series 27, Marcel Dekker, New York, 1988, chapter 13.
151 L.G.J. Fokkink, A. De Keizer, and J. Lyklema, *J. Colloid Interf. Sci.*, 135, 118, 1990.
152 R.O. James and T.W. Healy, *J. Colloid Interf. Sci.*, 40, 42, 1972.
153 R.O. James and T.W. Healy, *J. Colloid Interf. Sci.*, 40, 65, 1972.
154 G. Sposito, *The Surface Chemistry of Soils*, Oxford University Press, New York, 1984.
155 D.L. Dugger, J.H. Stanton, B.N. Irby, B.L. McConnell, W.W. Cummings, and R.W. Maatman, *J. Phys. Chem.*, 68, 757, 1964.
156 D.G. Kinniburgh, M.L. Jackson, and J.K. Syers, *Soil Sci. Soc. Am. J.*, 40, 796, 1976.
157 P.W. Schindler, B. Fürst, R. Dick, and R. Wolf, *J. Colloid Interf. Sci.*, 55, 469, 1976.
158 R.M. McKenzie, *Aust. J. Soil. Res.*, 18, 61, 1980.
159 F.J. Hingston, A.M. Posner, and J.D. Quirk, *J. Soil Sci.*, 23, 177, 1972.
160 J.F. Ferguson and J. Gravis, *J. Colloid Interf. Sci.*, 54, 391, 1976.
161 L. Sigg and W. Stumm, *Colloids Surf.*, 2, 101, 1980.
162 D.G. Kinniburgh and M.L. Jackson, in *Adsorption of Inorganics at Solid/Liquid Interfaces*, M.A. Anderson and A.J. Rubin, Eds., Ann Arbor Science, Ann Arbor, MI, 1981, chapter 3.
163 E.A. Jenne Ed., *Adsorption of Metals by Geomedia*, Academic Press, San Diego, USA, 1998.
164 E.A. Jenne, in *Adsorption of Metals by Geomedia*, E.A. Jenne, Ed., Academic Press, San Diego, USA, 1998, chapter 1.
165 T.E. Payne, G.R. Lumpkin, and T.D. Waite, in *Adsorption of Metals by Geomedia*, E.A. Jenne, Ed., Academic Press, San Diego, USA, 1998, chapter 2.
166 R.T. Pabalan, D.R. Tuner, F.P. Bertetti, and J.D. Prikryl, in *Adsorption of Metals by Geomedia*, E.A. Jenne, Ed., Academic Press, San Diego, USA, 1998, chapter 3.
167 F.P. Bertetti, R.T. Pabalan, and M.G. Almendarez, in *Adsorption of Metals by Geomedia*, E.A. Jenne, Ed., Academic Press, San Diego, USA, 1998, chapter 4.
168 R. Crawford, D.E. Mainwaring, and I.H. Harding, in *Surfaces of Nanoparticles and Porous Materials*, J.A. Schwarz and C.I. Contescu, Eds., Surfactant Science Series 78, Marcel Dekker, New York, 1999, p. 675.

169 E. Baumgarten, in: *Surfaces of Nanoparticles and Porous Materials*, J.A. Schwarz and C.I. Contescu, Eds., Surfactant Science Series 78, Marcel Dekker, New York, 1999, p. 711.

170 J.S. Geelhoed, T. Hiemstra, and W.H. van Riemsdijk, *Geochim. Cosmochim. Acta*, 61, 2389, 1997.

171 J.W. Bowden, S. Nagarajah, N.J. Barrow, A.M. Posner, and J.P. Quirk, *Aust. J. Soil Res.*, 18, 49, 1980.

172 N.J. Barrow, J.W. Bowden, A.M. Posner, and J.P. Quirk, *Aust. J. Soil Res.*, 19, 309, 1981.

173 L.G.J. Fokkink, A. De Keizer, and J. Lyklema, *J. Colloid Interf. Sci.*, 118, 454, 987.

174 N.J. Barrow, *Reactions with Variable-Charge Soils*, Kluwer, (M Nijhoff), Dordrecht, 1987, chapter 3.

175 F.J. Hingston, in *Adsorption of Inorganics at Solid/Liquid Interfaces*, M.A. Anderson and A.J. Rubin, Eds., Ann Arbor Science, Ann Arbor, MI, 1981, chapter 2.

176 A.N. Frumkin, *Z. Phys. Chem.*, 116, 466, 1925.

177 J. Westall, in *Particulates in Water*, M.C. Kavanaugh and J.O. Leckie, Eds., Advances in Chemistry Series, No 189, American Chemical Society, Washington, 1980, chapter 2.

178 M. Borkovec and J. Westall, *J. Electroanal. Chem.*, 150, 325, 1983.

179 J.C. Westall, J.L. Zachary, and F Morel, MINEQL Technical Note no 18, Ralph M Parsons Laboratory, MIT, Cambridge, MA, USA, 1976.

180 M. Keizer and W.H. van Riemsdijk, ECOSAT Manual 4.4, Department of Environmental Sciences, Section Soil Quality, Wageningen University, Wageningen, NL, 1998.

181 J.C. Westall, FITEQL Manual 2.1, Department of Chemistry, Oregon State University, Corvallis, OR, USA, 1982.

182 A. Breeuwsma and J. Lyklema, *Trans. Faraday Soc.*, 52, 324, 1971.

183 M. Schudel, S.H. Behrens, H. Holthoff, R. Kretzschmar, and M. Borkovec, *J. Colloid Interf. Sci.*, 196, 241, 1997.

184 M. Gunnarsson, A.M. Jakobsson, S. Ekberg, Y. Albinsson, and E. Ahlberg, *J. Colloid Interf. Sci.*, 231, 326, 2000.

185 M. Gunnarsson, M. Rasmusson, S. Wall, E. Ahlberg, and J. Ennis, *J. Colloid Interf. Sci.*, 240, 448, 2001.

186 M.M. Benjamin and J.O. Leckie, *J. Colloid Interf. Sci.*, 79, 209, 1981.

187 P. Venema, T. Hiemstra, and W.H. van Riemsdijk, *J. Colloid Interf. Sci.*, 181, 45, 1996.

188 D.R. Nagaraj, in *Reagents in Mineral Technology*, P. Somasundaran and B.M. Moudgil, Eds., Surfactant Science Series 27, Marcel Dekker, New York, 1988, chapter 9.
189 S. Xu and J.B. Harsh, *Soil Sci. Soc. Am. J.*, 54, 357, 1990.
190 R.P.J.J. Rietra, T. Hiemstra, and W.H. van Riemsdijk, *Geochim. Cosmochim. Acta*, 63, 3009, 1999.
191 T. Hiemstra and W.H. van Riemsdijk, *J. Colloid Interf. Sci.*, 179, 488, 1996.
192 W.H. van Riemsdijk and T. Hiemstra, in *Metals in Groundwater*, D. Brown, M. Perdue, and H.E. Allen, Eds., Lewis Publishers, Chelsea, MI, 1993, p. 35.
193 P. Venema, T. Hiemstra, and W.H. van Riemsdijk, *J. Colloid Interf. Sci.*, 183, 515, 1996.
194 P. Venema, T. Hiemstra, and W.H. van Riemsdijk, *J. Colloid Interf. Sci.*, 192, 94, 1997.
195 T. Hiemstra and W.H. van Riemsdijk, *J. Colloid Interf. Sci.*, 210, 182, 1999.
196 R.P.J.J. Rietra, T. Hiemstra, and W.H. van Riemsdijk, *J. Colloid Interf. Sci.*, 218, 511, 1999.
197 R.P.J.J. Rietra, T. Hiemstra, and W.H. van Riemsdijk, *J. Colloid Interf. Sci.*, 240, 384, 2001.
198 R.P.J.J. Rietra, T. Hiemstra, and W.H. van Riemsdijk, *Environ. Sci. Technol.*, 35, 3369, 2001.
199 J.D. Zachara, D.C. Girvin, L. Schmidt, and C.T. Resch, *Environ. Sci. Technol.*, 21, 589, 1987.
200 R.L. Parfitt, *Adv. Agron.*, 30, 1, 1978.
202 M.A. Burgess. *Metal Ions in Solution*, Ellis Horwood Publisher (Wiley), Chichester, UK, 1978.
203 B.E. Conway, *Ionic Hydration in Chemistry and Biophysics*, Elsevier, Amsterdam, 1981, chapter 29.
204 W. Stumm, H. Hohl, and F. Dalang, *Croat. Chem. Acta.*, 48, 491, 1976.
205 S. Golberg and G. Sposito, *Soil Sci. Soc. Am. J.*, 48, 772, 1984.
206 S. Golberg and G. Sposito, *Commun. Soil Sci. Plant Anal.*, 16, 801, 1985.
207 S. Golberg and R.A. Glaubig, *Soil Sci. Soc. Am. J.*, 49, 1374, 1985.
208 S. Golberg and R.A. Glaubig, *Soil Sci. Soc. Am. J.*, 50, 1442, 1986.
209 G. Sposito, J.C.M. De Wit, and R.H. Neal, *Soil Sci. Soc. Am. J.*, 52, 947, 1988.
210 K.F. Hayes and J.O. Leckie, *J. Colloid Interf. Sci.*, 115, 564, 1987.

211 J.A. Davis and J.O. Leckie, *J. Colloid Interf. Sci.*, 74, 32, 1980.

212 K.F. Hayes, C. Papelis, and J.O. Leckie, *J. Colloid Interf. Sci.*, 125, 717, 1988.

213 L.E. Katz and K.F. Hayes, *J. Colloid Interf. Sci.*, 170, 491, 1995.

214 Z. Xu, Q. Zhang, and J.A. Finch, in *Surfaces of Nanoparticles and Porous Materials*, J.A. Schwarz and C.I. Contescu, Eds., Surfactant Science Series 78, Marcel Dekker, New York, 1999, p. 593.

215 K.J. Farley, D.A. Dzombak, and F.M.M. Morel, *J. Colloid Interf. Sci.*, 106, 226, 1985.

216 V.A. Tertykh and V.V. Yanishpolskii, in *Adsorption on New and Modified Inorganic Sorbents*, A. Dabrowski and V.A. Tertykh, Ed., Studies in Surface Science and Catalysis 99, Elsevier, Amsterdam, 1996, p. 705.

217 N. Kallay, S. Zalac, and I. Kobal, in *Adsorption on New and Modified Inorganic Sorbents*, A. Dabrowski and V.A. Tertykh, Eds., Studies in Surface Science and Catalysis 99, Elsevier, Amsterdam, 1996, p. 857.

218 N. Kallay, T. Preocanin, and S. Zalac, in *Interfacial Dynamics*, N. Kallay, Ed., Surfactant Science Series 88, Marcel Dekker, New York, 2000, p. 225.

219 N. Kallay, T. Preocanin, and S. Zalac, in *Interfacial Dynamics*, N. Kallay, Ed., Surfactant Science Series 88, Marcel Dekker, New York, 2000, p. 249.

220 C.I. Contescu and J.A. Schwarz, in *Surfaces of Nanoparticles and Porous Materials*, J.A. Schwarz and C.I. Contescu, Eds., Surfactant Science Series 78, Marcel Dekker, New York, 1999, p. 51.

221 Th.F. Tadros and J. Lyklema, *J. Electroanal. Chem.*, 17, 267, 1968.

222 L.K. Koopal and J. Lyklema, *J. Electroanal. Chem.*, 100, 895, 1979.

223 A. De Keizer and J. Lyklema, *Can. J. Chem.*, 59, 1969, 1981.

224 J.H. Hildebrand, J.M. Prausnitz, and R.L. Scott, *Regular and Related Solutions*, Van Nostrand Reinhold, New York, 1970.

225 R. Defay, I. Prigogine, A. Bellemans, and D.H. Everett, *Surface Tension and Adsorption*, Longmans, London, 1986.

226 D.H. Everett, in *Colloid Science*, Vol 1, D.H. Everett, Ed., Specialist Periodical Reports, The Chemical Society, London, 1973, chapter 2.

227 D.H. Everett, *Trans. Faraday Soc.*, 61, 2478, 1965.

228 L.K. Koopal, in *Colloid Chemistry in Mineral Processing*, J.S. Laskowski and J. Ralston, Eds., Elsevier, Amsterdam, 1991, chapter 2.

229 P.J. Flory, *Principles of Polymer Chemistry*, Cornell University Press, Ithaca, NY, 1953, chapter 12.

230 D.H. Everett, *Trans. Faraday Soc.*, 60, 1803, 1964.

231 J.M. Prausnitz, R.N. Lichtenthaler, and E. De Azevedo, *Molecular Thermodynamics of Fluid-Phase Equilibria*, 2nd ed., Prentice-Hall, Englewood Cliffs, NJ, 1986.

232 R.H. Fowler and E.A. Guggenheim, *Statistical Thermodynamics*, Cambridge University Press, Cambridge, 1965, p. 429.

233 R.H. Ottewill and A. Watanabe, *Kolloid Z.*, 170, 132, 1960.

234 P. Somasundaran, T.W. Healy, and D.W. Fuerstenau, *J. Phys. Chem.*, 68, 3562, 1964.

235 A. Mpandou and B. Siffert, *J. Colloid Interf. Sci.*, 102, 138, 1984.

236 L.K. Koopal and L. Keltjens, *Colloids Surf.*, 17, 371, 1986.

237 L.K. Koopal, in *Fundamentals of Adsorption*, I., A.L. Myers and G. Belfort, Eds., American Institute of Chemical Engineers, Engineering Foundation, New York, 1983, p. 283.

238 A. De Keizer and L.G.J. Fokkink, *Colloids Surf.*, 51, 323, 1990.

239 L.K. Koopal, in *Structure–Performance Relationships in Surfactants*, K. Esumi and M. Ueno, Eds., Surfactant Science Series 112, Marcel Dekker, New York, 2003, p. iii; Idem Surfactant Science Series 76, 1997, p. 395 (1st ed).

240 L.K. Koopal and T. Goloub, in *Surfactant Adsorption and Surface Solubilization*, R, Sharma, Ed., ACS Symposium Series 615, American Chemical Society, Washington, DC, 1995, p. 78.

241 L.K. Koopal, *Z. Wasser Abwasser Forsch.*, 16, 91, 1983.

242 B.S. Kim and R.A. Hayes, *J. Ralston. Carbon*, 33, 25, 1995.

243 S.B. Haderlein, K.W. Weissmahr, and R.P. Schwarzenbach, *Environ. Sci. Technol.*, 27, 316, 1993.

244 S.B. Haderlein, K.W. Weissmahr, and R.P. Schwarzenbach, *Environ. Sci. Technol.*, 30, 612, 1996.

245 R. Kummert and W. Stumm, *J. Colloid Interf. Sci.*, 75, 373, 1980.

246 W. Stumm, R. Kummert, and L. Sigg, *Croat. Chem. Acta.*, 53, 291, 1980.

247 L. Sigg and W. Stumm, *Colloids Surf.*, 2, 101, 1981.

248 A.R. Bowers and C.P. Huang, *J. Colloid Interf. Sci.*, 105, 197, 1985.

249 A.R. Bowers and C.P. Huang, *J. Colloid Interf. Sci.*, 110, 575, 1986.

250 N. Kallay and E. Matijevic, *Langmuir*, 1, 195, 1985.

251 B. Muller and L. Sigg, *J. Colloid Interf. Sci.*, 148, 517, 1992.

252 B. Nowack and L. Sigg, *J. Colloid Interf. Sci.*, 177, 106, 1996.

253 M.A. Ali and D.A. Dzombak, *Environ. Sci. Technol.*, 28, 2357, 1996.

254 C.R. Evanko and D.A. Dzombak, *Environ. Sci. Technol.*, 32, 2846, 1998.

255 C.R. Evanko and D.A. Dzombak, *J. Colloid Interf. Sci.*, 214, 189, 1999.

256 J.F. Boily, P. Persson, and S. Sjoberg, *Geochim. Cosmochim. Acta.*, 64, 3453, 2000.

257 J.D. Filius, T. Hiemstra, and W.H. van Riemsdijk, *J. Colloid Interf. Sci.*, 195, 368, 1997.

258 J.D. Filius, J.C.L. Meeussen, T. Hiemstra, and W.H. van Riemsdijk, *J. Colloid Interf. Sci.*, 244, 31, 2001.

259 I. Langmuir, *J. Am. Chem. Soc.*, 40, 1361, 1918.

260 M. Jaroniec and R. Madey, *Physical Adsorption on Heterogeneous Solids*, Elsevier, Amsterdam, 1988.

261 W. Rudzinski and D.H. Everett, Adsorption of Gases on Heterogeneous Surfaces, Academic Press, London, 1992.

262 M. Jaroniec, *Adv. Colloid Interf. Sci.*, 18, 149, 1983.

263 M. Jaroniec and P. Bräuer, *Surf. Sci. Rep.*, 6, 65, 1986.

264 W.A. House, *Colloid Science* 4, D.H. Everett, Ed., Specialist Periodical Reports, Royal Society of Chemistry, London, 1983, chapter 1.

265 W.H. van Riemsdijk, L.K. Koopal, and J.C.M. De Wit, *Neth. J. Agric. Sci.*, 35, 241, 1987.

266 W.H. van Riemsdijk, G.H. Bolt, and L.K. Koopal, in *Interactions at the Soil Colloid–Soil Solution Interface*, G.H. Bolt, M.F. de Boodt, and M.H.B. Hayes, Eds., Kluwer Academic Publishers (M. Nijhoff), Dordrecht, 1990, chapter 3.

267 W.H. van Riemsdijk and L.K. Koopal, *Environmental Particles*, Vol. 1, J. Buffle and H.P. van Leeuwen, Eds., Lewis Publishers, London, 1992, p. 455.

268 W. Rudzinski, R. Charmas, S. Partyka, and J.Y. Bottero, *Langmuir*, 9, 2641, 1993.

269 R. Sips, *J. Phys. Chem.*, 16, 490, 1948.

270 R. Sips, *J. Phys. Chem.*, 18, 1024, 1950.

271 D.G. Kinniburgh, J.A. Barker, and M. Whitfield, *J. Colloid Interf. Sci.*, 95, 370, 1983.

272 D.A. Dzombak, W. Fish, and F.M.M. Morel, *Environ. Sci. Technol.*, 20, 669, 1986.

273 D.A. Dzombak, W. Fish, and F.M.M. Morel, *Environ. Sci. Technol.*, 20, 676, 1986.

274 G. Sposito, *CRC Crit. Rev.Environ. Control*, 16, 193, 1986.

275 A. Katchalski and P. Spitnik, *J. Polym. Sci.*, 2, 432,1947.

276 H.P. Gregor, L.B. Luttinger, and E.M. Loebl, *J. Phys. Chem.*, 59, 366, 1955.

277 M. Mandel, *Eur. Polym. J.*, 6, 807, 1970.

278 I.H. Suffet and M.J. McGuire, Eds., *Activated Carbon Adsorption of Organics from the Aqueous Phase*, Ann Arbor Science, Ann Arbor, MI, 1980.

279 J.F. Kuo and T.F. Yen, *J. Colloid Interf. Sci.*, 121, 220, 1988.

280 J.C.M. De Wit, W.H. van Riemsdijkm, and L.K. Koopal, in *Metal Speciation, Separation and Recovery*, Vol, II, J.W. Patterson and R. Passino, Eds., Lewis Publishers, Chelsea, MI, 1990, p. 329.

281 L.K. Koopal, T. Saito, J.P. Pinheiro and W.H. van Riemsdijk, colloids surfaces A, special issue 'Interfaces against Pollution', 2005 (in press). L.K. Koopal, W.H. van Riemsdijk, J.C.M. de Wit, and M.H. Benedetti, *J. Colloid Interf. Sci.*, 166, 51 1994.

282 L.K. Koopal, W.H. van Riemsdijk, and D.G. Kinniburgh, *Pure Appl. Chem.*, 73, 2005, 2001.

283 M.M. Nederhof, W.H. van Riemsdijk, and L.K. Koopal, *J. Colloid Interf. Sci.*, 135, 410, 1990.

284 L.K. Koopal, M.M. Nederhof, and W.H. van Riemsdijk, *Prog. Colloid Polym. Sci.*, 81, 1990.

285 M.M. Nederhof, W.H. van Riemsdijk, and L.K. Koopal, *Environ. Sci. Technol.*, 26, 763, 1992.

286 J.C.M. de Wit, W.H. van Riemsdijk, and L.K. Koopal, *Environ. Sci. Technol.*, 27, 2015, 1993.

287 T. Saito, S. Nagasaki, S. Tanaka, and L.K. Koopal, Colloids Surfaces A, special issue 'Interfaces against Pollution', 2005 (in press) M.J. Avena, L.K. Koopal, and W.H. van Riemsdijk, *J. Colloid Interf. Sci.*, 217, 37, 1999.

288 M.F. Benedetti, C.J. Milne, D.G. Kinniburgh, W.H. van Riemsdijk, and L.K. Koopal, *Environ. Sci. Technol.*, 29, 446,1995.

289 D.G. Kinniburgh, W.H. van Riemsdijk, L.K. Koopal, M. Borkovec, M.F. Benedetti, and M.J. Avena, *Colloids Surf. A*, 151, 147, 1999.

290 C.J. Milne, D.G. Kinniburgh, and E. Tipping, *Environ. Sci. Technol.*, 35, 2049, 2001.

291 J.P. Pinheiro, A.M. Mota, and M.F. Benedetti, *Environ. Sci. Technol.*, 34, 5137, 2000.

292 I. Christl and R. Kretzschmar, *Environ. Sci. Technol.*, 35, 2505, 2001.

293 I. Christl, C.J. Milne, D.G. Kinniburgh, and R. Kretzschmar, *Environ. Sci. Technol.*, 35, 2512, 2001.
294 B. Noble, in *The State of the Art in Numerical Analysis*, D. Jacobs, Ed., Academic Press, San Diego, 1979, p. 915.
295 L.K. Koopal and C.H.W. Vos, *Colloids Surf.*, 14, 87, 1985.
296 C.H.W. Vos and L.K. Koopal, *J. Colloid Interf. Sci.*, 105, 18b, 1985.
297 L.K. Koopal and C.H.W. Vos, *Langmuir*, 9, 2593, 1993.
298 D.L. Philips, *J. Assoc. Comput. Mach.*, 9, 84, 1962.
299 A.N. Tikhonov, *Sov. Math.*, 4, 1035 and 1624, 1963.
300 S. Twomey, *J. Assoc. Comput. Mach.*, 10, 97, 1963.
301 W.A. House, *J. Colloid Interf. Sci.*, 67, 166, 1978.
302 P.H. Merz, *J. Comput. Phys.*, 38, 64, 1980.
303 M. Von Szombathely, P. Brauer, and M. Jaroniec, *J. Comput. Chem.*, 13, 17, 1992.
304 J. Jagiello, *Langmuir*, 10, 2778, 1994.
305 J. Jagiello, T.J. Bandosz, K. Putyera, and J.A. Schwarz, *J. Colloid Interf. Sci.*, 172, 341, 1995.
306 M.M. Nederhof, W.H. van Riemsdijk, and L.K. Koopal, in *Heavy Metals in the Environment*, J.P. Vernet, Ed., Elsevier, Amsterdam 1991, p. 365.
307 M.M. Nederhof, J.C.M. de Wit, W.H. van Riemsdijk, and L.K. Koopal, *Environ. Sci. Technol.*, 27, 846, 1993.
308 S.S. Roginsky, *Adsorption and Catalysis on Heterogeneous Surfaces*, Academy of Sciences of the USSR, Moscow, 1948 (in Russian).
309 J.P. Hobson, *Can. J. Phys.*, 43, 1934 and 1941, 1965.
310 L.B. Harris, *Surf. Sci.*, 10, 129, 1968.
311 C.C. Hsu, B.W. Wojciechowski, W. Rudzinski, and J. Narkiewicz, *J. Colloid Interf. Sci.*, 67, 292, 1978.
312 W. Rudzinski, J. Jagiello, and Y. Grillet, *J. Colloid Interf. Sci.*, 87, 478, 1982.
313 K. Ninomiya and J.D. Ferry, *J. Colloid Interf. Sci.*, 14, 36, 1959.
314 D.L. Hunston, *Anal. Biochem.*, 63, 99, 1975.
315 A.K. Thakur, P.J. Munson, D.L. Hunston, and D. Rodbard, *Anal. Biochem.*, 103, 240, 1980.
316 C.M. Reinsch, *Numer. Math.*, 10, 177,1967; 16, 451, 1971.
317 H.J. Woltring, *Adv. Eng. Software*, 8(2), 104, 1986.
318 P. Craven and G. Wahba, *Numer. Math.*, 31, 377,1979.
319 M.M. Nederlof, W.H. van Riemsdijk, and L.K. Koopal, *Environ. Sci. Technol.*, 28, 1037, 1994.

6

Fundamentals of Homopolymers at Interfaces and their Effect on Colloidal Stability

G.J. Fleer and F.A.M. Leermakers

Laboratory of Physical Chemistry and Colloid Science, Wageningen University, Wageningen, The Netherlands

I. INTRODUCTION

Theoretical insights into the behavior of polymers at interfaces are to a large extent based on a self-consistent field (SCF) analysis of the problem. The basic equation that describes the interfacial layer is the Edwards equation [1–4], which considers the chains as walks where each step is weighted in a self-consistent potential field, wherein the excluded volume of the chains and nonideal interactions between the segments are (to first order) accounted for. The presence of the surface is incorporated essentially through the boundary condition. We have to mention, however, that in this approach the chains in the bulk are ideal (Gaussian) and do not change their conformation due to the presence of excluded-volume interactions; the swelling in good solvents is disregarded. Consequently, SCF-results for polymers at interfaces are only approximate. For example, the adsorbed amount is systematically underestimated slightly. Apart from this inherent drawback, mean-field results are internally consistent and thermodynamically sound. They allow for interpretations in terms of simple scaling behavior, which can be supplemented with the numerical prefactors. The predictions can readily be compared to the experimental results. Therefore it is worthwhile to investigate the problem in this simple scheme and consider the various intricacies. The results may, in the end, be extrapolated to mimic real systems by exchanging the mean-field power laws by scaling laws consistent with chain swelling. By doing so, we lose information on the numerical prefactors, however.

Starting some 25 years ago, exact SCF results were generated numerically by Scheutjens et al. [5–8] for a large variety of systems and regimes. For obtaining numerical solutions of the Edwards equation it is necessary to discretize the equations; these authors used a lattice. In the lattice approach the equations become rather transparent and in this review we will give a flavor of the method. More recently, the insights were advanced due to analytical work based upon the ground-state approximation and extensions of it [9–13]. In this analytical work, the way in which the boundary conditions are implemented is crucial for matching the analytical predictions with the numerical results. One of the key features is that the continuous description of polymers at interfaces breaks down on length scales smaller than the segment size; this is especially important for the case of adsorption. A discrete, Stern-layer type boundary condition solves this problem [12,13]. We will pay attention to some of the pertinent details. We discuss several regimes for polymer adsorption and show that the knowledge of the behavior of polymers in solution is a prerequisite to understand polymers at interfaces.

Polymers are used in colloidal systems to influence the colloidal stability. An insight into the physics of polymers at isolated surfaces is required to understand the effect of polymers in the confined space between two colloidal particles. There exists a general procedure, based on the free energy as a function of the order parameter, to predict how polymers influence the pair interaction between surfaces. Essential in this general procedure is the identification of the correct order parameter in the system. De Gennes [2] identified the ground-state end-point distribution function (which obeys the ground-state equivalent of the Edwards equation) as the order parameter. Because this order parameter is symmetric with respect to the midplane the result is attraction. When the ground-state approximation is sufficient, we thus expect always attraction between adsorbed polymer layers. However, the ground-state approximation fails to include the effect of tails. Tails occur at the periphery of the layer and therefore one might expect nonmonotonic force curves, with repulsion due to tails at weak overlap, whereas at shorter separation the attraction must dominate. This repulsive contribution is most relevant for those cases that tails are important, i.e., for finite chain lengths around the overlap concentration.

We take the numerically exact predictions as a source of inspiration to find analytical approximations. As a result our approach to the colloidal interactions deviates from the De

Gennes approach [2] on some subtle aspects. For example, in the superposition approximation needed to get an estimate of the concentration profile between two particles. We do not superimpose the ground-state eigenfunctions, but the volume fractions. This choice is made because much better agreement with the numerical SCF data (where no ground-state approximation is involved) is obtained. This has no consequences for the sign of interactions; both the ground-state eigenfunction and the concentration are symmetric 'order parameters'. Anyhow, from this review it will become clear that much is left to do on the issues of polymers in confined spaces.

The equilibrium behavior of homopolymers at interfaces is basically well understood as to the generic behavior, where the chemical details of the chains do not matter. There are at least three important issues for applications that will not be covered in this review. (i) Intuitively, one expects that nongeneric effects, i.e., as a result of the chemistry on the monomer length scale, become important for the colloidal interaction for not too long chains. Experimentally this aspect is potentially important, but it remains largely unexplored from a theoretical point of view. (ii) It remains an important challenge in the field to go beyond the mean-field SCF approximations. Here computer simulations undoubtedly are important. (iii) Any off-equilibrium effect is expected to add an extra repulsion to the pair interaction. This may well be of significant importance in practical applications and is the basis of the so-called 'restricted equilibrium Ansatz' for interacting polymer layers.

The basic concepts of polymers in solution are briefly reviewed in Section II, using the governing Edwards equation as the starting point. It is not generally known that this equation reproduces the main scaling features for the bulk solution, provided the excluded volume is included properly in the molecular field.

In Section III we discuss polymers near an isolated solid–liquid interface. We present the basis of the numerically exact lattice method. Numerical results are useful, but should not be the final goal. Analytical work can help to analyze and visualize the numerical results and to interpret the trends. For isolated chains at the interface there is an exact partition function, which gives insight in the adsorption–desorption transition. For mutually interacting chains, for which no exact description is available, we will discuss results obtained from the ground-state approximation in various regimes (good and theta solvents, dilute and semidilute solutions, adsorption and depletion). We then show the importance of tails, by going (slightly) beyond the ground-state approximation.

In the last Section IV of this review, we discuss how homopolymers influence the interaction between two particles. For depletion the situation is rather good: very accurate (and simple) analytical predictions are obtained which closely follow the numerical results. For adsorption the situation is less advanced. The numerical results show that the effects of tails are important, giving rise to a small repulsion at relatively large separation. Here the ground-state approximation fails. Using heuristic arguments the effect of tails is included in our analysis, but we still rely on the numerical predictions to obtain the full picture.

Many interesting further developments of the SCF method deal with extensions such as copolymers [8,14–17] which in selective solvents self-assemble [18–20]), polyelectrolytes [21] (both annealed [22,23] and quenched [24–26]), polymer mixtures (polydispersity effects [27], displacement [28]), effects of chain architecture (rings [29], combs [30], stars [23,25]), adsorption at curved surfaces [31], adsorption of (semiflexible) chains at liquid–liquid interfaces [32,33], adsorption on inhomogeneous surfaces [34], and last but not least polymer brushes [26,35–40]. These extensions will not be treated here, and the reader is referred to the literature.

II. HOMOPOLYMERS IN SOLUTION

A system of many (long) polymer chains in a monomeric solvent is actually very complex. The hardest issue to deal with is the excluded-volume problem. The polymer chains should not only be internally nonoverlapping (self-avoiding), but should not intersect with neighboring chains either. We may distinguish *dilute* solutions where the polymer chains are basically isolated and float freely in solution, *semidilute* solutions where the chains partially overlap, and the *concentrated* regime where there is only little solvent in between the segments. It is instructive to review the effects of excluded-volume interactions. In Section II.A we first consider a dilute polymer solution. In this case there are only intramolecular excluded-volume effects. In Sections II.B and II.C, we discuss the effect of intermolecular excluded-volume interactions.

A. Dilute solutions

To appreciate the effect of the finite volume of the chain segments, it is useful to realize that a chain without volume can be described exactly. This model of an ideal chain is known as the Gaussian chain. There exists a strong analogy between the diffusion of a Brownian particle and the path followed by such a chain. We are not interested *per se* in one particular conformation, but rather in the average of all possible conformations. They follow from the diffusion equation (for a zero field)

$$\frac{\partial G(r,s|0,0)}{\partial s} = \frac{1}{6} \nabla_r^2 G(r,s|0,0).$$ (1)

The notation r, s indicates the spatial position r of a point s on the chain, where s is the distance (measured along the contour) from the starting point of the walk; s runs from 0 (start) to N (end point of the chain), where N is the contour length or chain length. The quantity $G(r,s|0,0)$ is the end-point distribution for the chain section from zero to s, giving the statistical weight that point s ends up at coordinate r given that the start $(s=0)$ is at the origin $(r=0)$. In spherical coordinates $\nabla_r^2 = \frac{2}{r}\frac{\partial}{\partial r} + \frac{\partial^2}{\partial r^2}$. The end-point distribution that satisfies this differential equation is a Gaussian. For $s=N$ the solution is

$$G(r,N|0,0) = C \cdot \exp\left(-\frac{3r^2}{2N}\right),$$ (2)

where the normalization constant C is proportional to $N^{-3/2}$ such that the normalized partition function, given by the volume integral over the end-point distribution, is unity and the corresponding Helmholtz energy is zero. The reference of the free energy is thus chosen to be the random coil; this has serious consequences for the adsorption problems as we will see below. As an example how Equation (2) can be used, we note that the Helmholtz energy F (in units kT) for a particular end-to-end distance r is given by $-\ln G(r, N|0, 0)$ so that

$$F(r) = F(0) + \frac{3}{2} \frac{r^2}{N} \tag{3}$$

from which it follows that the force $-\partial F/\partial r$ needed to stretch the chain is proportional to r. From the Gaussian end-point distribution it can be derived that the coil size is given by $R_g = \sqrt{N/6}$ where N is the chain length (which for large N is nearly the same as the number of segments in the chain). Here and below all length scales are normalized to the segment size and all energies to the thermal energy kT.

Flory [41] showed that the effect on the coil size of volume interactions is a function of the solvent quality. It is instructive to first discuss the case of a good solvent. The first step is to calculate the number of segment–segment contacts in the coil. Let us assume that the chain assumes a size R_F (which is larger than the Gaussian radius R_g). The number of contacts is proportional to N^2/R_F^3; each of them gives an energy contribution of order kT. The extension reduces the entropy, according to Equation (3), by an amount proportional to R_F^2/N. This gives a free energy as a function of R_F of the following form:

$$F \cong \frac{N^2}{R_F^3} + \frac{R_F^2}{N}. \tag{4}$$

Minimizing this equation leads directly to the famous result $R_F \propto N^{3/5}$.

To go beyond this scaling argument, it is necessary to modify the diffusion equation such that the segments have volume. The result is known as the Edwards equation that adds a potential energy term to Equation (1):

$$\frac{\partial G(r, s|0, 0)}{\partial s} = \frac{1}{6} \nabla_r^2 G(r, s|0, 0) - u(r)G(r, s|0, 0), \tag{5}$$

where $u(r)$ is the potential field generated by the excluded-volume interactions in the system. Below (Equations (18) and (31)) we will find a general (mean-field) expression for this field. For homopolymers the field does not depend on s because all segments are identical and experience the same 'external' field. As we will show, in a good solvent one can replace $u(r)$ by the volume fraction of segments $u(r) = \varphi(r)$ (see Equation (54) in Section III.E3). At this point the problem is well defined. The procedure to solve the equation is first to obtain the correct end-point distributions and then relate these to the segment densities. We will explicitly go through this method in Sections III.C1 and III.C2.

The result for a central chain in a spherical coordinate system may be summarized as follows [42]. The radial concentration distribution, with the first segment in the center of the spherical coordinate system, assumes a power-law form [2] $\varphi(r) \propto r^{-4/3}$ for relative small values of r, with an exponential tail for larger values of r. The integral $\int_0^{R_F} r^2 \varphi(r) dr$ over this concentration profile scales as $R_F^{5/3}$ and should correspond to the number or segments N. Hence, also in this way we find $R_F \approx N^{3/5}$, in agreement with the Flory result. Interestingly, the crossover from the power-law to the exponential part occurs precisely at this length scale. The polymer concentration φ_{co} at the crossover point between the power-law and the exponential part of the profile thus scales as $\varphi_{co} \propto (R_F)^{-4/3} \propto N^{-0.8}$. In the next section we shall see that this is the same scaling as for the overlap concentration φ_{ov} which describes the crossover from dilute to semidilute solutions.

There exists a particular solvent quality for which the pair interactions vanish, called the theta condition. We may use the Edwards equation again to prove that for these conditions the coil size scales as $N^{1/2}$, as for Gaussian chains. Theta conditions can be considered in Equation by using $u(r) \approx \varphi(r)^2$ (see Equation (54) in Section III.E3). The radial concentration distribution has again a self-similar profile with a slightly smaller exponent [2] $\varphi(r) \propto r^{-1}$. The crossover to the exponential part occurs at a distance $r = R_g$ where the concentration is $\varphi_{co} \propto N^{-1/2}$.

In a poor solvent the chain collapses and the coil assumes a homogeneous polymer concentration that depends slightly on the solvent quality. The radius now scales with the chain length to the power 1/3.

Summarizing, polymers in dilute solutions are characterized by a coil size, or the radius of gyration. For flexible chains $R_g \propto N^\alpha$, where α depends on the solvent quality: $\alpha = 3/5$ in a good solvent and $\alpha = 1/2$ under theta conditions. In poor solvents the chain is completely collapsed and $\alpha = 1/3$.

B. Semidilute Solutions

Flory [41] was the first to realize that the intramolecular excluded-volume effect can be screened by intermolecular excluded-volume interactions. To understand this we must consider the effect of finite polymer concentrations. The first issue is what happens when the space is completely filled with polymer coils of radius R_g. Do the chains shrink in the sense of reducing their size or do they interpenetrate upon a further increase in concentration? We will see that in a good solvent they do both. The polymer concentration at overlap can be estimated from the condition that the volume fraction of polymer segments within the coil becomes similar to the overall volume fraction in the system. The average volume fraction in the coil is given by N/R_g^3 and thus the overlap concentration in a good solvent is $\varphi_{ov} \propto N^{-0.8}$. Note that this overlap condition coincides with the crossover concentration found for the excluded-volume coil.

What happens next, when the concentration of polymer exceeds the overlap concentration, may again be estimated from the Edwards equation. The way to do this is to consider a central chain surrounded by free chains. Now the potential field becomes $u(r) = \varphi(r) - \varphi_b$. It turns out that in this case only the power-law radial volume fraction profile remains, which crosses over to the bulk concentration at a distance identified by the correlation length ξ. At the crossover $\varphi_b = \varphi(\xi) \approx \xi^{-4/3}$. This means that the correlation length $\xi \approx \varphi_b^{-3/4} < R_g$ becomes independent of the molar mass, see Figure 6.1(A). This correlation length is sometimes called

(a)   (b)

Figure 6.1 (A) The semidilute polymer solution. The spheres are the blobs and are only drawn to guide the eye. The chain connects a series of blobs that are in a Gaussian configuration. Inside the blob the chain is swollen and excluded-volume statistics apply. With increasing polymer concentration the blob size decreases. (B) A semidilute polymer solution in the mean-field approximation. Now the imaginary blobs overlap, which means that the there are many chain fragments in each blob.

the blob size or mesh size in semidilute solution. The physical picture [2] is that the chain in a semidilute solution splits up into n blobs of N_b segments each: $n = N/N_b$. Within a blob the chain is locally swollen. This implies for good solvents $\xi = N_b^{3/5}$ or $N_b = \varphi_b^{-5/4}$. Hence $n = N\varphi_b^{5/4}$. The blobs themselves feel no self-consistent potential field ($u = 0$) and are thus Gaussian. As a result the chain can be considered as a Gaussian sequence of n blobs of size ξ each: $R_g \approx \xi\, n^{1/2} \propto \varphi_b^{-1/8} N^{1/2}$. Note that setting this equal to $R_g = N^{3/5}$ gives again $\varphi_{ov} = N^{-4/5}$. It follows that the coil gradually goes from a swollen state ($R_g = \varphi_{ov}^{-1/8} N^{1/2} = N^{3/5}$) near the overlap concentration to the Gaussian dimensions ($R_g = N^{1/2}$) in the melt.

In a theta-solvent $\xi \approx \varphi_b^{-1}$. When the correlation length equals the coil size we obtain the overlap concentration: $\varphi_{ov} \propto N^{-1/2}$. The coil size remains unaltered upon increasing the concentration, all the way up to the melt.

C. Mean-Field Picture of Semidilute Solutions

The credit for the scaling picture for polymer solutions outlined above is usually given to De Gennes [2]. The fact that this scaling picture follows completely from the Edwards diffusion equation (with the proper molecular field and constraints) is not generally known. Having said this, it is necessary to introduce yet another set of scaling results, which will be relevant when the Edwards equation is applied to the problem of polymer at interfaces. Typically, the application of the Edwards equation implies the equilibration of the chains in the adsorbed layer with chains in the bulk. In the bulk the potential field is, by definition, zero and the chains in the bulk solution are thus Gaussian (reference state). As this zero field and this Gaussian distribution apply for any solvent quality, it is clear that the physics for such mean-field solutions is not the same as outlined above. As a consequence there exists a mean-field correlation length in semidilute solutions.

Let us consider again a semidilute solution of chains that now have Gaussian dimensions and interact with the solvent. For theta solvents there is no problem because the coil size in the bulk is handled correctly: $\xi = \varphi_b^{-1}$ as found above. The inconsistency only occurs for good solvents. Let there be a mean-field correlation length ξ^{mf}. Again a chain will donate N_b segments to this blob with size ξ^{mf}. As the chain is Gaussian we have the relation $N_b \approx (\xi^{mf})^2$. The volume fraction of polymer segments in the blob is equal to the overall segment volume fraction φ_b. The interaction energy felt by this chain fragment amounts to $N_b \varphi_b$ in units of kT. Now the blob size is defined by the Ansatz that this energy is equal to unity (i.e., $1 kT$). We thus find $\xi^{mf} = \varphi_b^{-1/2}$. Indeed, it is possible to show that within one-blob segments of more than one chain come together, c.f. Figure 6.1B. Below (Equation (65) in Section III.E4) we will introduce a more general expression for the mean-field correlation length which depends on the solvency parameter χ (zero in good solvents and 0.5 in theta conditions) and which includes the limits $\xi^{mf} \sim \varphi^{-1/2}$ in a good solvent and $\xi \sim \varphi_b^{-1}$ in a theta solvent. This expression also includes the prefactors: $\xi = (3v\varphi_b)^{-1/2}$ for good solvents and $\xi = \sqrt{2/3}\varphi_b^{-1}$

for theta conditions. Here $v = 1 - 2\chi$, where χ is the Flory-Huggins solvency parameter (which is zero in a thermal good solvents).

In passing we note that for semiflexible chains the mean-field scaling is expected to be relevant in the regime where the blob size becomes of the order of the chain persistence length (the marginal regime). Also in the mean-field world there exists an overlap concentration. The overlap concentration is defined here by the value of φ_b where the blob size ξ^{mf} reaches the coil size R_g. This leads to $\varphi_{ov} \approx 2/N$, which should be compared to $N^{-4/5}$ as found above for chains with excluded-volume interactions.

III. HOMOPOLYMERS AT AN ISOLATED SURFACE

A. General Aspects

Polymers in solution accumulate at an interface when their segments have a higher affinity for surface sites than the solvent, and they tend to avoid the surface region when the solvent adsorbs preferentially. The former case is known as adsorption, and it is accompanied by drastic changes in the conformation of the polymer molecules. In the latter case, denoted as depletion, the entropy loss for chains close to the surface is not compensated by segmental adsorption energy, and the interfacial concentration is lower than in the bulk solution. For long chains the crossover from depletion to attraction occurs in a very narrow range of adsorption energies. The adsorption transition is known to be second order, the transition remains continuous [43]. In Section III.E2 we will discuss the critical region in more detail.

Experimentally it is possible to measure adsorption isotherms, for example, by monitoring the loss of material from the bulk solution. Such an isotherm gives the amount of polymer associated to the interface as a function of the equilibrium concentration. More difficult to measure is the concentration profile of polymer segments near the surface; useful techniques are neutron reflection or neutron scattering [8]. The theoretical equivalents are computed in an idealized setting. More explicitly, we will consider the surfaces to be ideally flat (no height fluctuations or surface roughness) and laterally homogeneous (no gradients in surface affinity). Then we may expect that parallel to the surface the volume fraction is rather homogeneous and that it is not a serious approximation to ignore lateral fluctuations. One can then implement a mean-field approximation and define a local volume fraction. In this case, the volume fraction profile varies only in one dimension and may be expressed as $\varphi(z)$, where z is the distance from the surface (measured again in units of the bond length). Far away from the surface, the concentration gradients have relaxed and the polymer concentration is homogeneous. This region of the system is the bulk solution where the volume fraction is φ_b. The excess amount of polymer per unit area adsorbed at the interface is found by integrating over the volume fraction profile

$$\theta^{ex} = \int_0^\infty (\varphi(z) - \varphi_b)\mathrm{d}z. \tag{6}$$

The excess number of chains per united area is found by dividing by the chain length N: $n^{ex} = \theta^{ex}/N$. The adsorption isotherm is then simply the excess amount θ^{ex} as a function of the concentration φ_b of polymer in the bulk.

Alternatively, one may choose to define a molecule to be adsorbed if it makes physical contact with the surface. The adsorbed amount defined in this way may be denoted as Γ. In adsorption from relatively dilute solutions, both measures for the adsorbed amount turn out to be virtually the same. The excess amount is conceptually less complicated and will be discussed below.

Basically, we may distinguish three regimes in an isotherm. At extremely low concentrations the adsorbed amount is so low that the adsorbed chains are isolated. In this regime the adsorbed amount increases linearly with the polymer concentration (Henry regime). At higher concentrations the adsorbed chains start to overlap and the adsorbed amount saturates. Typically, when the segments have sufficient affinity for the surface, the crossover from the Henry to the saturation regime (semipateau) occurs at a very low concentration because of the cooperativity of the adsorption process. The third regime occurs in semidilute and concentrated solutions, where the increasing bulk concentration causes the excess to decrease (now there is a significant difference between Γ and θ^{ex}). In the limit $\varphi_b \rightarrow 1$, θ^{ex} is necessarily zero whereas Γ keeps increasing with polymer concentration and reaches its maximum in the melt ($\varphi_b = 1$).

Not all the segments belonging to adsorbed polymer necessarily touch the surface. This is only the case in the Henry regime (unless the adsorption energy is very small): the adsorbed amount is then extremely low, and the amount of polymer is too low to cover all the available surface area. Segments in contact with the surface are known as *train* segments. Chain parts in between two train fragments are called *loops*. The remainder of the segments is in *tails* at the ends of the chain. The fact that the number of tails is limited to two and that the number of loops typically increases with increasing chain length could suggest that tails may be ignored for long chains. This is not necessarily the case as the tails may be very long and the fraction of segments in tails does not go to zero with increasing N (see also the final part of Section III.E6).

The importance of tails has long been an issue of debate [5,6,44], which would be not very relevant if the tails essentially would have the same contribution to the segment density profile as the loops. We know now that this is not the case and tails are important, especially for the colloidal stability for adsorbed layers in the plateau of their adsorption isotherm, in dilute solutions. The analysis of the adsorbed layer in terms of the train, loop, and tail contributions was extensively done in the early literature [6,8] of the (numerical) SCF theory and will not be repeated here. We note, however, that information on train, loop, and tail contributions may become experimentally observable with atomic force microscopy (AFM) [45].

Before discussing any calculations we make some remarks about the structure of the adsorbed layer. Due to the loops and tails, in the (semi)plateau the layer is much thicker than the segment size. On the other hand, the layer thickness cannot exceed the coil size R_g, because then the chain would have to stretch perpendicular to the surface. At weak adsorption, very close to the adsorption–desorption transition, the coils will hardly deform and we expect a thickness of the order of the coil size. When the adsorption energy is higher, the crowding of chains near the interface is important and when the bulk solution is dilute the adsorbed layer has in fact rather complex features. There is then a concentrated polymer region near the interface. This part of the layer is often called the *proximal* region; it is dominated by the adsorption energy. Then the concentration drops to semidilute values. This region is called the *central* part of the profile. It is expected to have universal characteristics as will be discussed next (Section III.B). Finally, in the periphery of the layer we find the dilute part of the profile where the volume fraction relaxes exponentially to the bulk value; this part is referred to as the *distal* region.

B. Proximal, Central, and Distal Parts
in a Scaling Picture

The surface is often already a crowded place at bulk concentrations well below overlap. The high degree of polymerization and some finite adsorption energy lead to a (semi)plateau in the adsorption isotherm for bulk concentrations $\varphi_b \ll \varphi_{ov}$. Indeed this is the most challenging case to tackle theoretically because the layer has a concentrated, a semidilute, and a dilute region; these are the proximal, central, and distal regions referred to above. A pictorial view of the structure of the adsorbed layer is given in Figure 6.2; it is based largely on concepts introduced by De Gennes [2,46].

The proximal part is roughly the train layer and is dominated by the affinity of the segments for the surface. It is expected to depend on specific interactions and has therefore no universal features, apart from the fact that for strong adsorption the volume fraction in the train layer approaches unity. In Equation (75) (Section III.E5) and Equation (86) (Section III.E6) we give explicit expressions for the dependence of the thickness p of the proximal region on the adsorption

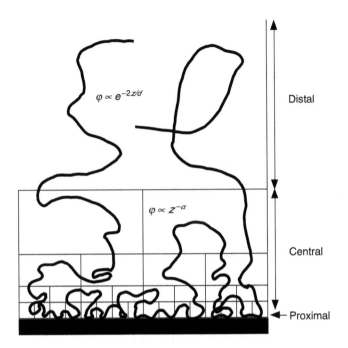

$\varphi \propto e^{-2z/d}$

Distal

$\varphi \propto z^{-\alpha}$

Central

Proximal

Figure 6.2 Schematic illustration of the structure of the adsorbed layer from a dilute solution. The self-similar grid introduced by De Gennes is drawn and the scaling behavior in the central and distal regimes are indicated. The parameter d is the distal length, equal to R_g/y where y is in the range 3–5, depending on φ_b (see Equation (78) in Section III.E6). The exponent α in the central region is $+1$ in a theta solvent, $+4/3$ in a good solvent with chain swelling due to intramolecular excluded-volume interactions, and $+2$ in a mean-field good solvent. The thickness of the proximal region is the proximal length p, which depends on the adsorption energy χ_s but is of order unity (unless χ_s is very close to the critical value χ_{sc}). An analytical expression for p is given in Equations (75) and (86).

energy χ_s; this relation is basically the same for dilute and semidilute solutions.

For the central part De Gennes launched a very interesting idea. He realized that in a semidilute solution the correlation length is the central length scale: $\xi \propto \varphi^{-3/4}$. He considered a distance z from the surface in the region of the adsorbed layer where the polymer concentration is in the semidilute range: the local concentration $\varphi(z)$ is above the overlap concentration φ_{ov}. The correlation length belonging to this concentration is given by $\xi(z) \propto \varphi(z)^{-3/4}$. De Gennes realized that the local correlation length should 'fit into the system'. This means that it cannot exceed the distance to the wall z. He proposed $z = \xi$, which directly leads to the self-similar result $\varphi(z) \propto z^{-4/3}$. In the literature one does not find further motivations for this choice. We may note that when $\xi < z$, the profile drops slowly and we have $\varphi(z) > z^{-4/3}$. The slow decay is expected to be a profile with a high-interfacial free energy. Indeed, the system will try to minimize the interfacial free energy (at fixed chemical potentials). As a result the profile drops as fast as it can ($z = \xi$).

Along the same lines in a theta solvent ($\xi \sim \varphi^{-1}$) one expects $z = \xi \sim \varphi(z)^{-1}$, or $\varphi \sim z^{-1}$.

We will see that an analysis using the Edwards equation supports the self-similarity concept, with the correct exponent -1 in a theta solvent but the mean-field exponent -2 (rather than $-4/3$) in good solvents. The latter difference can fully be accounted for by the difference between the mean-field and the excluded-volume expression for the mesh size ξ as a function of the polymer concentration.

The distal part of the profile should, again according to general considerations, decay exponentially. We do not repeat the arguments here, because the Edwards route, which will be discussed in some detail, fully recovers this result.

C.   SCF Modeling

In an SCF theory we typically consider the properties of a test polymer chain in an 'external' potential field. The effects of the interactions between the segments, sometimes called the volume interactions, are reflected in the properties of the field. The field can thus not be chosen as an arbitrary external potential, because it depends on the local concentrations. One refers to this field as the self-consistent potential. The approach boils down to the problem of finding the potential field $u[\varphi]$ as a function of the volume fractions, and the polymer volume fraction profile $\varphi[u]$ as a function of the potential field, such that the two profiles are mutually consistent. In the next Sections (III.C1–III.C3) we will see in more detail how such problems are treated properly. In general the exact equations can only be solved numerically. In some cases rather accurate analytical approximations exist, which will be discussed in Section III.E.

1.   The Scheutjens–Fleer Discretization Scheme

All numerical methods to solve differential equations need to discretize the, in principle, continuous functions. Scheutjens and Fleer [5,6,8] suggested to discretize both the space and the polymer chains by using a lattice and subdividing the chains into segments of a size compatible with the lattice unit, just as in the Flory–Huggins [41] treatment of a bulk solution. We will follow this approach, basically because of its intrinsic transparency for counting properly the conformations. We will show that the lattice approach has some advantages over the continuum route, especially when the near-surface behavior is considered where the size of the segments matters. The relationship between lattice and continuum descriptions is discussed in Section III.D.

The space next to an adsorbing plane is subdivided in discrete lattice layers with a thickness of the segment size. The layers are numbered $z = 0, 1, 2, \ldots, M$, where M is the last layer considered in the system (see Figure 6.3). The lattice is characterized by the fraction λ_0 of neighboring sites in the same layer. The fraction of neighbors in one of the adjacent layers is then $\lambda_1 = (1 - \lambda_0)/2$. In this review, we will choose a simple cubic lattice with $\lambda_0 = 4/6$ (four out of six neighbors in the same layer) and $\lambda_1 = 1/6$.

2.   From the Potential Fields to the Concentrations

Solvent molecules (label S) and polymer segments (no label) are assigned the self-consistent potentials $u_S(z)$ and $u(z)$, respectively, with respect to the homogeneous bulk solution. In this section we will assume that the potential fields $u_S(z)$ and $u(z)$ are known and calculate $\varphi[u]$. In Section III.C3 we address the opposite question: how to find the field from known

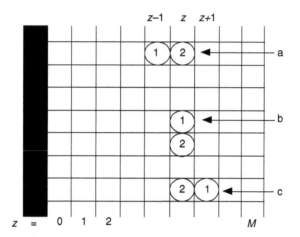

Figure 6.3 Two-dimensional representation of the system with its discretization in a lattice. Layers $z < 0$ are in the adsorbent, layers $z > 0$ in the solution. Conformations a, b, c of dimers with their end segment $(s = 2)$ in layer z are indicated. The first segment $(s = 1)$ can be in layers $z - 1$ (a), z (b), or $z + 1$ (c).

concentrations, i.e., $u[\varphi]$. It is assumed that every segment in layer z experiences the same potential (homopolymers). The first step is to introduce segmental weighting factors. The segmental weighting factors are simple Boltzmann factors

$$G_A(z) = e^{-u_A(z)}, \tag{7}$$

where it is understood that u is a dimensionless potential, i.e., normalized to kT. The label A refers to the segment type, where A is either the solvent S or the polymer unit, so we have two potential fields. The complete sets $\{G_S(z)\}$ and $\{G(z)\}$ are known when the potential fields $\{u_S(z)\}$ for the solvent and $\{u(z)\}$ for the segments are available.

For a monomeric solvent, the volume fraction profile is related to the weighting factors $G_S(z)$ in a very simple way because the solvent "segments" are free (not connected by chemical bonds). According to the Boltzmann distribution

$$\varphi_S(z) = \varphi_S(\infty)G_S(z) = \varphi_{Sb}G_S(z), \tag{8}$$

where the concentration $\varphi(\infty) = \varphi_b$ because the field is zero in the bulk solution. For a binary mixture of polymer and solvent $\varphi_{Sb} = 1 - \varphi_b$.

For the polymeric component, the relation between $\varphi(z)$ and φ_b is more complicated because of the chain connectivity, and involves in general two end-segment distributions $G(z,s|1)$ (starting from $s = 1$) and $G(z,s|N)$ (starting from $s = N$) where $s = 1, 2, \ldots, N$, with N the number of segments per chain. Note that we use a slightly different notation from that in Equations (1) and (5): the start of the walk is at $s = 1$ (first segment of the discretized chain) instead of at $s = 0$ (start of a continuous space curve), and we place no restriction on the spatial position of the start of the walk (i.e., we integrated over these spatial positions).

Let us first consider dimers ($N = 2$). We want the statistical weight $G(z,2|1)$ that the end segment ($s = N$) of a dimer ($N = 2$) is in layer z, with $s = 1$ free (see Figure 6.3). Because of the connectivity, $s = 1$ must be in one of the layers $z-1$, z, or $z+1$. The statistical weight of the upper conformation [$(z-1,1)$ $(z,2)$] in Figure 6.3 is $G(z-1)\lambda_1 G(z)$, that of the middle conformation [$(z,1)(z,2)$] is $G(z)\lambda_0 G(z)$, and that of the lower one [$(z+1,1)(z,2)$] is $G(z+1)\lambda_1 G(z)$. In these products, the first weighting factor is that of the first segment ($s = 1$) and the last that of $s = 2$ (the end segment is in z). The factors $\lambda_1 = 1/6$ and $\lambda_0 = 4/6$ express the a priori probabilities of 'perpendicular' and 'parallel' conformations, respectively. Summation (i.e., integration over the starting point) of the above weights gives $G(z,2|1) = G(z)[\lambda_1 G(z-1) + \lambda_0 G(z) + \lambda_1 G(z+1)]$, where the vertical bar and the 1 following it indicate that we sum over all positions of the first segment. In fact, the factors $G(z)$ in the right-hand side of this equation are equal to $G(z,1|1)$. If $G(z) = 1$ for all z (i.e., zero field according to Equation (7)), $G(z,2|1)$ is also unity: in that case the probability for a dimer to end in layer z is the same for all layers (except $z = 0$), and equal to that in the bulk solution. When we use the abbreviated notation $\langle Q(z) \rangle$ for the weighted average of a quantity Q over the three layers $z-1$, z, and $z+1$:

$$\langle Q(z) \rangle = \frac{1}{6}Q(z-1) + \frac{4}{6}Q(z) + \frac{1}{6}Q(z+1) \tag{9}$$

we may write $G(z,2|1) = G(z)\langle G(z,1|1)\rangle$, where the notation $G(z,1|1) = G(z)$ is used for reasons of generality.

Similarly, for trimers we have $G(z,3|1) = G(z)\langle G(z,2|1)\rangle$. Here we thus propagate from the dimer to the trimer, taking the ends of the dimer as the starting point to extend the chain in all possible directions for the next step. Note that this propagator implies a so-called (first-order) Markov approximation. Although the chain connectivity is precisely maintained in this propagation scheme, the chain can fold back on previously visited sites. It is very hard to rigorously correct for these step reversals because the histories are added up to get $G(z,2|1)$, and details of previous steps are no longer available. It is possible to systematically improve on the local excluded-volume correlations by implementing higher-order Markov approximations [18,30], which we do not consider here.

Generalizing the first-order scheme for sequences of s segments long ($s - 1$ steps) we obtain the forward propagator

$$G(z, s|1) = G(z)\langle G(z, s - 1|1)\rangle, \tag{10}$$

which can be applied for $s = 2, \ldots, N$, generating all end point distributions $G(z,s|1)$ that have in common that the evaluation of it started with segment $s = 1$, for which $G(z,1|1) = G(z)$; these segment weighting factors $G(z)$ are fully defined by the field $u(z)$. The complementary end-point distributions $G(z,s|N)$ follow from the backward propagator

$$G(z, s|N) = G(z)\langle G(z, s + 1|N)\rangle, \tag{11}$$

which is started from $G(z,N|N) = G(z)$. Because we discuss here only homopolymers composed of segments that are chemically identical, the segment weighting factor $G(z) = e^{-u(z)}$ does not depend on the ranking number, and $G(z,s|1) = G(z,N-s+1|N)$ (inversion symmetry). We note, however, that for copolymers or even for end-attached homopolymers the inversion symmetry is broken: $G(z,1|1) \neq G(z,N|N)$ or, more generally, $G(z,s|1) \neq G(z,N-s+1|N)$ because the segment types are not the same (copolymers) or the starting conditions are different (e.g., $s = 1$ attached to a surface and $s = N$ free).

The distribution of end segments for an N-mer is simply proportional to the end-point distribution found by the forward propagator, i.e.,

$$\varphi(z,N) = CG(z,N|1), \tag{12}$$

where C is a normalization factor. For equilibrium with a bulk solution, $C = \varphi_b/N$ is the proper normalization because in the bulk all G's are unity. For a homopolymer we could also apply $\varphi(z,N) = \varphi(z,1)$, with $\varphi(z,1) = (\varphi_b/N)G(z,1|N)$.

The distribution of an arbitrary segment s in the chain requires both end-point distribution functions. It is found by considering the joint probability that a walk $s' = 1, \ldots, s$ from segment 1 connects with the complementary walk $s' = N, \ldots, s$ from segment N. The resulting equation is known as the composition law (for a pictorial representation see Figure 6.4)

$$\varphi(z,s) = C\frac{G(z,s|1)G(z,s|N)}{G(z)} = C\frac{G(z,s|1)G(z,N-s+1|1)}{G(z)},$$
$$\tag{13}$$

where the second equality is only valid for homopolymers (or completely symmetric chains). Again, for equilibrium with the bulk $C = \varphi_b/N$. The division by the free segment weighting factor $G(z)$ for segment s corrects for the fact that this factor is included in both end-point distributions, whereas in the final result this weight should occur just once. Indeed, it is easily seen that $(G(z,s|1)G(z,s|N)/G(z)) = \langle G(z,s-1|1) \rangle G(z) \langle G(z,s+1|N) \rangle$ for $1 < s < N$. Here again the first-order Markov approximation is evidently present: nothing prevents segments $s-1$ and $s+1$ to have the same position in space. For $s = N$, Equation (13) reduces to the end segment distribution given in Equation (12), for monomers ($s = N = 1$), Equation (8) is recovered.

Obviously, the total polymer volume fraction is given by

$$\varphi(z) = \sum_{s=1}^{N} \varphi(z,s). \tag{14}$$

For symmetric dimers, $\varphi(z) = 2\varphi(z,1) = 2\varphi(z,2)$.

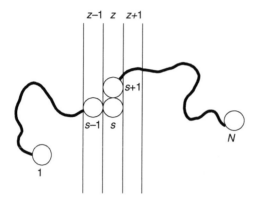

Figure 6.4 The volume fraction in layer z due to segments with ranking number s is found from the joint probabilities that the walks $(1,2,\ldots, s-1,s)$ and $(N,N-1,\ldots, s+1,s)$ both end in z. The solid curves connecting the two end-points with the interior segment s must be interpreted to represent all possible and allowed conformations with the end points free, given that the two walks end up with segment s in layer z.

The combined statistical weight to find the chain in the system is given by

$$G(N|1) = \sum_z G(z,N|1). \tag{15}$$

The quantity $G(N|1)$ serves as the chain partition function.

3. From the Concentrations to the Potential
    Field

In Section III.C2 we have seen how the volume fractions $\varphi_s(z)$ of the solvent and $\varphi(z)$ of the polymer follow from the segment potentials $u_S(z)$ and $u(z)$, respectively. In turn, these potentials depend on the segment volume fraction profiles because the field must be self-consistent. The potentials follow from differentiation of the (energetic) part of the Helmholtz energy with respect to the segment concentration under the constraint that the lattice remains completely filled. Typically such a constraint can be handled by introducing a Lagrange parameter $u'(z)$ coupled to the local condition $\varphi(z) + \varphi_S(z) = 1$ or, in general

$$\sum_A \varphi_A(z) = 1 \tag{16}$$

for all z-coordinates. In addition to the Lagrange contribution we account for segment–solvent and segment–surface interactions. The result may be expressed as

$$u_A(z) = u'(z) + u_A^{int}(z). \tag{17}$$

Physically, $u'(z)$ is the volume work (in units of kT) per segment to remove a segment from the bulk and to insert it in layer z. In other words, because of the finite size of the segments one needs to invest energy to create space for the segment. This is an important term to the segment potential because it corrects in part for the backfolding problem introduced by using the Markov approximation. Moreover, it accounts for the crowding of segments in an adsorbed layer.

A corresponding physical interpretation can be given for the interaction term. There may be an energetic effect (again in units of kT) needed to move a segment from the bulk to coordinate z because of the different chemical surroundings. In a lattice model it is customary to account for nearest-neighbor interactions by Flory–Huggins parameters $\chi_{AB} = Z(2U_{AB} - U_A - U_B)/(2kT)$, where Z is the lattice coordination number ($Z = 6$ for a cubic lattice) and U_{AB} is the contact energy per AB-bond. Here, U_{AA} and U_{BB} define the reference potential energies in the system, i.e., the reference state is pure unmixed A and B. When segments A and B effectively repel each other $\chi_{AB} > 0$, and for attraction the reverse is true. Realizing that the adsorbent (indicated by the index W of wall) has a fixed concentration profile, i.e., $\varphi_W = 1$ for $z < 0$ and zero otherwise, we find for the solvent $u_S^{int}(z) = \chi_{SW}$ $\langle \varphi_W(z) \rangle + \chi(\langle \varphi(z) \rangle - \varphi_b)$ and for the polymer units $u^{int}(z) = \chi_{PW}$ $\langle \varphi_W(z) \rangle + \chi(\langle \varphi_S(z) \rangle - \varphi_{Sb})$, where the index P is used here to refer to polymer segments. It is also understood that the χ without labels represents the polymer–solvent interaction parameter: $\chi_{PS} = \chi$. The angular brackets are necessary to count accurately the contacts in a system in which there are concentration gradients (i.e., near the surface).

It is obvious that the interactions with the surface are only present when the segment is in layer $z = 0$ and thus $\langle \varphi_S(z) \rangle = (1/6)\delta(z)$, where $\delta(z) = 1$ if $z = 0$ and zero otherwise. Therefore the adsorption contribution could also be written as $u_S^{ads} = (1/6)\chi_{SW}$ (solvent) and $u^{ads} = (1/6)\chi_{PW}$ (polymer segments). The notation with angular brackets seems a bit overdone, but it allows us to write the general equation:

$$u_A(z) = u'(z) + \sum_B \chi_{AB}(\langle \varphi_B(z) \rangle - \varphi_{Bb}), \qquad (18)$$

where the summation is over $B = S,P,W$. This equation applies also to multicomponent systems (e.g., copolymers): the summation then contains more terms, one for each segment type. The segment potential can easily be extended to account for, e.g., electrostatic contributions, polarization contributions, etc. [15,42]. We will not do this here.

The method outlined above for the surface interactions through Flory–Huggins nearest-neighbor contact energies is an alternative for using the classical Silberberg [47] adsorption parameter $\chi_s = -(u^{ads} - u_S^{ads})$, which is typically assigned only to the polymer segments and not to the solvent units. Indeed, it can be shown [8,15] that only the difference $\chi_{PW} - \chi_{SW}$ is important and the Silberberg parameter is given by

$$\chi_s = -(\chi_{PW} - \chi_{SW})/6, \qquad (19)$$

Note that, due to the minus sign, a positive value for χ_s indicates that the surface prefers contacts with polymer segments to that with solvent, which is only for historical reasons. This should be contrasted to the Flory–Huggins parameter where negative values for χ_{AB} reflect net attraction between A and B.

4.   Fixed Point of SCF Equations

The SCF equations are routinely solved numerically up to high precision (typically seven significant digits are needed).

This means that the potentials are found that generate the segment distributions on the one hand, and follow from the segment distributions on the other: $u[\varphi]$ and $\varphi[u]$ should be consistent. Such a fixed point of the equations has optimized the free energy in the system without the need to imply other approximations. For this reason we refer to these solutions as the numerically exact SCF results. The results can easily be generated in the dilute, semidilute, and concentrated systems, for strong and weak adsorption, near the critical point and for depletion, and for good, marginal, and poor solvent conditions. Also very bad solvency conditions (phase separation, wetting) can be handled. In the analytical work to be discussed in Section III.E we always have to make additional approximations and specify the regime.

The numerical method implies the use of a lattice. For some problems a discrete picture may have distinct advantages to unravel the physics of a problem. This is for example the case to understand the critical adsorption energy. This is the topic of the next section.

5. Critical Adsorption Energy

Polymers near a surface cannot penetrate the solid phase and thus experience an entropy loss. One of the important advantages of a lattice model is that this entropy loss is easily visualized and quantified. In a cubic lattice, each bond has six optional directions from each lattice site except from $z = 0$ where there are only five options. The implication is the existence of a critical adsorption energy χ_{sc} for adsorption to take place; it is needed to compensate this entropy loss. To quantify this critical adsorption energy it is instructive to focus first on very dilute systems where the polymer units do not feel each other ($u^{int}(z) = 0$ for $z > 0$). In such dilute systems $u'(z)$ is also zero and Equation (17) reduces to $u(z) = -\chi_s \delta(z)$: for $z > 0$, $u(z) = 0$ and $G(z) = 1$, for the surface layer we have $u(0) = -\chi_s$ and $G(0) = e^{\chi_s}$.

Let us assume that there exists a field due to the surface at $z = 0$ (i.e., a value of χ_s) such that the end-point distributions remain unity. This means that also the concentrations do not deviate from the bulk value (no adsorption or depletion). For this (hypothetical) situation we focus on the forward propagator $s \to s+1$ for the case that $s+1$ is at $z=0$

$$G(0,s+1|1) = G(0)\left(\frac{4}{6}G(0,s|1) + \frac{1}{6}G(1,s|1)\right) \qquad (20)$$

since the term $G(-1,s|1)$ is zero. When $G(0,s+1|1)$, $G(0,s|1)$, and $G(1,s|1)$ are unity, we thus need to take $G(0)=6/5$ to obtain our target. This implies that $u(0)=-\ln(6/5)$ and that the critical adsorption strength χ_{sc} is given by $\chi_{sc}=\ln(6/5)$ in a good solvent $(\chi=0)$. When $\chi>0$, a term $\chi/6$ has to be subtracted because of the missing contact of an adsorbed segment with solvent

$$\chi_{sc} = \ln\frac{6}{5} - \frac{\chi}{6}. \qquad (21)$$

There is, however, a slight inconsistency in this argument. As $G(0)$ is not unity, the end-point distribution for the starting segment does not obey our target value. When we analyze $G(0,s|1)$ using $G(0)=6/5$ and $u(z)=0$ for all layers $z>0$, we find the following for $s=1, 2, \dots$. Monomers have $G(0)=6/5=1.2$, and have a higher concentration than the bulk solution. For dimers we find $G(0,2|1)=1.16$, which is already closer to unity. It turns out that $G(0,s|1)$ quickly converges to unity for higher s, as does $G(z,s|1)$ for all other (positive) z. Hence, for long chains (where the difference between the number of bonds and the number of segments does not matter) the excess amount θ^{ex} is zero, corresponding to the adsorption–desorption transition taking place at $\chi_s = \chi_{sc} = \ln(6/5)$ when $\chi=0$.

The consequence of the mentioned inconsistency is that the critical adsorption strength depends (slightly) on the length of the polymer chain [48]. For infinite chain length the critical value is indeed $\chi_{sc}=\ln(6/5)$ (the start-up problems do not matter), but for finite chain length the critical affinity is slightly lower. In the limit $N=1$ (monomers) χ_{sc} is zero as there is no entropy loss for any bond. We note that the result for the critical adsorption energy given in Equation (21) depends on the lattice type, i.e., the lattice coordination number Z, and as such must be a typical lattice result. We return to this below.

The general boundary condition may now be given by replacing the weighting factor $G(0)$ in Equation (20) by $e^{-u(0)}$. Multiplying both sides by 6/5 and rearranging gives

$$\frac{6}{5}e^{u(0)} = \frac{4G(0,s|1) + G(1,s|1)}{5G(0,s+1|1}. \tag{22}$$

This form of the boundary condition (propagator for the surface layer $z = 0$) will be used frequently below, where the finite size of the segments is important. It enables us to make a connection between continuum and lattice models by introducing a 'Stern layer' in the continuum equations. For example, for very dilute layers in a good solvent $u(0) = -\chi_s$ and the left-hand side of Equation (22) reduces to $e^{-\Delta\chi_s}$, where $\Delta\chi_s = \chi_s - \chi_{sc}$; this form will be used in Equation (45) (Section III.E2). For more crowded surface layers an additional factor $(1 - \varphi_b)/(1 - \varphi(0))$ enters, as shown in Equation (72) (Section III.E5).

6. Thermodynamics

An important characteristic of self-consistent-field models is that once the potentials are consistent with the volume fractions, one has explicit expressions for the thermodynamic variables. For example, the dimensionless (in units of kT) Helmholtz energy per unit area (in units of the segmental area) is for a homopolymer in a monomeric solvent given by [8,15]

$$F = n\ln(NC) + n_S \ln \varphi_{Sb} + \varphi(0)\chi_S$$
$$- \sum_z [\varphi(z)u(z) + \varphi_S(z)u_S(z)] + \sum_z \varphi(z)\chi\langle\varphi_S(z)\rangle, \tag{23}$$

where n is the number of chains per unit area and C the normalization constant for the polymer. In general this quantity is found as the ratio between n and the single-chain partition function $G(N|1) = \Sigma_z\, G(z,N|1)$ as defined in Equation (15): $C = n/G(N|1)$. For homopolymers in equilibrium with a bulk with volume fraction φ_b, the normalization simplifies to $C = \varphi_b/N$. In the second term of Equation (23) $n_S = M - nN$ is the number of solvent molecules per unit area in the system. The χ_S and χ terms in Equation (23) give the contributions due to the interaction energy. The remaining fourth term is a sum over the local volume fraction times the local potential.

In the bulk the potentials $u(z)$ are zero and Equation (23) reduces to the Flory–Huggins (regular solution) Helmholtz energy. It is convenient to introduce the Helmholtz energy density $f = F/V$, where V is the volume (in number of lattice sites). For the bulk solution we obtain

$$f = \frac{\varphi_b}{N}\ln\varphi_b + (1 - \varphi_b)\ln(1 - \varphi_b) + \chi\varphi_b(1 - \varphi_b), \qquad (24)$$

which is completely expressed in the polymer concentration in the bulk.

In general one can define a bulk system which is in equilibrium with the interfacial system. Typically this bulk is composed of polymer chains and solvent molecules. Once the composition of this bulk is known, it is possible to evaluate the chemical potentials of these (mobile) components; they essentially follow from differentiation of Equation (24) with respect to $\varphi_{Sb} = 1 - \varphi_b$ (solvent) or with respect to φ_b (segments). The chemical potential of the solvent (directly related to the osmotic pressure π_b of the bulk solution) is given by

$$
\begin{aligned}
\mu_S = -\,\pi_b &= \ln(1 - \varphi_b) + \left(1 - \frac{1}{N}\right)\varphi_b + \chi\varphi_b^2 \\
&\approx -\left(\frac{\varphi_b}{N} + \frac{1}{2}v\varphi_b^2 + \frac{1}{3}\varphi_b^3 + \cdots\right)
\end{aligned}
\qquad (25)
$$

and the change of π_b with concentration (inversely proportional to the osmotic compressibility) may be written as:

$$\frac{\partial \pi_b}{\partial \varphi_b} = \frac{1}{N} + \frac{\varphi_b}{1 - \varphi_b} - 2\chi\varphi_b \approx \frac{1}{N} + v\varphi_b + \varphi_b^2 + \cdots \qquad (26)$$

In Equations (25) and (26) $v \equiv 1 - 2\chi$ is the segmental second virial coefficient, often also denoted as the Edwards excluded-volume parameter.

The chemical potential of a polymer chain of N segments is N times the segment chemical potential μ. For a solution of one type of homopolymer in a monomeric solvent μ is found by differentiating Equation (24) with respect to φ_b:

$$\mu = \frac{1}{N} \ln \varphi_b - (1 - \frac{1}{N})(1 - \varphi_b) + \chi(1 - \varphi_b)^2. \tag{27}$$

Polymers at interfaces influence the surface free energy of the system. In general, when the adsorption occurs spontaneously this will reduce the surface free energy, also known as the grand potential. The grand potential Ω is defined as the Helmholtz energy per unit area minus the chemical work μn of both the polymer and the solvent, $\Omega = F - \Sigma_i n_i \mu_i$. It is equal to the interfacial Gibbs (or Helmholtz) energy with respect to that of the surface in contact with pure solvent: $\Omega = \gamma a$, where a is the area of a site and γ the surface tension difference. It can be shown that for a homopolymer at an S/L interface in a monomeric solvent the result is [5,7,8,15]

$$\Omega = \left(1 - \frac{1}{N}\right)\theta^{ex} - \sum_z u'(z) - \sum_z \chi[\varphi(z)\langle\varphi_S(z)\rangle - \varphi_b\varphi_{Sb}]. \tag{28}$$

It is useful to mention that in the limit that the concentration gradients are not very large, one can approximate $\langle\varphi(z)\rangle \approx \varphi(z)$. Then one can write the grand potential as a summation over a well-defined grand potential density $\omega(z)$, i.e., $\Omega = \Sigma_z\omega(z)$, where the grand potential density ω is the difference between the bulk osmotic pressure π_b and the local osmotic pressure $\pi(z)$:

$$\Omega = \sum_z \omega(z), \qquad \omega(z) = \pi_b - \pi(z). \tag{29}$$

The bulk osmotic pressure is given in Equation (25) and the local osmotic pressure is found by replacing in Equation (25) φ_b by $\varphi(z)$.

We will use the exact grand potential Equation (28) in the numerical evaluation of the colloidal stability issues and

Equation (29) (with the expansion for π as given in Equation (25)) in some of the analytical approximations used to interpret the trends.

D. The Equivalence of the Propagator and the Edwards Equation

So far we did not make any attempt to relate the Scheutjens–Fleer propagators (Equations (10) and (11)) to the Edwards equation given in Equation (5). For gradients perpendicular to a flat interface we may replace ∇_r^2 by $\nabla_z^2 = \partial^2/\partial z^2$.

It is instructive to elaborate first on the segment potentials, given in Equations (17) and (18). It is easily seen that the angular brackets translate in continuous language to

$$\langle Q(z) \rangle \approx Q(z) + \frac{1}{6}\frac{\partial^2 Q(z)}{\partial z^2}. \tag{30}$$

When the second derivatives of the polymer concentration profile are small, which is typically a good approximation for not too small values of z, we can approximate $\langle \varphi(z) \rangle \approx \varphi(z)$. This approximation thus makes the exact nonlocal expression for the segment potentials (Equation (18)) local, i.e., only a function of one value of z. For $z > 0$ the interactions with the surface are absent and we have from Equation (18) for the solvent $u_S = u' + \chi(\varphi - \varphi_b)$. According to Equations (7) and (8) we also have $u_S = -\ln G_S = -\ln(\varphi_S/\varphi_{Sb}) = -\ln[(1 - \varphi)(1 - \varphi_b)]$. Hence, $u'(z)$ may be expressed in the polymer concentration profile $\varphi(z)$ as $u' = -\ln[(1 - \varphi)/(1 - \varphi_b)] - \chi(\varphi - \varphi_b)$. Inserting this in the segment potential $u = u' + \chi(\varphi_S - \varphi_{SB}) = u' - \chi(\varphi - \varphi_b)$ for the polymer units gives a rather accurate expression for the self-consistent potential of the polymer segments:

$$u(z) = -\ln\frac{1 - \varphi(z)}{1 - \varphi_b} - 2\chi(\varphi(z) - \varphi_b) \tag{31}$$

for $z > 0$. For $z = 0$ this expression applies as well, but there is an additional contribution $-(\chi_S + \chi/6)$ due to the presence of the surface. An expanded version of Equation (31) is given in Equation (54) (Section III.E3).

To obtain the Edwards equation from the propagators (Equation (10)) we assume that the second derivative of the end-point distribution function is small. Starting with Equation (10), we can first introduce a parameter Δs as the step increment used in the propagator; in the Scheutjens–Fleer approach $\Delta s = 1$. In the continuum approach we allow very small increments of s, which implies breaking up a segment in many partial segments. The energy associated with such a partial segment is $u(z)\Delta s$, since u is the energy per segment.

Upon rearranging Equation (10), using Equation (30) for the angular brackets, we may write

$$\frac{G(z, s + \Delta s|1)e^{u(z)\Delta s} - (G(z, s|1))}{\Delta s} = \frac{1}{6}\frac{\partial^2 G(z, s|1)}{\partial z^2}. \tag{32}$$

In the limit $\Delta s \to 0$ we can safely expand the exponent, $e^{u(z)\Delta s} = 1 + u(z)\Delta s$, and the Edward equation is recovered exactly by realizing that the left-hand side reduces to $\partial G(z,s|1)/\partial s + u(z)G(z,s|1)$:

$$\frac{\partial G(z, s|1)}{\partial s} = \frac{1}{6}\frac{\partial^2 G(z, s|1)}{\partial z^2} - u(z)G(z, s|1). \tag{33}$$

This equation is a good starting point for analytical approximations. It may also serve for obtaining numerical results. But in order to get such numerical solution Equation (33) has to be discretized again. There is then hardly any advantage of using Equation (33) above Equation (10), as the latter takes into account the finite segment size, which always gives problems for the continuum variant close to the surface. Moreover, the discrete propagators are expected to remain accurate even where Equation (33) breaks down, e.g., when the second derivative of the end point distributions is not small.

E. Analytical Descriptions for Homopolymers Near a Single Surface

1. General

For a variety of situations, analytical descriptions are available. In this section we will pay attention to dilute ideal chains interacting with a surface (Section III.E2), to depletion

(Section III.E4), to adsorption from semidilute solutions (Section III.E5), and to adsorption from dilute solutions (Section III.E6). In all cases we restrict ourselves to homodisperse homopolymers in a monomeric solvent.

As explained in Section III.C, the starting point is the Edward equation (33) and the end point distribution function $G(z,s|1)$ is the key function of interest. Unless specified otherwise we will write (in the continuum language) either $G(z,s)$ or simply G for this quantity. It is the statistical weight of all walks that end with segment s at coordinate z, where the starting point (segment $s = 1$) can assume any position. The walks take place in a field $u(z)$, which is defined such that u is zero in the bulk solution. The general expression for the potential field is Equation (31), which is already an approximation because the nonlocal energy effects are ignored as $\langle \varphi \rangle$ was replaced by φ.

In the potential field the adsorption energy contribution is omitted; this will be accounted for in the boundary condition at the surface. We use the Silberberg convention (Equation (19)) for the adsorption energy and take χ_s positive (and higher than χ_{sc}) for adsorption and negative for depletion.

The adsorption energy causes $G(z,s)$ to have a negative slope $\partial G/\partial z$ at the surface for adsorption and a positive slope in depletion. De Gennes [2,46] expressed the boundary condition as a given extrapolation length $1/c$, defined by

$$\frac{1}{G}\frac{\partial G}{\partial z}\bigg|_{z=0} = \frac{\partial \ln G}{\partial z}\bigg|_{z=0} = -c. \tag{34}$$

The meaning of c is illustrated qualitatively in Figure 6.5.

We will return in Section III.E2 to the precise relation between c and χ_s (Figure 6.8). For the moment it suffices to note that c increases monotonically with $\Delta\chi_s \equiv \chi_s - \chi_{sc}$ where χ_{sc} is the critical adsorption energy, which for long chains and $\chi = 0$ equals $\chi_{sc} = \ln(6/5)$ (Equation (21)). For a strongly repulsive surface, both $\Delta\chi_s$ and c are large negative and $1/c$ approaches zero from the negative side. For strong adsorption, $\Delta\chi_s$ and c are large positive and the extrapolation length $1/c$ is small and positive. In between there is a point (the adsorption–desorption transition, see Section III.C5) where

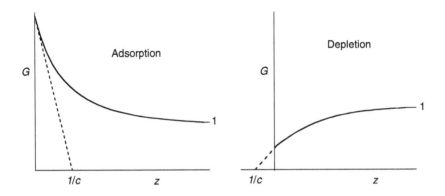

Figure 6.5 The extrapolation length $1/c$, which is positive in adsorption and negative in depletion. The function G is unity in the bulk solution. At the adsorption-desorption transition ($\chi_s = \chi_{sc}$, $G = 1$ everywhere), $1/c = \infty$ and $c = 0$.

$\Delta\chi_s = c = 0$; here the extrapolation length is infinite. In Figure 6.5 this corresponds to $G = 1$ at any distance from the surface.

With a given value of c, Equation (33) may be solved. When the full form for the segment potential Equation (31) is used, this has to be done numerically. Analytical solutions for some cases may be found when some simpler (expanded) form of $u(z)$ is taken. When $G(z,s)$ is known, the relative concentrations ρ_e of end points and ρ for all segments are found from

$$\rho_e \equiv \frac{N\varphi_e}{\varphi_b} = G(z,N), \tag{35}$$

$$\rho \equiv \frac{\varphi_e}{\varphi_b} = \frac{1}{N} \int_0^N G(z,s)G(z,N-s)\mathrm{d}s. \tag{36}$$

This latter expression differs from the discrete composition law Equation 13 because the chain is now seen as a space curve and s is the length along this curve; that is why the factor $1/G(z) = e^{u(z)}$ needed in the lattice approach does not appear. Note that we have dropped the information from which side the walk has started. This is justified for homopolymers as inversion symmetry applies. The quantities ρ_e, ρ, and G are normalized such that they are unity in the bulk solution.

2. Exact Solutions for $u=0$

Only when the last term of Equation (33) may be neglected are exact solutions available. That is the situation where the concentrations in Equation (31) are so low that $u=0$: the segments do not feel each other. There is only interaction with the surface as expressed by the parameter c. Apart from the length scale $1/c$, the natural length scale in this case is the gyration radius $R_g = \sqrt{N}/6$. Eisenriegler [43,49,50] derived for this case

$$G = \text{erf}(Z) + e^{-Z^2}Y(Z - C),\tag{37}$$

where Z is the distance expressed in units $2R_g$, and C is the ratio between R_g and the extrapolation length $1/c$:

$$Z \equiv \frac{z}{2R_g},\quad C \equiv cR_g.\tag{38}$$

The function $Y(x)$ in Equation (37) is defined as:

$$Y(x) \equiv e^{x^2}\text{erfc}(x) \approx \begin{cases} \frac{1}{\sqrt{\pi}}\frac{1}{x} & x \gg 1 \\ 1 - \frac{2}{\sqrt{\pi}}x + x^2 & |x| \ll 1 \\ 2e^{x^2} & x \ll -1 \end{cases}\tag{39}$$

We consider first three limiting cases of Equation (37).

Strong repulsion (depletion).   In this case $-C = -cR_g$ is large, and the second term of Equation (37) is small. Then $G = \text{erf}(Z)$ or

$$\rho_e = \text{erf}\frac{z}{2R_g}.\tag{40}$$

We may define the depletion thickness δ as the zeroth moment of the end-point profile

$$\delta = \int (1 - \rho_e)dz.\tag{41}$$

This equation applies generally, also when the field is non-zero and ρ_e has a more complicated structure. For the present case (extremely low concentration) we denote δ as δ_0. Substitution of Equation (40) into Equation (41) gives

$$\delta_0 = \frac{2}{\sqrt{\pi}} R_g. \tag{42}$$

The depletion thickness is thus slightly larger than R_g ($2/\sqrt{\pi} = 1.13$).

By inserting $G = \mathrm{erf}\, Z$ into Equation (36) an exact expression for the overall concentration $\varphi = \varphi_b \rho$ may be derived. We do not give this exact result (which is rather complicated) but present only the simple form proposed by Fleer et al. [51] which turns out to be an accurate approximation

$$\rho = \tanh^2 \frac{z}{\delta_0}. \tag{43}$$

A plot of ρ_e and ρ as a function of z is given in Figure 6.6. The zeroth moment of both profiles is the same (namely δ_0).

Critical point. When $C = c = 0$, $G(z,N) = 1$ for every z and we have a flat profile. It is possible to derive expressions for small

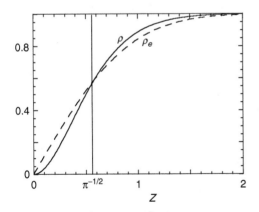

Figure 6.6 The distribution ρ_e of end segments and the overall distribution ρ as a function of $Z = z/(2R_g)$ according to Equations (40) and (43). The value of $\delta_0/2R_g = \pi^{-1/2}$ is indicated. The extrapolation length $1/c$ is zero, so that $\rho_e(0) = \rho(0) = 0$.

$|C|$ using the expansion (39b), but we do not go into these details here; we refer to Van der Gucht et al. [52]. It is useful to mention, however, that there is a one-to-one relation between the profile found at the surface and the adsorption parameter c. Upon increasing c from zero to some small finite value, the conformation of an adsorbed chain changes from an undeformed coil (size R_g) gradually towards flatter states: the chain gradually goes from a three-dimensional to a two-dimensional coil. The longer the chain, the smaller the range in c over which the transition occurs. However one cannot identify metastable situations and the transition is not first-order.

Strong adsorption. Now $C = cR_g$ is large positive, and Equation (37) reduces to

$$G = 2e^{c^2N/6}e^{-cz} \qquad (z \ll R_g), \qquad (44)$$

where $c^2/6$ is of order $\Delta\chi_s$, as will be shown below. The integral over Equation (44) gives the partition function and hence the free energy of the chain. The result is $F \approx Nc^2/6$. In a $1kT$ per blob Ansatz there are $n = Nc^2/6$ blobs per chain. In each blob there are $N_b = N/n = 6/c^2$ segments and as the chain is Gaussian ($N_b = 6r^2$, where r is blob radius) the blob size is of order $1/c$. A graphical representation is given in Figure 6.7.

From Equation (44) it is seen that the adsorption profile $\rho \sim e^{-2cz}$ is exponential, with a decay length $0.5/c$ which is small. This implies that the three-dimensional coils in solution (with radius R_g) change at the surface into a two-dimensional flattened shape with thickness $0.5/c$. Again, this shape may also be visualized as a string of blobs (which may be called adsorption blobs) with diameter $1/c$. The blobs are made up only of loops: tails are absent in the strong adsorption regime (dilute chains).

Relation between c and χ_s. The remaining problem is to relate c to the adsorption energy χ_s. Following Gorbunov et al. [48], we do this by inserting the continuum expression (which contains c) into the lattice boundary condition Equation (22) (which contains χ_s). In the present case ($u = 0$ except for the adsorption energy χ_s for train segments), Equation (22) reads

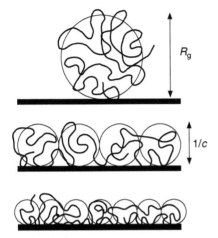

Figure 6.7 The conformation of an isolated adsorbed chain. The adsorbed chain splits up in a string of adsorption blobs with size $1/c$. With increasing surface affinity the blobs become smaller as indicated going from top to bottom. The top diagram represents the critical case ($c = 0$) where there is just one blob left. The blob size is in this case given by the radius of gyration.

$$\mathrm{e}^{-\Delta\chi_s} = \frac{4G(0,s) + G(1,s)}{5G(0,s+1)}. \tag{45}$$

Substitution of Equation (37) into Equation (45) gives in the long-chain limit

$$\mathrm{e}^{-\Delta\chi_s} = \begin{cases} 1 - \frac{c}{5} & c \leq 0 \\ \frac{1}{5}(4 + \mathrm{e}^{-c})\mathrm{e}^{-c^2/6} & c \geq 0 \end{cases}. \tag{46}$$

Figure 6.8 gives the relation between $\Delta\chi_s$ and c according to Equation (46). Around the critical point $c \approx 5\Delta\chi_s$, at high χ_s we find $c^2 \approx 6(\Delta\chi_s - \ln\frac{5}{4}) = 6(\chi_s - \ln\frac{3}{2})$.

The excess amount θ^{ex} is found from $\theta^{\mathrm{ex}} = \varphi_b \int (1 - \rho_e)\mathrm{d}z$ with $\rho_e = G$ given by Equation (37). The result is

$$\theta^{\mathrm{ex}} = \frac{\varphi_b}{c}\left[\mathrm{Y}(-C) - 1 - \frac{2C}{\sqrt{\pi}}\right] \approx \begin{cases} -\varphi_b\delta_0 & \text{depletion} \\ 0 & \text{critical point}. \\ \frac{2\varphi_b}{c}\mathrm{e}^{Nc^2/6} & \text{adsorption} \end{cases} \tag{47}$$

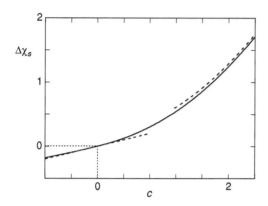

Figure 6.8   The relation between $\Delta\chi_s$ and c according to Equation (46) (full curve). The dashed curves are the asymptotes $c = 5\,\Delta\chi_s$ and $c^2 = 6\,(\Delta\chi_s - \ln(5/4))$, respectively.

The outcome for depletion coincides with the result that follows from Figure 6.6. For high χ_s the excess amount is roughly proportional to $\varphi_b e^{N\Delta\chi_s}$. We note that Equation (47c) and Equation (44) are expected to hold only as long as $\varphi_b e^{N\Delta\chi_s}$ is small, which means that φ_b should be (extremely) small when $N\Delta\chi_s$ is high. We are then in the Henry region of the adsorption isotherm. When θ^{ex} becomes of order 0.05, crowding effects play a role, so that the assumption $u = 0$ breaks down and we enter a different regime (see Sections III.E5 and III.E6).

When comparing Equations (44) and (47c) with lattice results, it is useful to treat the train layer in a discrete way and the loops in a continuous fashion. The argumentation is that strong variations on a length scale smaller than the segment length are treated irrealistically in a continuum theory. This would imply $\varphi_{tr} \approx 2\varphi_b\, e^{N\Delta\chi_s}$ for the trains $(-1/2 < z < 1/2)$ and an exponential loop decay for $z > 1/2$. We do not elaborate this point for the Henry regime, but we return to such a Stern-layer approach in Sections III.E5 and III.E6.

3.   Ground-State Approximation (GSA)

When $u \neq 0$ in Equation (33), no exact analytical solutions are known. One can find an approximate solution by using only the first term of an exact eigenfunction expansion

$$G(z,s) = \sum_k g_k(z)e^{\varepsilon_k s} \approx g(z)e^{\varepsilon s}, \tag{48}$$

where e^ε is the largest (ground-state) eigenvalue and $g(z)$ the corresponding (ground-state) eigenfunction. With this approximation the distribution $\varphi(z,s) \sim G(z,s)G(z,N-s) = g^2 e^{\varepsilon N}$ for an interior segment s in the chain becomes independent of s: all segments are assumed to have the same spatial distribution. This implies that end effects (tails) are neglected. In Section III.E6 we will see that it is still possible to describe tails, but then we have to go beyond the standard ground-state approach. When using only Equation (48), essentially only loops (and trains) are accounted for.

Upon substituting Equation (48) into Equation (33), we obtain an ordinary differential equation in g:

$$\frac{\partial^2 g}{\partial z^2} = 6(\varepsilon + u)g. \tag{49}$$

This equation may be solved when ε is known and when u is expressed in terms of g. Since u is a function of φ (Equation (31)), we need the relation between φ and g. Substitution of Equation (48) into Equation (35) and (36) gives

$$\begin{aligned} \rho_e &= e^{\varepsilon N} g \\ \rho &= e^{\varepsilon N} g^2 \end{aligned} \tag{50}$$

which shows that in GSA $\rho_e \sim g$ and $\rho \sim g^2$.

For depletion or adsorption from semidilute solutions, where ρ and g^2 correspond to the total concentration profile, we may apply Equation (49) to the bulk solution (where $\partial^2 g/\partial z^2$ and u are zero) and find that

$$\varepsilon = 0 \quad \text{(semidilute)}. \tag{51}$$

Consequently, in semidilute solutions Equation (50b) reduces to

$$\begin{aligned} \rho &= g^2 \\ \varphi &= \varphi_b g^2. \end{aligned} \tag{52}$$

Hence, in the bulk solution $\rho(\infty) = 1$ and $g(\infty) = 1$. Because in semidilute solutions ε is taken zero, the term $e^{\varepsilon N}$ is unity and there is no chain-length dependence for the interfacial properties. As discussed in Section II.B, the same applies for the solution properties.

For adsorption from dilute solutions, $\varepsilon \neq 0$. This implies that g cannot be coupled to the overall concentration profile because with $\varepsilon \neq 0$ the boundary condition $\partial^2 g / \partial z^2 = 0$ and $g \neq 0$ in the bulk solution cannot be satisfied. Instead, we may relate g to the adsorbed part φ^a of the volume fraction profile, which vanishes at large distance. Hence we use Equation (50b) in the form $\varphi^a / \varphi_b = e^{\varepsilon N} g^2$. It turns out that $\varphi_b e^{\varepsilon N}$ is of order unity [10,12,13]. Hence, in dilute solutions we have

$$\varepsilon \approx \frac{1}{N} \ln \frac{1}{\varphi_b}$$

$$\varphi^a \approx g^2. \tag{53}$$

In the bulk solution φ^a and g are zero, unlike the overall concentration, which includes free chains and becomes equal to φ_b.

We can solve Equation (49) analytically when we have a simple relation between u and g. Following De Gennes, we expand the logarithm in Equation (31) in terms of φ:

$$u \approx v(\varphi - \varphi_b) + \frac{1}{2}(\varphi^2 - \varphi_b^2) + \cdots$$

$$\approx v\varphi_b(g^2 - 1) + \frac{1}{2}\varphi_b^2(g^4 - 1) + \cdots \tag{54}$$

where $v \equiv 1 - 2\chi$ is Edwards' excluded-volume parameter. For the case of good solvents we use only the term linear in φ, whereas for a theta solvent ($\chi = 0.5$, $v = 0$) this linear term cancels and we are left only with the φ^2-term.

For adsorption from dilute solutions, we neglect φ_b with respect to $\varphi(z)$, and we disregard the difference between φ^a and φ in the adsorbed layer

$$u \approx v\varphi + \frac{1}{2}\varphi^2 \approx vg^2 + \frac{1}{2}g^4. \tag{55}$$

4. Depletion

Good solvents According to Equation (54) we have $u = v\varphi_b$
$(g^2 - 1)$ in semidilute solutions. The factor $v\varphi_b$ may be written
as the inverse square of a correlation length (blob size) ξ, as
first realized by De Gennes [2]:

$$\xi^{-2} = 3u_b = 3v\varphi_b. \tag{56}$$

The field u_b is the segment potential in the bulk solution with
respect to infinite dilution. As noted before (Section II.C), in
mean-field good solvents $\xi \sim \varphi_b^{-1/2}$. Here we also know the
prefactor. When we substitute $u = (\xi^{-2}/3)(g^2 - 1)$ into Equation (49) with $\varepsilon = 0$, we get

$$\frac{\partial^2 g}{\partial(z/\xi)^2} = 2g^3 - 2g. \tag{57}$$

This equation may be solved by multiplying both sides by
$\partial g/\partial z$. When we abbreviate $Z \equiv z/\xi$ we find $\partial[(\partial g/\partial Z)^2 - g^4 + 2g^2]/\partial Z = 0$. Integration gives $(\partial g/\partial Z)^2 - g^4 + 2g^2 = 1$, where
the integration constant 1 follows from $\partial g/\partial Z = 0$ and $g = 1$
in the bulk solution. Hence

$$(\partial g/\partial Z)^2 = (g^2 - 1)^2. \tag{58}$$

We thus find $\partial g/\partial Z = 1 - g^2$ with solution $(z+p)/\xi = \text{atanh } g$,
where p is an integration constant determined by the bound-
ary condition (which implies a choice for the extrapolation
length $1/c$ or, equivalently, for the segment-wall repulsion as
expressed in χ_s). The usual choice is $p = 0$ ($1/c = 0$). Then

$$g = \tanh\frac{z}{\xi}$$
$$\rho = \frac{\varphi}{\varphi_b} = \tanh^2\frac{z}{\xi}. \tag{59}$$

Note the analogy with Equation (43) for the dilute limit. The
only difference is the length scale: δ_0 (which depends only on
R_g) in the dilute limit, and ξ (which does not depend on R_g and
is a function of χ and φ_b) in the semidilute case.

From Equation (59) we may calculate the depletion thickness δ in semidilute solutions, using $\delta = \int(1-\rho)\,\mathrm{d}z$. This is the equivalent of Equation (41) but now in terms of the overall concentration profile instead of only the end points. The result is simple:

$$\delta = \xi, \tag{60}$$

which expresses that the depletion thickness at the surface is identical to the bulk solution correlation length ξ. We note that one could also compute δ from Equation (41), using $\rho_e = g = \tanh Z$. The result is $\delta = \xi \ln 2$, which is 30% lower than Equation (60). This demonstrates the approximative nature of the GSA solution in Equation (59): in an exact solution (such as the Eisenriegler result in Equation (37)) the integrals over ρ and ρ_e are the same.

Theta solvent. Now Equation (54) reduces to $u = (1/2)\varphi_b^2$ $(g^4 - 1)$. We may again introduce a bulk solution correlation length ξ, which is now defined by

$$\xi^{-2} = 3u_b = \frac{3}{2}\varphi_b^2. \tag{61}$$

In a theta solvent the correlation length scales inversely with the concentration: $\xi \sim \varphi_b^{-1}$ (see also Section II.B); Equation (61) gives also the prefactor.

Along the same lines as for a good solvent we find instead of Equation (57)

$$\frac{\partial^2 g}{\partial(z/\xi)^2} = 2g^5 - 2g. \tag{62}$$

The solution is [51,53]:

$$\frac{\varphi}{\varphi_b} = g^2 = \frac{\cosh\left(\sqrt{8}z/\xi\right) - 1}{\cosh\left(\sqrt{8}z/\xi\right) + 2}. \tag{63}$$

Again the zeroth moment of the profile may be determined. The result is $\delta/\xi = \sqrt{\frac{3}{2}}\mathrm{atanh}\left(\frac{1}{\sqrt{3}}\right) = 0.806$, or

$$\delta = 0.81\xi. \tag{64}$$

Again the depletion thickness is about equal to the bulk correlation length, but the numerical prefactor is slightly smaller.

Generalization. As shown by Fleer et al. [51], it is possible to combine Equations (56) and (61) into one equation by defining ξ through

$$\xi^{-2} = 3u_b = -3\ln(1-\varphi_b) - 6\chi\varphi_b \qquad (65)$$

Note the analogy with Equation (31): the φ_b-terms in this equation may be written as $-u_b$. With Equation (65) a smooth crossover between good and theta solvents is obtained, with the correct (mean-field) asymptotes for the limiting cases. A plot of ξ for three solvencies as a function of φ_b is given in Figure 6.9.

When, furthermore, we approximate $\delta \approx \xi$ throughout (which implies a systematic error of around 20% in a theta solvent) we can use $\rho = \tanh^2 \frac{z}{\xi}$ for all concentrations and solvencies in the semidilute range. In Equation (43) we found $\rho = \tanh^2 \frac{z}{\delta_0}$ in the (very) dilute limit. Fleer et al. [51] showed that the two limits may be combined by writing

$$\rho = \tanh^2 \frac{z}{\delta}, \qquad (66)$$

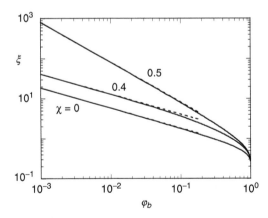

Figure 6.9 Double-logarithmic plot of $\xi(\varphi_b)$, for $\chi = 0$, 0.4 and 0.5 according to Equation (65). The limiting forms $\xi = (3v\varphi_b)^{-0.5}$ for $\chi = 0$ and 0.4 (Equation (56)) and $\xi = \sqrt{\frac{2}{3}}\varphi_b^{-1}$ for $\chi = 0.5$ (Equation (61)) are shown as the dashed lines.

where the depletion thickness δ is a function of N, χ and φ_b according to

$$\frac{1}{\delta^2} = \frac{1}{\delta_0^2} + \frac{1}{\xi^2} \tag{67}$$

with δ_0 given by Equation (42) and ξ by Equation (65). We may thus also write

$$\frac{1}{\delta^2} = \frac{\pi}{4R_g^2} - 3\ln(1 - \varphi_b) - 6\chi\varphi_b. \tag{68}$$

As shown in Figure 6.10, this equation for the depletion thickness is in excellent agreement with numerical calculations

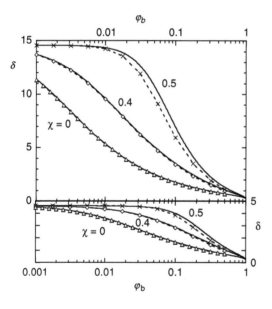

Figure 6.10 The depletion thickness as a function of the bulk concentration φ_b according to Equations (67) and (68) as compared to SCF results (symbols), for two chain lengths (upper part $N = 1000$, lower part $N = 100$) and three solvencies: $\chi = 0$, 0.4, and 0.5. The dotted curves for a theta solvent correspond to Equation (67) with ξ replaced by 0.81ξ, according to Equation (64). The numerical SCF data were computed with $\chi_s = -(1 + \chi)/6$.

using the SCF model. In this comparison, we should take a lattice boundary condition which ensures $g = 0$ and $\varphi = 0$ at the surface, which is the condition for which Equation (67) was derived. Fleer et al. [51] showed that this condition is met when in the lattice model the surface is made weakly repulsive: $\chi_s = -1/6$ for $\chi = 0$. For $\chi \neq 0$ we should add a term $\chi/6$ to correct for the missing contact with solvent (compare Equation (21)). Hence, the SCF-results in Figure 6.10 were computed with $\chi_s = -(1 + \chi)/6$.

For $\chi = 0$ and 0.4 in Figure 6.10 the analytical model describes the numerical data quantitatively. The small deviations that occur for a theta solvent are fully accounted for by the approximation to insert Equation (64) ($\delta = 0.81\xi$) into the simple form of Equation (60) ($\delta = \xi$).

5. Adsorption from semidilute good solvents

For depletion from good solvents (in the previous section) we had for the field $u = v\varphi_b(g^2 - 1)$, with g increasing from zero at the surface to unity in the bulk solution, so that close to the surface the field is negative. For adsorption, we can use the same expression, but now the field is positive, with g decreasing from a high value at the surface to unity in the bulk. The differential equation (57) does still apply, but the different boundary condition has to be taken into account. The solution is now in terms of coth z/ξ:

$$g = \coth \frac{z + p}{\xi}$$

$$\rho = \frac{\varphi}{\varphi_b} = \coth^2 \frac{z + p}{\xi}, \tag{69}$$

which may be compared to Equation (59). In this case we cannot take the integration constant zero, as this would mean that the concentration would diverge at $z = 0$. Instead, we have to find p, which may be called the proximal length, from the boundary condition, which takes into account the adsorption energy plus the requirement that the concentration in the train layer should not exceed unity. We will see that

p, which is essentially the extrapolation length $1/c$ in Equation (34), is of order unity.

For small $(z+p)/\xi$, Equation (69b) simplifies to

$$\varphi = \frac{1}{3v(z+p)^2},\tag{70}$$

where we have used $\varphi_b\xi^2 = 1/(3v)$ according to Equation (56). Applying this equation to the train layer ($z=0$) gives the relation between p and the train concentration φ_0: $p^{-2} = 3v\varphi_0$.

The meaning of the proximal length p is visualized in Figure 6.11. This figure corresponds to a situation where $\xi = 4$ ($\varphi_b \approx 0.02$) and $p = 0.632$ (or $\chi_s \approx 2$, as shown below). We have placed $z = 0$ in the middle of the train layer $(-1/2 < z < 1/2)$, so that the surface is at $z = -1/2$. It is clear that the continuum description predicts a very strong

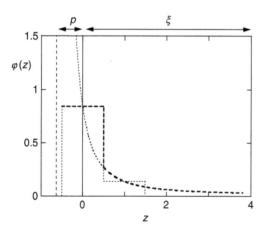

Figure 6.11 Volume fraction profile according to Equation (69) for $\xi = 4$ and $p = 0.632$ (corresponding to $\varphi_b \approx 0.02$ and $\chi_s \approx 2$). The distance z is measured from the middle of the train layer; the surface is thus situated at $z = -0.5$ (vertical dotted line). The vertical dashed line at $z = -p$ is the asymptote where the coth diverges. The length scales p and ξ are indicated. In the Stern-layer concept, the concentration $\varphi_0 = \varphi(0)$ in the train layer is homogeneous in the range $-1/2 < z < 1/2$, which replaces the (irrealistically diverging) dotted part of the continuum curve.

variation of $\varphi(z)$ over a distance which is much smaller than the segment length. In the example of Figure 6.11, $\varphi_0 = \varphi(0)$ is 0.84 (which is close to unity, as expected for relatively high χ_s). At the surface ($z = -1/2$), Equation (69) predicts the unrealistic volume fraction of more than 19! This problem is very similar to the classical problem of applying the continuum Poisson–Boltzmann equation to electrical double layers at high surface potentials. This problem was solved as early as 1924 by Stern [54], who proposed to treat the surface layer, with a thickness corresponding to a hydrated ion, in a discrete way; this layer is now commonly called the Stern layer.

We do the same for an adsorbed polymer layer, where now the appropriate length scale is the segment length. Segments in this 'Stern layer' may be called train segments and are treated in a discrete way, and loop (and tail) segments are in the region $z > 1/2$ and are described using the continuum Equation (69). Consequently, we distinguish trains and loops:

$$\varphi_0 = \varphi_{tr} = \varphi_b \coth^2 \frac{p}{\xi} \qquad (-\frac{1}{2} < z < \frac{1}{2})$$

$$\varphi_{lp} = \coth^2 \frac{z+p}{\xi} \qquad (z > \frac{1}{2}). \tag{71}$$

We are now in a position to find the relationship between p and χ_s. For convenience, we restrict ourselves to $\chi = 0$, so that the field $u_0 = u(0)$ in the surface layer is given by $e^{-u_0} = e^{\chi_s}(1 - \varphi_0)/(1 - \varphi_b)$, which is Equation (31) complemented with the χ_s-term. Inserting this in the lattice boundary condition Equation (22), we find

$$e^{-\Delta\chi_s}(1 - \varphi_b) = (1 - \varphi_0)\left(\frac{4}{5} + \frac{1}{5}\frac{g_1}{g_0}\right), \tag{72}$$

where $\varphi_0 = \varphi_b g_0^2$, $g_0 = \coth\frac{p}{\xi}$, and $g_1 = \coth\frac{p+1}{\xi}$. In terms of Figure 6.11, p is found by shifting the continuum curve horizontally until the relationship between g_0 and g_1 (or, equivalently, between φ_0 and φ_1) is satisfied.

For given φ_b and χ_s, Equation (72) constitutes an equation with one unknown parameter p, which equation may be solved numerically; for $\varphi_b = 0.02$ and $\chi_s = 2$, as in

Figure 6.11, the result is $p = 0.632$. Figure 6.12 gives a plot of $p(\varphi_b)$ for three values of $\Delta\chi_s$ and of $p(\Delta\chi_s)$ at three values of φ_b. The dependence on φ_b is only weak (unless $\Delta\chi_s$ is small), and p decreases monotonically with increasing χ_s, to reach a limit $p \approx 0.6$ for strong adsorption.

In order to see the trends, we consider the last factor in Equation (72), which varies from 4/5 (at relatively low φ_b and high χ_s) to 1 (at high φ_b or small $\Delta\chi_s$). We do not make a big error if we take 1 throughout. We then find a simple equation for the train concentration φ_0:

$$1 - \varphi_0 = (1 - \varphi_b)e^{-\Delta\chi_s}. \tag{73}$$

In this approximation φ_0 is linear in φ_b, increasing from $1 - e^{\Delta\chi_s}$ at low φ_b to 1 at $\varphi_b = 1$. At given φ_b, $1 - \varphi_0$ decays exponentially with $\Delta\chi_s$. In most cases we may approximate $1 - \varphi_b \approx 1$ so that $\varphi_0 = 1 - e^{-\Delta\chi_s}$, which is only a function of χ_s.

The proximal length is now found by combining this with Equation (71) for φ_0:

$$\coth^2\frac{p}{\xi} \approx \frac{1 - e^{-\Delta\chi_s}}{\varphi_b} = 3\xi^2(1 - e^{-\Delta\chi_s}) \tag{74}$$

from which p may be calculated from any φ_b and χ_s. For $p \ll \xi$ we have the simple explicit form

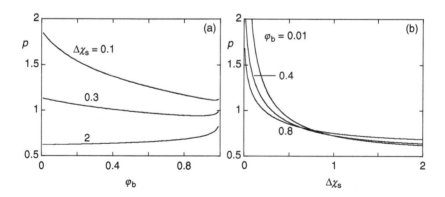

Figure 6.12   The dependence of $p(\varphi_b)$ and $p(\Delta\chi_s)$ for $\chi = 0$.

$$p \approx \frac{1}{\sqrt{3(1 - e^{\Delta\chi_s})}}. \tag{75}$$

which has the limit $p = 1/\sqrt{3} \approx 0.58$ at high χ_s (see Figure 6.12). For $\Delta\chi_s \to 0$, p diverges. However, for very small $\Delta\chi_s$ GSA breaks down.

The excess amount θ^{ex} is found by integrating $\varphi - \varphi_b$. As in Equation (71), we treat the trains discretely. We give both the full form and the simplified versions, which apply to relatively dilute solutions ($\varphi_b \leq 0.01$):

$$\theta^{ex}_{tr} = \varphi_b(\coth^2\frac{p}{\xi} - 1) \approx \frac{1}{3vp^2}, \tag{76}$$

$$\theta^{ex}_{lp} = \frac{2\varphi_b\xi}{e^{\frac{2p+1}{\xi}} - 1} \approx \frac{2}{3v(2p+1)}. \tag{77}$$

The second form for θ^{ex}_{tr} in Equation (76) is identical to φ_0 according to Equation (70). This result is valid as long as φ_b remains above the overlap concentration, which depends on the chain length. At overlap $\xi \approx R_g$ or $\varphi_{ov} \approx 2/N$; this is the lowest concentration where the above (semidilute) treatment is (approximately) valid. When φ_b is low enough for the simplified versions of Equations (76) and (77) to hold, θ^{ex} as a function of φ_b has a plateau for $\varphi_b > \varphi_{ov}$. However, for higher concentrations θ^{ex} decreases according to the full expressions in Equations (76) and 77; in the limit $\varphi_b \to 1$, θ^{ex} approaches zero. Later (Figure 6.14) we will give a plot of $\theta^{ex}(\varphi_b)$, but first we discuss the adsorption from dilute solutions, where $\varphi_b < \varphi_{ov}$.

6. Adsorption from Dilute Solutions

Good solvents. We can now use Equation (49), $\partial^2 g/\partial z^2 = 6(\varepsilon + u)g$, with $\varepsilon \approx N^{-1}\ln(1/\varphi_b)$ (Equation (53)) and $u \approx v\varphi^a \approx vg^2$ (Equation (55)). For convenience, we write $\varepsilon = y^2/N$, where the concentration parameter y is defined through

$$y \equiv \sqrt{\ln\frac{1}{\varphi_b}}. \tag{78}$$

In the range $10^{-10} < \varphi_b < 10^{-3}$, y decreases gradually from about 5 to slightly below 3.

In Equation (49), ε may be related to the inverse square of a length (in units of the bond length). This is most easily seen for large z where $u = 0$ and $\partial^2 g/\partial z^2 = 6\varepsilon g$. The solution of this differential equation is $g \sim e^{-z\sqrt{6\varepsilon}}$, which we may write as $g \sim e^{-z/d}$ where d is the exponential decay length of $g = \sqrt{\varphi}$ at large z. We call d the distal length. It is defined as

$$d = \frac{1}{\sqrt{6\varepsilon}} = \frac{1}{y}\sqrt{\frac{N}{6}} = \frac{R_g}{y}. \tag{79}$$

The parameter d may also be seen as the thickness of the adsorbed layer. From Equation (79) it is clear that d is of order of the gyration radius R_g, but it is smaller by a factor y which is 3 to 5, depending on φ_b. Through the parameter d the chain-length dependence (which is absent in the semi-dilute regime), enters into the dilute regime. Basically the N-independent length scale ξ in the semidilute regime is replaced by the N-dependent length scale d in dilute solutions.

In terms of d, Equation (49) becomes

$$\frac{\partial^2 \tilde{g}}{\partial(z/d)^2} = 2\tilde{g}^3 + \tilde{g}, \tag{80}$$

where we introduced, for convenience, a normalized eigen-function $\tilde{g} \equiv d\sqrt{3v}g$. Note the analogy (and the difference) with Equation (57). After multiplying with $\partial \tilde{g}/\partial z$ we may write Equation (80) as $\partial[(\partial \tilde{g}/\partial Z)^2 - \tilde{g}^4 - \tilde{g}^2]/\partial Z = 0$, where $Z = z/d$:

$$\left(\frac{\partial \tilde{g}}{\partial Z}\right)^2 = \tilde{g}^4 + \tilde{g}^2. \tag{81}$$

In this case the integration constant is zero as $\partial g/\partial Z$ and g vanish in the bulk solution. Hence, $\partial \tilde{g}/\partial Z = \tilde{g}\sqrt{1 + \tilde{g}^2}$. The solution is $(z + p)/d = \text{atanh}(1/\sqrt{1 + \tilde{g}^2})$, which may be re-arranged to $\tilde{g} = 1/\sinh[(z + p)/d]$. Consequently,

$$g = \frac{1}{\sqrt{3v}d} \frac{1}{\sinh[(z+p)/d]}$$

$$\varphi = g^2 = \frac{1}{3vd^2} \frac{1}{\sinh^2[(z+p)/d]}. \tag{82}$$

Again, the proximal length p is an integration constant that is related to χ_s (see Equations (85) and (86) below). We recall that φ in Equation (82) is the adsorbed part, which vanishes at large distances. This is an obvious difference with the equation for the semidilute regime, where φ becomes a constant (φ_b) at large z, see Equation (69). However, for small z the same form is obtained as in Equation (70):

$$\varphi \approx \frac{1}{3v(z+p)^2}. \tag{83}$$

As for the semidilute case, we adopt the 'Stern layer' approach, and we separate Equation (82b) into train and loop contributions

$$\varphi_0 = \varphi_{tr} = \frac{1}{3vd^2} \frac{1}{\sinh^2[p/d]} \quad \left(-\frac{1}{2} < z < \frac{1}{2}\right)$$

$$\varphi_{lp} = \frac{1}{3vd^2} \frac{1}{\sinh^2[(z+p)/d]} \quad \left(z > \frac{1}{2}\right). \tag{84}$$

As before, the middle of the train layer is at $z = 0$.

In order to find p, we start from Equation (72) for $\chi = 0$ or its simplified form $\varphi_0 = 1 - e^{-\Delta\chi_s}$, which is Equation (73) with $1 - \varphi_b = 1$. (In principle, in Equations (72) and (73) we should now replace $-\Delta\chi_s$ by $\varepsilon - \Delta\chi_s$, but since ε is small we disregard this refinement; for details we refer to Fleer et al. [12].) Combining this with Equation (84a), we find an expression for p:

$$\sinh^2\frac{p}{d} = \frac{1}{3d^2\varphi_0} = \frac{1}{3d^2(1 - e^{-\Delta\chi_s})}. \tag{85}$$

This equation looks quite different from Equation (74). However, for small p/d (relatively large N) we have the same form as Equation (75):

$$p \approx \frac{1}{\sqrt{3(1 - e^{-\Delta\chi_s})}}. \tag{86}$$

A plot of $p(\Delta\chi_s)$ is therefore nearly the same as the curve for low φ_b in Figure 6.12(b), and we do not give a separate figure.

When $\chi \neq 0$, a similar set of equations may be derived to find p. These are slightly more involved [12,13] because the field experienced by train segments depends also on χ. As a first approximation, we may use Equation (86) with the factor 3 replaced by $3v$. Hence, when $\chi > 0$ $(v < 1)$, the proximal length is larger than for $\chi = 0$.

Equations (84a) and (84b) are the contributions of trains and loops. Tails do also contribute to the adsorbed amount as will be discussed at the end of this section, but this contribution is small. To a good approximation we may find the adsorbed amount Γ (which in the dilute case is essentially the same as the excess amount θ^{ex}) by considering only trains and loops.

From Equation (84) we find

$$\theta_{tr}^{ex} = \frac{1}{3vd^2} \frac{1}{\sinh^2(p/d)} \approx \frac{1}{3vp^2}, \tag{87}$$

$$\theta_{lp}^{ex} = \frac{1}{3vd} \left(\coth\frac{2p+1}{2d} - 1 \right) \approx \frac{1}{3v} \left(\frac{2}{2p+1} - \frac{1}{d} \right). \tag{88}$$

The limiting forms of Equations (87) and (88) are those for small p/d, i.e., large N. They may be compared to Equations (76) and (77) for the semidilute regime. For trains, the limiting forms are identical. For loops, the leading terms are the same, but there is a small difference in the term $1/(3vd)$. For very large N this difference vanishes, but for realistic chain lengths it is significant (though small). For example, for $N = 1000$ (corresponding to a relative molar mass of order 10^5), d is of order 3 or 4 bond lengths in dilute solutions, so that in Equation (88) the contribution $1/(3d)$ for $\chi = 0$ is not negligible (though it is relatively small). Whereas we found that θ^{ex} is a decreasing function of φ_b in the semidilute regime, Equations (87) and (88) predict θ^{ex} ($\approx \Gamma$) to increase (weakly) with φ_b. Below, we consider these effects in more detail.

Adsorption isotherms. Equations (87) and (88) give θ^{ex} ($\approx \Gamma$) in the dilute regime ($\varphi < \varphi_{ov} \approx 2/N$); here θ^{ex} increases (weakly) with φ_b and N. Similarly, Equations (76) and (77) give the excess amount θ^{ex} in semidilute solutions ($\varphi > \varphi_{ov} \approx 2/N$), where θ^{ex} decreases with increasing φ_b and is independent of N. We may combine these two regimes to construct an adsorption isotherm, which is done in Figure 6.13 for $\chi = 0$ ($v = 1$).

The semidilute branch has a plateau $1/(3p^2)$ (trains) + $(2/3)/(2p+1)$ (loops) at low concentrations (dashed in Figure 6.13), and decreases for high φ_b to zero. This branch is independent of N as is typical for the semidilute regime. The dilute branch increases (weakly) with φ_b and depends (weakly) on N. The N-dependence originates mainly from the term $1/3d$ in

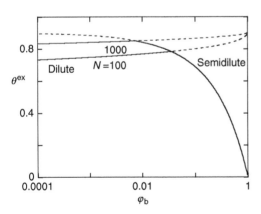

Figure 6.13 Adsorption isotherms $\theta^{ex}(\varphi_b)$ in the range $\varphi_b = 10^{-4} - 1$ for two chain lengths, $\chi = 0$, and $\Delta\chi_s = \chi_s - \chi_{sc} = 1$. In the dilute regime θ^{ex} increases with N and φ_b according to Equations (87) and (88), in the semidilute regime θ^{ex} decreases with φ_b and is independent of the chain length (Equations (76) and (77)). The dashed sections are the extrapolation of the equations for the two regimes where these equations are no longer valid.

Equation (88). The intersection is indeed around (but above) the overlap concentration $\varphi_{ov} \approx 2/N$.

When we extend the concentration range downwards to more dilute solutions, at some (very) dilute concentration the dilute regime as described by Equations (87) and (88) is no longer applicable and we enter the Henry regime of isolated chains. In Figure 6.14 we give an example for moderate N ($= 100$) and χ_s ($= 0.4$). In this case we plot $\theta^{ex}(\varphi_b)$ double-logarithmically, so that the slope in the Henry regime is unity.

The crossover between the Henry and semiplateau regimes may be estimated by setting the volume fraction φ_0 in the train layer for the Henry regime, $\varphi_0 \approx \theta^{ex} = (2\varphi_b/c)e^{Nc^2/6}$ (Equation (47c)), equal to $\varphi_0 \approx 1 - e^{-\Delta\chi_s}$ (Equation (73)) for the semiplateau. Hence, $\varphi_b \sim e^{Nc^2/6}$ at this crossover, which corresponds to extremely low volume fractions when $Nc^2/6 \approx N\Delta\chi_s$ is large. For the example of Figure 6.14, where N ($= 100$) and $\Delta\chi_s = 0.218$ (or $c = 0.784$ as follows from Equation (46)) are relatively low, we find $\varphi_b = 2.7 \; 10^{-6}$ at this crossover.

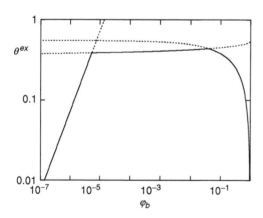

Figure 6.14 Double-logarithmic adsorption isotherm for $N = 100$, $\chi_s = 0.4$, and $\chi = 0$. The increasing branch at very low concentrations is the Henry regime (Equation (47c)), with slope unity. The weakly increasing middle part is the dilute regime (Equations (87) and (88)), the decreasing branch is the semidilute regime (Equations (76) and (77)). The dotted branches show how the various dependencies extrapolate outside the expected range of validity.

Obviously, around the two crossovers we expect a more realistic gradual transition between the various regimes than the abrupt changes in Figure 6.14. For the transition region between the Henry and dilute regimes, an analytical (though complicated) description is available [12], for the transition around $\varphi_b = \varphi_{ov} \approx 2/N$ this is, so far, not the case.

Theta solvent In a theta solvent the main difference with the good-solvent case as treated at the beginning of this section, is the field, which is no longer high and proportional to φ but much smaller and proportional to φ^2. According to Equation (55), $u \approx \varphi^2/2 = g^4/2$. This implies that Equation (80) is modified to

$$\frac{\partial^2 \tilde{g}}{\partial (z/d)^2} = \tilde{g}^5 + \tilde{g}, \tag{89}$$

where $\tilde{g} = 3^{1/4} d^{1/2} g$. The solution is

$$\varphi = g^2 = \frac{1}{d \sinh \frac{2(z+p)}{d}}. \tag{90}$$

Again, the exponential decay at large distances is $\varphi \sim e^{-2z/d}$, where the distal length is (approximately) given by Equation (79). The proximal length p has to be found from χ_s and χ ($= 1/2$) through an equation which is similar to Equation (85), but more complicated (and implicit). For details we refer to the literature [12].

For calculating the adsorbed amount, we split the profile of Equation (90) again into train and loop contributions. After integration we get

$$\theta_{tr}^{ex} = \frac{1}{d \sinh \frac{2p}{d}} \approx \frac{1}{2p}, \tag{91}$$

$$\theta_{lp}^{ex} = \frac{1}{2} \ln \coth \frac{2p+1}{2d} \approx \frac{1}{2} \ln \frac{2d}{2p+1}. \tag{92}$$

The train concentration can be shown to be higher than in a good solvent, which is not surprising because the segments experience less mutual repulsion so that they can accumulate more easily. More importantly, the loop contribution is now logarithmic in the chain length, since $d = R_g/y$ so that $\ln d$ increases as $(1/2)\ln N$ and θ_{lp}^{ex} as $(1/4)\ln N$. This feature is found also from numerical calculations as shown in Section III.F.

The structure of the adsorbed layer.   Equations (82) and (90) describe the concentration profile for good and theta solvents. Figure 6.15 illustrates these profiles both on a log–log and a log–lin scale. In this example a relatively high value for d was chosen: $d = 30$ or $N \approx 6y^2 30^2$, which is of order 10^5 in dilute solutions (y around 4). Hence the molar mass exceeds a million.

The semilogarithmic plot in Figure 6.15 highlights the distal region where the decay is exponential with decay length $d/2$; this distal region extends beyond $d \approx 30$, as expected. The double-logarithmic plot shows a central region where φ decays approximately as a power law in z. The exponent is -1 for a theta solvent and -2 in a good solvent. These results corroborate the self-similar argument of De Gennes, with in this case the mean-field scaling behavior. Indeed, with

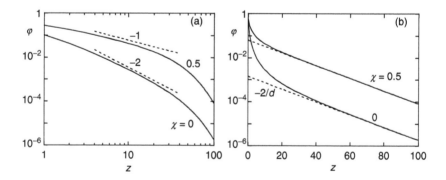

Figure 6.15   Adsorbed profiles for $p = 0.8$ and $d = 30$ for $\chi = 0$ and 0.5, both in a double-logarithmic presentation (a) and on a semilogarithmic scale (b). In the double-logarithmic plot the train layer (situated at $z = 0$) is omitted. The dashed lines to the left have slopes -1 ($\chi = 0.5$) and -2 ($\chi = 0$), those on the right are the limiting forms $[4/(3d^2)]e^{-2(z+p)/d}$ ($\chi = 0.5$) and $(2/d)e^{-2(z+p)/d}$ ($\chi = 0$), respectively.

$z \approx \xi \approx \varphi^{-1/2}$ or $\varphi \approx z^{-2}$ in a good solvent (see Section II.C) we find directly this mean-field self-similar behavior. For theta conditions $z \approx \xi \sim \varphi^{-1}$, and $\varphi(z) \sim z^{-1}$, as also shown in Figure 6.15.

We have to put in a warning, though, with respect to this scaling behavior. Figure 6.15 applies to a very high molar mass (N of order 10^5, R_g of order 120). For more realistic chain lengths ($N \sim 10^3$–10^4, d below 10), the central region is very narrow, extending over only a few segment lengths. In that case it will be difficult to find a scaling regime in experiments, even when it is realized that $R_g \propto N^{3/5}$ with swollen chains is larger than $R_g \propto N^{1/2}$ for Gaussian chains.

Tails. So far we have considered mainly trains and loops. It is well known that, for example, the hydrodynamic thickness of the adsorbed layer in dilute solutions is determined mainly by the tails [55]. These have a relatively low concentration, but protrude far in the solution. Their relative contribution increases with increasing φ_b, but in semidilute solutions they are more or less 'drowned' in the high concentration of free chains. That is why for the concentration profile tails play a role mainly in the dilute regime. Significantly more important is the role that tails play in the problem of colloidal stability. We will argue below (Section III) that tails add a repulsive contribution to the interaction between polymer-coated surfaces. This contribution is present both in dilute and in semidilute solutions but appears most prominent in the crossover region, i.e., around the overlap concentration.

Through the work of Semenov et al. [9–11] it has become possible to find an analytical description of tails. We saw before that in dilute solutions the volume fraction φ_{lp} of loops is given by the square of the eigenfunction g, which decays to zero at large z. This ground-state eigenfunction satisfies the differential equation $(1/6)(\partial^2 g/\partial z^2) = (\varepsilon + u)g$ (Equation (49)). Semenov et al. [10] showed that the volume fraction φ_{tl} of tails is proportional to the product of g (decaying) and another function f which increases from zero at the surface to a constant value in the bulk solution. This function f

satisfies the differential equation $1 + (1/6)\,(\partial^2 f/\partial z^2) = (\varepsilon + u)\,f$. Its solution is possible, but relatively complicated. We do not give the details and refer to the literature [10,12]. We just note that φ_{tl}, which is zero at $z = 0$, passes through a maximum around $z = d$ and decays at larger distances exponentially, with a decay length which is twice the decay length for the loops.

The analysis shows that the contribution of tails to the adsorbed amount is approximately given by

$$
\begin{aligned}
\theta_{tl}^{ex} &\approx 1/y^2 & (\chi = 0), \\
\theta_{tl}^{ex} &\approx N^{1/4}/y^{5/2} & (\chi = 0.5),
\end{aligned}
\tag{93}
$$

where $y^2 = \ln(1/\varphi_b)$ (Equation (78)). In a good solvent, where θ^{ex} is usually around unity, the tails thus contribute around 5% when $y = 4$ ($\varphi_b \approx 10^{-6}$), increasing to about 10% when $\varphi_b \approx 10^{-3}$; these contributions do not depend on the chain length. In a theta solvent the tails are relatively more important and their contribution increases (weakly) with the chain length. In all cases the tail length l_{tl} is roughly proportional to N, so that the tail fraction $\nu_{tl} \approx l_{tr}/N$ is nearly independent of N. For quantitative details we refer to the literature [9–13].

F. Numerical Results

As shown in Section III.C, numerically exact results can be obtained using the Scheutjens–Fleer SCF approach. In some limiting cases analytical results are available (Section III.E) which are usually based upon GSA. In this section we present numerical results focusing on quantities that are, in principle, measurable. For interpreting the trends, we will use some of the GSA predictions.

First we discuss the volume fraction of the train segments, i.e., the surface concentration (Figure 6.16). This quantity may be determined by NMR or by spectroscopic methods [8]. Then we present some volume fraction profiles (Figure 6.17 and Figure 6.18). Although it is not easy, one can obtain these profiles by neutron scattering or neutron reflectivity measurements [8,56,57]. Next, we give some predictions

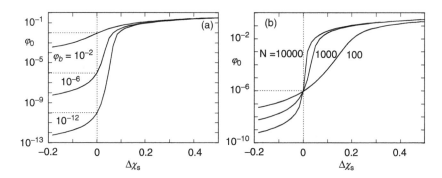

Figure 6.16 Numerical SCF results for the train volume fraction φ_0 in a good solvent ($\chi=0$) as a function of $\Delta\chi_s=\chi_s-\ln(6/5)$. Diagram (a) gives this dependence for three values of the bulk concentration φ_b at fixed chain length ($N=1000$). Diagram (b) shows the results for three chain lengths N at $\varphi_b=10^{-6}$. At $\Delta\chi_s=0$ (vertical dotted line) φ_0 equals φ_b (horizontal dotted lines). For $\Delta\chi_s<0$ we have depletion ($\varphi_0<\varphi_b$), for $\Delta\chi_s>0$ we are in the adsorption regime ($\varphi_0>\varphi_b$); in the latter case φ_0 approaches unity for large χ_s (outside the range shown).

for the adsorbed amount (Figure 6.19 and Figure 6.20) and the layer thickness (Figure 6.21). These quantities can in principle also be found from reflectivity measurements (e.g. ellipsometry). The adsorbed amount may also be found from classical techniques measuring the amount of polymer depleted from the solution [8].

One important conclusion obtained in Section III.C5 is that the adsorption transition is second order. Numerical data are presented in Figure 6.16, where we plot the train volume fraction φ_0 as a function of $\Delta\chi_s=\chi_s-\chi_{sc}$ in a good solvent ($\chi=0$). Here χ_s is the adsorption energy and χ_{sc} the critical adsorption energy. In Figure 6.16(a) results are given for a fixed chain length ($N=1000$) and the bulk concentration was varied from above overlap to well below overlap ($\varphi_{ov}\approx 2/N=2\times 10^{-3}$). At high adsorption energies the concentration of segments near the surface is nearly independent of the bulk concentration: the layer is in the semiplateau of the isotherm. From Equation (73) we have $\varphi_0\approx 1-e^{-\Delta\chi_s}$ as a first approximation in the adsorption regime.

At the critical point ($\Delta\chi_s = 0$) the surface concentration φ_0 equals the bulk concentration φ_b. For negative values of $\Delta\chi_s$ the surface concentration φ_0 is below the bulk concentration φ_b (depletion). In order to interpret the trends of Figure 6.16, we can use the Eisenriegler model (Section III.E2) as long as φ_0 is small so that crowding effects do not play a role. Hence, we can analyze, for example, the slope $\partial\varphi_0/\partial\chi_s$ at $\chi_s = \chi_{sc}$ and the value of φ_0 in the depletion regime.

Around the critical point we have $G(0,s) \approx 1 + 2c\sqrt{s/6\pi}$ according to Equations (37)–(39). Inserting this in the composition law (36) we find $\rho_0 = \varphi_0/\varphi_b = 1 + (4/9)c\sqrt{6N/\pi}$. From Equation (46) we have $c = 5\Delta\chi_s$ around the critical point, hence $\partial\varphi_0/\partial\chi_s = (20/9)\sqrt{6N/\pi}\varphi_b$ or $\partial\ln\varphi_0/\partial\chi_s = (20/9)\sqrt{6N/\pi}$. As a result, $\partial\log\varphi_0/\partial\Delta\chi_s \approx 1.33\sqrt{N}$, which is precisely the slope of the curves in Figure 6.16. It is clear that the adsorption transition becomes sharper with increasing chain length and diverges for infinite N. This means that the transition remains smooth and that the adsorption transition is second order.

In the depletion region ($\Delta\chi_s < 0$) we have $G(0,s) = c^{-1}(\pi s/6)^{-1/2}$ according to Equations (37)–(39). Now the composition law gives $\varphi_0 \approx 6/(Nc^2)$. With $c = 5(1 - e^{-\Delta\chi_s})$ according to Equation (46) we find $\rho_0 = (6/25)N^{-1}(1 - e^{-\Delta\chi_s})^{-2}$, which is in quantitative agreement with Figure 6.16.

Next we discuss the numerical predictions for the overall volume fraction profile in the adsorption case. An example is given in Figure 6.17 for strong adsorption, for a system in the semiplateau of the adsorption isotherm. This figure may be compared with Figure 6.15, which gives analytical GSA predictions for the adsorbed part φ^a of the profile. As predicted by the GSA analysis and in line with the arguments of De Gennes for the dilute regime, the layer splits up in a proximal part near the surface, a central part showing up as a (more or less) straight section in log–log coordinates (Figure 6.17a), and a distal part giving a straight line in semilog coordinates (Figure 6.17b). For this chain length ($N = 10\,000$) the central part can be identified even in theta-conditions. In the latter case the central region is just a few lattice layers wide. The main conclusion of Figure 6.17 is

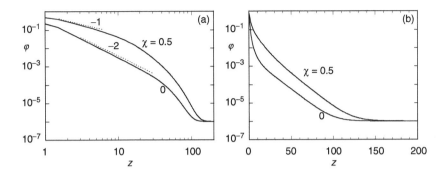

Figure 6.17 Numerical SCF results for the overall polymer concentration profile for $N = 10\,000$, $\chi_s = 1$, and $\varphi_b = 10^{-6}$. Diagram (a) is in log–log coordinates and shows the selfsimilar volume fraction profile in the central region with slope -1 for $\chi = 0.5$ and -2 for $\chi = 0$. Diagram (b) is in log–lin coordinates and illustrates the exponential decay in the distal region. The figure is the numerical analog of Figure 15, which gives analytical predictions for the adsorbed part φ^a of the profile.

that the overall picture of the adsorption layer is well established.

The semilogarithmic plot of Figure 6.15(b) shows a decay length $d/2$. For $N = 10^4$ ($R_g = 40.8$) and $\varphi_b = 10^{-6}$ ($y = 3.72$), $d = R_g/y$ is about 11. Hence, on the basis of simple GSA one would expect an inverse slope $\partial z/\partial \log\varphi$ in Figure 6.17(b) of $2.3d/2$: about 13 layers per decade in φ. The actual slope is 27 layers per decade, or a decay length d instead of $d/2$. This clearly shows the shortcomings of GSA, which describes only loops. The decay length seen in Figure 6.17(b) is that of tails, twice that of loops. Indeed, when in the numerical model the loop contribution is calculated separately, a loop decay length $d/2$ is found. In Figure 6.17(b) only the overall profile (loops plus tails) is shown, so that the loop decay length cannot be identified.

There is another small feature worth noting in the comparison of Figure 6.15 and Figure 6.17. The crossover of the concentration profile to the bulk concentration is not covered in the GSA analysis because this describes only adsorbed chains and neglects free polymer. Another interesting detail is the following. In the numerical profiles it is found that before the profile goes to the bulk value it shows a narrow dip. In other words, the concentration profile falls slightly below φ_b before it stabilizes at $\varphi(\infty) = \varphi_b$. This small dip is not visible on the scale of Figure 6.15(b). It is slightly more pronounced for good solvents. This dip can be attributed to the depletion of free chains: free chains are depleted slightly from the adsorption layer, which may be seen as a 'soft wall'. This depletion is not found from the GSA analysis.

In Figure 6.18 log–log plots of the overall profiles are shown for a series of polymer concentrations ranging from slightly below to well above overlap. From close inspection of Figure 6.18(a), which applies to adsorption ($\chi_s = 1$) from a good solvent ($\chi = 0$), one finds that the dip in the volume fraction profile is largest for $\varphi_b = 10^{-3}$, which is near overlap for $N = 1000$.

When $\varphi(z)$ is above overlap, i.e., above $\varphi_b = 10^{-3}$, the profiles closely follow the GSA prediction $\varphi = \varphi_b \coth^2[(z+p)/\xi]$. We do not give any quantitative detail here; this would require the mapping of p and χ_s along the lines as discussed in Section III.E6. The proximal region is hardly visible in Figure 6.18a as the train layer is omitted on the log–log plot. When

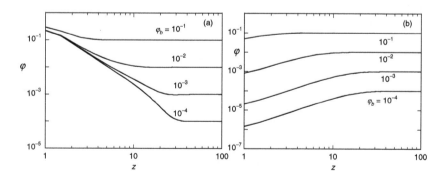

Figure 6.18 Numerical SCF results for the polymer concentration profile for $N = 1000$ and $\chi = 0$ on log–log coordinates for a range of polymer concentrations for adsorption ($\chi_s = 1$) to the left (a) and for depletion ($\chi_s = 0$) to the right (b).

ξ is large enough, there is a central region with slope -2, as seen for the lower concentrations in Figure 6.18(a). For the higher concentrations, ξ becomes too small for this exponent to show up.

The depletion regime is shown in Figure 6.18(b) ($\chi_s = 0$ and $\chi = 0$). Again the profiles are plotted on a double-logarithmic scale. From these profiles one can clearly see that the thickness of the depletion layer goes to a constant value δ_0 which is of order R_g ($= 12.9$) for bulk concentrations below overlap ($\varphi_b = 10^{-4}$ and 10^{-3}), and becomes a decreasing function of the polymer concentration above the overlap concentration (compare Figure 6.10). This behavior is completely in line with the GSA analysis of Section III.E4.

When the depletion profile is described as $\rho = \tanh^2(z/\delta)$, we would expect $\varphi \sim z^2$ when $z \ll \delta$. The slope on a log–log plot should then be $+2$. In Figure 6.18(b) there is a more or less constant slope for small φ_b, but it is smaller than 2. This is because $\chi_s = 0$ does not correspond to $\rho = \tanh^2(z/\delta)$ but to $\rho = \tanh^2[(z+p)/\delta]$, where $p = 0.5$ for $\chi_s = 0$ as shown recently by Fleer et al. [51]. Now the slope over the interval from $z = 1$ to $z = 2$ is given by $(2/\ln 2)\ln[(2+p)/(1+p)]$, which is 1.47 and quite close to the slope observed in Figure 6.18(b). This subtle difference shows the pitfalls of comparing scaling laws with numerical data. In the discussion of Figure 6.10 we saw that $p = 0$ is only obtained for a particular choice of χ_s ($\chi_s = -(1+\chi)/6$).

In Section III.A we discussed two measures for the adsorbed amount. The amount of chains that touch the surface (see Ref. [5] for details of the method to compute this quantity) is denoted by Γ, the excess is θ^{ex}. In Figure 6.19 we present adsorption isotherms for three chain lengths and a wide concentration range, in double-logarithmic coordinates. As predicted in Figure 6.14 the three regimes (Henry, semi-plateau, and concentrated) are easily identified. For very low concentration, in the Henry region, both θ^{ex} and Γ are proportional to φ_b. The adsorbed amount in this very dilute part depends significantly on N ($\theta^{ex} \propto e^{Nc^2/6}$ according to Equation (47)) whereas in semidilute and concentrated solutions the excess amount θ^{ex} becomes nearly independent of N.

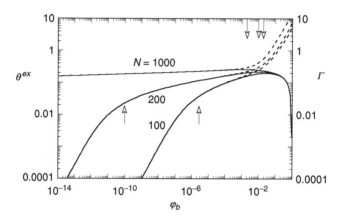

Figure 6.19 Adsorption isotherms calculated from the numerical SCF theory, for $\chi_s = 0.4$, $\chi = 0$, and three chain lengths. Both the excess amount θ^{ex} (solid curves) and the adsorbed amount Γ (dashed) are given as a function of the bulk polymer concentration in log–log coordinates. The arrows pointing upwards indicate the crossover from the Henry regime to the semiplateau region for $N = 100$ and 200. The arrows pointing down are the crossover concentrations $\varphi_{ov} = 2/N$.

In the discussion of Figure 6.14 we saw that the transition between the Henry and semiplateau regimes is approximately found at $\varphi_b = (c/2)(1 - e^{-\Delta\chi_s})e^{-Nc^2/6}$, with $\Delta\chi_s = 0.218$ and $c = 0.784$ (Equation (46b)) for $\chi_s = 0.4$ and $\chi = 0$. For $N = 100$ and 200 these concentrations are 2.7×10^{-6} and 9.5×10^{-11}, respectively. These concentrations are indicated in Figure 6.19 by the upward arrows. For $N = 1000$ this crossover is around 10^{-46}, way below the range shown in the figure. The other crossover towards the semidilute regime, at $\varphi_{ov} \approx 2/N$, is indicated in Figure 6.19 by the downward arrows.

The two measures for the adsorbed amount differ from each other only in the semidilute and concentrated regime, as expected. The number of chains that touch the surface grows as $\sqrt{N}$ at high concentrations [6,8].

As compared to Figure 6.14, where the GSA predictions were discussed, the main difference is that Figure 6.19 gives smooth transitions between the various regimes, unlike the simple GSA treatment.

Another interesting aspect is how, for a fixed polymer concentration and fixed surface affinity, the adsorbed amount depends on the chain length. From the GSA analysis we expect that the adsorbed amount diverges logarithmically for theta conditions (Equation (92)) and that it remains of order unity with only a weak N-dependence for good solvents (Equation (88)). The excess adsorbed amount is presented in Figure 6.20 for $\varphi_b = 10^{-4}$ and $\chi_s = 1$, in semilogarithmic coordinates. For $\chi = 0.5$ the logarithmic divergence shows up prominently as an increasing straight line, which sets in already at very low chain length $N \approx 100$. Below Equation (92) we argued that θ_{lp}^{ex} increases as $(1/4)\ln N$ or $0.58\log N$; this is indeed the slope found in Figure 6.20. Experimentally similar results were reported by Van der Linden and Van Leemput [58]. As expected the adsorbed amount levels off for good solvent conditions; for long chains the $1/d$-term in Equation (88) vanishes. The closer the solvent quality is to theta conditions, the longer the adsorbed amount follows the logarithmic behavior, but eventually it will level off at higher chain length.

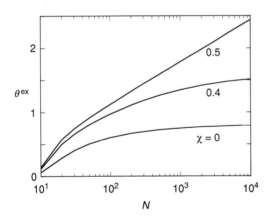

Figure 6.20 Numerical results for the excess adsorbed amount as a function of the chain length N for three values of the solvent quality $\chi = 0$, 0.4, and 0.5, as indicated. The adsorption affinity is $\chi_s = 1$ and the bulk concentration is $\varphi_b = 10^{-4}$.

The crossover chain length N^* may be estimated from evaluating when the intra-chain excluded-volume interactions become of order unity in one chain. Thus, the chain remains ideal for $N < N^*$ where $\sqrt{N^*}v \approx 1$. For $\chi = 0.4$ ($v = 1 - 2\chi = 0.2$) $N^* \approx 25$, and the deviations from $\chi = 0$ (N^* of order unity) are significant.

The layer thickness is easily measured experimentally by hydrodynamic or ellipsometric experiments. In the final Figure 6.21 of this numerical section we present the results for the layer thickness of the adsorbed polymer based on the ratio δ_1 between the first and zeroth moment over the concentration profile: $\delta_1 = D^{(1)}/D^{(0)}$. Here $D^{(i)} = \int z^i(\varphi(z) - \varphi_b)dz$ is the ith moment of the excess polymer volume fraction profile. The fact that some experimental data [8,56] have shown that $\delta \propto N^{0.4}$ is sometimes interpreted as evidence for the occurrence of the self-similar volume fraction profile of $\varphi \propto z^{-4/3}$. This is because $\int_0^{N^{3/5}} z(z^{-4/3})dz = N^{0.4}$, where it is assumed that the zeroth moment is almost independent of N.

Referring to Figure 6.21a it is seen that the numerical data for the layer thickness as a function of the chain length can be reasonably well represented by a power-law. The exponent is a rather strong function of the solvent quality: fitting $\delta_1 = N^\alpha$ gives values of $\alpha \approx 0.4$, 0.25, and 0.16 for $\chi = 0.5$, 0.4,

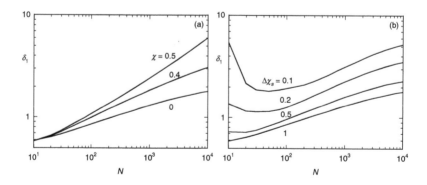

Figure 6.21 The layer thickness δ_1 as a function of the chain length N in double-logarithmic coordinates for $\varphi_b = 10^{-4}$. Diagram (a) gives results for $\chi_s = 1$ and three solvencies (as indicated), diagram (b) is for $\chi = 0$ and four values of $\Delta\chi_s$.

and 0, respectively. For a good solvent this exponent is considerably smaller than 0.4. Note that GSA ($\varphi \sim z^{-2}$ in a good solvent) would give δ_1 proportional to $\int_0^{N^{1/2}} z^{-1} \, dz$, which is logarithmic in N. For a theta solvent a similar argument leads to $\delta_1 \sim N^{0.5}$, which is not far from $\alpha = 0.4$ in Figure 6.21a. Nevertheless, these arguments are rather rough. Our preliminary conclusion is that the layer thickness is too sensitive a parameter to judge the failure of the GSA approximation, or the success De Gennes self-similar scaling for polymer adsorption.

Finally, in Figure 6.21(b) we show the dependence of $\delta_1(N)$ for various values of the effective adsorption strength $\Delta\chi_s$ in good solvent conditions. For very low values of $\Delta\chi_s$, the layer thickness goes through a minimum. This minimum reflects the crossover from weak to strong adsorption around $Nc^2/6 \approx 1$, where $c \approx 5\Delta\chi_s$. For $\Delta\chi_s = 0.1$ ($c \approx 0.5$) this predicts the minimum around $N = 24$, which is in fair agreement with Figure 6.21(b). The layer thickness is relatively large for small N and small $\Delta\chi_s$ because the chains gain little adsorption energy and are not strongly deformed. In this case the adsorbed amount is small, and δ tends towards the radius of gyration of the chains.

IV. COLLOIDAL STABILITY

A. General Aspects

Polymers are often applied to modify surface properties. An important aspect is that polymers influence the way in which particles interact with each other, which has direct consequences for the colloidal stability. In the remainder of this review we discuss how adsorbing or nonadsorbing homopolymers affect the interaction between two particles. Implicitly we will assume that the polymer contribution can be superimposed upon other contribution such as van der Waals forces, electrostatic repulsion, etc. [59]. Here we discuss only the effect of the polymer.

When calculating the Gibbs energy of interaction one needs to know to what extent exchange of material is allowed during particle approach. In most cases we will assume full equilibrium, i.e., constant chemical potentials of both solvent and polymer. This implies that the polymer can escape from the gap (or be drawn in) sufficiently fast: if it would not do so, its chemical potential would change.

For interacting depletion layers the assumption of full equilibrium is usually warranted since the polymer is not attached to the particles: no desorption or adsorption step is involved. For adsorption layers the main issue is whether the (de)sorption processes are fast enough. As a starting point, we assume that this is the case. Any deviation from this scenario implies off-equilibrium features ('restricted equilibrium'), with a higher Gibbs (or Helmholtz) energy than in full equilibrium. In restricted equilibrium the interaction is therefore necessarily more repulsive than in full equilibrium. One type of restricted equilibrium is that in which the amount of polymer in the gap is fixed (no transport of polymer is possible) but where the conformations of these interfacial chains are fully relaxed and adjust to a varying plate separation. Another option is to keep the adsorbed amount fixed during the interaction, but allow the free chains to exchange between the gap and the bulk solution. We will not go into these details and we concentrate on full equilibrium.

The general issue is sketched in Figure 6.22, which gives (numerical) volume fraction profiles in the gap between two particles at finite separation $2h$ (bottom) and at large separation (top), both for adsorption (left) and for depletion (right). At large separation the volume fraction in the middle of the gap equals the bulk concentration φ_b, for interacting surfaces the volume fraction φ_m in the middle differs from φ_b: $\varphi_m > \varphi_b$ for adsorption and $\varphi_m < \varphi_b$ for depletion.

In Section III.C we discussed how to calculate the profiles of Figure 6.22 numerically. In this numerical model the method to calculate the profiles at large and small h is essentially the same: the only difference is the number of layers taken into account. However, for analytical approximations there is a large difference because of the different boundary conditions. In Section III.E the equations for isolated surfaces (top diagrams in Figure 6.22) were given for various

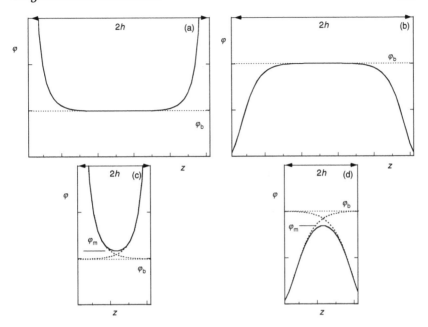

Figure 6.22 Volume fraction profiles in a slit between two surfaces at large separation (top; a,b) and at small separation (bottom; c,d), both for adsorption (left; a,c) and depletion (right; b,d). At large h the volume fraction φ_m in the middle of the slit equals the bulk volume fraction φ_b, at small separation $\varphi_m > \varphi_b$ for adsorption and $\varphi_m < \varphi_b$ for depletion. The dotted profiles in the bottom diagrams are the single-plate profiles.

situations; in these cases relatively simple solutions were possible because of the boundary condition $g = 1$ (semidilute) or $g = 0$ (dilute) in the bulk solution. For overlapping polymer layers the situation is more complicated, as discussed in the next section.

B.   Concentration Profiles in the Gap

As in Section III.E, we distinguish between the semidilute and dilute cases. The GSA equation for the semidilute case is Equation (57), $\partial^2 g / \partial Z^2 = 2g^3 - 2g$, with $Z = z / \xi$ and $\varphi = \varphi_b g^2$. For the dilute case we have Equation (80), $\partial^2 \tilde{g} / \partial Z^2 = 2\tilde{g}^3 + \tilde{g}$, with $\tilde{g}=(d\sqrt{3}v)g$, $Z=z/d$, and $\varphi=g^2=\tilde{g}^2/(3vd)^2$. These differential equations also apply to narrow slits, but because of the different boundary conditions the solutions are not as easy as for single (nonoverlapping) profiles.

1. Semidilute Solutions

For a single surface we found the solution $(\partial g/\partial Z)^2=(g^2-1)^2$ according to Equation (56). This simple form applies because $g=1$ when $\partial g/\partial Z=0$. For interacting surfaces we have $g=g_m \neq 1$ in the middle of the gap where $\partial g/\partial Z=0$ (see Figure 6.22). The value of g_m depends on the plate separation $2h$. Now Equation (56) transforms into

$$\left(\frac{\partial g}{\partial Z}\right)^2 = (g^2-1)^2 - (g_m^2-1)^2. \tag{94}$$

For $g_m=1$ Equation (56) is recovered. When $g_m \neq 1$, there is no analytical solution. We can find $Z(g)$ from numerical integration, according to

$$Z \equiv z/\xi = \int_{g_m}^{g} \frac{dg}{\sqrt{(g^2-1)^2-(g_m^2-1)^2}}$$

$$H \equiv h/\xi = \int_{g_m}^{g_0} \frac{dg}{\sqrt{(g^2-1)^2-(g_m^2-1)^2}}. \tag{95}$$

The procedure is to first choose a value of g_m. Numerical integration of Equation (95a) then gives the value of Z that corresponds to a certain g (i.e., the upper integration limit). In a similar way Equation (95b) gives the value of H corresponding to the chosen g_m; to that end a choice must be made for $g_0=g(0)$ in the train layer. A possible choice is to take the value for g_0 in a single adsorption layer. We note, however, that this choice is not strictly correct, as in full equilibrium the train concentration adjusts itself to a varying plate separation. For small overlap this adjustment is expected to be weak.

Equation (95) does not only apply to adsorption $(g \geq g_m \geq 1)$, but also to depletion $(g \leq g_m \leq 1)$, provided the upper and lower integration limits are interchanged. In the depletion case a reasonable choice for g_0 is $g_0 = 0$. However, also then no simple analytical solution is available.

2. Dilute solutions

In this case we had Equation (81) for $g = 0$ in the bulk solution $(\partial g / \partial Z = 0)$: $(\partial \tilde{g} / \partial Z)^2 = \tilde{g}^4 + \tilde{g}^2$. With $\partial g / \partial Z = 0$ at $g = g_m > 0$, this equation is extended to

$$\left(\frac{\partial \tilde{g}}{\partial Z}\right)^2 = \left(\tilde{g}^2 + \frac{1}{2}\right)^2 - \left(\tilde{g}_m^2 + \frac{1}{2}\right)^2. \tag{96}$$

Again there is no analytical solution. A numerical procedure analogous to Equation (95) is possible but we do not go into any detail.

We conclude that, in general, the route to calculate the profiles from Equation (94)–(96) is not very attractive. Yet, we need these profiles to calculate the grand potential Ω and the Gibbs energy of interaction G_a, as will be discussed in the next Section IV.C. For the profiles obtained with the numerical SCF model there is no problem. For obtaining analytical approximations we will fall back on alternatives. In the depletion case we shall use a step-function approach, see Section IV.D.

For interacting adsorption layers we try to get insight in the trends by using the superposition approximation (see Section IV.E). This implies that the (excess) volume fraction profile is assumed to be given by the sum of the two single-plate profiles. For the concentration φ_m in the middle of the gap we then have $\varphi_m - \varphi_b = 2[\varphi_s(h) - \varphi_b]$, where $\varphi_s(z)$ is the single-plate profile. In extremely dilute solutions (zero field), we expect that this superposition is exact: the chains do not feel each other so chains coming from one plate are not affected by chains adsorbed on the other plate. When the field is nonzero, this superposition does not hold. Nevertheless, we expect to capture the main trends for small overlap. Numerical SCF calculations show that the superposition is reasonable to first order, although it systematically overestimates the true equilibrium profile, which relaxes to give slightly lower concentrations than predicted by the superposition.

C.   Grand Potential and Gibbs Energy of Interaction

In full equilibrium the chemical potentials of all components do not depend on the distance between the two surfaces, and the interfacial Gibbs energy γ is determined by the grand potential $\Omega = \gamma a = F - n_p\mu_p - n_S\mu_S$, where a is the area per site and all quantities are defined per plate. The quantity μ_p is N times the segment chemical potential μ, defined in Equation (27), and μ_S is the solvent chemical potential (Equation (25)). As shown in Equations (28) and (29), Ω follows from an integration over the concentration profile, where the integration extends from $z = 0$ to $z = h$, which is half the separation between the two plates.

The Gibbs energy of interaction between the two plates is given by

$$G_a(h) = 2[\Omega(h) - \Omega(\infty)], \tag{97}$$

where $\Omega(h)$ corresponds to the bottom diagrams of Figure 6.22 and $\Omega(\infty)$ to the top diagrams. In a continuous representation, Equation (29) may be written as

$$\Omega(h) = \int_0^h \omega(z)\mathrm{d}z = \int_0^h [\pi_b - \pi(z)]\mathrm{d}z. \tag{98}$$

The full expression for Ω is given in Equation (28). Here we write it in terms of the grand potential density $\omega(z)$, which is a function of $\varphi = \varphi(z)$:

$$
\begin{aligned}
\omega &= \ln\frac{1-\varphi}{1-\varphi_b} + (1 - \frac{1}{N})(\varphi - \varphi_b) + \chi[\varphi\langle\varphi\rangle - \varphi_b^2] \\
&\approx -\frac{1}{N}(\varphi - \varphi_b) - \frac{1}{2}v(\varphi^2 - \varphi_b^2) - \frac{1}{3}(\varphi^3 - \varphi_b^3)
\end{aligned}
\tag{99}
$$

Obviously, $\omega = 0$ in the bulk solution. The first form of Equation (99) is the full SCF result, the second is obtained by approximating $\langle\varphi\rangle \approx \varphi$ and expanding the logarithm up to the cubic term.

For analytical purposes, for adsorbing polymer we will use slightly more simplified expressions for ω:

$$\omega \approx \begin{cases} -\frac{1}{2}v(\varphi^2 - \varphi_b^2) & \text{semidilute} \\ -\frac{\varphi}{N} - \frac{1}{2}v\varphi^2 & \text{dilute} \end{cases}. \tag{100}$$

In the semidilute case we thus assume dominance of the quadratic term $\varphi^2 - \varphi_b^2$, for dilute solutions we neglect φ_b with respect to $\varphi = \varphi(z)$.

Apart from G_a, also its (negative) derivative with respect to h is useful:

$$\Pi(h) = -\frac{\partial G_a(h)}{2\partial h} = -\frac{\partial \Omega(h)}{\partial h}, \tag{101}$$

where again $2h$ is the plate separation. The quantity $\Pi(h)$ is the force per unit area needed to keep the two surfaces at a specified distance; it is also known as the *disjoining pressure*.

D. Depletion

Equation (98) gives a way to calculate $\Omega(h)$ by integrating over the (local) osmotic pressure profile. We could call this the 'osmotic route'. We will use this below, but sometimes it is easier to follow a different route in which Ω is coupled to the (negative) adsorption. We first outline this 'adsorption route'.

1. The Adsorption Route to Calculate Ω

In this method the final system (two plates at a given separation $2h$ in equilibrium with a bulk concentration φ_b) is reached by starting from two plates with only solvent between them, and gradually adding the polymer. At any point the change in the interfacial Gibbs energy (or grand potential) is given by Gibbs' law:

$$d\Omega(h) = -\theta^{ex}(h)d\mu, \tag{102}$$

where θ^{ex} is the excess amount of polymer per plate (negative for depletion), and μ is the segment chemical potential of the polymer, given by Equation (27). The final grand potential is found as $\Omega = \int d\Omega$. In order to perform the integration, we have to account for the concentration dependence of both θ^{ex} and μ. The integration variable is φ, which runs from zero to the final value φ_b.

Let us start with a single surface ($h = \infty$). The excess amount θ^{ex} may be expressed in the depletion thickness δ as $\theta^{ex} = -\varphi_b\delta$, where δ is given by Equation (67), $\delta^{-2} = \delta_0^{-2} + \xi^{-2}$; its concentration dependence enters through ξ as given in Equation (65). We may relate the change $d\mu$ for the polymer to the change $d\mu_s$ for the solvent using the Gibbs–Duhem relation $\varphi_b d\mu + (1 - \varphi_b)d\mu_s = 0$. Since $d\mu_s = -d\pi_b$, we have $d\mu = (1/\varphi_b - 1)d\pi_b = (1/\varphi_b - 1)(d\pi_b/d\varphi_b)d\varphi_b$. Inserting θ^{ex} and $d\mu$ into Equation (102) and integrating gives

$$\Omega(\infty) = \int_0^{\varphi_b} \delta(1 - \varphi)\frac{\partial\pi_b}{\partial\varphi}\,d\varphi. \tag{103}$$

Here, $\partial\pi_b/\partial\varphi$ is the inverse of the osmotic compressibility; it is given by Equation (26).

Despite the quite different forms of Equations (98) and (103), they give identical results when the numerically exact values for the profile and for $\delta = -\theta^{ex}/\varphi_b$ are used, as they should. When approximations (such as Equation (67) for δ) are used, slight deviations may obviously occur.

For analytical approximations, both forms of $\Omega(\infty)$ can be used as a starting point, but for a single surface Equation (98) is easier because a rather accurate approximation for the profile is available: $\varphi = \varphi_b\tanh^2(z/\delta)$ according to Equation (66). Inserting the second form of Equation (99) into Equation (98), we can integrate the three terms separately, to give $\Omega(\infty) = \varphi_b\delta(1/N + (2/3)v\varphi_b + (23/45)\varphi_b^2)$. With $N = 6R_g^2 = (3\pi/2)\delta_0^2$ according to Equation (42), we may rewrite this as $\Omega(\infty) = (2/(3\pi))\varphi_b\delta(\delta_0^{-2} + \pi v\varphi_b + (23\pi/30)\varphi_b^2)$. To a good approximation we may replace the last two terms by $\xi^{-2} = 3v\varphi + (3/2)\varphi_b^2$, which is the expanded form of Equation (65). When we next replace $\delta_0^{-2} + \xi^{-2}$ by δ^{-2} according to Equation (67), we end up with an extremely simple result

$$\Omega(\infty) \approx \frac{2}{3\pi}\frac{\varphi_b}{\delta}. \tag{104}$$

This equation may be seen as the extension to finite concentrations of the exact expression derived Eisenriegler et al. [60] and Louis et al. [61] for ideal chains in the dilute limit, which is Equation (104) with δ_0 instead of δ.

For interacting surfaces we also need $\Omega(h)$ for finite h. In this case we have no analytical result for the profile, so that the osmotic route is not as easy as for $h \to \infty$. We now return to Equation (103), and apply the step-function approach by assuming that there is no polymer in the gap when $h < \delta$. Then the excess amount per plate is approximated as $\theta^{ex}(h) = -\varphi_b h$ for $h < \delta$. Now Equation (103) transforms into

$$\Omega(h) = \int_0^{\varphi_b} h(1-\varphi)\frac{\partial \pi_b}{\partial \varphi}\,\mathrm{d}\varphi \qquad h < \delta. \tag{105}$$

In the next section we will combine Equations (103) and (105) to find $G_a(h)$.

2. The Gibbs Energy of Interaction

For calculating $G_a(h)$ from Equation (97), we have to subtract $\Omega(\infty)$ according to Equation (103) from $\Omega(h)$ according to Equation (105). When $h > \delta$, we should take $G_a(h) = 0$ in the step-function approach. We incorporate this by writing $\delta - h = (\delta - h)H(\delta - h)$ where $H(x)$ is the Heavy-side step function which is zero for $x < 0$ and unity for $x \geq 0$. Hence

$$G_a(h) = -2\int_0^{\varphi_b} (\delta - h)(1-\varphi)\frac{\partial \pi_b}{\partial \varphi}H(\delta - h)\,\mathrm{d}\varphi. \tag{106}$$

The minus sign shows that there is attraction. When a constant value for $\delta = \delta(\varphi)$ would be used in Equation (106), $G_a(h)$ would be strictly linear in the range $0 < h < \delta$, and zero for $h > \delta$. In fact δ equals δ_0 for small φ (at the lower end of the integration interval), and δ decreases as the integration variable increases towards it highest value φ_b (lowest δ). Now there are contributions to $G_a(h)$ for $\delta(\varphi_b) < h < \delta_0$, and $G_a(h)$ shows a tail around $h = \delta(\varphi_b)$.

Figure 6.23 shows an example for $N = 1000$, $\chi = 0.4$, and four values of φ_b, both for the analytical expression (106) and the full SCF results. As in Figure 10, $\chi_s = -(1+\chi)/6$, with $\chi = 0.4$ in Figure 6.23. This value was chosen to ensure that $\varphi = 0$ at the surface, which is the prerequisite for Equation (67), $\delta^{-2} = \delta_0^{-2} + \xi^{-2}$, to hold. The agreement between Equation (106) and numerical SCF is nearly quantitative. However,

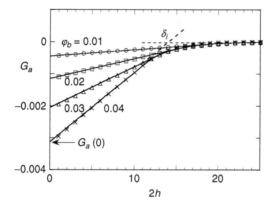

Figure 6.23 The Gibbs energy of interaction $G_a(h)$ as a function of the plate separation $2h$, for $N = 1000$, $\chi = 0.4$, and four bulk concentrations φ_b as indicated. Symbols are the SCF results for $\chi_s = -(1 + \chi)/6$; the curves were obtained from Equation (106). The minimum value $G_a(0)$ and the characteristic interaction distance δ_i for $\varphi_b = 0.04$ is indicated.

Equation (106) is not that simple, as it requires a numerical integration.

As can be seen from Figure 6.23, the basic features of $G_a(h)$ are a linear dependence over most of the range, characterized by a minimum $G_a(0)$ and an interaction distance δ_i which determines the slope $\partial G_a/\partial h = -G_a(0)/\delta_i$ of this linear part. For this linear part we may write $G_a(h) = G_a(0)[1 - h/\delta_i]$, where $G_a(0)$ is negative. It is relatively easy to find expressions for $G_a(0)$ and δ_i.

For $h \rightarrow 0$ there is no polymer in the slit, so $\Omega(0) = 0$. According to Equation (97) we have then $G_a(0) = -2\Omega(\infty)$. With Equation (104):

$$G_a(0) = -\frac{4}{3\pi}\frac{\varphi_b}{\delta}. \tag{107}$$

Figure 6.24a shows $G_a(0)$ according to Equation (107) for $N = 1000$ and three values of χ ($\chi = 0$, 0.4, and 0.5) as a function of φ_b in a double-logarithmic representation, both according to Equation (107) and for the numerical SCF-data. For

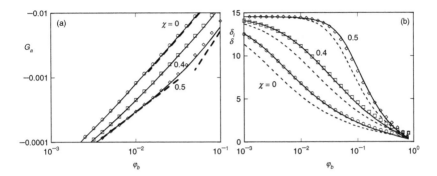

Figure 6.24 The characteristic quantities $G_a(0)$ and δ_i (see Figure 6.23) as a function of φ_b for $N = 1000$ and three values of χ. Symbols are the SCF result for $\chi_s = -(1+\chi)/6$, the solid curves were computed from (a) Equation (107) (left) and (b) Equation (108) (right). In the left diagrams the asymptotes $(4/\pi\sqrt{3})\varphi_b^{3/2}$ $(\chi=0)$ and $(\sqrt{8/3}/\pi)\varphi_b^2$ $(\chi=0.5)$ for high φ_b are shown, as well as the low-φ_b asymptote $\sqrt{8/3\pi N}\varphi_b$. In the right-hand side diagram the value of the single-plate depletion thickness δ (Figure 6.10) is indicated as the dashed curves.

comparison, the limiting forms for high concentration $(\delta \approx \xi)$ for $\chi = 0$ and $\chi = 0.5$ are given as well (dashed lines). For $\chi = 0$, $\xi^{-2} \approx 3\varphi_b$ and this asymptote is given by $G_a(0) = -(4/\pi\sqrt{3})\varphi_b^{3/2}$, for $\chi = 0.5$ we have $\xi^{-2} = \frac{3}{2}\varphi_b^2$ and $G_a(0) = -(\sqrt{8/3}/\pi)\varphi_b^2$. For small φ_b $(\delta = \delta_0 = \sqrt{2N/3\pi})$ all the curves converge to $\sqrt{8/3\pi N}\varphi_b$; this limiting behavior is also indicated in Figure 6.24(a). The agreement between the analytical predictions and the numerical SCF results is quite good.

Equation (107) for $G_a(0)$ followed from Equation (104), which was derived from the osmotic route. In a similar way we find an expression for δ_i. For narrow slits the gap contains only solvent, so $\pi(z) = 0$ and $dG_a/dh = 2d\Omega/dh = 2\pi_b$ according to Equation (98): G_a is linear in h for small h (see Figure 6.23).

When we extrapolate this linear dependence to $G_a = 0$ with $h = \delta_i$, we find $-G_a(0)/\delta_i = 2\pi_b$ or, using $G_a(0) = -2\Omega(\infty)$:

$$\delta_i = \frac{\Omega(\infty)}{\pi_b}. \tag{108}$$

Intuitively, one expects δ_i (the zeroth moment of the osmotic pressure profile) to be of order δ (the zeroth moment of the volume fraction profile). Indeed, $\Omega(\infty) = \varphi_b \delta((1/N) + (2/3)v\varphi_b + (23/45)\varphi_b^2)$, (see above Equation (104), and $\pi_b \approx \varphi_b(1/N + (1/2)v\varphi_b + (1/3)\varphi_b^2)$ according to Equation (25), so that the ratio δ_i/δ is given by

$$\frac{\delta_i}{\delta} \approx \frac{1/N + (2/3)v\varphi_b + (1/2)\varphi_b^2}{1/N + (1/2)v\varphi_b + (1/3)\varphi_b^2}. \tag{109}$$

Hence, δ_i/δ is close to (but higher than) unity.

Figure 6.24(b) gives a plot of $\delta_i(\varphi_b)$ according to Equation (108) (curves) and the numerical SCF data (symbols) for $\chi = 0$, 0.4, and 0.5. Again the agreement is quite good, which supports the analytical description given above. For comparison, this figure gives also $\delta(\varphi_b)$ taken from Figure 6.10 (dashed curves). As expected, δ_i/δ is above unity, but not much so.

We conclude this depletion section with a qualitative illustration of a slit that is void of polymer for $h < \delta_i$, with an outside osmotic pressure π_b (Figure 6.25). The physical picture is that there is an overlap volume $V_{ov} = 2(\delta_i - h)A$ of the two depletion zones, giving rise to an attraction $G(h) = -\pi_b V_{ov}$.

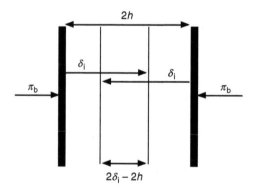

Figure 6.25 Two plates of area A at distance $2h$ (with $h < \delta_i$) without polymer in between. The outside osmotic pressure π_b pushes the plates together which leads to an attractive Gibbs energy of interaction $G(h)$ which is equal to $-\pi_b V_{ov}$, where $V_{ov} = 2(\delta_i - h)A$ is the overlap volume.

This is the same result as found above for the linear part of $G_a(h)$. Usually, a figure like Figure 6.25 is presented with the interaction distance δ_i replaced by the single-plate depletion thickness δ. The analysis given in this section shows that this is qualitatively correct, but that the picture may be refined somewhat by using δ_i as defined by Equation (108), which is slightly larger than δ, see Equation (109) and Figure 6.24(b).

E. Adsorbing Polymer

The numerical procedure to calculate $G_a = 2[\Omega(h) - \Omega(\infty)]$ uses Equations (98) and (99), which are valid for adsorption and depletion; the only difference is the value of χ_s, which is above χ_{sc} for adsorption and below it for depletion. However, we also want analytical approximations. To that end we use Equation (100) for the grand potential density ω for adsorbing polymer: $\omega = -(1/2)v(\varphi^2 - \varphi_b^2)$ in semidilute solutions, $\omega = -\varphi/N - (1/2)v\varphi^2$ in the dilute case. When $\varphi > \varphi_b$, ω is negative. We now need information about the concentration profile $\varphi(z)$ in the gap between two plates. As discussed in Section IV.B, the relevant GSA equations have no easy analytical solutions. We therefore have to make further approximations.

It turns out that it is easier to calculate first the disjoining pressure $\Pi(h) = -\partial\Omega/\partial h$ and not $G_a(h)$ itself; once an expression for $\Pi(h)$ is obtained, $G_a(h)$ can be reconstructed as

$$G_a(h) = -\int_\infty^h \Pi(h')dh'. \tag{110}$$

However, also for obtaining $\Pi(h)$ we need information about the profile in the gap; to that end we shall use the superposition approximation in the analytical approach.

1. The Disjoining Pressure

Since $\Pi(h) = -\partial\Omega(h)/\partial h = -\int(\partial\omega/\partial h)dz$, we can find $\Pi(h)$ by comparing the ω-profiles at half-separations h and $h + dh$. Figure 6.26 gives a discretized example, obtained from the numerical SCF model, for $h = 9$ and $dh = 1$. When $\varphi(z) > \varphi_b$, as is typical for adsorption, ω is negative as is clear from

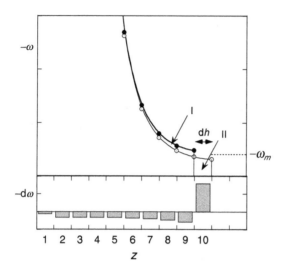

Figure 6.26   Numerical example of the profile $-\omega(z)$ at two plate separations $2h$ and $2h + dh$. Only the left half of the slit is indicated. The curve connecting the filled circles is for $h = 9$, that with the open circles is for $h = 10$. The bar diagram at the bottom shows $-d\omega(z) = \omega(z,9) - \omega(z,10)$ for $z = 1, \ldots, 9$; these differences are negative and their sum corresponds to area I in the main figure. The positive value for $-d\omega$ in layer 10 (approximately equal to $-\omega_m$) corresponds to area II.

Equations (99) and (100). That is why $-\omega$ is plotted in Figure 6.26. The function $-\omega$ decays from a high value in the train layer (outside the range shown in Figure 6.26) to a lowest value $-\omega_m$ in the middle of the slit. In the discrete model ($dh = 1$) there is a difference between the values of ω_m for the two plate separations; when a continuum approach ($dh \rightarrow 0$) is used, this difference vanishes.

When the plate separation is increased ($h \rightarrow h + dh$), $\varphi(z)$ and $-\omega(z)$ decrease: the curve for $h = 10$ (open circles) lies below that for $h = 9$ (filled circles). From Figure 6.26 it is clear that there are two contributions to the disjoining pressure Π:

$$\Pi = \Pi_a + \Pi_m. \qquad (111)$$

Here, Π_a is the contribution due to the *adjustment* of the profile in the region $0 < z < h$, and Π_m is the *midplane* contribution due to the increase of the integration range from h to $h + dh$.

This procedure to split up $\Pi(h)$ into two contributions is analogous to that used by De Gennes [62]. However, De Gennes separated the total Ω in a surface contribution (which is negative) and an integral over the remaining local term ω^{loc} (which is positive). Hence, the signs of ω (which we use) and ω^{loc} (used by De Gennes) are different, as are the signs of Π_a and Π_m. De Gennes' conclusion that in GSA there is always attraction (because ω_m^{loc} at the midplane dominates) is the same as that obtained in our procedure, as shown below.

In Figure 6.26, $\Pi_a dh = -\int_0^h (d\omega) dz$ corresponds to area I (which is the sum of the 9 negative values for $-d\omega$ in the bottom part of Figure 6.26; since $-d\omega$ is negative this constitutes a negative (attractive) part of Π_a). Analogously, $\Pi_m dh = -\omega_m dh$ corresponds to area II, which is the positive value of $-d\omega$ in layer 10. Hence, Π_m constitutes a positive (repulsive) contribution to Π. We can thus write

$$\Pi_a = -\int_0^h \frac{\partial \omega}{\partial h} dz, \tag{112}$$

$$\Pi_m = -\omega_m. \tag{113}$$

The net effect (attraction or repulsion?) is determined by the relative magnitudes of areas I and II in Figure 6.26, which depend very sensitively on details of the profile. In the example of Figure 6.26, both areas are of the same order of magnitude but area I is slightly larger, so that the net effect is attraction. This is the general trend for strong overlap. However, at weak overlap this trend is reversed: the two contributions still compensate each other largely, but then the net effect is a slight repulsion. This illustrates the subtle balance between Π_a and Π_m. In Section V we will present several numerical examples.

In the ground-state approximation, we can evaluate Π_a and Π_m from Equations (112) and (113) only when we use the superposition approximation for the volume fraction profile.

In the next section we shall see that the net effect is again compensation of the leading terms of Π_a and Π_m, with a net attraction due to the higher-order terms. The repulsion at weak overlap as found from the numerical data can be interpreted analytically as well, but then we have to go beyond standard GSA and take end effects into account (Section III).

2. Attraction Due to Bridge Formation

In order to use Equations (112) and (113) for calculating Π_a and Π_m, we have to know how $\omega(z)$ and $\varphi(z)$ depend on h. We approximate the (excess) profile in the slit as a superposition of the two individual single-plate (excess) profiles: $\varphi(z) - \varphi_b = \varphi_s(z) + \varphi_s(2h - z) - 2\varphi_b$, where φ_s is the single-plate profile. Figure 6.27 illustrates the principle. In semidilute solutions, $\varphi_s = \varphi_b \coth^2(Z + P)$ for the left plate. Hence, the profile in the gap is approximated as

$$\frac{\varphi}{\varphi_b} = \coth^2 x + \coth^2 y - 1, \tag{114}$$

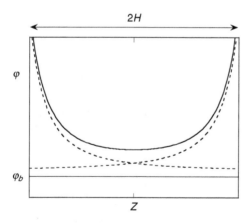

Figure 6.27 Superposition approximation according to Equation (114). The dotted curves are the single-plate profiles given by $\varphi_s(z) = \varphi_b \coth^2(Z + P)$ (left) and $\varphi_s(z) = \varphi_b \coth^2(-Z + 2H + P)$ (right), the solid curve is their sum minus φ_b.

where $x = Z + P$ and $y = -Z + 2H + P$. As before, all distances are normalized by the correlation length ξ: $Z = z/\xi$, $H = h/\xi$, $P = p/\xi$. We note that this superposition approximation does include some adjustment in the train layer: φ_0 is slightly larger than $\varphi_0 = \coth^2 P$ for a single plate.

For the evaluation of Π_a we use $\omega = -\frac{1}{2}v(\varphi^2 - \varphi_b^2)$. In Equation (112) we need $\partial\omega/\partial h$, hence $\partial\varphi^2/\partial H = 2\varphi\partial\varphi/\partial H$. From Equation (114) we have $\partial(\varphi/\varphi_b)/\partial H = 4\coth y(1 - \coth^2 y)$ $\approx -16e^{-2y}\coth y$. We will integrate over the left half of the slit where $\coth y \approx 1$ and $\coth^2 y - 1 \ll \coth^2 x$. Hence, $\partial\varphi^2/\partial H \approx -32e^{-2y}\coth^2 x = -32e^{-4(H+P)}e^{2x}\coth^2 x$. The integral of $e^{2x}\coth^2 x$ is given by $(1/2)\coth x(e^{2x} - 3) + 2[x + \ln\sinh x]$. Taking the leading terms at both boundaries, we obtain as a first approximation

$$\Pi_a \approx -8v\varphi_b^2\left\{e^{-2(H+P)} + \frac{2}{P}e^{-4(H+P)}\right\}. \tag{115}$$

With $\varphi_m/\varphi_b = [2\coth^2(H+P) - 1] - 1 \approx 16e^{-2(H+P)}$ we have for Π_m:

$$\Pi_m \approx 8v\varphi_b^2 e^{-2(H+P)}. \tag{116}$$

We thus find that the leading contributions of Π_m (repulsive) and Π_a (attractive) exactly compensate each other. We see also that $-\Pi_a$ is larger than Π_m: the attractive contribution due to the adjustment of the ω profile overcompensates the repulsive contribution of Π_m. This is fully in line with the findings in Figure 6.26. The net result is an attraction which is exponential as $\Pi \approx -(16v\varphi_b^2/P)e^{-4h/\xi}$. Integration of $\Pi(h)$ to find G_a according to Equation (110) implies multiplication by $\xi/4$: $G_a \approx -(4\varphi_b/3p)e^{-4h/\xi}$ where we used $1/P = \xi/p$ and $\xi^2 = (3v\varphi_b)^{-1}$. This result is only approximate in view of the rather rough derivation. We write it as

$$G_a(h) = -Ae^{-2h/L}, \tag{117}$$

where $L/2$ is the decay length of the attraction and A the amplitude. In this case $L/2 = \xi/4$. We note that in the exponential region of a single adsorbed layer the decay length of $\varphi - \varphi_b$ is $\xi/2$; since G_a is determined by $\omega \sim \varphi^2 - \varphi_b^2$, the decay length here is $\xi/4$, a factor of 2 smaller. In the present example $A \approx \varphi_b/p$, so the attraction increases with φ_b and with increasing χ_s (decreasing p). We do not want to attach too much value to the quantitative details. The most important finding is that GSA predicts attraction, with an exponential decay of G_a in the limit of weak overlap between adsorption layers.

For dilute solutions, a similar exercise can be done. Now the superposition approximation gives $\varphi(z) = \varphi_s(z) + \varphi_s(2h - z) = [1/(3vd^2)] \, [1/\sinh^2 x + 1/\sinh^2 y]$. Here x and y are defined as in Equation (114), but $Z = z/d$, $H = h/d$, and $P = p/d$; the characteristic length scale is the distal length d (Equation (79)) rather than the correlation length ξ. This $\varphi(z)$ has to be substituted into $\omega = -\varphi/N - (1/2)v\varphi^2$ and Π_a and Π_m can be evaluated from Equations (112) and (113). We skip all the mathematical details. The final result is again Equation (117) for G_a. As expected, now the decay length $L/2$ of the attraction is $d/4$.

In conclusion, both in dilute and in semidilute solutions we see that GSA leads to attraction with an exponential decay $G_a(h) = -Ae^{-2h/L}$, where L is of order $\xi/2$ in semidilute solutions and around $d/2$ in the dilute case.

As the GSA description neglects tails, the attraction must be attributed to the loops. Every time a conformation crosses the midplane it loses the information on which surface it visited last time. Essentially the chain can choose randomly between returning to the same plane and form a loop or proceed to the opposing surface and form a bridge (see Figure 6.28). As this choice is immaterial with respect to the energy (i.e., the number of polymer surface contacts), it implies just an increase in entropy by a factor of ln2 for each midplane crossing. This entropy increase leads to a reduction of Gibbs energy and thus to attraction. In other words, attraction arises because loops can transform into bridges.

3. Repulsion Due to Tails

In the previous sections we considered only ground-state dominance for "average segments," which for single surfaces describes only loops. In this situation we have only attraction

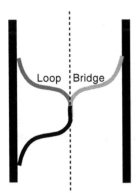

Figure 6.28 Conformational entropy gain occurring when a walk starting from the left surface (solid curve) reaches the midplane. It can follow some path which returns to the same surface to make a loop, or it follows with equal probability its mirror image to make a bridge.

between two polymer layers in close proximity, both for adsorption and for depletion.

For adsorbed polymer layers, end effects do play a role. This has been known since the 1980s due to the numerical work of Scheutjens et al. [5,6,8,63]. Analytical theories describing tails became only available after 1995 and are largely due to Semenov et al. [9–11]. The starting point is that, whereas the overall (loop) concentration is proportional to g^2, the end-point concentration is proportional to g (see Equation (50)). From Equation (50) we have $\rho_e \equiv N\varphi_e/\varphi_b = e^{\varepsilon N}g$. Here, $\varphi_e = \varphi_e(z) = \varphi(z,N)$ is the volume fraction of segments with ranking number N. The total volume fraction of end points is then $2\varphi_e$. With $\varepsilon = 0$ and $g = \coth(Z+P)$ in semidilute solutions, and $\varphi_b e^{\varepsilon N} \approx 1$ and $g \sim 1/\sinh(Z+P)$ in dilute solutions, we may approximate the excess of end points with respect to the bulk solution as

$$\rho_e - 1 = b(g-1) \sim \coth\frac{z+p}{\xi} - 1 \qquad \text{semidilute}$$

$$\rho_e = bg \sim 1/\sinh\frac{z+p}{d} \qquad \text{dilute}$$

(118)

The factor b is a normalization constant which ensures that the excess number of end points in the adsorbed layer equals the excess number of chains; it is defined through $b\int(g-1)dz = \int(g^2-1)dz$ for semidilute solutions and similar (without the term -1) for dilute systems. We do not consider the precise value of b; it suffices to note that b is of order unity.

In order to see the effect of chain ends on the interaction force, the GSA description as discussed in the previous sections has to be extended to include the end points. This was done by Semenov et al. [9] from first principles, along the lines of De Gennes' theorem of forces [62]. In this approach the disjoining pressure $\Pi_e(h)$ due to the ends is related to the end-point concentration φ_{em} at the midplane. The idea is that the end-points give an ideal gas-like contribution to the force, i.e., there is an additional contribution $\Pi_e(h) \sim +(\varphi_{em} - \varphi_{eb}) \sim (g_m - 1)$, where $\varphi_{eb} = \varphi_b/N$ is (half) the volume fraction of end points in the bulk solution. For dilute solutions we have $\Pi_e(h) \sim \varphi_{em} \sim g_m$.

The most important implication is the sign: end effects give rise to repulsion between adsorbed polymer layers. As mentioned this repulsion may be interpreted as originating from an ideal gas of end points. When the two surfaces are brought closer, the concentration of chain ends increases, the 'gas' pressure increases and the plates repel each other.

For making a quantitative estimate, we approximate $g_m - 1$ according to Equation (118) as $\coth(H+P) - 1 \approx 2e^{-2(h+p)/\xi}$ for semidilute solutions and $g_m \sim 1/\sinh(H+P) \approx 2e^{-(h+p)/d}$ for the dilute case. The result is $\Pi_e(h) \sim e^{-h/L}$. Upon integration (Equation (110)) this gives the Gibbs energy of interaction as

$$G_{ae}(h) = Re^{-h/L}, \tag{119}$$

where R is the amplitude of the repulsion. As before, the decay length L is $\xi/2$ is semidilute solutions and $d/2$ in dilute solutions.

From a comparison of Equations (117) and (119) it is clear that the decay length L for the repulsive contribution is twice the decay length $L/2$ for the attraction. The reason is that the attraction is due to the overall concentration profile $\varphi \sim g^2$, whereas the repulsion is due to the end points with concentration profile $\varphi_e \sim g$. For weak overlap the midpoint value g_m decays exponentially as $e^{-h/L}$ towards unity in semidilute solutions (where $L = \xi/2$) and towards zero in the dilute case (with $L = d/2$). Obviously, the decay length of g_m^2 is half that of g_m.

4. Total Gibbs Energy of Interaction

From Equations (117) and (119) we find the total Gibbs energy of interaction as

$$G_a = -Ae^{-2h/L} + Re^{-h/L}. \tag{120}$$

A sketch of $G_a(h)$ is given in Figure 6.29. The curve has a maximum G_m at $h = h_m$, and passes through zero at $h = h_0$. These three parameters follow from L, A, and R as

$$G_m = \frac{R^2}{4A}$$

$$h_m = L \ln \frac{2A}{R} \tag{121}$$

$$h_0 = h_m - L \ln 2.$$

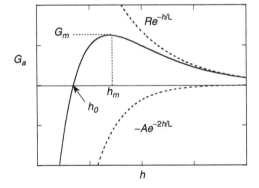

Figure 6.29 Plot of $G_a(h)$ according to Equation (120). The maximum G_m, the corresponding separation h_m, and the separation h_0 where $G_a = 0$ are indicated. The dotted curves give the two contributions to G_a.

Exact numerical SCF results provide also a dependence $G_a(h)$ with a maximum. When these curves are interpreted according to Equation (120), the parameters L, A, and R may be found from the numerical curves as

$$L = \frac{h_m - h_0}{\ln 2}$$
$$A = G_m e^{2h_m/L} \qquad\qquad (122)$$
$$R = 2G_m e^{2h/L}.$$

In the next section we will see that the exact numerical results can indeed be interpreted in this way. We will find that A is typically of the order 0.01–0.1 (in kT per lattice site), whereas R is much smaller, giving G_m values of order 10^{-4}–10^{-5}. Hence, the dominating effect is the attraction, with only a (very) small repulsive contribution. The decay length L is indeed of order $\xi/2$ (semidilute) or $d/2$ (dilute).

5.   Numerical examples of $G_a(h)$

In this section we present some numerical results for the Gibbs energy of interaction between two surfaces with adsorbed polymer layers in full equilibrium. We first show that indeed the dominating effect is attraction but that there is a weak repulsive effect, with interaction curves which can be analyzed along the lines of Equation (120) and Figure 6.29, using the parameters G_m (height of the maximum), h_m (position of the maximum), and L (decay length of the repulsion or twice the decay length of the attraction). We will see how these parameters depend on polymer concentration φ_b, chain length N, adsorption energy χ_s, and solvency χ.

Figure 6.30 shows $G_a(h)$ for $N = 1000$ and $\varphi_b = 10^{-2}$, which is in the semidilute range. The linear plot of Figure 6.30(a) shows an enlarged view of the region of weak overlap, around the maximum. The symbols in Figure 6.30a are the numerical SCF data; from interpolation $G_m = 1.71 \times 10^{-4}$, $h_0 = 4.96$, and $h_m = 7.0$. From these numbers Equation (122) gives $A = 1.41 \times 10^{-2}$, $R = 3.11 \times 10^{-3}$, and $L = 2.95$. The solid curve in Figure 6.30(a) is Equation (120) with these

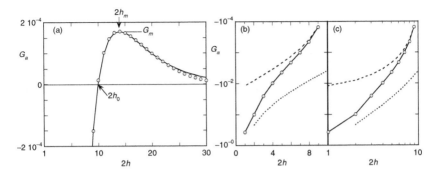

Figure 6.30 The Gibbs energy of interaction $G_a(2h)$ for $N = 1000$, $\varphi_b = 10^{-2}$, $\chi_s = 1$, and $\chi = 0$. In diagram (a) the region near the maximum is plotted with linear scales. The maximum G_m, the corresponding position h_m, and the position h_0, where $G_a = 0$, are indicated. The points are the SCF results, the curve is the fit with the two exponentials according to Equation (120), using $A = 1.41 \times 10^{-2}$, $R = 3.11 \times 10^{-3}$, and $L = 2.95$. The attractive part of the inter-action curves for the range $h < h_0$ (solid curve) is plotted in panels (b) and (c) on semilogarithmic and double-logarithmic coordinates, respectively. In these graphs the dashed curve is the extrapolation of the solid curve in diagram (a). The dotted curve was calculated from the numerical integration of the superposition profile accord-ing to Equation (114). In this calculation a Stern layer was taken into account, analogously to Figure 6.11; the curves start at $2h = 2$ where the two Stern layers touch.

parameters. In this region of weak overlap the fit is quite good, which corroborates the double-exponent Ansatz.

For $h < h_0$ there is strong attraction, as shown in Figure 6.30(b) and (c) where G_a is plotted logarithmically and $2h$ on a linear (Figure 6.30(b)) or logarithmic scale (Figure 6.30(c)). The dashed curves in these figures show the extrapolation of Equation (120). For h slightly below h_0 this extrapolation works quite well, but for smaller h the attraction is severely underestimated. The reason is that Equation (120) is only valid in the exponential part, where $\coth x \approx 1 + e^{-2x}$. For smaller h the full hyperbolic functions have to be used, which give rise to stronger attraction, as shown by the dotted curves in Figure 6.30(b) and (c). These curves correspond to the superposition approximation in GSA (only attraction) without any expansion, and were obtained by numerical integration of Equation (114). As mentioned (Section IV.B) the superposition approximation overestimates $\varphi(z)$ and, consequently, $-\omega(z)$ so that now the attraction is overestimated. For small plate separations this is a relatively small effect, for higher separations the discrepancy with the numerical integration of Equation (114) is larger because then the endpoint repulsion (not accounted for in Equation (114)) starts to play a role.

We conclude that the numerical data smoothly cross over from the dotted curve at low h (which constitutes an overestimation of $-G_a$) to the dashed (double-exponential) curve at higher h. Even though we have no accurate analytical description over the full range, this crossover from one analytical limit to the other is gratifying.

From Figure 6.30(b) and (c) we can see that the attractive part of the interaction curve has a complex structure. It is neither a true exponential nor a true power-law curve. In Figure 6.30(c) it would be possible to force a power-law through the data in the middle range with an exponent not far from -3. When $2h \approx 1$ an exponential decay of the interaction is found in Figure 6.30(b), which is not present in the GSA result: GSA is not expected to hold when the train layers start to overlap.

In Figure 6.31 a number of plots is collected similar to Figure 6.30(c), under various conditions. The attractive part of the interaction energy ($h < h_0$) is plotted on a double-logarithmic scale for three values of the bulk concentration φ_b (Figure 6.31(a)), the chain length N (Figure 6.31(b)), the effective adsorption energy $\Delta\chi_s$ (Figure 6.31(c)), and the solvency χ (Figure 6.31(d)).

The first thing to notice is that for narrow gaps only the adsorption energy has an effect: for small h the curves for different φ_b, different N, and different χ all come together. At the other end of the scale, for h close to h_0, the curves for different χ_s merge, whereas there is an effect of φ_b, N, and χ.

All the curves start at $2h = 1$, which corresponds to one (lattice) layer of solution between the plates. The attraction in

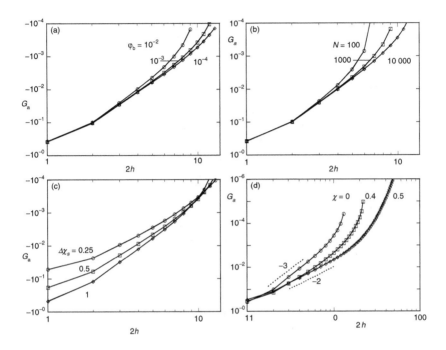

Figure 6.31 Numerical SCF data for the attractive part of $G_a(h)$, in the range $h < h_0$, for three values of φ_b (a), N (b), $\Delta\chi_s$ (c) and χ (d). In diagram (a) $N = 1000$, $\chi_s = 1$, $\chi = 0$, in (b) $\varphi_b = 10^{-2}$, $\chi_s = 1$, $\chi = 0$, in (c) $N = 1000$, $\varphi_b = 10^{-3}$, $\chi = 0$, and in (d) $N = 1000$, $\varphi_b = 10^{-3}$, $\chi_s = 1$.

this situation is proportional to the effective adsorption energy as will be discussed in Figure 6.35 (Section VI). For larger plate separations, the attraction becomes weaker in all cases. For h close to h_0 (where the loop-bridge attraction and the tail repulsion just compensate each other), the curves become approximately independent of χ_s (Figure 6.31(c)), but they do depend on the other parameters. The interparticle separation where the curves bend upwards is determined mainly by the value of h_0, which increases with decreasing φ_b (Figure 6.31(a)), increasing N (Figure 6.31(b)), and increasing χ (Figure 6.31(d)). The value of h_0 will be discussed in more detail in Figure 6.32–Figure 6.34.

As the range h_0 of the attractive interaction does not depend on the adsorption energy, and the Gibbs energy of interaction at contact is proportional to the effective adsorption energy (see also Figure 6.34 below), the slope in the interaction curve is a weak function of the adsorption energy. To a reasonable approximation the interaction curves remain power-law-like with an exponent, which decreases slightly with decreasing adsorption energy.

There is a very large effect of the solvent quality on the range of the attraction, as shown in Figure 6.31(d). Going from a good ($\chi = 0$) to a theta solvent, the attractive component becomes stronger because the attraction between (loop) segments reduces the excluded-volume effect; at $\chi = 0.5$ the second virial coefficient vanishes. The repulsive effect of the chain ends is then relatively less important. Apparently, GSA is qualitatively correct in this case and the interaction curve for $\chi = 0.5$ in the double-logarithmic coordinates of Figure 6.31(d) extends to very large h. Indeed, for large h the interaction is purely exponential (not shown). At shorter plate separations ($2h < 15$) we again find a power-law-like dependence of the interaction force. Now the exponent is to a good approximation equal to -2.

In the following sets of graphs we discuss the maximum in the interaction curves and present information on the dependencies of $G_m = G_a(h_m)$, the position h_0, and the characteristic interaction length L on φ_b, N, and χ_s. Roughly speaking, the magnitude of the repulsion G_m is of the same order of magnitude as (though usually smaller than) the minimum in the depletion interaction for the nonadsorbing surfaces. Therefore, we expect that this repulsion could be important in practical applications.

The interaction depends on the polymer concentration φ_b, as shown in Figure 6.32 which applies to $N = 1000$, $\chi_s = 1$, and $\chi = 0$. This figure covers the range $10^{-4} < \varphi_b < 10^{-1}$. As the overlap concentration is near $\varphi_{ov} \approx 0.002$, we thus cross over from the dilute to the semidilute regime. The position of the maximum h_m as well as the compensation point h_0 are decreasing functions of the polymer concentration; both quantities scale with the polymer concentration with an exponent

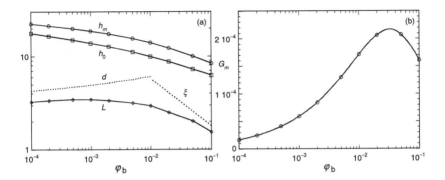

Figure 6.32 (a) The position h_m of the maximum in the interaction curve, the position h_0 where the interaction is zero, and the interaction length L as a function of the polymer concentration φ_b. For comparison the distal length d and the correlation length ξ are also plotted. (b) The maximum $G_m = G_a(h_m)$ in the interaction curve as a function of the polymer concentration φ_b. Both graphs are for $N = 1000$, $\chi_s = 1$ and $\chi = 0$.

close to -0.15. The interaction length L goes through a maximum and follows the distal length d in dilute solutions and the correlation length ξ in the semidilute range. In Figure 6.32(a) d and ξ are also indicated (dotted curve). These two lengths cross at a point well in the semidilute regime. The maximum in L, however, is much closer to the overlap concentration φ_{ov}.

In the dilute range L is close to $d/2$ and in the semidilute region to $\xi/2$, as expected from Equations (117) and (119). The value of G_m is a rather strong function of the polymer concentration, roughly as $G_m \sim \sqrt{\varphi_b}$ for low φ_b. Deep in the semidilute regime G_m goes through a maximum and drops sharply in the concentrated regime. The maximum repulsion is thus found at relatively high polymer concentrations.

The interaction depends also on the chain length (Figure 6.33). In this graph the polymer concentration was fixed to $\varphi_b = 0.001$ and again the solvent quality is good ($\chi = 0$). The position of the maximum of the interaction curve increases monotonically with the chain length. For short chains the system is below the overlap concentration and a chain-length

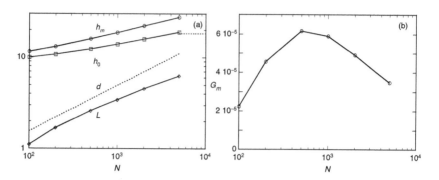

Figure 6.33   (a) The position h_m of the maximum in the interaction curve, the position h_0 where the interaction is zero, and the interaction length L as a function of the chain length N. For comparison the distal length d and the correlation length ξ are also plotted. (b) The maximum $G_m = G_a(h_m)$ in the interaction curve as a function of the chain length N. Both graphs are for $\varphi_b = 10^{-3}$, $\chi_s = 1$, and $\chi = 0$.

dependence is expected. However, for long chains the system is above the overlap concentration and the chain-length dependence should be weak. Inspection of Figure 6.33(a) shows that $h_m \sim N^{0.2}$ without any tendency to level off for high-chain lengths. In the range of N values presented in Figure 6.33 the interaction length L is a monotonically increasing function of the chain length which follows half the distal length $d/2$ in good approximation; for the largest N shown, where $\varphi_b/\varphi_{ov} = 10^{-3}$. $N/2$ is above unity, it is not far from $\xi/2$.

The height of the maximum of the interaction curve has a very interesting dependence on the chain length as shown in Figure 6.33(b). In this case a very pronounced maximum is found. The fact that the strength of the repulsion decreases for large N is easily explained. As the adsorbed amount is only a weak function of the chain length the number of end points in the layer drops inversely proportional to N. As the range of interaction is not a constant the effect is smaller and as a result a logarithmic dependence with N is found. On the other side of the maximum, the drop may be explained from the fact that the adsorbed amount goes down with decreasing chain length. Thus, when one is interested to use homopolymers to stabilize particles, one should choose an intermediate degree of polymerization. Very long polymers are not effective because they have not enough ends, short polymers do not adsorb strongly enough.

Figure 6.34 shows the effect of $\Delta\chi_s$ on the characteristics of the maximum. The adsorption energy does not influence the spatial characteristics of the maximum in the interaction curves dramatically, at least for relatively strong adsorption. Only when the adsorption energy is close to χ_{sc} do we observe that the position of the maximum shifts to higher h values. This signals the approach to the critical region. The height of the maximum increases with the adsorption energy. This can be understood from the fact that the adsorbed amount and thus the number of end points is an increasing function of the effective adsorption energy. To stabilize particles with homopolymers one should thus use strongly adsorbing polymer.

6. The Gibbs Energy of Interaction at Contact

Apart from a small repulsion due to tails which is mainly relevant in good solvent conditions, homopolymers generate

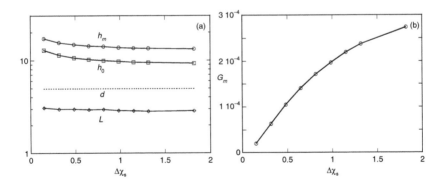

Figure 6.34 (a) The position h_m of the maximum in the interaction curve, the position h_0 where the interaction is zero, and the interaction length L as a function of the adsorption energy $\Delta\chi_s$. For comparison the distal length d is also plotted. (b) The maximum $G_m = G_a(h_m)$ in the interaction curve as a function of the adsorption energy $\Delta\chi_s$. Both graphs are for $N = 1000$, $\varphi_b = 10^{-3}$, and $\chi = 0$.

an attractive contribution to the pair potential between two
particles, both for depletion and for adsorption. To put all of
the above into perspective it is necessary to realize that depletion forces are significantly less effective to drive particles
together than the bridging attraction: the value of $-G_a(0)$ in
Figure 6.23 and Figure 6.24a (depletion) is much smaller than
in Figure 6.31 (adsorption). To illustrate this once more we
present the Gibbs energy of interaction at contact as a function of the adsorption energy in Figure 6.35. In Figure 6.35a
we show the result for a wide range of $\Delta\chi_s$ for two values of φ_b.
The numerical SCF data for small h were extrapolated to $h = 0$
to find $G_a(0)$. On the scale of Figure 6.35a we see only attraction for positive values of $\Delta\chi_s$, with hardly any φ_b-dependence.
For $\Delta\chi_s > 0.3$, $G_a(0)$ depends linearly on $\Delta\chi_s$, as was also found
in Figure 6.31(c). The dotted line in Figure 6.35a was drawn
according to $G_a(0) = 0.2 - \Delta\chi_s$ and describes the data quite
well for $\Delta\chi_s > 0.3$.

For small values of $\Delta\chi_s$ this linear behavior breaks down.
In fact, around the critical point there is actually a region
where $G_a(0)$ is positive. Figure 6.35(b) gives some details for
the range $-0.1 < \Delta\chi_s < 0.1$ and a more detailed scale for $G_a(0)$.

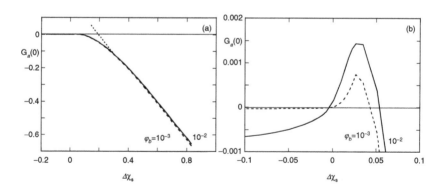

Figure 6.35 The Gibbs energy of interaction $G_a(0)$ at contact, as a
function of the adsorption energy $\Delta\chi_s$ for a wide range of $\Delta\chi_s$ (a) and
for small $\Delta\chi_s$ near the critical point (b). The dotted line in (a) is a
linear extrapolation according to $G_a(0) = 0.2 - \Delta\chi_s$. Parameters
$N = 1000$, $\varphi_b = 10^{-3}$ and 10^{-2}, and $\chi = 0$.

Now the attraction in the depletion range is visible. According to Equation (107) $G_a(0) \sim \varphi_b/\delta$, which would give a linear dependence on φ_b if δ would not depend on φ_b (which is the case for $\varphi_b < 10^{-4}$, where $\delta = \delta_0$). Since δ at $\varphi_b = 10^{-2}$ is about half of the value at $\varphi_b = 10^{-3}$ (Figure 6.10), the ratio of $G_a(0)$ between $\varphi_b = 10^{-2}$ and 10^{-3} is close to 20.

The repulsion for $0 < \Delta\chi_s < 0.05$ cannot be described by GSA and shows once again that GSA breaks down around the critical point. Probably, it is again the tails which are responsible for the repulsion. In $G_a(0) = 2[\Omega(0) - \Omega(\infty)]$, tails do not show up in the part $\Omega(0)$, but they do so for sure in the part $\Omega(\infty)$.

Note that $\chi_s = 0$ corresponds to $\Delta\chi_s = -\ln(6/5) \approx -0.18$. The interval plotted in Figure 6.35(b) thus correspond to positive values of χ_s. The dependence of the depletion force on $\Delta\chi_s$ (Figure 6.35(b) for $\Delta\chi_s < 0$) may become important in experimental systems because it is difficult to avoid all possible interactions with the surface. Some residual interaction to the surface may already result from nonideal solvency effects.

V. CONCLUDING REMARKS

To understand polymers between surfaces, it is necessary to know the physics of polymers at a single surface, for which in turn insight into the behavior of polymers far away from the surface (i.e., in the bulk solution) is required. For example, the bulk correlation length ξ characterizing a semidilute polymer solution appears to be also the relevant length scale defining the thickness δ of the depletion layer for nonadsorbing polymers in semidilute solutions; it is approximately equal to the interaction length δ_i. The scaling of the bulk correlation length with the polymer concentration may be used to rationalize the power-law exponent in the self-similar (central) part of the profile for adsorbing polymers. In the nonmonotonic interaction curves for adsorbing polymers an interaction length L occurs, which in semidilute solutions is again determined by the bulk correlation length ξ.

In dilute solutions the relevant length scale in the bulk is the radius of gyration R_g. This length gives also the depletion layer thickness of the depletion zone in very dilute systems and is then the length scale at which the interaction appears. For adsorption from dilute solutions R_g is still important since the length scale is the distal length $d = R_g/y$, where y depends logarithmically on the concentration. For quantitative estimates this logarithmic factor cannot be neglected, though, since it takes a value between 5 (in extremely dilute solutions) and 3 (around overlap).

In some situations both length scales ξ (which depends only on concentration and solvency) and R_g (which depends only on the chain length N) play a role. This is, for instance, the case in depletion at intermediate concentrations, where the typical length scale is the depletion thickness δ, given by $\delta^{-2} \approx R_g^{-2} + \xi^{-2}$.

Apart from the fundamental length scales ξ and d which describe the extension of the interfacial layers, in the adsorption problem another length scale enters: the proximal length p (essentially equal to the extrapolation length $1/c$). This length is determined by the adsorption energy and by crowding effects in the train layer, and is in most cases below unity (i.e., smaller than the segment size). It is therefore problematic to determine this length from a continuum description only, as variations on a length scale smaller than the bond length are unphysical. Indeed, using only a continuum model predicts a very strong and irrealistic variation over (part of) the train layer, with local volume fractions far above unity, see for example Figure 6.11. The method to deal with this problem is to use a surface boundary condition taken from the lattice model, in which the finite segment size is accounted for. This boils down to considering the train layer as a discrete Stern layer, and enables a quantitative evaluation of the proximal length (which largely determines the adsorbed amount) as a function of the adsorption energy (and other quantities).

The Edwards equation with a self-consistent segment potential may be used to obtain rather accurate description of polymers in solution, for both an isolated surface and for two interacting surfaces. There exist numerically exact solutions, where a computer is used to solve the equations. In such an approach it is necessary to discretize the equations. Here the scheme of Scheutjens and Fleer is particularly suitable.

Numerical analysis is useful but should not be the final goal. To interpret the trends and to consider limiting behavior, one needs analytical descriptions. For many problems we have rather accurate analytical predictions based on the ground-state approximation, especially for single surfaces. As stated above, for adsorption a discrete train layer should be incorporated. Unfortunately, so far there are not many accurate predictions for the case of polymers between two surfaces; here further work is needed.

There is relatively little literature on detailed SCF predictions about the effects of polymers on the pair potential between particles. In this review we have attempted to cover both depletion and adsorption in the full equilibrium case. The non-monotonic interaction curves, where there is repulsion due to tails and attraction due to bridge formation, are analyzed in more detail than was done before. A rather complex picture emerges. For example, one and the same type of (adsorbing) polymer–solvent system can be used to stabilize or to flocculate a particular system, using only the polymer concentration and the molecular weight as the control parameters. For optimal stabilization the polymer tails should be prominently present. This means that one should have an intermediate molecular weight (enough chain ends) around the overlap concentration (relatively large contribution of tails). Very long polymers or suboverlap concentrations may lead to bridging attraction.

The SCF predictions may be compared to computer simulation efforts such as molecular dynamics or Monte Carlo. Our predictions are sufficiently detailed to put them next to experimental data. Unfortunately in this review space did not allow for a detailed evaluation of the simulation-SCF confrontation [64,65]. We also apologize for the fact that we did not pay any attention to the experimental validation. However, this review gives, to our modest opinion, a good picture of what is known about polymers in full equilibrium between two surfaces. We hope that our analysis will give an impulse to perform new and better-defined experiments, initiate new simulations, and promote further theoretical developments.

REFERENCES

[1] S.F. Edwards, *Proc. Phys. Soc.*, 85, 613, 1965.

[2] P.G. De Gennes, *Scaling Concepts in Polymer Physics*, Cornell University Press, Ithaca, NY, 1979.

[3] M.V. Volkenstein, *Configurational Statistics of Polymeric Chains*, Interscience Publishers, New York, 1966.

[4] A.Y. Grosberg and A.R. Khokhlov, *Statistical Physics of Macromolecules*, AIP Press, New York, 1976.

[5] J.M.H.M. Scheutjens and G.J. Fleer, *J. Phys. Chem.*, 83, 1619, 1979.

[6] J.M.H.M. Scheutjens and G.J. Fleer, *J. Phys. Chem.*, 84, 178, 1980.

[7] J.M.H.M. Scheutjens and G.J. Fleer, *Macromolecules*, 18, 1882, 1985.

[8] G.J. Fleer, M.A. Cohen Stuart, J.M.H.M. Scheutjens, T. Cosgrove, and B. Vincent, *Polymers at Interfaces*, Chapman & Hall, London, 1993.

[9] A.N. Semenov, J.-F. Joanny, and A. Johner, Polymer adsorption: Mean field theory and ground state dominance approximation, in *Theoretical and Mathematical Models in Polymer Research*, Academic Press, New York, 1998.

[10] A.N. Semenov, J. Bonet-Avalos, A. Johner, and J.-F. Joanny, *Macromolecules*, 29, 2179, 1996.

[11] A. Johner, J. Bonet-Avados, C.C. van der Linden, A.N. Semenov, and J.-F. Joanny, *Macromolecules*, 29, 3629, 1996.

[12] G.J. Fleer, J. van Male, and A. Johner, *Macromolecules*, 32, 825, 1999.

[13] G.J. Fleer, J. van Male, and A. Johner, *Macromolecules*, 32, 845, 1999.

[14] O.A. Evers, J.M.H.M. Scheutjens, and G.J. Fleer, *J. Chem. Soc. Faraday Trans.*, 86, 1333, 1990.

[15] O.A. Evers, J.M.H.M. Scheutjens, and G.J. Fleer, *Macromolecules*, 23, 5221, 1990.

[16] O.A. Evers, J.M.H.M. Scheutjens, and G.J. Fleer, *Macromolecules*, 24, 5558, 1991.

[17] C.M. Wijmans, F.A.M. Leermakers, and G.J. Fleer, *J. Colloid Interf. Sci.*, 167, 124, 1994.

[18] F.A.M. Leermakers and J.M.H.M. Scheutjens, *J. Chem. Phys.*, 89, 3264, 1988.

[19] F.A.M. Leermakers and J.M.H.M. Scheutjens, *J. Chem. Phys.*, 89, 6912, 1988.
[20] F.A.M. Leermakers and G.J. Fleer, *Macromolecules*, 28, 3434, 1995.
[21] R.R. Nertz and D. Andelman, *Phys. Rep.*, 380, 1, 2003.
[22] O.A. Evers, G.J. Fleer, J.M.H.M. Scheutjens, and J. Lyklema, *J. Colloid Interf. Sci.*, 111, 446, 1986.
[23] J. Klein Wolterink, J. van Male, M.A. Cohen Stuart, L.K. Koopal, E.B. Zhulina, and O.V. Borisov, *Macromolecules*, 35, 9176, 2002.
[24] M.A. Cohen Stuart, G.J. Fleer, J. Lyklema, and W. Norde, *Adv. Colloid Interf. Sci.*, 34, 477, 1991.
[25] F.A.M. Leermakers, J.M.P. van den Oever, and E.B. Zhulina, *J. Chem. Phys.*, 118, 969, 2003.
[26] R. Israëls, F.A.M. Leermakers, G.J. Fleer, and E.B. Zhulina, *Macromolecules*, 27, 32491, 1994.
[27] S.P.F.M. Roefs, J.M.H.M. Scheutjens, and F.A.M. Leermakers, *Macromolecules*, 27, 4810, 1994.
[28] M.A. Cohen Stuart, G.J. Fleer, and J.M.H.M. Scheutjens, *J. Colloid Interf. Sci.*, 97, 515, 1984.
[29] B. van Lent, J.M.H.M. Scheutjens, and T. Cosgrove, *Macromolecules*, 20, 366, 1987.
[30] C.C. van der Linden, F.A.M. Leermakers, and G.J. Fleer, *Macromolecules*, 29, 1000, 1996.
[31] C.M. Wijmans, F.A.M. Leermakers, and G.J. Fleer, *Langmuir*, 10, 4514, 1994.
[32] C.C. van der Linden, F.A.M. Leermakers, and G.J. Fleer, *Macromolecules*, 29, 1172, 1996.
[33] M.C.P. van Eijk and F.A.M. Leermakers, *J. Chem. Phys.*, 109, 4592, 1998.
[34] C.C. van der Linden, B. van Lent, F.A.M. Leermakers, and G.J. Fleer, *Macromolecules*, 27, 1915, 1994.
[35] T. Cosgrove, T. Heath, B. van Lent, F.A.M. Leermakers, and J.M.H.M. Scheutjens, *Macromolecules*, 20, 1692, 1987.
[36] S. Alexander, *J. Physique (Paris)*, 38, 983, 1976.
[37] P.G. De Gennes, *J. Physique (Paris)*, 37, 1443, 1976.
[38] S.T. Milner, T.A. Witten, and M. Cates, *Macromolecules*, 22, 853, 1989.
[39] C.M. Wijmans, J.M.H.M. Scheutjens, and E.B. Zhulina, *Macromolecules*, 25, 2657, 1992.

[40] C.M. Wijmans, E.B. Zhulina, and G.J. Fleer, *Macromolecules*, 27, 3238, 1994.

[41] P.J. Flory, *Principles of Polymer Chemistry*, Cornell University Press, Ithaca, New York, 1953.

[42] J. van Male, *Self-Consistent-Field Theory for Chain Molecules: Extensions, Computational Aspects, and Applications*, PhD thesis, Wageningen University, the Netherlands, 2003.

[43] E. Eisenriegler, K. Kremer, and K. Binder, *J. Chem. Phys.*, 77, 6296, 1982.

[44] J.M.H.M. Scheutjens and G.J. Fleer, *Colloids Surf.*, 21, 285, 1986.

[45] X. Chatellier, T.J. Senden, J.F. Joanny, and J.M. di Meglio, *Europhys. Lett.*, 41, 303, 1998.

[46] P.G. De Gennes, *Rev. Prog. Phys.*, 32, 187, 1969.

[47] A Silberberg, *J. Chem. Phys*, 48, 2835, 1968.

[48] A.A. Gorbunov, A.M. Skvortsov, J. van Male, and G.J. Fleer, *J. Chem. Phys.*, 114, 5366, 2001.

[49] E. Eisenriegler, *Polymers Near Surfaces*, World Scientific, Singapore, 1993.

[50] E. Eisenriegler, K. Kremer, and K. Binder, *J. Chem. Phys.*, 79, 1052, 1983.

[51] G.J. Fleer, A.M. Skvortsov, and R. Tuinier, *Macromolecules*, 36, 7857, 2003.

[52] J. van der Gucht, N.A.M. Besseling, and G.J. Fleer, *Macromolecules*, 35, 2810, 2002.

[53] J. van der Gucht, N.A.M. Besseling, J. van Male, and M.A. Cohen Stuart, *J. Chem. Phys.*, 112, 2886, 2000.

[54] O. Stern, *Z. Elektrochem.*, 30, 508, 1924.

[55] M.A. Cohen Stuart, F.H.W.H. Waajen, T. Cosgrove, B. Vincent, and T.L. Crowley, *Macromolecules*, 17, 1825, 1984.

[56] T. Cosgrove, T.L. Crowley, K. Ryan, and J.R.P. Webster, *Colloids Surf.*, 51, 255, 1990.

[57] L Auvray and J.P. Cotton, *Macromolecules*, 20, 202, 1987.

[58] C. van der Linden and R. van Leemput, *J. Colloid. Interf. Sci.*, 67, 48, 1978.

[59] J.N. Israelachvili, *Intermolecular and Surface Forces*, Academic Press, New York, 1992.

[60] E. Eisenriegler, A. Hanke, and S. Dietrich, *Phys. Rev. E*, 54, 1134, 1996.

[61] A.A. Louis, P.G. Bolhuis, E.J. Meijer, and J.-P. Hansen, *J. Chem. Phys.*, 116, 10547, 2002.

[62] P.G. De Gennes, *Macromolecules*, 15, 492, 1982.

[63] J.M.H.M. Scheutjens and G.J. Fleer, *Adv. Colloid Interf. Sci.*, 16. 361, 1982. (Erratum: *Adv. Colloid Interf. Sci.*, 18, 309–310, 1983.)

[64] M. Rangarajan, J. Jimenez, and R. Rajagopalan, *Macromolecules*, 35, 6020, 2002.

[65] J. De Joannis, J. Jimenez, R. Rajagopalan, and I. Bitsanis, *Macromolecules*, 34, 4597, 2001.

7

Flotation as a Heterocoagulation Process: Possibilities of Calculating the Probability of the Microprocesses, Rupture of the Intervening Thin Liquid Film, and Progress in Modeling of the Overall Process

Hans Joachim Schulze[†] and Werner Stöckelhuber

Max-Planck-Research Group for Colloids and Interfaces at the TU Bergakademie Freiberg, D-09599 Freiberg, Chemnitzer Str 40, Germany

Institute of Polymer Research, Hohe Str. 6, D-01069 Dresden, Germany

[†] Deceased.

I. INTRODUCTION

Flotation is a separation process, in which one component
of a heterogeneous mixture of solid or liquid particles in an
aqueous medium selectively attaches to gas bubbles or oil
droplets, forming aggregates of lower specific gravity, which
in turn rise to the surface of liquid in a flotation cell and

form a froth layer there. Thus, it can be separated from the remaining components. Flotation is therefore a macro process composed of a very large number of individual physical, hydrodynamic, and physicochemical (adsorption) micro processes taking place simultaneously in space and in time. Recovery of particles by flotation is most successful in the 10 to 200 μm particle size range. However, a number of flotation-related techniques are available below and above this size range. Some of these are as follows:

- Carrier or ultra flotation
- Oil or emulsion flotation
- Agglomerate or flocculation flotation
- Liquid–liquid or two-liquid extraction.

For small and for large particles in particular, special techniques such as thin film flotation and flotation in the froth layer are available. In thin film flotation, particles float on the pulp surface rather than attach to bubbles as in conventional flotation. In flotation in the froth layer particles are feted into the froth. But we will discuss only the conventional flotation process here.

From the perspective of colloid science, flotation is a *heterocoagulation process* between solid particles and bubbles or droplets. The stability of the bubble (b) and particle (p) system or, alternatively, the rate of aggregation ("flotation kinetics") therefore depends on the interaction forces between approaching particles and the external field forces. The following kinds of interaction forces are known to occur:

1. Electrostatic repulsion or attraction resulting from the overlap of electrical double layers
2. van der Waals forces always present between particles
3. Steric and hydrophobic interaction due to adsorbed molecules of solvent, surfactant, or macromolecules, and
4. Capillary forces in the case of the existence of a three-phase contact (TPC).

The net superposition of all the interactions and external field forces determines the energy change associated with aggregation and hence stability or instability of the system. Details are described in the literature [1–3].

The total-interactions energy and distance curve has, in principle, two minima, one primary and one secondary, and one maximum. If direct contact occurred (i.e., the separation distance between the particles is in the order of atomic radii), then a primary minimum exists and this would be very deep. But this situation is unlikely to occur, since the distance of the closest approach is determined by adsorbed layers. A secondary minimum exists at a greater distance. It is characterized by the fact that an intervening thin liquid film exists between the particles, which keeps the flocculated particles at a certain distance.

In flotation, attachment of very small particles, the weight of which are extraordinarily small, can take place through heterocoagulation in this secondary minimum of energy without formation of a three-phase contact (*contactless flotation*) [4]. The binding force is very weak in this case. With an increase in the particle size it will generally become insufficient to compensate weight and external stress forces as well. Therefore, formation of a TPC between the bubble and the particle is necessary for the attachment of bigger particles of normal size according to particle density. Here, capillary forces occur, which hold particle and bubble together. Capillary forces are always stronger than attraction forces at the secondary minimum even at small contact angles.

Electrostatic interaction between particles can, as is known, take place at constant charge as well at constant potential. If the exchange of charge occurs within the interaction time in the course of the mutual penetration of the electric double layers during the interaction, the charge at the surface changes depending on mutual penetration, whereas the potential remains constant (ionic crystals, oxides, and metals). If, on the contrary, the charge exchange is retarded and, as in the case of an air bubble, controlled by slow diffusion of surfactant molecules, then the charge keeps constant during interaction and the potential changes. Consequently, in flotation we are always faced with the so-called mixed electrostatic interactions, constant charge, and constant potential, where sign reversal of interaction forces can occur at small distances. This influence of the gas bubble on flotation of fine particles is, in principle, not unimportant. But it cannot be dealt with in the following discussion.

II. AIM OF THIS SURVEY

The general aim of physical modeling of the flotation process is to precalculate floatability of one component as a function of flotation time on the basis of physical, physicochemical, and hydrodynamic parameters. The simplest form of a kinetic equation describing floatability is the variation of particle number per time unit in a given volume of apparatus. It depends on the number of bubble–particle collisions per time unit, Z, on the number of particles, N_p, and of bubbles, N_B, and on the probabilities of interaction. At this, it is commonly assumed that these probabilities are independent of each other, so that the product law of the individual probabilities may be used. Then and only then, the following equation holds for the variation of particle number:

$$\frac{dN_p}{dt} = ZN_pN_BP_cP_aP_{tpc}P_{stab},\qquad(1)$$

where P_c is probability of collision, P_a is probability of adhesion, P_{tpc}, is probability of TPC formation, and P_{stab} is probability of the stability of particle–bubble aggregates against external stress forces. Nevertheless, it can be seen from Equation (1) that the probability of the overall process of flotation is a combination of the individual probabilities. The latter ones then depend on both the interparticle interaction forces and the hydrodynamic forces in the vicinity of the bubble. P_c is mainly determined by the hydrodynamics of the flow field. P_a and P_{tpc} are influenced essentially by the surface forces. P_{stab} depends on both the hydrodynamics in the turbulent flow field and the surface (respectively interface) forces.

The aim of the present work is therefore to

- describe the interplay between interface forces and hydrodynamic forces

- derive equations for calculating the individual probabilities mentioned above and
- illustrate and discuss essential factors influencing the individual probabilities.

For this reason the most important equations characterizing the flow field close to the bubble are given first, from which the behavior of particle motion can be derived. Then the individual probabilities are described in detail. The probability of formation of a three-phase contact is only briefly discussed because there exist no utilizable investigations as yet. Special attention will be focused also on the present knowledge of thin film instability and rupture process as a key to understanding of p–b heterocoagulation. In the last part the rapid progress during the last decade in modeling of the overall batch process on the basis of the available micro process probabilities will be described.

III. FLUID MECHANICS OF BUBBLES

Solid particles moving in the vicinity of a bubble are subjected to the tangential and radial component of fluid flow. Particle motion in the radial (normal) direction to the bubble surface leads to collision with the bubble, whereas in the case of tangential motion the particle slides across the bubble [5]. There exists a comprehensive literature, e.g., [6–8], which describes the hydrodynamics of bubbles. The following facts are substantial for the problem of b–p interaction:

1. For very small diameters of the bubble, the Stokes law for solid particles is valid, which is based on the conception that the bubble–liquid interface is rigid, i.e., completely immobile. The surface of the bubble in this case is covered with surfactant molecules and is said to be completely retarded. In the interior of bubble no vortex occurs.
2. In the opposite case, when the surface of bubble is completely mobile, a flow occurs in its external envir-

onment, with the formation of a circular vortex within the bubble itself. Equations of motion of spherical bubbles with internal circulation have been described by Haas et al. [9]; for example, Hadamard and Rybczynski (in Ref. [6], p. 30) dealt with the mobility of the phase boundary in the case of an incompletely retarded bubble in a classical work.

3. According to the type of the surfactant, adsorption on the bubble surfaces, rising velocity of bubble, and locality on the bubble surface, each condition from completely immobile (quasi-rigid) to completely free mobile is possible. Normal and tangential components of fluid flow across the bubble surface change their values depending on the degree of mobility (Figure 7.1).

Therefore, we assume ad hoc that the two approximated boundary cases for the flow field in the vicinity of bubble are the Stokes flow at bubble Reynolds numbers $Re_B \ll 1$ and the potential flow at $Re_B \gg 1$. They differ with regard to the tangential component of fluid flow, which is important to the attachment process, in such a way that tangential component of liquid flow u_ϕ, is equal to zero at $r = R_B$ in the case of Stokes flow (no slip condition on the surface) and finite in the case of potential flow. The radial component u_r is always finite.

The most important equations describing the field of fluid flow in the vicinity of bubble in flotation are summarized considering the facts mentioned above.

A. Stokes Flow

Hadamard and Rybczynski equation is as follows [7]:

$$
\begin{aligned}
u_r &= -v_B \left(1 - \frac{(2\eta + 3\gamma)R_B}{2(\eta + \gamma)r} + \frac{\gamma R_B^3}{2(\eta + \gamma)r^3} \right) \cos \Phi, \\
u_\phi &= v_B \left(1 - \frac{(2\eta + 3\gamma)R_B}{4(\eta + \gamma)r} - \frac{\gamma R_B^3}{4(\eta + \gamma)r^3} \right) \sin \Phi
\end{aligned}
\tag{2}
$$

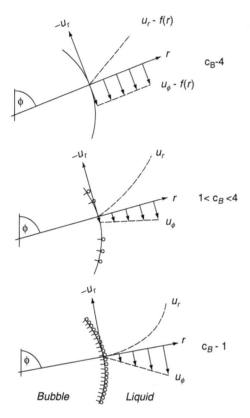

Figure 7.1 Schematic representation of the fluid flow field near the bubble surface with several degrees of coverage with surfactant molecules, i.e., with several steps of surface mobility: $C_{\mathrm{B}} = 1$, quasi-rigid; $C_{\mathrm{B}} = 4$, completely mobile.

where η is the dynamical viscosity and γ is a coefficient depending on mobility of the bubble surface.

The Stokes equation for gas bubbles in water ($\gamma \gg \eta$ and η is approximately zero) is as follows:

$$u_{\mathrm{rSt}} = -v_{\mathrm{B}}\left(1 - \frac{3R_{\mathrm{B}}}{2r} + \frac{R_{\mathrm{B}}^3}{2r^3}\right)\cos \Phi,$$

$$u_{\Phi\mathrm{St}} = v_{\mathrm{B}}\left(1 - \frac{3R_{\mathrm{B}}}{4r} - \frac{R_{\mathrm{B}}^3}{4r^3}\right)\sin \Phi. \tag{3}$$

B. Potential Flow

The potential flow equation is as follows:

$$u_{rpot} = -v_B \left(1 - \frac{R_B^3}{r^3}\right) \cos \Phi, \qquad u_{\Phi pot} = v_B \left(1 + \frac{R_B^3}{2r^3}\right) \sin \Phi$$

$$(4)$$

C. Intermediate Reynolds Number Flow

For intermediate Reynolds numbers of the bubble $(1 < Re_B < 100)$, which can be frequently found in flotation machines, we get from a modified stream function given by Yoon and Luttrell [10]:

$$u_r = u_{rSt} - v_B \frac{3Re_B^*}{2} \left(\frac{R_B^4}{r^4} - \frac{R_B^3}{r^3} - \frac{R_B^2}{r^2} + \frac{R_B}{r}\right) \cos \Phi,$$

$$u_\Phi = u_{\Phi St} + v_B \frac{3Re_B^*}{4} \left(-\frac{2R_B^4}{r^4} + \frac{R_B^3}{r^3} + \frac{R_B}{r}\right) \sin \Phi,$$

$$(5)$$

where

$$Re_B^* = \frac{Re_B^{0.72}}{15}.$$

$$(6)$$

For modeling the interaction probabilities in flotation we have made the following additional assumptions:

1. At a mobile bubble $(C_B = 4)$ the tangential component u_ϕ can be described approximately by the potential flow.
2. For the immobile bubble $(C_B = 1)$, the equation for intermediate flow $(1 < Re_B < 100)$ is suitable as an approximation.

IV. PROBABILITY OF PARTICLE–BUBBLE COLLISION

Generally, the collision probability (P_c) is defined by the ratio of the number of particles with encountering a bubble per time unit to the number of all particles approaching the bubble in a stream tube having a cross section equal to the projected area of the bubble. Usually this value is less than 1. The reason is that not all the particles within the bubble projection path can collide with the bubble, but only particles within a limiting radius of collision of the flow tube R_c. Therefore the collision probability is given by the ratio (Figure 7.2):

$$P_c = \left(\frac{R_c}{R_B}\right)^2. \tag{7}$$

The fraction of particles reaching a bubble surface, the shape of the particle trajectory in the fluid flow, and the interactions possible between particle and bubble surface depend on particle mass and its inertia and on the flow field itself. The related characteristic quantity is the Stokes number. It is the ratio between the particle size and the bubble diameter, or more generally, the ratio between the inertia force of the particle and the viscous drag force of the bubbles:

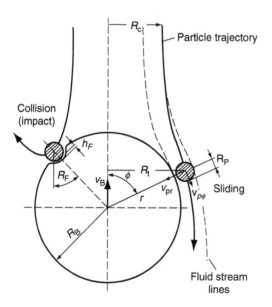

Figure 7.2 Interaction of particles with the radius R_p and a gas bubble with the radius R_B by impact and sliding. R_c, collision radius; V_{pr} and $V_{p\phi}$, radial and tangential components of the particle velocity; V_B bubble-rising velocity.

$$St = \frac{Re_{\text{B}} \rho_{\text{p}} d_{\text{p}}^2}{9 \rho_{\text{fl}} d_{\text{B}}^2} = \frac{\rho_{\text{p}} d_{\text{p}}^2 v_{\text{B}}}{9 \eta d_{\text{B}}}, \tag{8}$$

where ρ_{p} is the particle density, v_{B} is the bubble-rising velocity, and η is the dynamic viscosity.

Depending on St, the following regimes of particle attachment to a bubble can be distinguished:

1. At $St < 0.1$, inertia forces do not influence the particle motion in practice. Particles following the fluid streamlines slip freely. Collisions due to impact with a bubble are not possible.
2. At $0.1 < St < 1$ inertia forces play a role in the course of attachment. Therefore, bubble–particle collisions can take place with the particle smaller than the bubble. Particle trajectories deviate more or less from the streamlines of liquid.
3. At $St > 1$ particle trajectory is next to a straight line. Particles predominantly collide with the bubble.

But the exact calculation of the overall collision probability is rather complicated. More details will be described in the textbook *Colloidal Science of Flotation* by Nguyen and Schulze [11]. Many approximation equations of different kinds are available. One possibility of calculation is to split of the whole process into single ones on the basis of the acting forces in the flow field at different Stokes numbers. The overall collision probability than can be written (Figure 7.3) as the algebraic sum of the

*   Interception collision efficiency
*   Gravity collision efficiency and
*   Inertial collision efficiency.

In the following, we will distinguish generally between these three different probabilities in order to calculate the overall probability in a simple way [12]:

$$P_{\text{c}} = E_{\text{ic}} + E_{\text{g}} + \left[1 - \frac{E_{\text{ic}}}{(1 + (d_{\text{p}}/d_{\text{B}}))^2} \right] E_{\text{in}}. \tag{9}$$

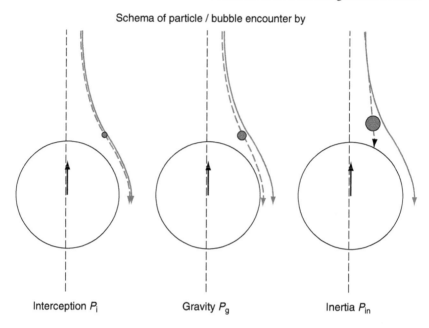

Figure 7.3 Schema of interception collision efficiency, gravity collision efficiency, and inertial collision efficiency.

Here E_{ic} is the interception collision efficiency, i.e., is the gravity collision efficiency, and E_{in} is the inertial collision efficiency. For p–b diameter ratios of approximately < 0.1, which is always valid in flotation, the following equations are applicable:

(i) *For the interceptional effect*:

$$E_{ic} = \frac{2\Psi_c^*}{[1 + (v_p/v_B)]} \tag{10}$$

with

$$\Psi_c^* = \exp\left\{\ln\Psi^*\left\{\frac{1}{\xi_0^*}\right\} + \left[\frac{(2.6328 - \ln\Psi^*\{1/\xi_0^*\})}{(1.4569 - \ln\Psi^*\{1/\xi_0^*\})}\right]\right.$$

$$\left.\left(\ln\left(\frac{d_p}{d_B}\right) - \ln\frac{1}{\xi_0^*}\right)\right\} \text{ at } R_p/R_B > 1/\xi_0^*, \tag{11}$$

where $\Psi^*\ \{1/\xi_0^*\}$ has to be substituted by ψ^* according to Equation (12) with $d_p/d_B = 1/\xi_0^*$

$$\Psi^* = \frac{3}{4}\left(\frac{d_p}{d_B}\right)^2\left[1 + \frac{(3/16)\,Re_B}{(1 + 0.249Re_B^{0.56})}\right] \text{ at } R_p/R_B \leq /\xi_0^*,$$
(12)

$$\xi_0^* = \frac{3}{2}\left[1 + \frac{(3/16)\,Re_B}{(1 + 0.249\,Re_B^{0.56})}\right]\frac{1}{\sin\,\Phi_c},$$
(13)

$$\Phi_c = 78.1 - 7.37\log Re_B \quad \text{for } 20 < Re_B < 400.$$
(14)

(ii) *For the gravitational effect:*

$$E_g = \left(1 + \frac{d_p}{d_B}\right)^2\left[\frac{(v_p/v_B)}{(1 + (v_p/v_B))}\right]\sin^2\Phi_c.$$
(15)

(iii) *For the inertial effect:*

$$E_{in} = \frac{(1 + (d_p/d_B))^2(St/(St + a))^b}{(1 + (v_p/v_B))},$$
(16)

where a and b are parameters, according to Ref. [13], at Re_B 100–250, $a = 0.8$, $b = 2.0$; at Re_B 50–100, $a = 1.12$, $b = 1.84$; at Re_B 25–50, $a = 2.06$, $b = 2.07$.

Bloom and Heindel [21] have derived the following relations for the probability of collision at intermediate flow conditions, which can be handled more easily for the modeling of the overall batch process:

$$P_c = \frac{1}{1 + |G|}\left\{\frac{1}{2[(R_p/R_B) + 1]^3}\left[2\left(\frac{R_p}{R_B}\right)^3 + 3\left(\frac{R_p}{R_B}\right)^2\right]\right.$$
$$\left. + \frac{2Re_B^*}{[(R_p/R_B) + 1]^4}\left[\left(\frac{R_p}{R_B}\right)^3 + 2\left(\frac{R_p}{R_B}\right)^2\right]\right\} + \frac{|G|}{1 + |G|},$$

where

$$G = v_{ps}/v_B, \quad Re_B^* = Re_B^{0.72}/15 \quad Re_B = \rho_{F1}v_Bd_B/\eta.$$

V. PROBABILITY OF PARTICLE ADHESION

After a particle has approached a bubble closely, interaction leads either to an attachment or to driftage of the particle in the immediate vicinity of the bubble surface. Two kinds of interaction may occur:

>   *Collision* (impact) processes, in which the bubble surface is strongly deformed and the particle is repelled unless attachment takes place during the first collision,
>   *Sliding* processes along the bubble surface with a weak deformation of the surface.

Hence the adhesion probability P_a consists, in principle, of two parts: the probability of adhesion by collision, P_{ac}, and the probability of adhesion by sliding, P_{asl}. Which of the interactions predominate in a special case depends on many factors and has not been investigated experimentally in detail. It may be stated, however, that collision occurs predominantly with large or heavy particles, high radial particle velocities, and chiefly in the region of the stagnation point flow of the bubble surface at touching angles less than 30°. Sliding occurs at small relative velocities between particles and bubbles and at larger touching angles independently of the particle size. Both probabilities are, however, not independent of each other because collision processes can pass over in sliding processes, depending on the strength of the tangential flow carrying away the particle after recoil [14]. Here the product rule for individual probabilities is therefore not fully valid.

The most important point is that during the contact between particle and bubble by collision or by sliding, a thin liquid film of the thickness, h_F, is formed (see Figure 7.2). Whether the particle can be attached or not depends crucially on the drainage time of the thin liquid film up to its critical thickness of rupture. Thus, the time requirement for attachment is that the contact time must be longer than the drainage time. The contact time is either the impact time or the sliding time of the particles. Therefore, the adhesion probability is connected with the ratio of the contact time and the film drainage time.

At aeration conditions usually found in technical flotation process the number of air bubbles per unit volume is normally very large. Therefore, we assume that the sliding interaction is the determining process for adhesion and the following relation holds true:

$$P_{asl} = f(\text{contact time/sliding time}).$$

A. Probability of Adhesion by Sliding, P_{asl}

Sliding processes occur if the Stokes number is less than 0.1, if particles are rebounded and carried away by the flow after the first collision, or, generally, if V_{pr} is very small. The probability of whether a particle can attach to a bubble by sliding depends again on the circumstance of whether the thin film ruptures or remains stable during the sliding time. The probability of the rupture is greater the smaller the sliding velocity, i.e., the greater the sliding time is. The latter depends on its part on the tangential particle motion in the flow field close to the bubble surface and consequently on the tangential component of the fluid flow. The probability of adhesion by sliding can therefore be derived from the equation of particle motion in the flow field of the bubble. Set-ups made by several authors, e.g., Sutherland [15], Dobby and Finch [16], can be found. By solving the complete equation of particle motion with the balance of the acting forces in radial and tangential direction of motion we have been also derived a first approach on the sliding probability [5]. The considered forces are:

(i) *In the radial direction*:

$$-F_{gr} + F_T + F_c - F_{ur} \pm F_L = 0. \tag{18}$$

They are given by:

- *the particle weight*

$$F_{gr} = \frac{4}{3}\pi R_p^3 \Delta \rho g \cos \Phi. \tag{19}$$

- *the resisting force during drainage of the thin film*

In this case, the Taylor equation is used since no extended thin film but only a *point contact* exists during sliding as was observed by us:

$$F_T = \frac{6\pi\eta R_p^2 v_{pr}}{h_F c_B}, \tag{20}$$

$$v_{pr} = -\frac{dh_F}{dt}.$$

- *the centrifugal force:*

$$F_c = \frac{4}{3}\pi R_p^3 \Delta\rho g \frac{v_{p\Phi}^2}{r}, \tag{21}$$

which occurs if the particle moves on a circular path with

$$r = R_p + R_B + h_F. \tag{22}$$

- *the flow force acting on a particle close to the bubble surface (wall)*

$$F_{ur} \approx 6\pi\eta R_p |u_r|. \tag{23}$$

- *the lift force*

This force according to Ref. [17] results from the following reasons: Within a tangential flow, in which a velocity gradient exists, a force acts on the particle perpendicularly to the flow direction, if the particle moves with a relative velocity: $v_{prel} = u_\Phi - v_{ps\Phi}$. The sign of this force depends on v_{prel} [18]. If $v_{prel} < 0$, then F_L directed to the bubble, if $v_{prel} > 0$, then F_L is a driftage force (*lift force*) directed away from the bubble. According Clift [6, p. 259]

$$F_L = 1.62\eta d_p v_{prel} \sqrt{Re_G}. \tag{24}$$

The characterizing value is the Reynolds number of shear, Re_G:

$$Re_G = \frac{|G| d_p^2 \rho_{Fl}}{\eta}, \tag{25}$$

where the shear gradient G is given by

$$G = \frac{\partial u_\Phi}{\partial r}. \tag{26}$$

For a potential flow it follows for G from Equation (4) that

$$G_{\text{pot}} = v_{\text{B}} \left(\frac{-3R_{\text{B}}^4}{2r^4} \right) \sin \Phi \tag{27}$$

and in the case of intermediate flow (Equation (5)) G may be written as:

$$G_{\text{in}} = v_{\text{B}} \left[\left(\frac{3R_{\text{B}}^2}{4r^2} + \frac{3R_{\text{B}}^4}{4r^4} \right) + \frac{3}{4} Re^* \left(-\frac{R_{\text{B}}^2}{r^2} - \frac{3R_{\text{B}}^4}{r^4} + \frac{8R_{\text{B}}^5}{r^5} \right) \right] \sin \Phi. \tag{28}$$

(ii) *In the tangential direction*:

$$F_{\text{g}\Phi} - F_{\text{w}\Phi} = 0 \tag{29}$$

with

• *the particle weight*

$$F_{\text{g}\Phi} = \frac{4}{3} \pi R_{\text{p}}^3 \Delta\rho g \sin \Phi. \tag{30}$$

• *the resisting force in the vicinity of the wall F_W*

In the vicinity of a wall, the resisting force acting on a moving particle is increased. It depends on the kind of flow field and consequently also on the degree of covering of the bubble surface with surfactant molecules. There is a very comprehensive literature on this subject that is too extensive to consider here. A short summary is given in Duchin and Rulev [7], for example. In a general form a correction function $f(h_\text{F}/R_\text{p})$, depending on both the particle size and the wall distance, is introduced to consider this increase of resistance within the equation for the resisting force of a particle according to Stokes [in Ref. 19]:

$$F_{\eta\Phi} = 6\pi\eta R_\text{p} v_{\text{prel}} f(h_\text{F}/R_\text{p}). \tag{31}$$

Thus we can write for the force near to a completely retarded bubble at $(h_F/R_P) > 10^{-3}$

$$F_{\eta\Phi} \approx \frac{16}{5} \pi\eta v_{p\Phi} R_p \ln\left(\frac{h_F}{R_p}\right). \tag{32}$$

Equations (18)–(32) follow a complete set of equations of particle motion by resolving $d\phi/dt$ and dh_F/dt:

$$\frac{d\Phi}{dt} = \frac{[u_\Phi + (v_{Ps} \sin \Phi/cor)]}{(R_B + R_p + h_F)},$$

$$\frac{dh_F}{dt} = -h_F c_B \left[g_0 \cos \Phi + \frac{|u_r|}{R_p} - \frac{g_1 v_{p\Phi}^2}{(R_B + R_p + h_F)} - g_2 v_{prel} \sqrt{Re_G}\right], \tag{33}$$

where $g_0 = 2\Delta\rho g R_p/9\eta$, $g_1 = 2\Delta\rho R_p/9\eta$, $g_2 = 0.1714/R_p$, and $cor = (8/15) \ln(h_F/R_p)$. Re_G is calculated according to Equations (25) and (27) or Equation (28); u_R and u_ϕ according to Equation (4) or Equation (5).

For $(h_F/R_P) < 10^{-3}$ the constant factor $cor = 2.1$ is set instead of $cor = (8/15) \ln(h_F/R_p)$. In the case of an unretarded bubble in a potential flow we assume $F_{W\phi} = 0$.

Solutions of Equation (33) can be found by numerical integration. That gives the thickness of the thin liquid film h_F in terms of the particle positions r and ϕ using the following initial conditions at $t = 0$ and $\phi(t = 0)$ according to [20]:

$$h_{F(t=0)} = h_0 = d_p \left(\frac{3\eta v_{Pr}}{8\sigma c_B}\right)^{1/2}. \tag{34}$$

This transition thickness is in the order of few micrometers. Attachment is possible if the critical film thickness, $h_F = h_{crit}$, is reached during sliding in the range of touching angle between ϕ at $t = 0$ and $\phi = 90°$. The touching angle at which h_{crit} is marginally reached is denoted as critical angle ϕ_{crit}. Particles touching the bubble at $\phi < \phi_{crit}$ are able to adhere; particles touching the bubble at $\phi > \phi_{crit}$ do not adhere. Therefore, attachment probability may be defined as follows:

$$P_{asl} = \left(\frac{R_t}{R_B + R_p}\right)^2. \tag{35}$$

with

$$R_t = r \sin \Phi \approx (R_B + R_p) \sin \Phi, \tag{36}$$

we get

$$P_{asl} = \sin^2 \Phi_{crit}. \tag{37}$$

Figure 7.4 shows, as an example, values of P_{asl} for immobile ($C_B = 1$) and mobile bubbles ($C_B = 4$) in intermediate as well as in potential flow as a function of the particle size, bubble-rising velocity, and critical film thickness. It can be seen that four factors influence the probability of adhesion:

1. The main influence arises from the bubble size, i.e., from the real bubble-rising velocity. P_{asl} is greater the smaller the bubble size.
2. The smaller the particles, the greater the P_{asl}. Decrease of P_{asl} with increasing particle size is small in the case of large particles but big in the case of small particles.
3. The influence of bubble mobility manifests itself in an increase of P_{asl} in the case of a mobile bubble.
4. With increasing critical thickness (increasing of hydrophobicity of particle surface) of rupture the intervening liquid film enlarges the probability of adhesion by sliding considerably.

Unfortunately the numerical procedure of integration the equation of particle motion over bubble surface is unsuitable for calculating the kinetics of the overall process. Therefore, Bloom and Heindel [21] have derived a new equation (38), which involves the flow conditions at intermediate

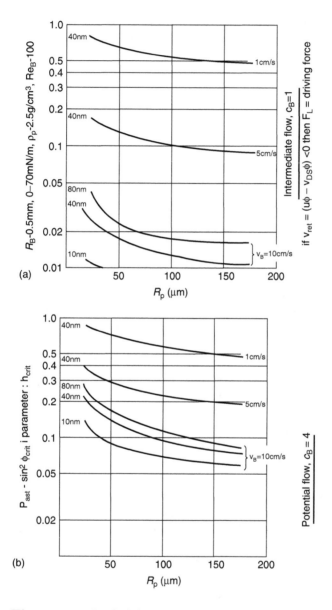

Figure 7.4 Probability of the particle adhesion P_{asl} on bubbles by the sliding process as a function of particle radius, the bubble-rising velocity V_B, and critical film thickness h_{crit}. $C_B = 4$, mobile bubbles, potential flow; $C_B = 1$, immobile bubble surface, intermediate flow regime.

Reynolds numbers, the transition film thickness, and the critical thickness of intervening film:

$$P_{asl} = \exp\left\{-2\left(\frac{\bar{\lambda}}{c_B}\right)\left(\frac{R_p}{R_B + R_p}\right)\left[\frac{g(r) - G}{|k(r)| - G}\right]\left(\frac{h_0}{h_{crit}} - 1\right)\right\}$$

$$g(r) = \left(1 - \frac{3R_B}{4r} - \frac{R_B^3}{4r^3}\right) + Re_B^*\left(\frac{R_B}{r} + \frac{R_B^3}{r^3} - \frac{2R_B^4}{r^4}\right)$$

$$k(r) = -\left[\left(1 - \frac{3R_B}{2r} + \frac{R_B^3}{2r^3}\right) + 2Re_B^*\left(\frac{R_B}{r} - \frac{R_B^2}{r^2} - \frac{R_B^3}{r^3} + \frac{R_B^4}{r^4}\right)\right]$$

$$\bar{\lambda} = 6\pi\mu R_p/f,$$

$$(38)$$

where f is a fluid flow friction factor according to Ref. [21].

Some other possibilities to calculate the adhesion probability on the basis of the induction time will be presented in the text book "Colloidal Science of Flotation" [11, Chapter 4].

B. Probability of Adhesion by Particle Impact

Geidel [22] and Rasemann [23] showed that the attachment probability at an impact is distributed exponentially. If the film drainage time is denoted as induction time τ_i according to Sven-Nilsson [24] the attachment probability by collision may be expressed by the following equation:

$$P_{ac} = 1 - \exp\left(-\frac{\tau_c}{\tau_i}\right). \tag{39}$$

The collision time, τ_c, depends on the depth of deformation, the surface tension of the liquid–gas interface, and the particle mass. The following relationship was derived from the equation of radial motion of a single particle by deformation of the bubble surface during the contact by us [25]:

$$\tau_c = \left[\frac{\pi^2 R_p^3(\rho_p + 1.5\rho_{Fl})}{3\sigma}\right]^{1/2} f, \tag{40}$$

where f is a nonlinear empirical function of the depth of deformation, surface tension, and particle density:

$$f \approx 1.39 - 0.46 \ln R_\mathrm{p}\{\mathrm{cm}\}. \tag{41}$$

The induction time τ_i (Equation (39)) depends essentially on the critical thickness of rupture of the thin film, i.e., on the degree of hydrophobicity of the particle surface, its dimensions, and the bubble retardation as well. The equation of Stefan–Reynolds holds for the case of formation of an extended plane-parallel liquid film:

$$\tau_\mathrm{i} = \frac{3\eta R_\mathrm{F}^2 R_\mathrm{p}}{8\sigma h_\mathrm{crit}^2 c_\mathrm{B}},$$

$$R_\mathrm{F} \approx \pi R_\mathrm{p}(-299.89 - 177.1 \ln R_\mathrm{p}\{\mathrm{cm}\})(0.32 v_\mathrm{Pr} \tau_\mathrm{c})^{0.62}/180. \tag{42}$$

Equation (42) applies to smooth, spherical particles. As a model for broken particles with a rough surface we employ a big sphere, whose surface radius is covered with micro hemispheres of the radius R_mic [26]. Then we may write

$$\tau_\mathrm{i}^* = \frac{3\eta \pi^2 R_\mathrm{mic}^2 R_\mathrm{p}}{128\sigma h_\mathrm{crit}^2 c_\mathrm{B}}. \tag{43}$$

VI. RUPTURE OF THE INTERVENING THIN LIQUID FILM

The crucial step of all involved microprocesses of the *classical* flotation process is the formation and rupture of the intervening thin liquid film between the interacting particle and the air bubble as a target. A short time ago we assumed that only the interparticle forces between particles and bubbles are responsible for the rupture process, which can be macroscopically characterized by the contact angle of the particle surface and the surface tension of the liquid [27]. These facts of the matter could be described by means of a physically determined empirical equation. An abundance of observations in conflict with the conceived ideas and a many inexplicable phenomena led during the last decade to the recognition that other mechanisms, which will be described following, are probably essential for the rupture process on hydrophobic surfaces. From these results we are not able to create an equation for calculating the critical thickness at present.

Dewetting of metastable thin liquid films from a solid surface is a topic of greatest interest since more than three decades because of the crucial importance of the film rupture between air bubbles and solid particles in mineral and deinking flotation, in modern technologies as polymer coatings, and some other applications. Also from a fundamental point of view there are many unsolved questions concerning the interpretation of experimental phenomena with respect to the behavior predicted theoretically.

We know today that two different mechanisms can be responsible for the destabilization and rupture of the liquid film depending on the nature of the solid and liquid, the degree of hydrophobicity, the kind of adsorption layer, and its morphological and chemical heterogeneity. The two mechanisms are the growing capillary wave-mechanism and nucleation.

The first mechanism is based on instability against thermal fluctuations in the presence of any kind of an attractive force which increases the amplitude of the fluctuation. According the theory firstly developed by Scheludko [28] and Vrij [29] this instability leads to rupture of the film during its drainage. A certain example recently described in literature [30] is the dewetting of molten gold films from fused silica substrates which occurs after melting with a laser pulse.

The second mechanism was firstly proposed by Derjaguin and Gutop [31]. Density fluctuations inside the film in the vicinity of a hydrophobic solid or tiny gas cavity at defects could be the reason for this, but no kind of an attractive force is necessary. An example in literature is the rupture of polystyrene layers on silicon wafers [32].

As well known [33], the van der Waals force in the chosen system: silica (or glass)–water-film–air is repulsive. Because of the repulsive electrostatic disjoining pressure between the negatively charged air bubbles and the negatively charged silica surface, the sum of interaction forces is also repulsive. Hence, thin water films on silica surfaces must be stable at its equilibrium thickness independently of the ionic strengths in water [34].

Due to methylation the surface becomes strongly hydrophobic. But neither their charge (or surface potential) nor the Hamaker constant of the system is influenced significantly by methylation of the surface hydroxyl groups [35,36]. Therefore, wetting films should remain stable upon such surfaces. But everybody knows that on hydrophobic surfaces no stable wetting films can exist. In order to explain this discrepancy and describe their large instability, additionally, the so-called long-range hydrophobic force (LRHF) was postulated in the literature during the last decade [37,38].

Another possibility to change the surface properties of hydrophilic silica surfaces is to change their surface charge to a positive value by adding of Al^{3+} cations. In this case an attractive electrostatic disjoining pressure occurs. Although the surface remains nearly hydrophilic, thin liquid wetting films rupture at a thickness which depends on the range of the electrostatic attractive double-layer force [39].

We are able to demonstrate experimentally that also water films on silica surfaces can be destabilized by both mechanisms, nucleation, and growing capillary waves: When the surface is, in principle, hydrophilic but an attractive electrostatic double-layer force is present, then the capillary wave rupture occurs. In the case of hydrophobic surface nucleation is responsible for the rupture. No LRHF must be introduced in order to explain the rupture on such strong hydrophobic surfaces.

To prove the nonexistence of LRHF during the time of thinning of the wetting film until its critical rupture thickness h_{crit} we had to generate metastable wetting films on hydrophobic surfaces. Formerly it had been proved that the heterogeneity of the hydrophobic surface causes a dramatic influence on lifetime of the film and its h_{crit}; the larger the heterogeneity, the shorter is the lifetime and the larger is h_{crit}. The goal was to generate very homogeneous hydrophobic surfaces. Gas-phase silanization of the silanol groups of the glass surface was carried out with hexamethyldisilazane (HMDS), where different reaction times lead to different hydrophobicities, expressed by contact angles between 20° and 90°.

On the other hand, positively charged silica surfaces by addition of a small amount of Al^{3+} ions (10^{-4} mol/l) [39] were prepared. In this case the surface remains rather hydrophilic, but the sum of the surface forces becomes attractive, which destabilizes the wetting film and leads to film rupture.

On these modified surfaces, microscopic wetting films on these modified surfaces were established using the well-known method of Derjaguin and Scheludko (D–S-film balance) [40]. Film thickness is measured versus time by micro-interferometric means at a wavelength of 470 nm.

If the film is formed at a distance of less than 150 nm between the bubble and the surface, it is possible to generate a flat parallel film without a central dimple. In this case the hydrodynamic equation of Reynolds describing the drainage can be used without restrictions. The capillary pressure P_σ in the gas bubble is measured separately and was adjusted to 250 Pa constantly in our experiments. Microscopic cover glasses made of soda-lime-floating-glass (Marienfeld Superior No. 1 20 mm^2) were used as substrates. Their roughness is less than 2 nm. They are cleaned by boiling them in a 70:30 H_2SO_4–H_2O_2 mixture prior to a long rinsing under running Milli-Q or Plus-water.

To evaluate the film drainage the well-known Reynolds equation was used, which describes the thinning of a parallel thin film [28]:

$$\frac{1}{h^2} - \frac{1}{h_0^2} = \frac{4}{3} n \frac{P_\sigma - \Pi_\Sigma(h(t))}{\eta R_F^2} t. \tag{44}$$

where h_0 is the initial thickness at $t = 0$ (last interferometric minimum); n is a factor for the interfacial mobility of the film ($n = 4$ in the case of a flow between a solid wall with nonslip condition and a completely mobile liquid–gas interface without adsorption); η is the dynamic viscosity of the liquid. The addition of the electrostatic term Π_{el} and the van der Waals term Π_{vdW} leads to the disjoining pressure Π_Σ, which is in sum with the capillary pressure in the bubble P_σ, the driving force for film drainage.

Numerical solution of Equation (44) leads to three different regimes of film thinning depending from the sign of the disjoining pressure Π_Σ (see Figure 7.5):

1. In the case when $\Pi_\Sigma = 0$ the capillary pressure in the bubble P_σ is the only driving force, the normal Reynolds equation is valid: the film is thinning, until after some time the thickness $h = 0\,nm$ will be reached.

2. For a repulsive disjoining pressure ($\Pi_\Sigma > 0$), the film drainage velocity is reduced and the film thickness reaches a constant value h_{eq}, which is governed by Π_Σ/h isotherm.

3. An attractive disjoining pressure ($\Pi_\Sigma > 0$), leads initially to an increased thinning rate of the film; after reaching a certain thickness, determined by the range of the attractive force, the film ruptures. Hereby, the numerical evaluation sometime shows two solutions of the Reynolds equation, rupture, and a period of metastability.

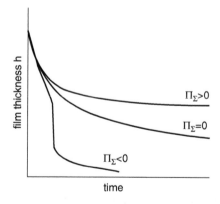

Figure 7.5 Calculated drainage curves according to the Reynolds equation for flat films, with respect to the acting surface forces.

1. Equilibrium Films on Hydrophilic Silica

To ensure the validity of our calculations, as well as the chosen parameters charge, surface potential, and Hamaker constant in the DLVO theory, we measured the drainage of equilibrium films with complete wetting hydrophilic glass. It is well known that their equilibrium thickness is controlled only by the repulsive electrostatic double-layer forces and the van der Waals forces [34, p. 100].

The kinetics of drainage and the critical rupture thickness on hydrophobic surfaces was measured in 10^{-3} mol/l KCl solution. The van der Waals term of the forces does not play any role in this case because of its short range (about 10 nm).

Figure 7.6 shows the mean and the standard deviation of experimentally investigated drainage curves for the case of a

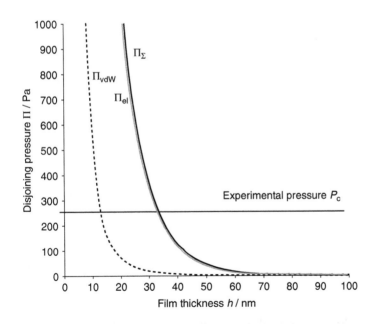

Figure 7.6 Disjoining pressure–distance isotherms of the system silica/electrolyte (10^{-3} mol/l KCl)–air according to the DLVO theory; charge of the glass surface $\sigma = -1$ mC/m^2 potential of the gas bubble $\Psi = -35$ mV; $A_{slv} = -1.0 \times 10^{-20}$ J.

stable, flat parallel film (without dimple). The mean value of the film radius for all experiments was $R_F = 90 \pm 15\,\mu m$.

The theoretical drainage curve is calculated by solving the Reynolds equation numerically. Within the experimental error, it is identical to the experimentally determined drainage curve. Because of our former extensive investigations [34] as well as literature data [41], charge and potential in a $10^{-3}\,m$ KCl solution at neutral pH value is known and therefore the electrostatic disjoining pressure term of the theoretical curve is calculated corresponding to the constant charge/constant potential approximation of Kar et al. [42]: charge of the glass surface $\sigma = -1\,mC/m^2$; potential of the gas bubble $\Psi = -35\,mV$.

The repulsive van der Waals term of the disjoining pressure is calculated using a Hamaker constant for the wetting film of $A_{slv} = -1.0 \times 10^{-20}\,J$ [1].

The equations used are

$$\Pi_{vdW} = -\frac{A_{slv}}{6\pi h^3}, \tag{45}$$

$$\Pi_{el} = -\frac{1}{2}\left[\kappa\left(\frac{\sigma_2^2}{\varepsilon_0\varepsilon\kappa} - \varepsilon_0\varepsilon\kappa\Psi_1^2\right)\frac{1}{\cosh^2(\kappa h)} - 2\kappa\Psi_1\sigma_2\frac{\sinh(\kappa h)}{\cosh^2(\kappa h)}\right], \tag{46}$$

where κ is the Debye–Hückel parameter, ε_0 is the absolute dielectric constant, ε is the relative dielectric constant, and h is the film thickness.

The isotherms of the disjoining pressure of the system silica–electrolyte (10^{-3} mol/l KCl-solution)–air are shown in Figure 7.7. According to them, a stable wetting film exists at the capillary pressure of 250 Pa with an experimentally corroborated equilibrium thickness of 34 nm. Therefore, we can consider our calculation and the chosen parameters for the applicability of the DLVO theory as correct.

2. Metastable Wetting Films on Methylated Silica

The hydrophobization of a surface changes the properties of a wetting film dramatically. It becomes unstable and during thinning rupture takes place at a certain critical thickness,

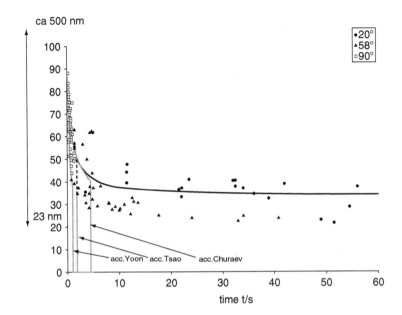

Figure 7.7 Drainage and rupture of metastable flat wetting films on gas phase methylated glass surface at different advancing contact angles between 20° and 90° realized by different reaction times with HMDS. Arbitrary chosen start time at the film thickness 89 nm. Larger rupture thicknesses are indicated by the arrow outside the coordinate system. Full line: calculated drainage according the Reynolds equation if only repulsive electrostatic and van der Waals forces are present. Broken lines: calculated drainage curves if additionally attractive LRHF is present (denoted according different approaches [37,44,45]).

h_{crit}. Due to the dewetting of the solid surface the film brakes apart into single droplets.

The parameters used in the DLVO theory like Hamaker constant or potential of the solid surface are not changed significantly due to the methylation reaction. In Derjaguin et al. [36] it has been shown that A_{slv} does not change its sign at thicknesses larger than 20 nm [34, p. 89]. The negative zeta potential of the silica surface remains virtually unchanged due to hydrophobization, the electrostatic interaction

between bubble and solid is essentially unaltered by the introduction of surface trimethylsilyl groups, since this reaction
can only eliminate a small amount of the superficial hydroxyl
groups [43]. Laskowski and Kitchener [35] proved experimentally that the potential of methylated silica surfaces remains
unchanged. Therefore, the surface forces summarized in the
DLVO-theory — only repulsive forces are acting — does not
provide any explanation for film rupture. Figure 7.7 shows the
h_{crit} values of metastable wetting films at different degrees of
hydrophobicity, expressed by the three selected advancing
contact angles of 20°, 58°, and 90°.

The two most important experimental observations on
this system are:

1. The rupture takes place along the theoretical drain-
    age curve for a system where only repulsive DLVO
    forces are present, with no kind of any attractive force.
2. Hence the smallest observed rupture thickness cor-
    responds to the equilibrium thickness, to which the
    film could thin, if no rupture would occur (in our
    described model system at 34 nm).
3. The largest observed critical thickness comes up to
    several hundreds of nanometers and is the higher,
    the larger the degree of hydrophobicity is.
4. The lifetime depends on the degree of hydrophobicity:
    The higher the contact angle, the shorter is the life-
    time.
5. High-speed video frames of hole formation and its
    expansion (Figure 7.8) show that only one formed
    hole is sufficient for destabilization and dewetting of
    the whole film.
6. Another remarkable fact is the rather isometric, cir-
    cular expansion of the newly formed three-phase con-
    tact perimeter.

It is obvious that for the kinetics of drainage the same
law is valid on hydrophilic as well as on hydrophobic surfaces.
If the film would not rupture occasionally it would be able to
thin until it reaches equilibrium thickness according to the
DLVO theory, which in this special case is 34 nm. Because

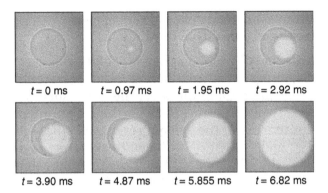

$t = 0$ ms	$t = 0.97$ ms	$t = 1.95$ ms	$t = 2.92$ ms
$t = 3.90$ ms	$t = 4.87$ ms	$t = 5.855$ ms	$t = 6.82$ ms

Figure 7.8 High-speed video sequence of the rupture of an aqueous wetting film on methylated glass (contact angle: 59°). Film diameter: 200 μm. The rupture occurs at 0.97 msec.

the DLVO theory does not assume any further attractive forces, the capillary pressure in the gas bubble is the only driving force for the thinning of the film.

On the assumption that attractive LRHF would exist on hydrophobic surfaces, the drainage process should be accelerated considerably, and the lifetime of the film could not be longer than few seconds depending on the strengths of the attractive force. In Figure 7.7, three different approaches for LRHF were given [37,44,45] and their accelerating influence on the drainage kinetics. It is clearly visible that the actual observed lifetimes are much longer. That demonstrates unambiguously that there are no LRHF present. Therefore, it is evident that in contrast to what is widely believed, not the capillary wave mechanism can be responsible for rupture, but nucleation must be the dominating process. The cause of the rupture is, with high probability, the appearance of tiny gas bubbles in nanometer range, adhering at solid surface. Recently, there are some publications that give clear evidence for the existence of such nano-bubbles, by means of IR-spectroscopy, force measurements, and also by image-giving methods like tapping-mode atomic force microscopy (AFM) (see the review article of Ralston et al. [46]).

3. Electrostatically Destabilized Wetting Films on Recharged Silica (Due to Al^{3+} Ions)

The addition of Al^{3+} ions in the electrolyte changes the sign of the disjoining pressure, acting in our system. Due to adsorption of the aluminium ions, the silica surface becomes charged positively. Figure 7.9 shows the disjoining pressure–distance isotherms in the system recharged silica–electrolyte (KCl: 10^{-3} mol/l, $AlCl^{3}$: 10^{-5} mol/l)–air. Due to the ionic strength of the electrolyte, the range of the electrostatic force can be controlled. In Figure 7.10, the calculated drainage behavior (lines) and rupture thicknesses (dots) for three different electrolyte concentrations are given; it is remarkable that the calculation can produce two solutions: fast rupture and a branch of metastability. The wetting film ruptures because of the attractive surface forces, although the surface remains rather hydrophilic.

The behaviour of wetting films on this recharged glass surfaces differs unambiguous from that on the methylated surface:

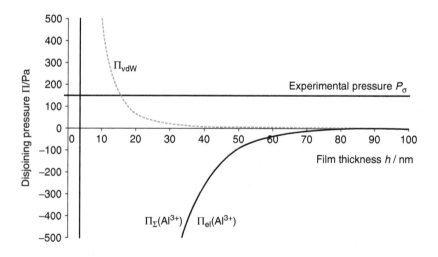

Figure 7.9 Disjoining pressure–distance isotherms of the system recharged silica electrolyte.

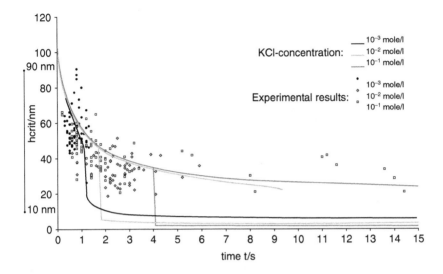

Figure 7.10 Drainage and rupture of a aqueous wetting film of $AlCl_3$ solution on glass at different KCl concentrations. Full lines: calculated curves according Reynolds law; dots: rupture thicknesses.

1. The rupture takes place simultaneously at many places. (Figure 7.11).
2. The critical thickness is never larger than the range of attractive electrostatic double-layer force, i.e., not larger than 90 nm at the given experimental conditions.
3. The holes do not enlarge. The pinning of the newly formed three-phase contact line on the solid surface can be easily visualized by careful receding of the pressed air bubble by means of slow pressure degradation inside the bubble.
4. The distance between the holes is remarkably constant.
5. Although many holes are formed simultaneously, the whole wetting film remains stable for a long time.
6. Such a kind of partially ruptured film could be named as perforated wetting film.

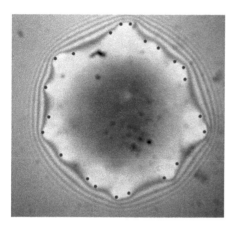

Figure 7.11 Ruptured wetting film of electrolyte solution with 10^{-2} mol/l KCl and 10^{-4} mol/l AlCl$_3$ on opposite charged glass surface. Pinning of the three-phase contact line during removing the bubble on multiple holes (here marked as black dots) indicates the wavelength of the critical fluctuation.

The observed critical thickness values scatter regularly around the calculated drainage curves for different electrolyte concentrations.

But the most amazing observation here is the formation of constant hole distance at the thinnest part of the film. We believe that it is a good evidence for the existence of a dominating fluctuation wave. In this case thermal fluctuations grow as a consequence of attractive electrostatic force, leading to hole formations as soon as the amplitude of the dominating wave reaches the size of the film thickness. If one assumes that every trough of wave leads with large probability to the formation of a hole, the distance between them should be scaled by the wavelength.

According to theory [28,47] this critical wavelength is inversely proportional to the square root of first derivative of the attractive disjoining pressure with respect to the film thickness and is given by:

$$\lambda_c = \frac{2\pi}{\sqrt{(1/\sigma)(\partial \Pi / \partial h)}}. \tag{47}$$

The evaluation of this relationship for the given disjoining pressure leads to theoretical critical wavelength λ_c of approximately 25 μm, which is in the same order as the measured distance of the holes. We are considering this as a good evidence of the existence of the capillary wave mechanism in this system.

Our experiments with thin water films on glass surfaces, which were either homogeneously hydrophobized by gas phase methylation with hexamethyldisilazane, or oppositely charged with aluminium chloride solution, has been detected for the first time in both mechanisms of destabilization of wetting films: nucleation and the capillary wave mechanism. In the case of hydrophobic surfaces the so-called long-range hydrophobic force could not observed. Its existence is even rather questionable not only in the investigated system, but also on other hydrophobic surfaces. If the existence of tiny gas cavities is assumed to be the reason for wetting film destabilization and its rupture, then a further consequence is that the estimated critical thickness should be commensurable to the size of the nanobubbles in a range between 30 nm and few hundred nanometers.

On recharged silica surfaces, a dramatical difference takes place in rupture process: the wetting film ruptures only at thicknesses where the attractive forces are acting. The several holes appearing are regularly ordered in a distance, which is calculated as the critical distance of capillary waves for this system. This leads us to the opinion that the capillary wave mechanism is responsible for rupture in this case.

As a result, we could show that both possible rupture mechanisms described in literature — nucleation and the capillary wave mechanism — can occur in the system silica–water–air, depending on the acting surface forces and the condition (hydrophobicity, heterogeneity etc.) of the surface.

VII. PROBABILITY OF EXTENSION OF THE THREE-PHASE CONTACT

The formation of the primary hole during the rupture of the thin film takes place in the microsecond or even nanosecond range and it is therefore not so important for the kinetics of the flotation process. But the probability of further TPC extension is an important step [48,49, p. 421–445] and depends on many influencing factors, above all the roughness of the particle surface, the chemical heterogeneity, solution effects, evaporation of the liquid along the contact line, and the gradient of surface tension [50]. The driving power for both the enlargement of the hole and the expansion of the three-phase contact is the difference of the interface energies. It is given by the difference between the instant value of the dynamic receding contact angle, depending on time and velocity, and the receding contact angle of equilibrium. Frictional as well as inertia forces decelerate the movement such that it is very fast at the beginning and approaching zero on the end.

Hence, after rupture of the thin liquid films a sufficiently large TPC must be formed in a sufficiently quick time, such that external forces can be received by the p–b aggregate. Therefore, some authors are looking at TPC expansion as the most important step of particle adhesion [51].

The external stress forces on aggregates are caused by turbulent vortices within the flotation machine. The lifetime of such vortices is in the millisecond range. Therefore, both sufficiently large TPC radius and contact angle must be formed within a time interval to ensure that a sufficiently strong force of attachment prevents the aggregate from getting detached. Consequent to this is also a time condition for stable attachment, as for particle adhesion, namely, $\tau_{tpc} > \tau_v$ lifetime of laminar vortices (dissipation region) ([52, p. 71, Equation (3.68)] and [53, p. 68, Equation (3.9), p. 113, Equations (1.11) and (1.12)]) is in the order of:

$$\tau_{vl} \approx 0.6 r_v^2 / \nu, \tag{48}$$

where $r = R_p + R_B$, and ν is the kinematic viscosity.

The lifetime of no laminar vortices in the inertial region is

$$\tau_{vnl} \approx 13 r_v^{2/3} / \varepsilon^{1/3}, \tag{49}$$

with the local energy dissipation ε.

The velocity of the expansion of a circular hole of the radius r after the rupture of the thin liquid film is represented in a general form similar to the expansion of holes in free-foam films [54]:

$$\frac{dr}{dt} = a_{tpc}\sigma(\cos\theta_{Rdyn} - \cos\theta_R). \tag{50}$$

Here a_{tpc} is the instantaneous mobility coefficient of the TPC. Experiments carried out by Tschaljowska [55] showed that it depends in a very complicated manner on surface properties, on external and friction forces, which act on the TPC.

Another problem for the precalculation of the TPC expansion time is the very strong dependence of a_{tpc} on the radius of the TPC, i.e., on the time itself because during movement of the TPC line a transfer of surfactant from the liquid–gas to the solid–liquid interface and vice versa is possible. Related to that, there are changes in both the interface energy and surface tension of the liquid in the region of moving liquid meniscus, which depend on the diffusion rate of the surfactant molecules. The measurable phenomenon of these complicated processes is the change of contact angle depending on the velocity of the TPC line. In contrast to expectation, for example, the advancing contact angle decreases with increasing TPC velocity [56]. One possible explanation should be the following: during TPC movement one part of the adsorption layer from the solid–gas interface is transferred to the liquid–gas interface of the advancing liquid meniscus. But because of the great velocity of movement no diffusion interchange with the bulk phase can occur there. This results in a decreasing dynamic surface tension in the region of meniscus, which must lead — according to the equation of Young and Dupre — to an increasing $\cos\theta$ and consequently to a decreasing θ.

One example for the strong influence of the surface heterogeneity on the TPC expansion has been observed by Mahnke et al. [57]. They have been able to demonstrate by high-speed video technique that already defects on a molecular length scale in the depth of an adsorption layer influencing the TPC expansion dramatically. In order to show that, a regular stripe patterns has been formed in skeletonized LB films of two monolayers of arachidic acid deposited on silica wafers from Cd^{2+} containing aqueous subphase after monolayer transfer at pH 5.7. Atomic force microscopy (AFM) and phase shift interference microscopy (PSIM) studies have shown that these stripes are grooves of about 6 nm depth aligned in definite distances along the meniscus and perpendicular to the dipping direction of the monolayer transfer in a regular distance. Figure 7.12 shows a sequence of images during the rupture process on such skeletonized LB system. It can be clearly seen that the rupture is strongly accelerated at once when the TPC reaches a defect stripe. The TPC expansion velocity is usually enlarged by a factor of 10–15 (111 µm/msec). In comparison the TPC expansion velocity on a completely homogeneous methylated silica surface is in the order of 60 µm/msec. It should be mentioned here that unlike other cases of heterogeneous LB-layers, the critical thickness of the wetting films at which rupture commences, is not affected by the skeletonization. So far however, it is not clear whether the presence of the strips initiates the formation of a first hole in the film. The reason for the influence of such only molecular depth structures on the TPC expansion velocity is completely unknown as jet and therefore should be investigated in more detail in future.

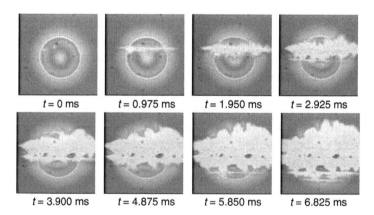

$t = 0$ ms $t = 0.975$ ms $t = 1.950$ ms $t = 2.925$ ms

$t = 3.900$ ms $t = 4.875$ ms $t = 5.850$ ms $t = 6.825$ ms

Figure 7.12 TPC expansion after hole formation at $t = 0$ msec in a thin-liquid wetting film on a skelettonized Langmuir–Blodgett arachidic acid bilayer.

When we assume that the TPC expansion probability is again exponentially distributed, than it may be written

$$P_{\mathrm{tpc}} = 1 - \exp(-(\text{TPC expansion time})/(\text{vortex lifetime})).$$

Experimentally determined times of TPC expansion on spherical glass particles are collected in Stechemesser and Nguyen [48]. One example and a schema of the experimental conditions are given in Figure 7.13 and Figure 7.14.

VIII. PROBABILITY OF AGGREGATE STABILITY

The probability of aggregate stability, P_{stab}, depends on the attachment force between the bubble and the particle in relation to the external stress forces in the flotation cell. If the attachment force is greater than the sum of all stress forces, then this aggregate remains stable on its long way from the place it is formed to the foam layer.

The force balance for spherical particles in the liquid–gas interface can be described in a physically clear manner:

$$F_{\mathrm{ca}} + F_{\mathrm{hyd}} + F_{\mathrm{b}} - F_{\mathrm{g}} - F_{\mathrm{d}} - F_{\sigma} = 0. \tag{51}$$

- *The force of gravity:*

$$F_{\mathrm{g}} = \frac{4}{3} \pi R_{\mathrm{p}}^{2} \rho_{\mathrm{p}} g. \tag{52}$$

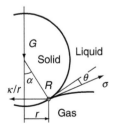

Figure 7.13 The gas–solid–liquid TPC between a solid sphere and an air–liquid interface in an experimental set-up for determination of TPC expansion according to Stechemesser et al. [87].

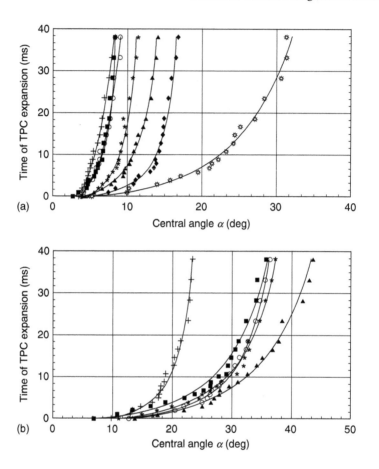

Figure 7.14 Time of TPC expansion on silanated glass spheres. Points: experimental values, curves: theoretical predictions according to Stechemesser and Nguyen [48] plus sign: $R_p = 151\ \mu m$; $\theta = 30°$; filled square: 163 μm, 34°; circle: 161 μm, 28°; star: 150 μm, 32°; filled triangle: 159 μm, 54°; diamond: 172 μm, 31°; sun: 138 μm.

- *The static buoyancy force of the immersed part:*

$$F_b = \frac{\pi}{3} R_p^3 \rho_{Fl} g [(1 - \cos \omega)^2 (2 + \cos \omega)],\qquad(53)$$

for the whole sphere approximately:

$$F_{b0} \approx \frac{4}{3} \pi R_p^3 \rho_{Fl} g.\qquad(54)$$

- *The hydrostatic pressure of the liquid of height z_0 above of the TPC area with the radius r_0:*

$$F_{\text{hyd}} = \pi r_0^2 \rho_{\text{Fl}} g z_0 = \pi R_p^2 (\sin^2 \omega) \rho_{\text{Fl}} g z_0. \tag{55}$$

- *The capillary force on the TPC with the radius r_0:*

$$F_{\text{ca}} = -2\pi \sigma R_p \sin \omega \sin (\omega + \theta). \tag{56}$$

- *Additional detaching forces which are represented globally as the product of the particle mass and the acceleration* a *in the external field of flow:*

$$F_d = \frac{4}{3} \pi R_p^3 \rho_p a. \tag{57}$$

- *The capillary pressure in the gas bubble which act on the contact area of the attached particle:*

$$F_\sigma = \pi r_0^2 P_{\text{bubble}} \approx \pi R_p^2 (\sin^2 \omega) \left(\frac{2\sigma}{R_B} - 2R_B \rho_{\text{Fl}} g \right). \tag{58}$$

A complete calculation of the force balance is only possible by numerical integration of the Laplace equation of capillary menisci, which give the meniscus deformation z_0 as a function of particle size. Actually this analysis has shown [34, p. 182] that under real flotation conditions, i.e., with contact angles $\theta < 90°$ and particle sizes smaller than 300 μm, the hydrostatic term is negligible and the maximum attachment force occurs, if the centre-angle between the rear part of the attached sphere and the TPC projection area on the sphere is

$$\omega = \omega^* = \pi - \frac{\theta}{2}. \tag{59}$$

The ratio of the attachment to the detachment forces characterizes the aggregate stability. This ratio is a dimensionless similarity parameter analogous to the Bond number

$$Bo' = \frac{F_{\text{detach}}}{F_{\text{adhesion}}} = \frac{F_g - F_b + F_d + F_\sigma}{F_{\text{ca}} + F_{\text{hyd}}}. \tag{60}$$

Therefore, the following equation is valid:

$$Bo' = \frac{4R_p^2(\Delta \rho g + \rho_p a) + 3R_p(\sin^2 \omega^*)((2\sigma/R_B) - 2R_B \rho_{\text{Fl}} g)}{|6\sigma \sin \omega^* \sin (\omega^* + \theta)|}. \tag{61}$$

The *additional acceleration* a, which determines the detachment forces, here depends on the structure and the intensity of the turbulent flow field; hence, finally on the local energy dissipation ε in a given volume of the apparatus. We assume, according to Schubert et al. [52, p. 65] and Albring [53], that aggregates of the dimension which corresponds to those of the turbulent vortices in the inertial region are moved mainly by the centrifugal acceleration $a = a_c$ present in the vortex. Moreover, this quantity is experimentally easily measurable by means of a centrifuge method [58]. If r_v is the radius of such a vortex then in the inertial region is

$$a_{cin} = 1.9 \frac{\varepsilon^{2/3}}{r_v^{1/3}}. \tag{63}$$

For aggregates, where the particle size is smaller than the bubble size, the vortex radius should be set equal to the aggregate radius:

$$a_{cin} \approx 1.9\varepsilon^{2/3}/(R_B + R_p)^{1/3}. \tag{64}$$

Much smaller aggregates obey a stress in the dissipation region of a vortex flow and it is valid:

$$a_{cdis} \approx 0.52\varepsilon^{3/4}/\nu^{1/4}. \tag{65}$$

In flotation machines the mean energy dissipation is in the range from 10^{-3} to 10^{-1} kW/kg or 10^4 to 10^6 cm^2/s^3, respectively. Accordingly, the acceleration involved is of the order of 2–200 g units.

For a bubble flowing laminar, i.e., for conditions realized in a column flotation for example, a_c depends on a first approximation only on the tangential velocity of particle motion across the bubble surface.

$$a_c = \frac{(u_\Phi + v_{Ps\Phi})^2}{R_B + R_P}. \tag{66}$$

On a mobile in a potential flow, we obtain the following approximation using Equation (4):

$$u_\Phi \approx \frac{3}{2} v_B \sin \Phi. \tag{67}$$

Thus it is

$$a_c \approx \frac{[((3/2)v_B + v_{Ps}) \sin \Phi]^2}{R_B + R_p}. \tag{68}$$

The maximum of acceleration at $\phi = 90°$, which is only decisive for the detachment force, is therefore:

$$a_{c\,max} \approx \frac{9(v_B + v_{Ps})^2}{4(R_B + R_p)}. \tag{69}$$

We assume again that P_{stab} is exponentially distributed

$$P_{stab} = 1 - \exp\left(-\frac{F_{ca\,max} - F_{det}}{F_{det}}\right) = 1 - \exp\left(1 - \frac{1}{Bo'}\right), \tag{70}$$

where the maximum capillary force is after a little algebra

$$F_{ca\,max} = \pi \sigma R_p (1 - \cos \theta). \tag{71}$$

One example for P_{stab} as function of particle size, energy dissipation, and contact angle is given in Figure 7.15. Bloom and Heindel [21] used the following equation for describing the aggregate stability:

$$P_{stab} = 1 - \exp\left[A_s\left(1 - \frac{1}{Bo'}\right)\right], \tag{72}$$

where an additional stability constant A_s was introduced to match the experimental results with flat disk-shaped toner particles. We assume that this parameter reflects the particle shape because the theory mentioned above has been developed for spherical particles only. Other improved equations for calculating the probability of stability will be given in Nguyen and Schulze [11, part 5].

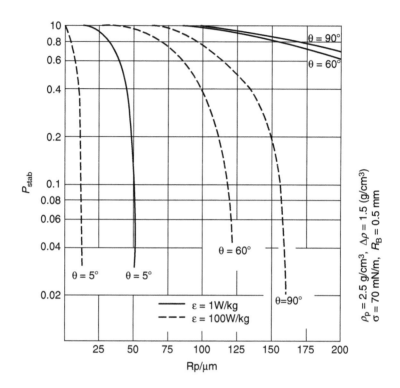

Figure 7.15 Probability of aggregate stability as function of contact angle θ, energy dissipation ε and particle size R_p.

IX.   MODELLING OF A SEMIBATCH PROCESS ON THE BASIS OF THE MICROPROCESS PROBABILITIES

A considerable effort has been made to mathematically model the flotation separation process because experimental studies are very costly and time consuming [59]. The complexity of the flotation process, however, has prevented the development of a flotation model based on first principles. Bloom and Heindel have developed a model for a semibatch deinking flotation process based on available microprocess probabilities, and compared them to experimental data obtained using a WEMCO laboratory flotation cell during the last decade. In general, the model predicts the correct experimental trends. In some cases, the model also predicts removal efficiency very well. Parametric studies reveal that the model predictions are sensitive to a stability parameter, the turbulent energy density in the flotation cell, the contact angle between the solid particle and fluid, and the ratio of the initial to the critical film thickness.

In the following expositions on that subject we are undertaking the work of Bloom and Heindel. Experimental observations of the flotation process have revealed that it can be likened to a chemical reactor [60] and be described by an ordinary differential equation of the form [33]:

$$\frac{dn_p^f(t)}{dt} = -k' \left(n_B^f(t)\right)^m \left(n_p^f(t)\right)^n, \tag{73}$$

where $n_B^f(t)$ and $n_p^f(t)$ are the concentrations of free bubbles and particles, respectively, t is the flotation time, m and n are the respective orders of reaction, and k' is a pseudo rate constant. Assuming that the reaction is first order [61–65], that the bubble concentration is constant, that the removed particles represent a small volume [33,66], and that the pseudo-rate constant can be expressed as a product of various microprocess probabilities [34,67–79], Equation (73) can be rewritten as

$$\frac{dn_p^f(t)}{dt} = -kn_p^f(t), \tag{74}$$

where $n_p^f(t)$ is the free particle concentration in the flotation cell. The rate constant, k, has the form

$$k = ZP_cP_{asl}P_{tpc}P_{stab}n_B^f(t), \tag{75}$$

where Z is related to the bubble–particle collision frequency (Equations (94)–(96)) and $n_B^f(t)$ is the concentration of bubbles without particles attached to them. The necessary probabilities of the microprocess can be calculated according Equations (17), (38), and (72). P_{tpc} is assumed to be equal to 1. Additional information can be found in the literature [80–82].

Equation (74) represents a kinetic- or population bal-
ance-type model of the evolution of free particles. Bloom and
Heindel [72,73] extended the idea of a population balance
model to include a forward and a reverse reaction (i.e., the
birth and death of free particles). This model has the form

$$\frac{dn_p^f(t)}{dt} = -k_1 n_p^f(t) + k_2 n_B^a(t), \qquad (76)$$

where $n_B^a(t)$ represents the concentration of bubbles with attached particles. The first term on the right-hand side of
Equation (76) represents the formation of bubble–particle aggregates, while the second term monitors the dissolution of
those bubble–particle aggregates which become unstable before they reach the froth layer and split to yield *new* free
particles. The kinetic rate constants, k_1 and k_2, are positive
numbers with k_1 as in Equation (75) and

$$k_2 = Z^* P_{\text{destab}} = Z^*(1 - P_{\text{stab}}), \qquad (77)$$

where P_{destab} is the probability a bubble–particle aggregate will
become unstable (Equation (72)) and Z^* is the detach-
ment frequency of particles from bubbles according to
Equation (97).

A. Model Development

The experimental apparatus, to which the model described
below applies, consists of a vessel into which a desired
liquid–particle mixture is loaded and air is injected. Air at a
known constant volumetric flow rate Q is bubbled through the
liquid–particle mixture. It is assumed that the air bubbles
formed at the air injector move through the suspension at
their terminal rise velocity v_B. Bubble coalescence is assumed
to be negligible and the entire bubble surface is assumed to be
rigid because of the presence of surface-active substances. In
the assumed semibatch process, a fixed volume of the liquid–
particle mixture is maintained while air is continuously bubbled through the suspension. The total mixture volume (including liquid and gas) in this steady-state operation is
denoted by V. The agitator provides the necessary agitation
for bubble–particle mixing.

Assuming that all bubbles have a uniform size, the average number of bubbles in the vessel can be calculated by measuring the gas hold-up (gas percent by volume) and the average bubble size. The total number of bubbles consists of bubbles free of particles and those with particles attached to them. Both of them are functions of time, for a constant airflow rate the total bubble concentration (n_B) is constant at any given time with

$$n_B = n_B^f(t) + n_B^a(t). \tag{78}$$

If V_B is the average volume of a single bubble, then the total number of air bubbles entering the vessel per unit time is

$$\dot{N}_B = Q/V_B. \tag{79}$$

Assuming negligible bubble coalescence and ignoring the influence on the individual bubble-rising velocity, which can be modified by the presence of one or more attached under steady-state conditions, the total number of bubbles entering the vessel is identical with the total number of bubbles leaving the vessel. Let $n_B^a(t)/n_B$ denote the fraction of bubbles inside the vessel with attached particles; as the contents of the vessel are assumed to be well mixed, this same fraction also represents the fraction of bubbles with attached particles in the exit stream. Therefore, the number of particles leaving the vessel per unit time may be calculated as the product of this fraction, the total number of bubbles leaving the vessel per unit time, and the (time-averaged) average number of particles attached to a bubble. Hence, the rate of decrease of particles inside the vessel is given by

$$V\frac{dn_p(t)}{dt} = -\frac{Q}{V_B n_B} n_B^a(t) n_p^e, \tag{80}$$

where n_p^e represents the (time-averaged) average number of particles, which are attached to each bubble in the exit stream (i.e., n_p^e is defined as the total number of particles on each bubble in the exit stream divided by the total number of bubbles that have attached particles in the exit stream).

Since $n_p^e n_B^a (t) = n_p^a (t) = n_p (t) - n_p^f(t)$, where $n_p^a (t)$ is the number of particles attached to bubbles, $n_p (t)$ is the total particle concentration, and $n_p^f (t)$ is the number of particles free of bubbles, Equation (80) can be rewritten as

$$\frac{dn_p(t)}{dt} + \Lambda n_p(t) = \Lambda n_p^f(t) \tag{81}$$

with

$$\Lambda = \left(\frac{Q}{V}\right) \frac{1}{V_B n_B} = \frac{Q}{\varepsilon_g V}, \tag{82}$$

where $\varepsilon_g = V_B n_B$ is the gas hold-up in the flotation cell. Note that Λ can be determined based on known experimental conditions.

In Equation (81), both $n_p (t)$ and $n_p^f (t)$ are unknown functions of time. To determine the particle removal efficiency

$$\text{Eff}(t) = 1 - \frac{n_p(t)}{n_{p0}}, \tag{83}$$

where n_{p0} is the initial (free) particle concentration, and a second model equation is required to determine $n_p(t)$.

The original flotation model developed by Bloom and Heindel [72,73] and described by Equation (76) assumes that only bubbles which do not already have a particle attached to them are capable of picking up a particle and transporting it to the froth layer (i.e., the average number of particles on a bubble leaving the flotation cell with particles attached to it was identically equal to 1).

The new model removes this assumption be replacing $n_B^f (t)$ in Equation (75) by $n_B^A (t)$, where $n_B^A (t) > n_B^f(t)$ represents the concentration of bubbles, which are available to pick up particles. Therefore, to allow for the capture of more than one particle by a bubble during the flotation process, Equation (76) must be modified to

$$\frac{dn_p^f(t)}{dt} = -\bar{k}_1 n_p^f(t) + k_2 n_B^a(t), \tag{84}$$

where k_2 is still given by Equation (77), but k_1 is now given by

$$k_1 = ZP_c P_{asl} P_{tpc} P_{stab} n_B^A(t) = \bar{k}_1 n_B^A(t). \tag{85}$$

Equation (84) provides the second governing equation needed to determine $n_p(t)$. The following relations are obvious from the definitions

$$n_B^f(t) \leq n_B^A(t) \leq n_B, \tag{86}$$

$$n_B^a(t) \leq n_p^a(t) = n_p(t) - n_p^f(t), \tag{87}$$

$$-n_B^A(t) \leq -n_B^f(t). \tag{88}$$

Using these relations and Equation (84), a differential inequality can be developed which has the form

$$\frac{dn_p^f(t)}{dt} \leq -\bar{k}_1 n_p^f(t) n_B + \left(\frac{\bar{k}_1}{n_p^e}\right) n_p^f(t) \left[n_p(t) - n_p^f(t)\right]$$

$$+ k_2 \left[n_p(t) - n_p^f(t)\right]. \tag{89}$$

With the initial conditions

$$n_p(0) = n_p^f(0) = n_{p0}, \tag{90}$$

the mathematical procedure gives the global semibatch flotation model process, which has the form

$$\text{Eff}(t) \geq \left[1 - \left(\frac{B}{B-A}\right) \exp^{-At}\right] - \frac{1}{n_p^e}\left(\frac{n_{p0}}{n_B}\right)$$

$$\left[\frac{k_1' B(2\Lambda - A)}{(B-A)(k_1' + k_2 + \Lambda - 2A)}\right]\left(1 - \frac{1}{B-A}\right) \exp^{-At}, \tag{91}$$

where n_{p0} is the initial concentration of (free) particles in the vessel, n_B is the (constant) bubble concentration in the vessel, n_p^e is the (time-averaged) measure of the average number of particles attached to bubbles which have attached particles, Λ is given by Equation (91), k_2 is given by Equation

(77), and $k_1' = kn_B$. The two additional parameters, a and b, are defined by

$$A = \frac{\Lambda k_1'}{k_1' + k_2 + \Lambda},$$ (92)

$$B = k_1' + k_2 + \Lambda - A.$$ (93)

To obtain removal efficiency predictions using Equation (82), the microprocess probabilities must be evaluated. The collision frequency Z and the bubble–particle detachment frequency Z^* have the following forms: according to Schubert et al. [52]

$$Z = 5(R_p + R_B)^2 \sqrt{u_p^2 + u_B^2} \exp\left[-\frac{1}{2}\frac{(|v_{ps}| + v_B)^2}{u_p^2 + u_B^2}\right]$$

$$+ \pi(R_p + R_B)^2 \left\{ \frac{(|v_{ps}| + v_B)^2 + u_p^2 + u_B^2}{(|v_{ps}| + v_B)} \mathrm{erf} \right.$$ (94)

$$\left. \left[\frac{(|v_{ps}| + v_B)}{\sqrt{2(u_p^2 + u_B^2)}}\right] \right\},$$

$$u_p = 0.4 \frac{\varepsilon^{4/9} d_p^{7/9}}{\nu^{1/3}} \left(\frac{\rho_p - \rho_{Fl}}{\rho_{Fl}}\right)^{2/3},$$ (95)

$$u_B = 0.4 \frac{\varepsilon^{4/9} d_B^{7/9}}{\nu^{1/3}} \left(\frac{|\rho_B - \rho_{Fl}|}{\rho_{Fl}}\right)^{2/3},$$ (96)

$$Z^* = \sqrt{2}\varepsilon^{1/3}/(d_p + d_B)^{2/3}.$$ (97)

B. Experimental Methods

Batch bench-top flotation deinking experiments were completed for the purpose of obtaining experimental particle removal efficiencies to be used for the flotation model validation [83–86]. The stock used in these flotation trials consisted of fused toner particles that were added to unprinted copy paper and than repulped. The pH of the system was neutral. A frother (Triton X-100, henceforth referred to as TX-100) was added to the fibre suspension at a level of 20 mg/(g o.d. fiber) to create stable foam during flotation. TX-100 is an octyl ethoxylate (9.5 moles EO) nonionic *generic* surfactant. It acts as a moderate foaming agent over a wide range of pH conditions. No additional chemicals were added to the stock.

The experimental flotation deinking efficiency is defined as Eff(t) = (Feed Concentration – Accept Concentration)/Feed Concentration, where the feed and accept concentration for each particle size range are determined from the average values from the three flotation trials.

Power input into the WEMCO flotation cell was also measured. It was assumed that all the applied power could be equated to the turbulent energy density, and, therefore, ε was 6.7 W/kg. By recording the rise in the fluid level when the air was turned on, the gas hold-up was estimated to be $\varepsilon_g = 0.1$. From this value and the estimated bubble diameter, a constant bubble concentration was determined to be $n_B = 6.28 \times 10^7$ bubbles/(m^3 of 1% slurry). The feed particle size concentration was used to estimate the initial number of particles in the given size range (e.g., $n_{P0} = 1.41 \times 10^7$ particles/(m^3 of 1% slurry) for the $d_p = 125$–150 size range).

C. Model Validation

The data obtained in the semibatch flotation experiments have been compared to model predictions. Table 7.1 summarizes

Table 7.1 Parameter values used in the initial model calculations for $d_p = 125$–150 μm

$R_p = 75 \, \mu$m	$h_0/h_{crit} = 4$
$\rho_p = 1.2 \, \text{g/cm}^3$	$C_B = 1$
$R_B = 0.75 \, \text{mm}$	$\theta = 60°$
$\rho_l = 1 \, \text{g/cm}^3$	$\varepsilon_g = 0.10$
$\mu_l = 1 \, \text{cP}$	$V = 3 \, l$
$\nu_B = 12 \, \text{cm/sec}$	$Q = 3 \, \text{Lpm}$
$\sigma = 35 \, \text{dynes/cm}$	$A_s = 0.5$
$\varepsilon = 6.7 \, \text{W/kg}$	$n^c_p = 1$
$n_{p0} = 1.41 \times 10^7$ particles/(m^3 of 1% stock)	$n_B = 6.28 \times 10^7$ bubbles/(m^3 of 1% stock)

all the parameter values used in the initial calculations. The ratio of initial-to-critical film thickness, h_0/h_{crit}, must be known to determine P_{asl}, which, in turn, influences k_1. h_0 is a function of particle diameter, fluid viscosity, particle settling velocity, surface tension, and surface mobility, and this function depends on the particular system of interest. Rulev and Dukhin [20] concluded that both h_0 and h_{crit} are functions of the surface tension and collision process. They determined that for quasielastic collisions, $(St > 1)$, $h_0/h_{crit} \sim 3$, and for inelastic collisions, $(0.1 < St \leq 1)$, $h_0/h_{crit} \sim 4$. Therefore, although the specific values of h_0 and h_{crit} are system dependent, the ratio h_0/h_{crit} is typically on the order of 3 to 4 for mineral flotation systems. Since mineral flotation and flotation deinking are assumed to follow similar microprocesses, assuming that h_0/h_{crit} for flotation-deinking systems is also within this range [75].

D. Model Parametric Variations

The data obtained in semibatch experiments have been compared to the model predictions. In these calculations all parameters, except the one variable parameter, are assumed to be fixed at the values given in Table 7.1. Additionally, the varying parameter is assumed to be independent of the other parameters in Table 7.1. Some examples for the main influencing parameters are given in the following.

Figure 7.16 displays the effect on the predicted removal efficiency in which efficiency increases as the *surface tension* increases. The surface tension of the stock filtrate was measured during the flotation trials and was shown to increase as flotation time increased. This is due to some of the surfactant removed with the foam during flotation. An estimated value of $\gamma = 35$ mN/m was used in the calculations. However, the effect over the range of experimental surface tension measurements is not too significant.

Variations in the *turbulent energy density*, ε, will influence the predicted removal efficiency through the collision rates as well as the probability of stability. Figure 7.17 shows that as the turbulent energy density increases, the

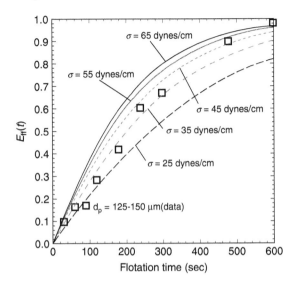

Figure 7.16 Effect of surface tension on the predicted removal efficiency. The remaining parameters are fixed at those given in Table 7.1, according to Bloom and Heindel [21].

predicted removal efficiency decreases. This should be expected because turbulence can destabilize bubble–particle aggregates.

The *contact angle* will influence the predicted flotation efficiency through P_{stab}. For hydrophobic ink particles, found in flotation deinking applications, θ will typically be large. However, chemicals added to the system during paper recycling could depress this value. Figure 7.18 displays the effect of contact angle on predicted flotation efficiency. High contact angles yield better removal rates, and low contact angles produce extremely poor removal rates.

The ratio of *initial-to-critical film thickness* will affect the predicted flotation efficiency through the probability of adhesion. Increasing h_0/h_{crit} (Figure 7.19) reduces the flotation efficiency. But this research area requires in future a detailed study to fully understand the influence of h_0/h_{crit} on the particle removal efficiencies.

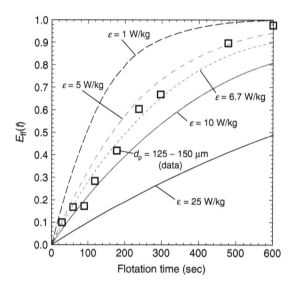

Figure 7.17 Effect of turbulent energy density on the predicted removal efficiency. The remaining parameters are fixed at those given in Table 7.1, according to Bloom and Heindel [21].

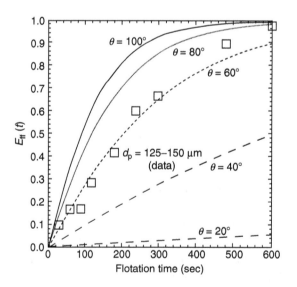

Figure 7.18 Effect of contact angle on the predicted removal efficiency. The remaining parameters are fixed at those given in Table 7.1, according to Bloom and Heindel [21].

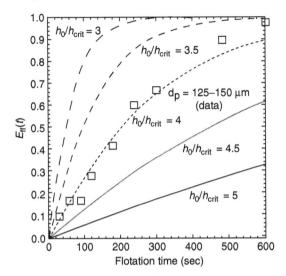

Figure 7.19 Effect of the relation between initial and critical-film thickness on the predicted removal efficiency. The remaining parameters are fixed at those given in Table 7.1, according to Bloom and Heindel [21].

In conclusion one can estimate that the computer simulation underlines at least the determining interplay of physicochemical and hydrodynamic process engineering parameters for successful flotation as a heterocoagulation process, This interplay is described by three essential parameters, the *contact angle*, the *critical thickness* of film rupture, and the *energy dissipation* into the flotation apparatus.

X. SUMMARY

Flotation is a heterocoagulation process. While interparticle interaction forces are often sufficiently strong for very small particles ($d < 20\,\mu m$) to form stable bubble–particle aggregates (contactless flotation), big particles can only be kept attached to the bubble by capillary forces at the three-phase contact. A prerequisite for that is the destabilization of the thin liquid film between particle and bubble, associated with formation of a TPC. To precalculate kinetics of p–b coagulation, the probability of p–b collision, the probability of adhesion, the probability of expansion of TPC, and the probability of stability of aggregates must be known. For that, available methods of calculation have been summarized in this chapter.

An important elementary step of the flotation process is the forming, drainage, and rupture of a thin intervening liquid film between the bubble and the mineral particle. After encounter of the bubble with the solid surface a wetting film is formed. The initial thinning of this film is driven by the capillary pressure inside the bubble. When the film has reached a certain thickness, rupture can take place. There are two possible mechanisms of this process:

(a) Growing fluctuation waves (*spinodal dewetting*) on fluid interfaces under the influence of any kind of attractive forces (electrostatic, van der Waals).

(b) Nucleation inside the film.

Metastable wetting films on negatively charged, hydrophobic glass surfaces (gaseous phase methylated), or on hydrophilic, positively charged glass surfaces (with Al^{3+} ions), have been analysed by means of the kinetics of film thinning according to the Reynolds law. These experiments demonstrate that both mechanisms can be really responsible for the rupture of a thin wetting film. In the case of hydrophobic surfaces the nucleation mechanism is preferred; in the case of opposite charged silica surface the capillary waves mechanism due to an attractive electrostatic double-layer force between silica and air bubble takes place. Hereby, the existence of a 'long-range hydrophobic force' on a hydrophobic surface can be excluded. The apparent interaction can be explained by the presence of gas nuclei formed on heterogeneous surface sites.

A model for a semibatch flotation separation process has been developed recently in the literature, based on available microprocess probabilities, and compared to experimental data obtained using a WEMCO laboratory flotation cell, is described. In general, the model predicts the correct experimental trends. In many cases, the model also predicts removal efficiency very well. Parametric studies reveal that the model predictions are sensitive to the turbulent energy density in the flotation cell, the contact angle between the solid particle and fluid, and the ratio of initial-to-critical film thickness, also to the stability of the intervening liquid film.

REFERENCES

1. J.N. Israelachvili, *Intermolecular and Surface Forces*, Academic Press, New York, 1992.
2. H. Sonntag and K. Strenge, *Coagulation Kinetics and Structure Formation*, Deutscher Verlag der Wissenschaften, Berlin, 1987.
3. G. Lagaly, O. Schulz, and R. Zimehl, *Dispersionen und Emulsionen*, Steinkopff, Darmstadt, 1997.
4. B.V. Derjaguin, S.S. Duchin, and N.N. Rulev, *Mikroflotacija*, Chimija Moskva, Moskva, 1986.
5. H.J. Schulze, *Adv. Colloid Interf. Sci.*, 40, 283, 1992.
6. R. Clift, J.R. Grace, and M.E. Weber, *Bubble, Drops and Particles*, Academic Press, New York, 1978.
7. S.S. Duchin, N.N. Rulev, and A.S. Dimitrov, *Koaguljacija i dinamika tonkich plenok*, Naukova dumka, Kiev, 1986.
8. V.G. Levich, *Physicochemical Hydrodynamics*, Prentice-Hall, Englewood Cliffs, NJ, 1962, p. 395.
9. U. Haas, H. Schmidt-Traub, and H. Brauer, *Chemie Ingen Technik*, 44, 1060, 1972.
10. R.H. Yoon and G.H. Luttrell, *Miner. Process. Extractive Metal. Rev.*, 5, 101, 1989.
11. A.V. Nguyen and H.J. Schulze, *Colloidal Science of Flotation*, Marcel Dekker, New York, 2003.
12. H. Plate, *Modellierung der Flotation auf der Grundlage der Mikroprozesse*, Forschungsbericht FIA der Akademie der Wissenschaften der DDR, Freiberg, 1989.
13. G. Schuch and Fr. Löffler, *Verfahrenstechnik*, 12, 302, 1978.
14. S.S. Duchin and H.J. Schulze, *Kolloidnyj Sh.*, 49, 644, 1987.
15. K.L. Sutherland, *J. Phys. Chem.*, 52, 394, 1948.
16. G.S. Dobby and J.A. Finch, *Int. J. Mineral Process.*, 21, 241, 1987.
17. P.G. Saffman, *J. Fluid. Mech.*, 22, 385, 1965.
18. R.C. Jeffrey and J.R.A. Pearson, *J. Fluid Mech.*, 22, 721, 1965.
19. A.J. Goldmann, R.G. Cox, and H. Brenner, *Chem. Eng. Sci.*, 22, 63, 1967.

20. N.N. Rulev and S.S. Duchin, *Kolloidnyi Zhurnal*, 48, 302, 1986.
21. F. Bloom and T.J. Heindel, *Chem. Eng. Sci.*, 58, 353, 2003.
22. Th. Geidel, Aufbereitungstechnik, Wiesbaden, 26, 287, 1985.
23. W. Rasemann, *Int. J. Miner. Process.*, 24, 247, 1988.
24. J. Sven-Nilsson, *Kolloidzeitschrift*, 69, 230, 1934.
25. H.J. Schulze, B. Radoev, Th. Geidel, H. Stechemesser, and E. Töpfer, *Int. J. Mineral Process.*, 27, 263, 1989.
26. H.J. Schulze, in *Frothing in Flotation*, J. Laskowski, Ed., Gordon and Breach, New York, 1989, 43.
27. H.J. Schulze and J.O. Birzer, *Colloids Surf.*, 24, 209, 1987.
28. A. Scheludko. *Adv. Colloid Interf. Sci.*, 1, 391, 1967.
29. A. Vrij, *Disc. Faraday Soc.*, 42, 23, 1966.
30. J. Bischof, D. Scherer, St. Herminghaus, and P. Leiderer, *Phys. Rev. Lett.*, 77, 1536, 1996.
31. B.V. Derjaguin and J.U.V. Gutop, *Doklad Akad Nauk SSSR*, Moscow, 153, 859, 1963.
32. K. Jacobs and S. Herminghaus, *Langmuir*, 14, 965, 1989.
33. N. Ahmed and G.J. Jameson, *Miner. Process. Extractive. Metall. Rev.*, 5, 77, 1989.
34. H.J. Schulze, *Physico-Chemical Elementary Processes in Flotation*, Elsevier, Amsterdam, 1984.
35. J.A. Laskowski and J. Kitchener, *J. Colloid Interf. Sci.*, 29, 670, 1969.
36. B.V. Derjaguin, N.V. Churaev, and V.M. Muller, *Surface Forces* (in Russian), Isdatelstvo Nauka, Moskva, 1985.
37. R.H. Yoon, *Int. J. Miner. Process.*, 58, 129, 2000.
38. R.J. Pugh and M.W. Rutland, *Prog. Colloid Polym. Sci.*, 98, 284,1995.
39. H.J. Schulze, *Colloid Polym. Sci.*, 253, 730, 1975.
40. J. Mahnke, H.J. Schulze, K.W. Stöckelhuber, and B. Radoev, *Colloids Surf. A*, 157, 1, 1999.
41. J. Sonnefeld, *J. Colloid Polym. Sci.*, 273, 926, 1995.
42. G. Kar, S. Chander, and T.S. Mika, *J. Colloid Interf. Sci.*, 44, 347, 1973.
43. F. Grieser, R.N. Lamb, G.R. Wiese, D.E. Yates, R. Cooper, and T.W. Healy, *Radiat. Phys. Chem.*, 23, 43, 1984.
44. Y.H. Tsao, D.F. Evans, and H. Wennerström, *Langmuir*, 9, 779, 1993.
45. N.V. Churaev, *Adv. Colloid Interf. Sci.*, 58, 87, 1995.
46. J. Ralston, D. Fornasiero, and N. Mishchuk, *Colloids Surf. A*, 192, 39, 2001.

47. D.A. Edwards, H. Brenner, and D.T. Wasan, *Interfacial Transport Processes and Rheology*, Butterworth, Boston, 1991.
48. H, Stechemesser and A.V. Nguyen, *Int. J. Miner. Process.*, 56, 117, 1999.
49. J.S. Slattery, *Interfacial Transport Phenomena*, Springer, Berlin, 1990.
50. P.G. de Gennes, *Rev. Modern Phys.*, 573, 827, 1985.
51. R. Verbanov, *Colloid Polym. Sci.*, 263, 75, 1985.
52. H. Schubert, E. Heidenreich, F. Liepe, and Th. Neese, *Mechanische Verfahrentechnik. l.*, Deutscher Verlag für Grundstoffindustrie, Leipzig, 1977.
53. W. Albring, *Elementarvorgänge fluider Wirbelbewegungen*, Akademieverlag, Berlin, 1981.
54. A. Scheludko, Sl. Tschaljowska, and A Fabrikant, *Disc. Faraday Soc.*, 1, 112, 1970.
55. Sl. Tschaljowska, Metod sa isledvane javleniada tonkich tetschen filmi verchu tverda poverchnosti, thesis, Univ. Sofia, Fac. Chemie, 1988.
56. W. Hopf, Zur Geschwindigkeitsabhängigkeit des Rückzugsrandwinkels in tensidhaltigen Dreiphasensystemen, Doctoral thesis, Akademie der Wissenschaften, FIA Freiberg, 1987.
57. J. Mahnke, D. Vollhardt, K.W. Stöckelhuber, K. Meine, and H.J. Schulze, *Langmuir*, 15, 8220, 1999.
58. H.J. Schulze, B. Wahl, and G. Gottschalk, *J. Colloid Interf. Sci.*, 128, 57, 1989.
59. R.D. Crozier, *Flotation Theory, Reagents, and Ore Testing*, Pergamon Press, New York, 1992.
60. G.J. Jameson, S. Nam, and M Moo Young, *Miner. Sci. Eng.*, 9, 10318, 1977.
61. E.T. Woodburn, *Miner. Sci. Eng.*, 2, 3, 1970.
62. J. Ralston, in *Colloid Chemistry in Mineral Processing*, J.S. Laskowski and J. Ralston, Eds., Elsevier, Amsterdam, 1992, p. 203.
63. R.H. Yoon and L. Mao, *J. Colloid Interf. Sci.*, 181, 613, 1996.
64. A.V. Nguyen, J. Ralston, and H.J. Schulze, *Int. J.Miner. Process.*, 53, 225, 1998.
65. A.V. Nguyen, *Int. J. Miner Process.*, 56, 165, 1999.
66. R.J. Gochin, in *Solid–Liquid Separation*, Butterworths, London, 1990, p. 591.
67. R. Schuhmann, *J. Phys. Chem.*, 46, 891, 1942.
68. H.J. Schulze, *First Research Forum on Recycling*, CPPA Press, Toronto, 1991, p. 161.

69. H.J. Schulze, in *Coagulation and Flocculation*, B. Dobias, Ed., Marcel Dekker, New York, 1993, p. 321.
70. H.J. Schulze, Comparison of the Elementary Steps of Particle/ Bubble Interaction in Mineral and Deinking Flotation, Eighth International Conference on Colloid and Surface Science, Adelaide, 1994.
71. H.J. Schulze, *Wochenblatt für Papierfabrikation*. 122, 164, 1994.
72. F. Bloom and T.J. Heindel, *Math. Comput. Mod.*, 25, 13, 1997.
73. F. Bloom and T.J. Heindel, *J. Colloid Interf. Sci.*, 190, 182, 1997.
74. T.J. Heindel, *TAPPI J.*, 82, 115, 1999.
75. T.J. Heindel and F. Bloom, *J. Colloid Interf. Sci.*, 213, 10, 1999.
76. T.J. Heindel and F. Bloom, *J. Colloid Interf. Sci.*, 213, 10, 1999.
77. F. Julien Saint-Amand. *Int. J. Miner. Process.*, 56, 277, 1999.
78. F. Bloom and T.J. Heindel, *Colloid Interf. Sci.*, 218, 564, 1999.
79. F. Bloom and T.J. Heindel, *Chem. Eng. Sci.*, 57, 2467, 2002.
80. J. Ralston, S.S. Duchin, and N.A. Mishchuk, *Int. J. Miner. Process.*, 56, 207, 1999.
81. J. Ralston, D. Fornasiero, and R. Hayes, *Int. J. Miner. Process.*, 56, 133, 1999.
82. H. Schubert, *Int. J. Miner. Process.*, 56, 257, 1999.
83. D.A. Johnson and E.V. Thompson, *TAPPI J.*, 78, 41, 1999.
84. R. Pan, F.G. Paulsen, D.A. Johnson, D.W. Bousfield, and E.V. Thompson, TAPPI Pulping Conference, TAPPI Press, Atlanta, 1993, p. 1155.
85. E.V. Thompson, Review of flotation research by the cooperative recycled fiber studies program, in *Paper Recycling Challenge. II. Deinking Flotation*, M.R. Doshi and J.M. Dyer, Eds., Department of Chemical Engineering, University of Maine, Appleton. (W J. Doshi Account).
86. M.A. McCool, Flotation deinking, in *Secondary Fiber Recycling*, R.J. Spangenberg, Ed., TAPPI Press, Atlanta, 1993, p. 141.
87. H. Stechemesser, G Zobel, and H Partzscht, *Colloid Polym. Sci.*, 101, 172, 1996.

LIST OF SYMBOLS

A_s	stability parameter for the aggregate strength
a, a_c	acceleration
Bo'	modified Bond number
C_B	bubble surface mobility coefficient
d_B	bubble diameter
d_P	particle diameter
$\text{Eff}(t)$	particle removal efficiency
G	dimensionless particle settling velocity
g	acceleration due to gravity
h_0	initial film thickness
h_{crit}	critical film thickness
$h_{F,}, h$	thickness of intervening liquid film
k	flotation rate constant
k'	pseudo rate constant
k_1	flotation rate constant associated with the successful formation of a bubble–particle aggregate and its removal to the froth layer
k_2	flotation rate constant associated with the destabilization of a bubble-particle aggregate
m	order of reaction
N_B	total number of air bubbles entering the flotation cell per unit volume
n	order of reaction
n_B	total bubble concentration
$n_B^A(t)$	concentration of bubbles which are available to pick up particles
$n_B^a(t)$	concentration of bubbles with attached particles
$n_B^f(t)$	concentration of bubbles without particles
$n_p(t)$	total particle concentration
n_{p0}	initial (free) particle concentration
$n_p^a(t)$	number of particle attached to a bubble
n_p^e	average number of particles on bubbles in the exit stream which have attached particles
$n_p^f(t)$	concentration of free particles (i.e., particles not attached to bubbles)
P_{asl}	probability of bubble–particle attachment by sliding

P_c	probability of bubble–particle collision
P_{destab}	probability of bubble–particle aggregate destabilization
P_{stab}	probability of bubble–particle stability
P_{tpc}	probability of forming a three-phase contact
Q	volumetric gas flow rate
R_B	bubble radius
R_p	particle radius
Re_B	bubble Reynolds number
Re_G	shear Reynolds number
Re_p	particle Reynolds number
R_F	film radius
R_v	vortex radius $\sim R_p + R_B$
St	Stokes number
t	time
U_B	relative velocity between a bubble and the surrounding fluid
U_p	relative velocity between a particle and the surrounding fluid
u_ϕ	tangential component of fluid flow
u_r	radial component of fluid flow
V	mixture volume
V_B	average volume of a single bubble
Z	bubble–particle collision frequency
Z^*	bubble–particle detachment frequency
ε	turbulent energy density or relative dielectric constant
ε_g	gas hold-up
θ	contact angle
θ_A	advancing contact angle
θ_R	receding contact angle
Φ	polar angle
λ	wavelength of capillary surface wave
μ, η	liquid dynamic viscosity
ν	liquid kinematic viscosity
ρ_B	bubble density
ρ_l	liquid density
ρ_p	particle density
σ	surface tension or surface charge

V_B	bubble rise velocity
V_p	particle velocity
V_{ps}	particle settling velocity
Π	disjoining pressure
τ_i	induction time
τ_c	contact time
τ_v	vortex lifetime
κ	Debye–Hückel parameter
ψ	surface potential

8

From Clay Mineral Crystals to Colloidal Clay Mineral Dispersions

Gerhard Lagaly

Institute of Inorganic Chemistry, University of Kiel, Kiel, Germany

I. INTRODUCTION

Colloid scientists dream of colloidal dispersions to behave as demanded by theories. Probably the dreams can never be realized; neither the systems nor the theories are perfect.

Two systems are sometimes considered ideal: latex and clay dispersions. Latex particles can behave as ideal hydrophobic colloids (the DLVO theory describes the properties correctly) but strong deviations are often observed. The reputation of clay mineral dispersions as suitable model systems is based on the fact that the surface structure of these plate-like particles is well known and that minerals with widely differing properties are available. However, clay mineral dispersions never behave as ideal systems. Coagulation and flocculation processes are much more dependent on system parameters than for other dispersions. Several reasons must be mentioned that contribute to this behavior:

1. The particles are of irregular shape and of different thicknesses. The layers composing one particle have no common contour line and are frayed out.
2. The particles are often very thin and possess remarkable flexibility.
3. Particles with narrow size distribution cannot be obtained.
4. The crystals of smectites disarticulate into thinner lamellae or single silicate layers when certain experimental conditions are fulfilled.
5. The charges of the layers are not uniformly distributed.
6. The particles also carry charges at the edges, which change with the chemical parameters, in particular with pH.

Because of these facts, straightforward calculations of stability and coagulation conditions are difficult to obtain. Nevertheless, many practical uses of clay dispersions are just based on the variation of the colloidal stability with the system parameters.

II. CLAY MINERAL STRUCTURE

A. Clays and Clay Minerals

Clays are fine-grained sediments with particle sizes $<2 \mu m$. The clay minerals are the decisive components in clays. Properties, which are considered typical of clays (plasticity, thixotropy, water retention, swelling, ion exchange, adsorption of

inorganic and organic compounds), are related to the presence
of these minerals. Among the indifferent components, quartz
should be mentioned first because it is an abundant main
component besides the clay minerals. The yellow-brown color
of clays is mainly caused by iron oxides. The amount of organic materials in clays is often small but can have a strong
influence on color and the colloidal and flow properties.

The most important clay minerals are kaolinite, smectites, illites, and mixed-layer minerals. Clays containing appreciable amounts of kaolinite are called kaolins. Smectites,
mostly montmorillonites, are the main constituents in bentonites. Clays are formed by precipitation from fluids, by weathering or hydrothermal alteration of different types of parent
minerals and by hydrothermal neoformation.

Bentonites in technical applications are mainly used
without further processing as they are fine-grained materials
with particles mainly <2 μm. They may be transferred into the
sodium form (alkaline or soda activation) and into organic
derivatives (organic activation) or may be decomposed to
bleaching earths (acid activation) [1–3].

Kaolins normally contain coarser particles of quartz, feldspars, micas, unaltered granite, etc., and have to be benefited
by size separation or refining (hydrocycloning, sedimentation,
flotation, high-gradient magnetic separation) [4].

Common clays are dominant in ceramics, bricks, tiles,
and in the building industry including half-timbering and
contain kaolinite, illite–sericite, quartz, and illite–smectite
mixed-layer materials in very different ratios.

Synthetic clay minerals containing aluminum are difficult to obtain, and colloidal dispersions of these materials are
rarely studied. Phases that are similar to hectorite (a magnesium lithium smectite) are more easily obtainable. For instance, Laponite is produced by Laporte Industries on a
technical scale, and other companies also expand capacity
for the production of similar materials. An advantage of
these synthetic clay minerals is the purity and the ease with
which they form colloidal dispersions [5–11]. Magnesium lithium silicates can also be prepared from melts [10]. They are
often called synthetic fluoro hectorites but their layer charge

can be distinctly higher than for hectorite minerals and Laponite. Na-4-micas with high-Al content were also be prepared
by solid-state reactions [10,12]. For other clay minerals, see
Ref. [13].

B. Structure of Clay Minerals

The clay minerals considered here are composed of continuous
two-dimensional tetrahedral sheets of composition $(Si, Al)_2O_5$
and octahedral sheets which normally contain Mg^{2+}, Al^{3+},
Fe^{2+}, and Fe^{3+} cations (Figure 8.1) [2,14].

When one tetrahedral sheet is linked to one octahedral
sheet, the assemblage is known as 1:1 layer silicate (Figure

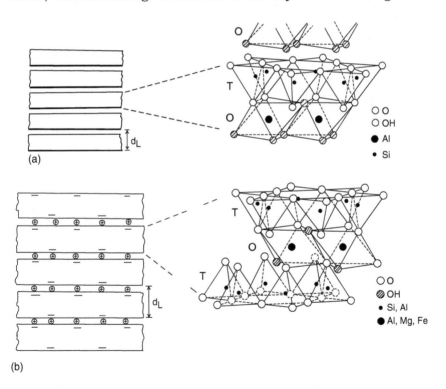

Figure 8.1 Structure of clay minerals. T, tetrahedral sheet; O,
octahedral sheet; C, interlayer cations $(+H_2O)$. (a) 1:1 clay minerals
(kaolinite); (b) 2:1 clay minerals (smectites, vermiculites, micas).
(From K. Jasmund and G. Lagaly, Tonminerale und Tone, Figure
1.2. Ref. 2. With permission.)

8.1(a)). The predominant species of this group of minerals is kaolinite. The layers are uncharged and held together by hydrogen bonds and van der Waals interactions. The positions of adjacent layers are governed by the requirement for pairing the OH groups on the octahedral sheet of the layer with the oxygen atoms of the tetrahedral sheet of one layer below. The general formula of kaolinite is $Al_2(OH)_4Si_2O_5$ and the dimensions of the unit cell are [2]: $a = 0.516$ nm, $b = 0.894$ nm, $c = 0.740$ nm, $\alpha = 91.7°$, $\beta = 104.9°$, and $\gamma = 89.8°$. The unit cell contains two formula units $[Al_2(OH)_4Si_2O_5]$.

The 2:1 clay minerals consist of two tetrahedral sheets linked to one central octahedral sheet (Figure 8.1(b)). The variety of minerals of this group (Table 8.1) arises from different types of isomorphous substitutions and cation vacancies in the octahedral sheets. The layers are uncharged (talc, pyrophyllite) or negatively charged (mica-type minerals). The charge density is mainly determined by the Al-for-Si substitutions in the tetrahedral sheets and the amount of divalent and trivalent cations within the octahedral sheets. The origin of the charges is expressed by the chemical formula

$$M_{(x+y)/m}^{m+}(H_2O)_n[(Me^{II},\ Me^{III})_{2-3}^{(6-x)+}(OH)_2(Si_{4-y}Al_y)O_{10}]^{(x+y)}$$

interlayer space   octahedral sheet   tetrahedral sheet

Clay minerals with about three divalent cations in the octahedral sheet per formula unit are called trioctahedral, and those with two trivalent cations are called dioctahedral. When the layers carry negative charges, cations are bound in the interlayer space for charge compensation. Generally, the interlayer space also contains water molecules, the amount of which depends on the layer charge and on the water vapor pressure around the clay particles or, in aqueous solution, on the type and concentration of salts (see Chapter 3). The unit cell data are [2,14]:

| Micas: | Biotite | $a = 0.533$ nm; $b \approx 0.923$ nm $c = 2.01$ nm $\beta = 95°$ |
| | Muscovite | $a = 0.519$ nm; $b \approx 0.900$ nm $c = 2.00$ nm $\beta = 96°$ |

Vermiculites:		$a = 0.534\,\text{nm}; b \approx 0.925\,\text{nm}$
Smectites:	Montmorillonite:	$a = 0.518\,\text{nm}; b \approx 0.900\,\text{nm}$
	Beidellite	$a = 0.518\,\text{nm}; b \approx 0.900\,\text{nm}$
	Hectorite	$a = 0.525\,\text{nm}; b \approx 0.918\,\text{nm}$

Table 8.1 Classification of clay minerals [15]

Layer type	Interlayer material	Group	Species[a]	Layer charge[b]
1:1	None or H$_2$O only	Serpentine– Kaolin	Kaolinite, Halloysite (di)	0
2:1	None, $(x + y) \sim 0^b$	Talc– Pyrophyllite	Talc (tri), Pyrophyllite (di)	0
	Hydrated exchangeable cations, $(x + y) \sim 0.2$–0.6	Smectite	Hectorite (tri)	0.2–0.25
			Montmorillonite (di)	0.25–0.4
			Saponite (tri)	
			Beidellite (di)	0.4–0.6
			Nontronite (di)	
	Hydrated exchangeable cations, $(x + y) \sim 0.6$–0.9	Vermiculite	Vermiculite	0.6–0.9
	Nonhydrated monovalent cations $(x + y) \sim 0.6$–1	Jllite Mica	Illite	0.8–1
			Biotite, Muscovite	1

[a] Only a few examples are given. Di, tri = di- or trioctahedral structure.
[b] Layer charge in charges/formula unit.

The unit cell contains two formula units (Si, Al)$_4$O$_{10}$. The c spacing of the vermiculites and the smectites is not listed because it depends on the interlayer cation and the relative humidity or the solvating liquid.

The interlayer cation in smectites is mainly calcium but a few bentonites (e.g., from Wyoming) contain sodium as the interlayer cation besides calcium ions. Bentonites known as sodium bentonites do not contain pure sodium montmorillonite but sodium calcium montmorillonite with sodium and calcium in varying ratios.

Mixed-layer minerals consist of two (or more) different types of layers within the same crystals (*layer* in connection with mixed-layer minerals comprises the silicate layer itself and the interlayer space). Very important for geological studies are illite–smectite (I/S) mixed-layer minerals [16]. Common clays often contain appreciable amounts of I/S materials. Their presence in ceramic masses improves plasticity and dry strength [2]. For x-ray powder diffraction of mixed-layer minerals see Refs. [17,18].

C.   Quality of Clay Mineral "Crystals"

The structure of clay minerals is characterized by different types and degrees of disorder [14,18,19]. The crystals, in particular of smectites, are never crystals in the strong sense. Many crystallographers are dismayed at the particles which clay scientists call *crystals*. In fact, a smectite *crystal* is more equivalent to an assemblage of silicate layers than to a true crystal (Figure 8.2). The particles seen in the electron microscope never have the regular shape of real crystals but look

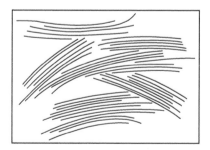

Figure 8.2   A schematic view of montmorillonite *crystals*. (From G. Lagaly and R. Malberg, *Colloids Surf.*, 49, 11–27, 1990. With permission.)

like paper torn into irregular pieces. Very instructive pictures of smectite particles were published by Vali and Köster [20]. The core of the particles is surrounded by disordered and bent silicate layers with frayed edges. Layers or thin lamellae of a few layers protrude out of the packets and enclose wedge-shape pores. The particles reveal many points of weak contacts between the stacks of the layers. At these *breaking points* the particles may easily break as under the influence of mechanical forces or during intracrystalline reaction.

In Ca^{2+} smectite crystals only a few layers form coherent domains in which all silicate layers have the same distance, for instance, corresponding to two layers of water. These domains are separated by zones composed of silicate layers in different distances because of the presence of one, three, four, or even more water layers. This structure could be clearly observed by transmission electron microscopy (TEM) inspection [21]. A few of the units composed of the coherent domains and the zones of differently spaced silicate layers are aggregated with almost parallel orientation to quasicrystals. The quasicrystals are arranged in a network with pores of different sizes, thus constituting the whole particle. The aggregated particles exhibit a very complex pore structure with lenticular pores between parallel oriented layers, pores within the network of quasicrystals and between the particles (see, electron micrographs in Ref. [22]). It seems to be possible to estimate the volumes of the different types of pores from the kinetics of water inhibition [22]. Nomenclature to name the different levels of organization is still lacking.

The structure of the smectite crystals is described here in detail because it greatly determines the type of units formed by dispersing smectites and determines the colloidal and rheological properties of the dispersions. It also governs the applicability of a bentonite in the practical uses.

The electrostatic attractions between the layers and the interlayer cations increase the stacking order in more highly charged 2:1 clay minerals. The domains with equally spaced layers become thicker and the influence of the defects on the shape and position of the (001) reflections decreases. Defects of unequally spaced silicate layers due to different degrees of hydration (probably a consequence of charge inhomogeneity) are still observable by delicate analysis of the x-ray reflection profiles of sodium beidellite with water monolayers [24]. The different types of layer stacking in the vermiculites were reviewed by Suquet and Pézérat [25]. Micas can be considered as true crystals, and many polymorphs were described [14]. Kaolinites also contain varying degrees of disorder in the layer stacking [14,26].

D. Layer Charge of 2:1 Clay Minerals

The 2:1 clay minerals are classified by the layer charge $(x+y)$ (Table 8.1). The most important minerals are the montmorillonites. The layer charge $(x+y)$ is given in charges/formula unit $(eq/(Si,Al)_4O_{10})$ and is, for montmorillonites, in the range $0.2-0.4$. The numerical value is identical to the interlayer cation density ξ (in equivalents of cations/mol formula unit).

The surface charge density is $(x+y)/2$ and may also be expressed as surface charge density σ_0 (in $C\,m^{-2}$). The unit cell contains two formula units $(Si, Al)_4O_{10}$ and has the dimensions a and b in the plane of the layer. The charge density is

$$\sigma_0 = 1.602 \times 10^{-19}(x+y)/ab \ (C \ m^{-2})$$

Typical surface charge densities are listed in Table 8.2.

An outstanding property of smectites and vermiculites (probably also of micas) is the nonuniform charge distribution. The density of the charges is not constant in all layers but varies from layer to layer within certain limits, and probably also from the core of the particles to the edge regions. Beidellites and vermiculites with considerably isomorphous substitution in the tetrahedral sheets can contain polar layers, i.e., both faces of a layer show different surface charge densities [27,28].

Due to the charge inhomogeneity the interlayer cation density ξ_i varies from interlayer space to interlayer space. The numerical value of ξ_i in the interlayer space i is the average of the surface charges of both layers enclosing this interlayer space. The interlayer cation density varies more or less randomly between certain limits ξ_i and ξ_j. The mean value of the

Table 8.2 Layer charge $(x + y)$, surface charge density σ_0, and equivalent area A_c of mica-type clay minerals[a]

Mineral	M^b	$(x + y)$ (charges/formula unit)	σ_0 (C/m^2)	ab (nm^2)	A_c (nm^2/charge)
Biotite	455	1	0.326	0.492	0.246
Muscovite	390	1	0.343	0.467	0.234
Vermiculites	390	0.8	0.259	0.495	0.309
		0.6	0.194	0.495	0.412
Beidellite	360	0.5	0.172	0.466	0.466
Montmorillonites	362	0.4	0.137	0.466	0.583
		0.3	0.103	0.466	0.777
		0.2	0.069	0.466	1.165
Hectorite	380	0.23	0.076	0.482	1.048

[a] The equivalent area is the area per monovalent interlayer cation: $A_c = ab/2(x+y)$.
[b] Mean molecular mass of the formula unit ($Si_4O_{10}(OH)_2$) without interlayer cations and water [2].

cation density $\bar{\xi}$ is equal to the mean value of the layer charge $(x + y)$. The frequency distribution of the interlayer cation densities $\xi_1, \xi_2, \ldots, \xi_j$ sometimes shows a maximum but often resembles to a bimodal distribution (cf. Figure 8.17) [27].

E. Determination of Layer Charge

The layer charge may be calculated from the chemical analysis and the cation exchange capacity. This method requires that the clay minerals be separated quantitatively from all other ancillary materials or that type, composition, and amount of these materials including the amorphous substances are exactly known. Pure microsized vermiculites are available but pure montmorillonites are never found. The amounts of accompanying minerals and amorphous silica in bentonites can be determined but the procedure is very time consuming [29]. Sometimes the content of montmorillonite in bentonites is estimated by (often somewhat suspect) methods, and the layer charge is calculated from the experimental cation exchange capacity. However, exchangeable cations are also located at the edges (see next section). If the amount of these counterions is not considered, a too high value of the layer charge is obtained.

The most reliable method for determining the layer charge is exchanging the interlayer cations by a series of primary n-alkylammonium ions $C_nH_{2n+1}NH_3^+$. These ions are quantitatively exchanged for the interlayer cations (and the cations at the external surfaces). The arrangement of these ions in the interlayer space depends on the chain length n and the charge density in the interlayer space. The different types of interlamellar arrangements are recognized by the basal spacing which is measured by simple x-ray powder technique. The exact layer charge is derived from the basal spacing measurements [27,30–32]. Primarily, the interlayer cation density $\bar{\xi}$ is obtained, which is numerically identical to the layer charge $(x+y)$, both in eq/mol. With an average molecular mass of the formula unit of smectites (without interlayer cations and water) of $\sim$360 the exchange capacity related to the interlayer cations is

$$C_i = \xi/360 \quad \text{(eq/g silicate)}$$

For smectites, the interlayer exchange capacity is about 80% of the total exchange capacity C_t. (For determinative methods for the total cation exchange capacity, see Ref. [33].)

The decisive progress brought about by the alkylammonium method is the evaluation of the charge heterogeneity. When the charge density changes from interlayer space to interlayer space, the arrangement of the alkylammonium ions in the interlayer spaces varies accordingly and, in characteristic ways, influences the position and shape of the (00l) reflections in the x-ray powder pattern. By a simple procedure [30] the charge heterogeneity is deduced from the x-ray powder patterns, and diagrams (as in Figure 8.17) are constructed which reveal the frequency distribution of the interlayer cation densities ξ_i. The mean interlayer cation density $\bar{\xi}$ is obtained from

$$\bar{\xi} = \frac{\sum \nu_i(\xi_i + \xi_{i+1})/2}{\sum \nu_i},$$

where ν_i is the frequency of interlayer spaces with a density between ξ_i and ξ_{i+1}. The mean cation density $\bar{\xi}$ is equal to the mean layer charge $(x+y)$.

F. Edge Charges

The sign and density of the charges at the crystal edges depend on the pH of the dispersion. Sometimes clay scientists still use the term *broken bonds*, which a colloid scientist never will use. The charging arises from adsorption or dissociation of protons as in case of oxides (Figure 8.3). In acidic medium, an excess of protons creates positive edge charges, and its density decreases with rising pH. Negative charges are produced by the dissociation of silanol and aluminol groups [34].

The interesting question concerns the condition that leads to virtually uncharged edges. When alkylammonium ions were exchanged at pH = 6.5, the total amount of alkylammonium ions bound by Wyoming montmorillonite was 1.07 meq/g silicate (average value, see Table 8.3). The interlayer cation exchange capacity was 78 meq/g silicate. Thus 0.29 meq alkylammonium ions were bound at the edges (or 27% of the

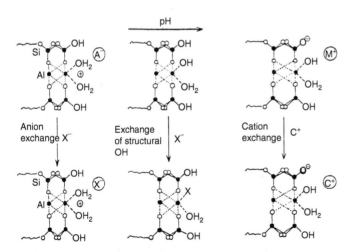

Figure 8.3 The pH-dependent ion and ligand exchange reactions at the edges of the clay minerals crystals. (From K. Jasmund and G. Lagaly, Tonminerale und Tone, Figure 3.1. Ref. 2. With permission.)

Table 8.3   Total (C_t) and interlayer exchange capacity (C_i) of Montmorillonite[a]. C_t from the carbon content of the alkylammonium derivatives, C_i by the alkylammonium method

n[b]	C_t (meq/g silicate)	C_i/C_t
6	1.06	0.74
8	1.06	0.74
10	1.10	0.71
12	1.07	0.73
14	1.01	0.77
16	1.10	0.71
18	1.12	0.70

[a] From Wyoming, type "Greenbond" M 40, supplied by Süd-Chemie AG in 1974.
[b] Number of carbon atoms of the alkylammonium ions.

total exchange capacity). A very similar result was obtained by surface charge measurements with the particle charge detector. This simple instrument indicates the point, at which all charges of a colloidal particle are compensated by macroions or, in our case, by the alkylammonium ions. The amount of alkylammonium ions required to uncharge the particles increased rapidly with pH because in acidic medium protons strongly competed with the alkylammonium ions also in the interlayer space. At pH = 6–8, a plateau was reached and, in agreement with the analytical data, a dosage of 1.03 meq alkylammonium ions/g montmorillonite uncharged the particles. The increase at higher pH was caused by the presence of alkylamine molecules, which were strongly adsorbed by alkylammonium montmorillonite.

The total amount of alkylammonium ions bound by 2:1 clay minerals is often slightly higher than the total cation exchange capacity determined by other methods [2,33]. This is a consequence of the charge regulation at the edges. Adsorption of surface-active agents on oxidic surfaces increases the surface charge density by adsorption (anionic surfactants) or desorption (cationic surfactants) of protons [35,36]. Thus, adsorption of alkylammonium ions exceeding the cation exchange capacity is accompanied by desorption of protons from surface OH groups. The analytical data clearly reveal that, at pH 6.5, a considerable amount of cations are bound at the edges. The point of zero charge of the edges must be below 6.5, probably around pH = 5.

The flow behavior of sodium montmorillonite dispersions in water or diluted NaCl solutions shows characteristic changes with pH (Section VII.A). The shear stress decreases strongly above pH = 4.5 (Figure 8.25). This reduction of viscosity is generally attributed to the destruction of the card-house when the amount of positive edge charges becomes too small to stabilize the card houses by stable edge($+$)–face($-$) contacts [37,38]. Thus, rheological measurements also indicate an apparent p.z.c. of the edges at pH = 5 or slightly above.

The point of zero charge at the edges of colloidal pyrophyllite particles ($x+y\sim0$, Table 8.1) was found at pH = 4.2 by titrations at various indifferent electrolyte concentrations [39].

G. Intracrystalline Reactions

Exchange of interlayer cations by other inorganic and organic ions is one of the outstanding properties of 2:1 clay minerals. A diversity of other reactions has to be mentioned including reversible hydration and solvation, displacement of interlamellar water by organic molecules, interlamellar complexation, and sorption of organic derivatives in the interlayer spaces (Figure 8.4). Electron transfer from or to the layers is the basis of several catalytic reactions [2,40–43].

Kaolinite intercalates several organic compounds [41]. Hydrated kaolinites are prepared by displacement of suitable organic guest molecules (e.g., dimethylsulfoxide) by water [44]. A variety of organic molecules can be intercalated in these hydrated kaolinites to give ordered or less ordered intercalates [45]. A refined displacement method consists of reacting kaolinite with NMF (giving a basal spacing of 1.08 nm), which is replaced by methanol (basal spacing 1.11 nm). The wet kaolinite–methanol intercalation compound is then treated with the methanolic solution of the final guest compound [46].

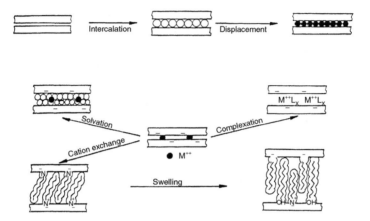

Figure 8.4   Intracrystalline reactions of 1:1 and 2:1 clay minerals.

III.   THE CLAY–WATER SYSTEM

A.   Hydrates of 2:1 Clay Minerals

The 2:1 clay minerals form hydrates with one, two, three, or four layers of water between the silicate layers. The state of hydration varies with the water vapor pressure, the water content, and, in salt solutions, with the type and concentration of salts, and is dependent on the layer charge and the interlayer cation density. Typical basal spacings are: 1.18–1.24 nm (water monolayers), 1.45–1.55 nm (water bilayers), and 1.9–2.0 nm (four water layers).

The changes of the basal spacing with the salt concentration are seen in Figure 8.5. The dotted fields comprise the spacings of a large collection of smectites and vermiculites. Sodium smectites change from a state indicated by a basal spacing $\rightarrow \infty$ (see Section III.C) into hydrates with four, and, at higher salt concentration, two layers of water [47]; vermiculites persist in the two-layer hydrate over the whole range of concentrations. Potassium ions in higher concentrations restrict the interlayer expansion of the smectites to water monolayers. The hydration of vermiculites is virtually impeded by KCl solutions at concentrations above $0.01\,M$. In the presence of calcium ions the four-layer hydrate of smectites and the two-layer hydrate of vermiculites persist over a wide range of concentrations.

The variation of the basal spacings with salt concentration (Figure 8.5) is of great interest to colloid scientists. To explain the reversibility of salt coagulation of many colloidal dispersions, Frens and Overbeek [48,49] introduced the concept of the distance of closest approach. They postulated that at the onset of coagulation the particles are not in direct contact but remain separated by a certain distance. A value of 0.4 nm (corresponding to two water layers) was shown by a number of calculations as the most probable minimum separation. That a distance of closest approach really exists is clearly seen by the behavior of sodium and calcium smectites as well as sodium and calcium vermiculites in NaCl and $CaCl_2$ solutions. Even far above the critical coagulation concentrations and up to concentrations as high as $5\,M$ NaCl the silicate layers remain separated by two water layers. The water layers are displaced from between the surfaces only when the counterions like potassium ions are attracted to the surface by specific interactions. The *fixation* of potassium ions (and cesium and rubidium ions) is well known to clay scientists and is explained by the reduced hydration energy and a better geometric fit of the potassium ions into the surface oxygen hexagons in comparison to sodium ions (cf. Ref. [50]).

B. Structure of the Hydrates

The hydrated forms of 2:1 clay minerals may be considered as quasicrystalline structures because there is a certain amount of order in the interlayer space. The counterions (interlayer cations) reside near the middle plane of the interlayer spaces [51–53]. The silicate layers of the more highly charged silicates (saponites, vermiculites) assume ordered or semiordered layer-stacking sequences [25]. In smectites, the superposition of the layers is generally random (turbostratic structure) but specimens with a certain periodic structure (e.g., beidellites) are also found [54,55]. The layers of Wyoming montmorillonite

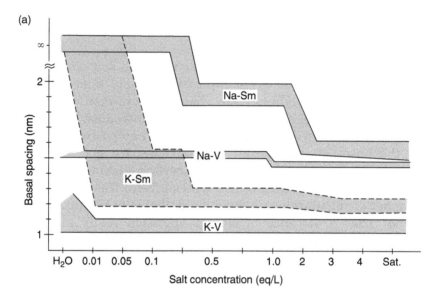

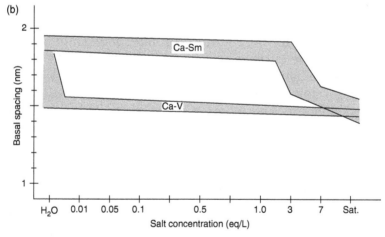

Figure 8.5   Basal spacing of 2:1 clay minerals as a function of salt concentration. Na–Sm, sodium smectites and NaCl; K–Sm, potassium smectites and KCl; Ca–Sm, calcium smectites and CaCl$_2$; Na–V, sodium vermiculites and NaCl; K–V, potassium vermiculites and KCl; and Ca–V, calcium vermiculites and CaCl$_2$. (From G. Lagaly and R. Fahn, in *Ullmann's Encyclopedia of Technical Chemistry*, 4th ed., Vol. 23, Verlag Chemie, Weinhein, 1983, pp. 311–326. With permission.)

with initially random superposition become more and more regularly stacked after wetting and drying cycles in the presence of potassium ions [56].

Organization, mobility, and dynamics of the interlayer water molecules were studied in great detail by Fripiat (see, for instance, Refs. [53,57–60]). Generally, a part of the water molecules form the hydration shells around the interlayer cations, the other water molecules fill the space between the hydration shells [61] and are in a state probably similar to the state around structure-breaking ions. There is no doubt that the properties of interlayer water are different from those of bulk water [62], e.g., showing increased acidity [53,59–65] and rotational correlation times about two orders of magnitude ($\sim 10^{-10}$ s) lower than in liquid water [59].

Grandjean and Laszlo [66] deduced from deuterium nuclear magnetic resonance (NMR) studies of montmorillonite–water dispersions (0.024 g/ml D_2O) that the water molecules are strongly polarized and are simultaneously bound to a negative center by a hydrogen bridge and to an interlayer cation by electrostatic forces. As a consequence, the acidity of interlayer water molecules is increased. In sodium montmorillonite the water molecules reorient predominantly around the hydrogen bond whereas they reorient around the metal–oxygen axis in the presence of calcium ions.

A question of importance concerns the influence of the surface force fields of the external surfaces onto the water structure around and between the particles. Fripiat et al. [59,60] could not detect long-range forces acting on the water molecules outside of the particles, neither at the thermodynamic scale (heat of immersion measurements) nor at the microdynamic range (NMR measurements). The number of water layers influenced by the surface forces is 3–4, i.e., a water film of thickness of about 1 nm. In contrast, Mulla and Low [67] concluded that the molecular dynamics of vicinal water as seen by infrared spectroscopy is affected by the particle surface to an appreciable distance of about 4 nm.

Molecular dynamics simulations of the montmorillonite hydrates mainly confirm the experimental results on the position of the interlayer cations and the interlayer water structure. A part of the lithium interlayer cations form inner-sphere surface complexes (the lithium ions are directly coord-inated to the surface oxygen atoms). The expansion of the interlayer space is accompanied by the conversion in outer-sphere surface complexes to diffuse layer species as a result of the strong Li^+–water interactions. However, a part of Li^+ ions still persist as inner-sphere surface complexes but are in ready exchange with the diffuse layer Li^+ [68]. Also, sodium and potassium ions have a significant coordination with surface oxygen atoms and exist in inner- and outer-sphere surface complexes [62,69–71]. Increasing tetrahedral substitution shows a trend of direct binding between Na^+ and surface oxy-gen atoms and a corresponding dissimilarity with the coordin-ation structure in bulk solution. The coordination structure of water molecules around K^+ ions is, as expected, not nearly so well defined as it is for Li^+ and Na^+ ions. Magnesium counter-ions on montmorillonite reside at the midplane of the inter-layer spaces. Nonsolvating water molecules move freely on planes above and below the midplane. In the case of beidellites, the motion of water molecules is more hindered because of the presence of negative charge sites close to the surface [72]. Surface energy studies by contact angle measurements also indicated that divalent cations are shielded from the silicate surface by water molecules whereas monovalent cations can be in direct contact with the surface oxygen atoms [73].

C.   Structure with Diffuse Ionic Layers

In Figure 8.5 the basal spacing $\to \infty$ is indicated for sodium (and several potassium) smectites in water or diluted salt solutions ($< 0.2 M$ for NaCl). This is an outstanding property of smectites with monovalent counterions. The crystals dela-minate into the individual silicate layers, or thin packets of them, and form a colloidal dispersion of very thin particles (Figure 8.6). The thickness and width of the individual layers and lamellae produced by the disarticulation of a large variety of smectites and illite–smectite mixed-layer crys-tals (see Section IV.D) were recorded by TEM [74,75]. The interlayer cations constitute the diffuse ionic layers around

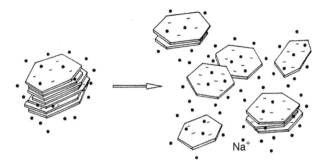

Figure 8.6 Disarticulation (delamination) of alkali smectite crystals in aqueous dispersions.

the silicate layers and silicate lamellae. The presence of discrete particles that do not interact strongly and flow independently was proven by light and small-angle neutron scattering (SANS) [9,76].

The average interparticle distances (obtained from small-angle scattering) respond to the addition of sodium salts in an almost linear decrease with $2/\sqrt{c}$ (c = salt concentration) until, at $c \approx 0.2$ mol/l, rearrangement into the quasicrystalline structure is seen by the sudden decrease of the spacing from about 4 to 2 nm [3,77].

With counterions other than lithium and sodium the distances between the individual layers are no longer equal (or, more correctly, varying around a mean value). Tactoids are formed by a columnar-like superposition of a few silicate layers in equal distances. The distances between the tactoids are larger than within these units. In Figure 8.7 the relative number N/N_{Li} of plates per tactoid ($N_{Li} = 1$ for lithium as counterion) when the calcium ions are progressively replaced by alkali and magnesium ions [76] is shown. In the presence of calcium ions the tactoids contain about seven silicate layers. Even small amounts (< 0.2 equivalent fractions) of alkali metal ions reduce the size of the tactoids dramatically to one to three layers.

Cebula et al. [78] concluded from small-angle neutron-scattering measurements that a considerable part of the

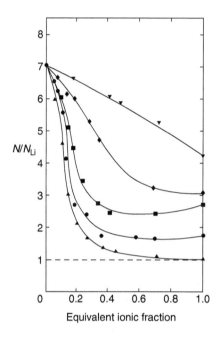

Figure 8.7 Relative number of layers per tactoid, N/N_{Li} ($N_{Li} = 1$) as a function of surface coverage when the calcium ions are exchanged by lithium (▲), sodium (●), potassium (■), cesium (◆), and magnesium (▼) ions. (From L.L. Schramm and J.C.T. Kwak, *Clays Clay Miner.*, 30, 40–48, 1982. With permission.)

layers is aggregated to units consisting of two silicate layers (potassium montmorillonite) or three silicate layers (cesium-montmorillonite) interleaved with bimolecular water layers.

D. Effect of Calcium Ions

The interaction between the dispersed clay mineral particles or silicate layers is described by the DLVO theory [79–81] even if reasons were put forward that other types of interactions must be considered [82–84]. One effect is difficult to explain by the DLVO theory. It is the pronounced sensitivity of the colloidal clay mineral dispersions against calcium ions (and other di- and trivalent metal ions). In the presence of these ions, the particles remain in coagulated state and cannot be dispersed even in pure water.

Fitzsimmons et al. [85] showed that, just after contact of sodium montmorillonite dispersions with calcium-saturated exchange resins, calcium montmorillonite existed as single platelets as in the sodium form. With time the individual silicate layers aggregated by overlapping of the edges and linked together to form flat, large sheets (band-like aggregation; see Section VII.A).

As revealed by the basal spacings (Figure 8.5), calcium ions maintain the quasicrystalline structure. They reside in the middle of the interlayer space, restrict the interlayer distance to 1 nm (basal spacing 2 nm), and impede transition into the structure with diffuse ionic layers. The effect of calcium ions is, therefore, related to the question of whether the quasicrystalline structure with electrostatic attraction between the layers can rearrange into the state with ionic double layers.

A first attempt to explain the attractive interactions in the presence of calcium ions was developed on the basis of the DLVO theory by Kleijn and Oster [86]. The counterions located between the layers are assumed to be in equilibrium with the bulk solution, so that their charge density was slightly different in magnitude from the charge density of the layers. The electrostatic contribution to the Gibbs energy, G_e, was calculated for constant surface charge density (see also Ref. [79]). The Gibbs energy was positive over a wide range of salt concentrations c_s and surface charge densities σ_0 when the counterions were monovalent. For smectites with $\sigma_0 = 0.07$–0.14 C/m^2 (Table 8.2) the electrostatic interaction was repulsive at $c_s < 10^{-1}$ M. Thus, sodium montmorillonite particles were coagulated by salt concentrations slightly above $0.1\,M$ as long as the surface charge density remained below 0.1 C/m^2 and by salt concentrations slightly below $0.1\,M$ for $\sigma_0 = 0.1$–0.15 C/m^2. For more highly charged clay minerals (vermiculites, micas) the Gibbs energy was negative even at very low salt concentrations, and the formation of colloidal dispersions was not expected. With divalent counterions the colloidal dispersions not only of more highly charged clay minerals but also of very low-charged smectites became un-stable at $c_s < 10^{-3}$ M, and colloidal dispersions were not formed.

More recently, Kjellander et al. [87] calculated the diffuse double-layer interactions with an advanced statistical mechanical method. In contrast to the predictions of the simple Poisson–Boltzmann theory this model gives strongly attractive double-layer interactions for divalent ions (Figure 8.8). The position of the potential minimum is in reasonable agreement with the x-ray basal spacing measurements. The most important reason for the occurrence of the potential minimum is the attraction due to the ion–ion correlation. In the Gouy–Chapman model of the diffuse ionic layer the correlation between the ions is entirely neglected, i.e., the ion density in the neighborhood of each ion is assumed to be unaffected by this ion. This neglect of the ion–ion correlation is a reasonable approximation when the electrolyte concentration *and* the surface charge density are sufficiently low. If any of these conditions are violated, the correlation between the ions must be considered and can lead to attractive double-layer interactions between equally charged particles at short separations [88]. This correlation influences the interaction by two different mechanisms: they make the ion concentration in the middle of the interlayer space change and they contribute to an attractive electrostatic fluctuation force [88].

IV.   HOW COLLOIDAL CLAY MINERAL
DISPERSIONS ARE PREPARED

A.   Purification of Clays

In most cases, natural products are used in studying the properties of clay mineral dispersions. Clay minerals having a certain degree of purity are separated from the clays by sedimentation techniques. The first step consists of removal of iron oxides and organic materials. These materials not only affect properties of the colloidal dispersions but also prevent optimal peptizing of the clay particles and successful fractionation by sedimentation. Iron oxides are removed by reduction

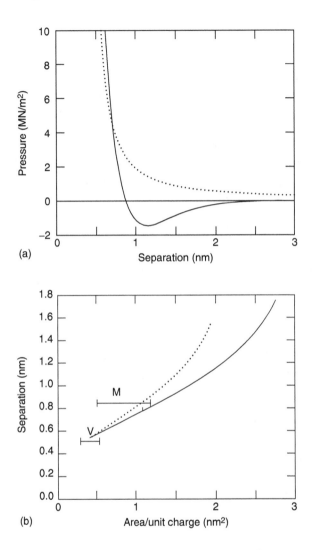

Figure 8.8 Calculation of the pressure between two silicate layers of calcium montmorillonite and vermiculite. (a) Total pressure, calculated for montmorillonite with a surface charge density $\sigma_0 = 0.12$ C/m^2 (full line); results of the DLVO theory shown as dotted line. (b) Position of the potential minimum as a function of the area per unit charge. Solid line: the total net pressure; dotted line: pressure without van der Waals interaction included. (M), montmorillonites; (V), vermiculites. (From R. Kjellander, S. Marčelja, and J.P. Quirk, *J. Colloid Interf. Sci.*, 126, 194–211, 1988. With permission.)

with sodium dithionate in the presence of citrate ions as complexing agents for the Fe^{2+} ions. Organic materials are decomposed to CO_2 by oxidation with H_2O_2. (For experimental details, see Refs. [89–92].)

There is a certain risk that during decomposition of the organic materials low molecular weight substances form which are adsorbed by the clay minerals. Significant amounts of oxalate were found when clays with high contents of organic matter were oxidized with H_2O_2. Most of it was present as soluble oxalato aluminate but a part was attached to the clay mineral [93]. Oxalate adsorption on sodium smectites showed a complex behavior. At low initial oxalate concentrations the amount adsorbed (in the range of 0.1–1 mg/g) decreased with pH. When the initial concentration of oxalate exceeded 0.003 mol/l, the amount adsorbed showed a sharp minimum at pH = 6–7 [94]. Other oxidizing agents, which may be used to decompose the organic materials, are sodium hypochlorite [95] and bromine water [96].

Sometimes the organic material is difficult to remove quantitatively, e.g., when polymeric materials have penetrated into the interlayer spaces and are protected from thermal and chemical degradation.

Other materials that generally have to be removed are calcium and magnesium carbonates and amorphous silica. Carbonate is decomposed by acids, and silica is dissolved by boiling in Na_2CO_3 solution [92]. The carbonates deliberate calcium or magnesium ions, which reduce the degree of peptization. Amorphous silica can act as cementing agent between the particles.

B. Fractionation of Clays

Particle size fractionation of clay is necessary to separate the clay minerals as well as possible from ancillary materials. A decisive step in preparing a stable dispersion of the clay is the replacement of divalent counterions by sodium (or lithium) ions. Large amounts of sodium ions are introduced during the removal of iron oxides and humic materials. Nevertheless, it is recommended to react the clay with a sodium chloride solution (about $1M$) several times further to remove all calcium ions (or other divalent ions). It should be noted that during and after the sodium for calcium exchange the clay cannot be intensively washed with water because the smectites begin to disperse into the colloidal state and are very difficult to filtrate or separate by centrifugation. Finally, the excess of salts must be removed. As dilution with water immediately leads to peptization of the sodium smectite, the only reliable way is to dialyze the dispersion against water. The pH of the dispersion should be maintained at 7–8 (adjusting with some drops of NaOH) because the clay particles aggregate at pH below 6.5.

The final dispersion obtained by dialysis and adjusted to pH 7.5 is stable and contains the particles in an optimal degree of dispersion. The addition of deflocculating agents like phosphates and polyphosphates is not required to attain the colloidal distribution. This is verified by Figure 8.9. The particle size (obtained by a Horiba particle size analyzer) of three fractions of two montmorillonites was independent of the amount of calgon (polyphosphate) added as peptizer. A similar plot was obtained for phosphate as the dispersing agent.

The stable dispersion of the clay minerals in homoionic form is fractionated by sedimentation in the gravitational or, for particle sizes below $<2\,\mu\text{m}$, in the centrifugal field (for details, see Ref. [92]). To reduce the particle–particle interaction during sedimentation, the volume fraction of the particles should be in the range 10^{-3}–2×10^{-3} (about $5\,\text{g}$ clay/ $1000\,\text{ml}$ water). The particle size of the fractions is expressed in Stokes equivalent spherical diameters.

The colloidal dispersions obtained may be used as such or concentrated by water vapor evaporation at about $60°\text{C}$, e.g., in a rotary evaporator. When these dispersions are stored for some time, they must be kept in the dark and refrigerated. Drying of the aqueous dispersions yields compact dense materials that are difficult to redisperse. Freeze-drying or washing with acetone and air-drying are recommended to obtain a fine powder. However, it is advisable to avoid the drying steps because the smectites may not be completely redispersed.

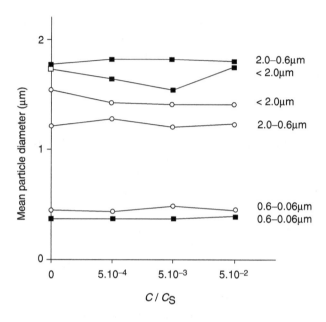

Figure 8.9   Mean particle size of different fractions of two montmorillonites versus the concentration c/c_s of polyphosphate added; c_s:
concentration at saturation at room temperature. ■, montmorillonite (Wyoming); ○, montmorillonite (Niederschönbuch, Bavaria). The
mean particle size of the different fractions was measured with the
Horiba centrifugal particle size distribution analyzer, CAPA-500.

C.   Dispersions of Kaolin

The plate widths of kaolinite particles vary from about 0.1 to
20 μm [4]. Cumulative particle size distribution curves of four
kaolins are shown in Figure 8.10. Kaolinite particles are often
found as largely composed of pseudohexagonal plates [97–99].
The content of ancillary minerals (feldspars, quartz, mica, and
smectites) varies with particle size. Figure 8.11 shows that
pure dispersions of kaolinite can be obtained from certain
kaolins by selecting the appropriate particle size fractions.
In fractions <0.1 μm, smectites are enriched; in fractions
>1 μm, quartz, feldspars, and micas become abundant. How-
ever, the variation of the composition with particle size

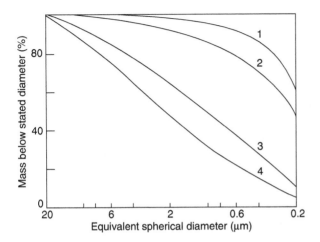

Figure 8.10 Cumulative particle size distribution curves of kaolins: curves 1 and 2, sedimentary kaolins from eastern Georgia; curve 3, sedimentary kaolin from central Georgia; curve 4, primary kaolin from Cornwall. (From I.E. Odom, *Philos. Trans. R. Soc., London A*, 311, 391–409, 1984. With permission.)

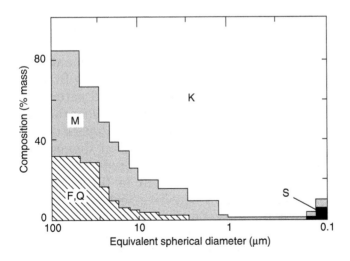

Figure 8.11 Mineral composition of Cornish kaolin with particle size. (F, Q) feldspars and quartz; (M) mica; (S) smectites; (K) kaolinite. (From W.B. Jepson, *Philos. Trans. R. Soc., London A*, 311, 411–432, 1984. With permission.)

depends on the deposit and, in many other cases, dispersions of pure kaolinite cannot be obtained by fractionation.

When the hydrogen bonds and the dipole interactions which hold together the silicate layers of kaolinite crystals are weakened by intercalation of suitable organic molecules, the crystals can be separated into thinner lamellae under the action of mechanical forces. This reaction has been used by Chinese ceramists to improve the quality of porcelain [100]. Stable colloidal dispersions of kaolinite lamellae can be prepared by partial delamination of the particles when kaolinite is treated with DMSO and ammonium fluoride [101]. Ammonium fluoride is added because exchange of some OH^- groups by F^- ions reduces the number of hydrogen bonds and the bonding energy between the layers [44].

D.   Dispersions of Smectites and Vermiculites

Smectite particles may be as large as $2\,\mu m$ and as small as $0.1\,\mu m$ with average sizes of about $0.5\,\mu m$ [3,102]. The morphology of individual particles ranges from lamellar to lath and even to fiber shapes, but mostly the particles are of irregular shape. Aggregates may be compact, foliated, or reticulated [99,103].

The singular processes proceeding during fractionation of smectites by sedimentation distinguishes smectites from all other clay minerals. Pretreatment reactions and saturation with sodium ions causes an increasing delamination: the particles disarticulate into the individual silicate layers during sedimentation. The particles, whose equivalent diameter is measured by sedimentation are artificial products and are not the same particles originally present in the bentonite. Nevertheless, fractionation is a sensitive tool for detecting differences between bentonites of different origin. A bentonite consisting of particles of various thickness and plate widths may be fractionated by sedimentation at conditions whereby all particles are delaminated. The mass content of the different particle size fractions is then representative of the number of silicate layers with a given diameter (mean spherical equivalent diameter), which originally were aggregated to thicker crystals. Naturally, nothing can be said about the thickness of the original particles. Thus, the particle size distribution obtained by fractionation represents the plate width distribution in the parent bentonite. A bentonite, which produces fractions of very fine particles, must contain particles of small diameters. In fact, bentonites of various deposits differ mainly in the particle size distribution below 2 μm (Figure 8.12).

As many montmorillonites are formed by alteration and weathering reactions, the differently sized particles may not be identical in layer charge. However, the mean layer charge of the particles of various fractions often changes only slightly [32]. For instance, the layer charge of montmorillonite of Wyoming only increases from 0.27 charges/formula unit (<0.06-μm fraction) to 0.28 charges/formula unit (2- to 63-μm fraction). The Bavarian montmorillonite (bentonite from Niederschönbuch) shows similar changes: 0.27 charges/formula unit for <0.06-μm particles and 0.29 charges/formula unit for 2- to 63-μm particles. However, distinct changes of the charge distribution curves are observed. During peptization and fractionation, the particles are completely disarticulated. Afterward the layers are reaggregated by coagulation, and the sequence of the differently charged layers (as a consequence of charge heterogeneity, see Section V.C) must not be the same as in the parent material. Dialysis can also change the charge distribution to some extent, not only because of the disaggregation–re-aggregation mechanism but also because of the increased risk of chemical attack on the thin silicate layers [32].

As a consequence of the disarticulation of smectite *crystals* into the individual silicate layers (Figure 8.6), illite–smectite mixed-layer *crystals* can disintegrate at the low-charged interlayer spaces. The type of the produced fundamental particles (Section V.C) depends on the charge distribution, that is the variation of the cation density from interlayer space to interlayer space (Figure 8.13). The particles do break up only at the interlayer spaces with cation densities typical of smectites. The way in which the I/S mixed-layer crystals delaminate has technical consequences because

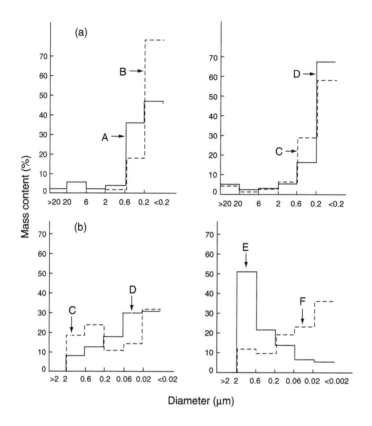

Figure 8.12 Particle size distribution of several bentonites. (a) Mass content, related to the original material; (b) Mass content, related to the fraction <2 μm. A, bentonite, Kumhausen, Bavaria; B, *Tixoton*, Kumhausen, Bavaria; C, bentonite, Belle Fourche, Dakota; D, bentonite, Amory, Mississippi; E, bentonite Cameron, Arizona; F, bentonite, Schwaiba, Bavaria. (From G. Lagaly and R. Fahn, in *Ullmann's Encyclopedia of Technical Chemistry*, 4th ed., Vol. 23, Verlag Chemie, Weinhein, 1983, pp. 311–326. With permission.)

many common clays contain I/S materials. The different types of particles produced by the break-up of the mixed-layer crystals of soda-activated clays determine the flow behavior of these dispersed clays (see also Ref. [38]).

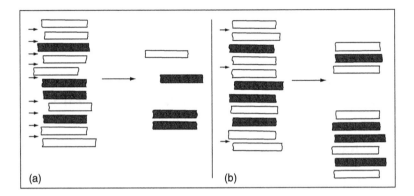

Figure 8.13 Delamination of I/S mixed-layer crystals. Different types of fundamental particles form depending on the variation of the interlayer cation density, i.e., the charge distribution. The particles are only split at the interlayer spaces with sufficiently low cation densities (arrows).

Submicron vermiculite particles could be prepared from macroscopic vermiculite flakes by ultrasound treatment [104].

E. H⁺-Saturated Smectites

Sometimes it is desirable to prepare dispersions of H^+-saturated smectites. Leaching of smectites with acids results in a high degree of H^+ saturation and is accompanied by a severe chemical decomposition of the layers. The aluminum ions (also magnesium and other divalent ions) deliberated by the decomposition are preferentially adsorbed and the remaining structure progressively transforms into the Al^{3+} form [34,105]. In the initial states, the dispersion is stable for a certain time [106]. With increasing alteration, the amount of deliberated aluminum ions causes coagulation. Barshad [107] recommended passing a sodium smectite dispersion (1–2%) rapidly (200 ml/1–5 min) through a train of three exchange resin columns arranged in the order of H^+ resin $\rightarrow$ HO^- resin $\rightarrow$ H^+ resin. An H^+-saturated, highly peptized smectite dispersion is obtained which remains stable for some time.

V. COAGULATION OF CLAY MINERAL DISPERSIONS

A. Stability Against Salts

Well-dispersed clay minerals in the sodium form are coagulated by very small amounts of salts. For NaCl, the critical coagulation concentration c_K varies between 3 and 15 mmol/l:

- Kaolinites, 7–12 mmol/l [108]; 5–10 mmol/l [109]; 16–40 mmol/l (for pH increasing from 4 to 10, Figure 8.14(a) [110]).
- Montmorillonites, 7.8 mmol/l [111]; 3.5 mmol/l [112]; 8 mmol/l (pH $\approx$ 7 [113]); 1–10 mmol/l (for pH increasing from 4 to 10, Figure 8.14(b) [110]); 7–20 mmol/l (pH = 7), 17–68 mmol/l (pH = 9.5) [114]; 8–15 mmol/l (for two montmorillonites and different particle sizes at pH = 6–7 [115]); 12 mmol/l [116]; 10–44 mmol/l (for pH increasing from 5 to 9.8 [117]); 15–32 mmol/l (for pH increasing from 6.4 to 9 [118]).
- Beidellite, 5–7 mmol/l (pH = 6–7 [115]).

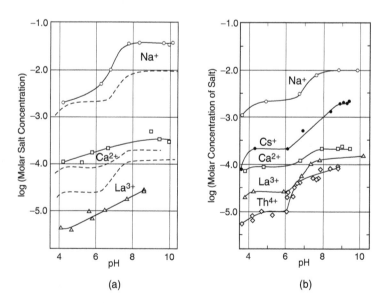

(a)   (b)

Figure 8.14   (*Continued*)

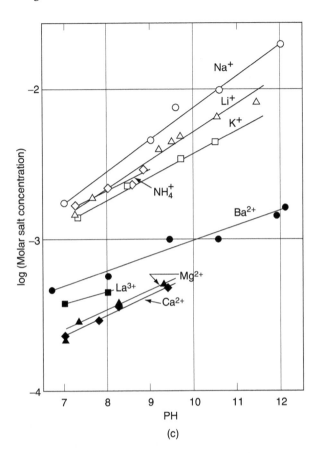

Figure 8.14 Critical coagulation concentration of various nitrates (counterions indicated) as a function of pH. (a) Dispersions of kaolinite (25 ppm); dotted line: sodium montmorillonite for comparison. (From L.S. Swartzen-Allen and E. Matijević, *J. Colloid. Interf. Sci.*, 56, 159–167, 1976. With permission.) (b) Dispersions of sodium montmorillonite (250 ppm). (From L.S. Swartzen-Allen and E. Matijević, *J. Colloid Interf. Sci.*, 56, 159–167, 1976. With permission.) (c) Dispersions of Laponite CP (0.2% by weight). (From R. Perkins, R. Brace, and E. Matijević, *J. Colloid Interf. Sci.*, 48, 417–426, 1974. With permission.)

- Laponite, 10 mmol/l (pH ≈ 8.5 (?) [7]); 2–20 mmol/l (pH increasing from 7 to 12 [119]; Figure 8.14(c)).

The data assembled above reveal the modest influence of the different types of montmorillonites; even beidellite and Laponite give c_K values similar to montmorillonites. As a function of pH, the c_K of sodium montmorillonite in the presence of salts showed a plateau between pH = 4 and 6 or 7 and increased at higher pH (Figure 8.14(b)) [110,117–119]. The dispersion coagulated spontaneously at pH<3.5 and was destabilized by the base necessary to rise the pH above 10.5. The c_K of Laponite increased linearly with pH (Figure 8.14(c)). A stepwise increase was also found for sodium kaolinite and $NaNO_3$, but not for sodium kaolinite and $Ca(NO_3)_2$ and $La(NO_3)_3$ (Figure 8.14(a)). Below pH = 6 the dispersions were destabilized by heterocoagulation between edges (+) and faces (−), and the critical coagulation concentration was very low. Recharging of the edges increased c_K of sodium montmorillonite from about 2 to 10 mmol/l $NaNO_3$. Coagulation may then occur by edge(−)– face(−) contacts. The importance of this type of contacts for the coagulation of pyrophyllite (no layer charge, only pH dependent edge charges) was clearly established [39]. It is also conceivable that band-like aggregates form with small contact areas between the faces. Coagulation by face–face contacts operating over large contact areas requires much higher salt concentrations.

The c_K values of Laponite CP at pH 7–10 were below the c_K values of the montmorillonite. The charge density of the Laponite layers is distinctly lower and probably negative edge charges gradually arise in the stronger alkaline medium only. As Laponite is very sensitive to acids, the dispersion could not be adjusted to pH below 7.

When sodium diphosphate is added to the dispersions, preferential binding of $H_2PO_4^-$ and HPO_4^{2-} ions by ligand exchange increases the edge charge density to a level that the salt concentration required to coagulate the dispersion by edge(−)/face(−) contacts approaches the value for face(−)– face(−) coagulation.[1] For instance, the c_K of NaCl for sodium

---

[1] In a similar way adsorption of the anionic surfactant sodium dodecylsulfate increased the C_K of NaCl to 136 mmol/liter at 10^{-1} mol/l surfactant [121].

montmorillonite dispersions increased to about 200 mmol/l when 0.1 mmol/l $Na_4P_2O_7$ was added [120]. The c_K value remained in the same order of magnitude when the concentration of diphosphate was increased (Table 8.4). As the area between two faces is much larger than between an edge and a face, coagulation occurs by face–face contacts. Keren et al. [117] suggested that face(−)–face(−) aggregation between two platelets is initiated by surface regions that, due to the charge heterogeneity (Section II.D, E), have lower specific charge densities than the average value of the montmorillonite surface.

Typical values of face–face coagulation are 0.25–0.30 M NaCl for beidellite, 0.33–0.38 M NaCl for montmorillonite from Cyprus, and 0.36–0.44 M NaCl for montmorillonite from Wyoming [115,122]. The more highly charged layers of beidellite (average surface charge density $\sigma_0 = 0.13$ C/m^2) are coagulated at distinctly lower NaCl concentrations than the montmorillonitic layers ($\sigma_0 = 0.10$ C/m^2). This is a consequence of the different charge distribution. Most of the negative charges in the layers of the beidellite are produced by the Al

Table 8.4 Effect of diphosphate addition on the critical NaCl concentration c_K of sodium beidellite dispersions (fraction 0.1–2 μm, 0.025% (w/w) beidellite) [115] and of polyphosphate addition on sodium montmorillonite (Wyoming) dispersions [116]

Beidellite			Montmorillonite	
$Na_4P_2O_7$ added (mmol/liter)	pH	c_K (mmol NaCl/l)	$NaPO_3$ added (mmol/l)	c_K (mmol NaCl/l)
0	6	6	0	12
1.25	6	230[a]	0.01	20
5.0	8.3	250[b]	0.1	80
10.0	9	270	1.0	120[b]
12.5	9.3	280		
25.0	9.7	310		

[a] Sodium montmorillonite dispersion at 0.1 mmol/l $Na_4P_2O_7$: $c_K = 195$ mmol NaCl/l [120].
[b] 0.1% (hydrogen, sodium) montmorillonite dispersion with 50 μmol $(NaPO_3)_6$/g montmorillonite: $c_K = 100$ mmol/l NaCl at pH = 2; $c_K = 200$–250 mmol/l NaCl at pH = 4; $c_K = 350$–400 mmol/l NaCl at pH = 8 [129].

for Si substitution and are therefore located in the tetrahedral sheets near the faces of the layer. The charges in montmorillonite created by substitutions and defects in the octahedral sheet are centered in the middle of the silicate layer. Thus, a larger number of counterions will reside in the Stern layer around the beidellite particles. Beidellite behaves as a colloidal particle with a constant surface charge density of about 0.07 C/ m^2 [115].

Coagulation of sodium montmorillonite with other than sodium as counterions showed the same trend with increasing pH as for $NaNO_3$ (Figure 8.14(b)). As expected, the c_K values of $CsNO_3$ were considerably lower than for $NaNO_3$. The critical concentrations of monovalent cations for dispersions of Laponite were in a strange order (Figure 8.14(c)).

The visual determination of coagulation concentrations [79,120,123] requires very diluted dispersions (< 0.025%). The c_K values of more highly concentrated sodium montmorillonite dispersions (< 2%) could be derived from viscosity measurements. The salt concentration at which the viscosity steeply increased was considered to be the critical coagulation concentration [124]. Noteworthy is the increase of the c_K values of most salts with the solid content [124,125]. It is the consequence of counterion adsorption in the Stern layer [123,126,127]. In the presence of phosphates, however, the c_K of the 2% sodium montmorillonite dispersion was smaller than for the 0.025% dispersion [125].

B. Calculation of Interaction Energies

The interaction energy between the dispersed particles of clay minerals with alkali metal ions as counterions is calculated in terms of the DLVO theory. As the charge density of the faces is determined by the isomorphous substitutions and defects within the layer, the calculations are started under the assumption that the charge density of the faces remains constant at variable salt concentrations. This may be done in an easy way by the method of van Olphen [79].

Lubetkin et al. [128] measured the pressure created by several alkali montmorillonites and beidellites as a function of the distance between the plates (calculated from the mass content of smectite assuming that the particles are completely delaminated) (Figure 8.15(a)). At separations above 5 nm the repulsive pressure arose solely from electrostatic repulsion. On the basis of diffuse layer interactions reasonably good agreement between theory and experiment was obtained (Figure 8.15(b)). The curves were calculated for a constant potential of $\psi = 150$ mV or in case of constant charge for $\sigma_0 = 0$ 0.118 C/m^2 (which is slightly above the typical charge density of Wyoming montmorillonite). The calculations did not take into account such factors as finite size of the ions and the particles and the charge heterogeneity. The silicate layers in the dispersion carry different charge densities, for Wyoming montmorillonite, for instance, varying between 0.08 and 0.12 C/m^2 (corresponding to interlayer cation densities of 0.24

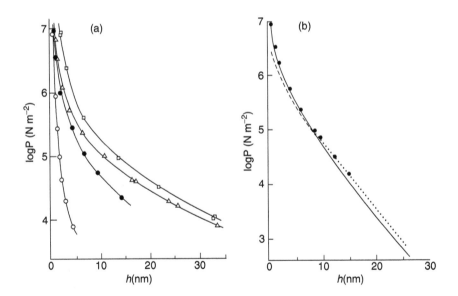

Figure 8.15 Pressure against plate distance h for Wyoming montmorillonite [128]. (a) In $10^{-4} M$ salt solutions of various counterions: □, Li$^+$; △, Na$^+$; ●, K$^+$; ○, Cs$^+$. (b) Experimental data for Li$^+$ montmorillonite in $10^{-2} M$ LiCl (●), and calculated curves for the constant potential model (---), constant surface charge model (—), and by approximate equation (Equation (7) in Ref. [128]) (....).

and $0.34\,eq/(Si, Al)_4\,O_{10})$. Ottewill [81] noted that the theory should be corrected for the fact that due to the finite volume of the dispersion a partition of the counterions between the clay dispersion and the external finite reservoir has to be considered.

The effect of tactoid formation is clearly seen in Figure 8.15(a). In the presence of alkali metal ions other than lithium and sodium ions the distances between the silicate layers are no longer equal; they are smaller within the tactoids than between the tactoids (Section III.C) and, therefore, the pressure decreases more strongly.

At separations of $\sim1\,nm$ the discontinuous decrease of the basal spacing indicates transition into the quasicrystalline structure. It is therefore no longer reasonable to calculate the interaction forces on the basis of the simple DLVO model. As discussed in Section III.B, the distribution of the interlayer cations differs considerably from that of two interacting double layers. The hydration shells around the interlayer cations resist further compression of the interlayer space.

Tombacz et al. [129] considered H^+-saturated montmorillonite as a solid acid with ionizable surface groups. They calculated the density of these sites and their intrinsic ionization constants from potentiometric and conductometric titrations [130,131]. Two groups of ionizable sites with intrinsic ionization constants $pK_{s1} = 2.6$ and $pK_{s2} = 6.4$ were distinguished. The total number of sites, $N_1 + N_2 = 4.8 \times 10^{17}$ (sites/m^2), was calculated from the cation exchange capacity of 0.59 meq/g (montmorillonite of Kuzmice, CSFR). The ratio of weak and strong acid groups was $N_2/N_1 = 0.41$, i.e., 29% of all sites were weak acidic centers. The most acidic sites were the H_3O^+ ions exchanged for the exchangeable sodium ions [34]. The surface potential as a function of pH was calculated from the charge densities. At $10^{-3}\,M$ NaCl and above pH 4, the surface potential showed a plateau at $185\,mV$. The surface potential decreased to $70\,mV$ in $10^{-1}\,M$ NaCl and to $30\,mV$ in $1\,M$ NaCl, and the plateau extended from pH ≈ 6. The total interaction curves (Hamaker constant $A = 0.5 \times 10^{-20}$ J) on the basis of these surface potentials were hypothetical at small distances (see remark above), but they clearly showed that the maximum of the total interaction energy disappears at about $0.1\,M$ NaCl (pH $= 2$), $0.3\,M$ NaCl (pH $= 4$), and $0.4\,M$ NaCl (pH $= 8$), which agrees with the experimental data ($0.1\,M$, 0.2–$0.25\,M$, and 0.35–$0.40\,M$ at pH $= 2$, 4, 8 [129] and 0.36–$0.44\,M$, pH $= 9$ [115]).

A quite different (and for a colloid scientist strange) view was put forward by Low on the basis of extensive studies on the clay–water system. The hydration of the clay mineral surface was seen as the primary cause of swelling. This non-specific interaction of water with the surfaces of clay particles could not be fully explained. Hydration of the interlayer cations was assumed to be of minor importance [82,83,132].

In contrast, Delville and Laszlo [62] showed that the Poisson–Boltzmann formalism correctly reproduces the relation between the interlamellar distance and the swelling pressure. The driving force is the stabilization of water molecules within the interlamellar force field. In all cases, the Poisson–Boltzmann approximation, modified to incorporate ion–polyion excluded volume effects, led to a concentration profile in agreement with Monte Carlo calculations. Recently, Quirk and Marčelja [133] examined published data for extensive swelling of Li^{+}-montmorillonite as revealed by $d_{(001)}$ spacings over the pressure 0.05–0.9 MPa and 1–$10^{-4}\,M$ LiCl. The Poisson–Boltzmann theory and DLVO double-layer theory satisfactorily predicted surface separations over the range 1.8–12.0 nm. The DLVO theory with a 0.55 nm thick Stern layer indicated Stern potentials of -58 to -224 mV (for 1–$10^{-4}\,M$ LiCl) and a constant Gouy plane charge of $0.038\,C/m^{2}$ (about 30% of the crystal lattice charge). There was no additional pressure contributed to hydration forces for surface separations of about 1.8 nm or larger. (However, the hydration force was considerable for muscovite with a surface charge density about three times that of montmorillonite [134].)

C. Coagulation of Dispersions Containing Two Smectites

The coagulation of dispersions containing differently charged particles was studied with colloidal smectites. When a dispersion of delaminated low and highly charged smectites is coagulated by addition of, for instance, sodium chloride, the particles can aggregate in different ways (Figure 8.16). When selective coagulation occurs, one sort of particles is formed first so that the coagulate consists of a mixture of crystals that contain the same layers as the starting crystals (but may differ in size and thickness). When low and highly charged layers aggregate within the individual crystals, mixed-layer crystals grow with random, regular, or zonal (segregation) layer sequences. The problem is to analyze the coagulate and find out which type of aggregation is predominant. The analysis is further complicated by the fact that the charge density varies to some extent within the individual crystals of the parent smectites (charge heterogeneity). There is only one method that allows a clear distinction between selective coagulation and random or regular mixed-layer formation. This

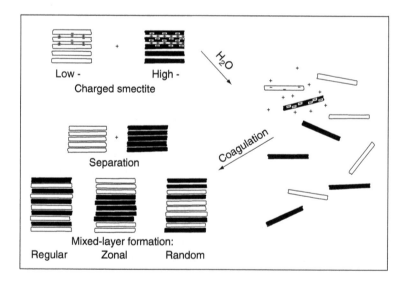

Figure 8.16 Disarticulation of sodium smectite crystals in water; formation of a colloidal dispersion and re-aggregation to crystals by coagulation (surface charges and interlayer cations not entirely shown). (From E. Frey and G. Lagaly, *J. Colloid Interf. Sci.*, 70, 46–55, 1979. With permission.)

is the measurement of the charge distribution by the alkylam-monium ion exchange (Section II.E) [30,32,115,135].

Two differently charged smectites were used: a montmor-illonite (Upton, Wyoming) with a mean surface charge density $\sigma_0 = 0.096 \, C/m^2$ (mean interlayer cation density $\bar{\xi} = 0.192 \, C/m^2$) and beidellite (Unterrupsroth, Germany) with $\sigma_0 = 0.13 \, C/m^2$ ($\bar{\xi} = 0.26 \, C/m^2$). The cation density in the interlayer spaces of the montmorillonite varied from 0.17 to 0.25 C/m^2 and of beidellite from 0.20 to 0.35 C/m^2 (Figure 8.17).

The homoionic sodium smectites were dispersed in $0.01 \, M$ sodium diphosphate solution at pH > 7 to give disper-sions with a mass content of 250 mg/l. Equal volumes of these

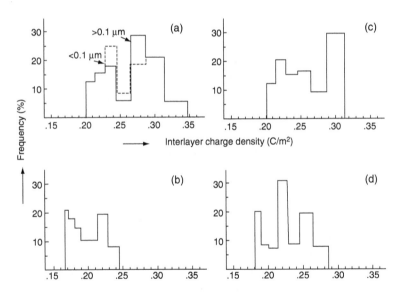

Figure 8.17 Formation of mixed-layer crystals by coagulation of dispersions containing two smectites. (a) Pure beidellite (Unter-rupsroth, Germany; particle size fraction $<0.1 \, \mu m$ and $0.1–2 \, \mu m$). (b) Pure montmorillonite (Wyoming, USA; particle size fraction $<2 \, \mu m$). (c) Coagulated material from the mixed colloidal dispersion, particle size fraction $0.1–2 \, \mu m$. (d) Coagulated material from the mixed colloidal dispersion, particle size fraction $<0.1 \, \mu m$. (From E. Frey and G. Lagaly, *J. Colloid Interf. Sci.*, 70, 46–55, 1979. With permission.)

colloidal dispersions were mixed and coagulated under different experimental conditions. The following results were
obtained:

1. The type of aggregation was determined by the par-
    ticle size and was less dependent on the experimental
    conditions.
2. Larger particles (fraction 0.1–2 μm) were selectively
    coagulated. Experimental conditions could be
    selected to separate beidellite from montmorillonite:
    a. Slow coagulation by adding NaCl solution grad-
        ually to rise the Na^+ concentration to $0.28\,M$,
        pH = 9. Coagulate: mostly beidellite-like par-
        ticles, remaining dispersion: dispersed montmor-
        illonite.
    b. Rapid coagulation by fast addition of NaCl, final
        concentration $0.5\,M$ NaCl. Coagulate: mixtures of
        low and highly charged particles. As discussed in
        Section V.A, the beidellite coagulated at a lower
        salt concentration. The particles composed of low
        or highly charged layers were not identical to the
        particles of the starting materials. The charge
        distribution curve of the highly charged particles
        (Figure 8.17(c)) was substantially different from
        that of the starting beidellite particles (Figure
        8.17(a)).
3. When the particles were smaller than 0.1 μm, the
    coagulate consisted of mixed-layer crystals. The
    charge distribution curve (Figure 8.17(d)) clearly
    proved that low and highly charged layers succeeded
    in the same crystal. The layer sequence was not com-
    pletely random; swelling tests revealed a certain
    amount of segregated layers [135].

In Figure 8.18 is explained why only very small particles
are mixed during coagulation. The maximum total interaction
energy $V_{t,m}$ is calculated for particles of 10^2 and $10^4\,nm^2$
and surface charge densities 0.069, 0.096, and $0.12\,C/m^2$.
The total interaction curves are obtained by the linear superposition approximation for constant surface charge density

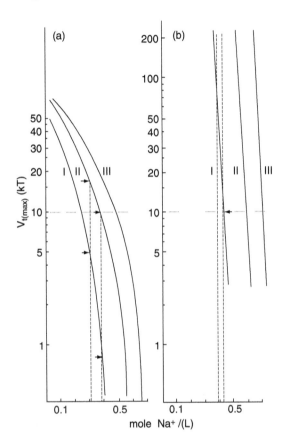

Figure 8.18 Maximum interaction energy (in kT/plate) as a function of Na^+ concentration for particles $10 \times 10\,nm^2$ (a) and $100 \times 100\,nm^2$ (b). Surface charge densities: I, 0.069; II, 0.096; III, 0.120 C/m^2; calculated for constant charge density, Hamaker value 5×10^{-20} J. (From E. Frey and G. Lagaly, *J. Colloid Interf. Sci.,* 70, 46–55, 1979. With permission.)

[79]. The curves are cut at half distances of 0.5 nm ($d_1 = 2\,nm$) because at smaller distances DLVO calculations are no longer appropriate. At $0.3\,M$ NaCl, $V_{t,m}$ is about $5\,kT$/particle for the small particles and $\sigma_0 = 0.069$ C/m^2 (representative of beidellite) and $17\,kT$/particle for $\sigma_0 = 0.096$ C/m^2 (montmorillonite) (Figure 8.18(a)). An increase of concentration to $0.38\,M$

NaCl leads to $V_{t,m} \approx 1\,kT$/particle for $\sigma_0 = 0.069\,C/m^2$ and $V_{t,m}$ $\approx 10\,kT$/particle for $\sigma_0 = 0.096\,C/m^2$. Thus, mixing of both types of layer is highly probable during coagulation. The curves calculated for larger particles (Figure 8.18(b)) clearly reveal that, at a salt concentration of about $0.4\,M$, only the beidellitic layers are coagulated whereas $V_{t,m}$ of the montmorillonitic layers remains so high that coagulation is prevented.

It is interesting to consider the depth of the secondary minimum. For larger layers with low charge density the depth is about $50\,kT$/particle and it is $\approx 20\,kT$/particle for the more highly charged layers. Thus, aggregation of the low-charged layers is initiated by preliminary demixing in the secondary minimum.

In more sophisticated calculations, the total interaction energies have to be calculated between differently charged plates. This reduces $V_{t,m}$ and makes the mixing of the small particles still more probable [136,137]. For large particles, the differences of $V_{t,m}$ remain large enough to prevent the layers from mixing during coagulation. The effect of particle size on the simultaneous coagulation of differently charged particles is also evident from the calculations of Pugh and Kitchener [138].

The delamination of illite–smectite mixed-layer crystals was studied in detail [74,75]. At conditions of delamination, only the smectitic interlayer spaces expand largely, and the particles break asunder into *fundamental particles* consisting of one, two, or a few illitic layers (Figure 8.13, Section IV.D). The physical dimensions of the fundamental particles were determined by TEM techniques. Once the clay minerals had been fully disarticulated, mixed colloidal dispersions could be prepared which, after re-aggregation, produced illite–smectite mixed-layer crystals substantially different from the starting materials. Aggregation may be performed not only by coagulation but also by other methods like simple air-drying, spray drying, or freeze-drying. Some possible applications are directed to the design of special heterogeneous catalysts and the preparation of very thin films and coatings [139,140].

VI. FLOCCULATION AND STABILIZATION OF CLAY DISPERSIONS BY MACROMOLECULES

A. Flocculation by Polyanions

Polymers are extensively used as flocculating agents for clay dispersions. In practical applications, polyanions are more effective in flocculating clay dispersions than polycations. Polyanions are attached to the particles at a few sites, and larger parts of the macromolecules remain free in solution to form bridges between neighboring particles.

Several studies are reported on the adsorption of polyacrylates and polyacrylamides on montmorillonite [94,141,142] and on kaolinite [143–147]. Polyacrylates and hydrolyzed polyacrylamides are bound by complex formation between carboxylate groups of the polyanion and aluminum ions exposed at the edges. Thus, polyacrylate adsorption on sodium montmorillonite reaches a maximum at pH 7.

Polyamides, as a consequence of hydrolysis, often contain carboxylate groups, so that they may be attached at the edges like polyacrylates. In addition, two further binding mechanisms between the clay mineral surface and the amide groups are operative:

1. *Hydrogen bonds between the amide groups and aluminol and silanol groups at the edges of kaolinite particles.* It seems that formation of these hydrogen bonds is competitive with hydrogen bond formation between neighboring surface OH groups (aluminol and silanol groups). Thus, hydrogen bonding of polyacrylamide to silanol groups as anchoring sites is promoted when, in acidic medium, neighboring aluminol groups are protonated. In alkaline medium, bonding to aluminol groups is favored, when the neighboring silanol groups are dissociated. The amount of polyacrylamide adsorbed by kaolinite therefore goes through a minimum at pH 6–8 [144,146].

2. Polyamide groups are protonated by the increased acidity of the interlayer water molecules of smectites and then bound by electrostatic forces [141].

The addition of polyacrylamide to sodium kaolinite dispersions (2% by mass) initiated different types of aggregation [148]. The fraction of flocculated kaolinite decreased first to a minimum, then increased to a maximum (Figure 8.19). Polyacrylamide that was adsorbed at the edges broke up edge(+)–face(−) contacts and increased the dispersibility of the kaolinite particles (B→C in Figure 8.19).[2] Bridging of the particles by the macromolecules at somewhat higher levels of polyacrylamide addition (C→E) increased the amount of flocculated kaolinite, which at high dosages of polymer again was reduced by steric stabilization (E→F). The floc density (determined from sediment volume and weight after freeze drying) changed in inverse direction to the amount flocculated. The presence of polymer on the edges of the particles prevented the formation of extended networks during sedimentation, and the particles were more densely packed (C in Figure 8.19). Bridging produced more loosely packed flocs (E in

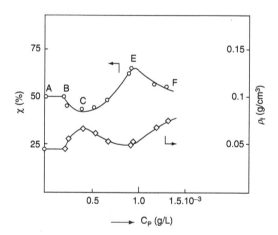

Figure 8.19 Flocculation of sodium kaolinite dispersions (2% by weight) with polyacrylamide ($M = 1.2 \times 10^6$) in $10^{-4} M$ NaCl (concentration c_P) at pH = 3.7. ○, fraction χ of kaolinite flocculated; ◇, density ρ_f of the flocs [148].

---

[2] On the edges of kaolinite particles pronounced adsorption of polyacrylamides was in fact observed [149].

Figure 8.19). The increase of floc density from E to F revealed that the most loosely packed aggregates were first redispersed by steric stabilization.

Bottero et al. [142] studied the effect of polyacrylamide on dispersions of sodium montmorillonite. The polyacrylamide did not flocculate montmorillonite particles but influenced the structure of the tactoids.

B. Flocculation by Polycations

Polycations are strongly adsorbed, and many segments are attached to the negative surface, so that bridging between particles does not occur as easily as in the presence of polyanions. In many cases flocculation occurs as a consequence of charge neutralization. This type of flocculation often requires a sophisticated adjustment of clay, polymer and salt concentration, and pH [150].

Durand–Piana et al. [151] examined the effect of cationic groups attached to polyacrylamide on the flocculation of sodium montmorillonite dispersions. Random copolymers of m units acrylamide (AM) and n units N,N,N-trimethyl aminoethyl chloride acrylate (CMA) were synthesized:

$$----(CH_2-CH)----(CH_2-CH)------$$

$$
\begin{array}{cc}
| & | \\
C=O & C=O \\
| & | \\
NH_2 & O-CH_2-CH_2-\overset{\oplus}{N}-CH_3 \quad Cl^{\ominus}
\end{array}
$$

with CH_3 and CH_3 substituents on N.

The cationicity $\tau = n/(n+m)$ was varied between 0 and 1. At low cationicities flocculation occurred by bridging of the particles by the weakly charged polyelectrolytes. The optimal flocculation concentration decreased with increasing molecular mass and, for $\tau \geq 0.01$, with increasing cationicity. Above ≥ 0.2 flocculation seemed to occur by charge neutralization, and the optimal flocculation concentration became independent of the molecular mass (Figure 8.20). It is interesting to note that the saturation value of adsorption decreased with the

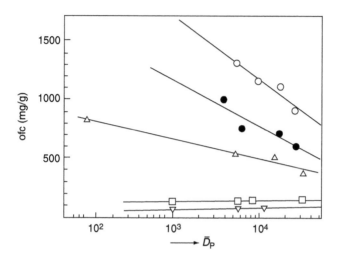

Figure 8.20 Optimal flocculation concentration (ofc, mg/g sodium montmorillonite) versus degree of polymerization, $\bar{D}_p$, for polycations (see formula) of various cationicities τ: $\bigcirc$, $\tau = 0.01$; $\bullet$, $\tau = 0.05$; $\triangle$, $\tau = 0.13$; $\square$, $\tau = 0.30$; $\triangledown$, $\tau = 1.0$ [151].

montmorillonite concentration when the solid content exceeded 2 g/l. Aggregation of montmorillonite platelets is strong enough to limit the surface area available to polymer adsorption.

Due to the opposite charges polycations penetrate into the interlayer spaces to a certain extent. However, an increasing number of contacts are formed between their positive centers and the surface charges that retard and eventually stop the diffusion of the polycations between the layers. However, complete coverage of the interlamellar surfaces by polycations may be achieved by a similar disaggregation–reaggregation process as proposed for the lysozym adsorption on sodium montmorillonite [152]. When a tactoid, in which the external faces are saturated with the polycations, collides with another tactoid, which has no polycations on its faces, strong interaction between the polymer-covered face and the bare face can peel an individual layer off the either tactoid. Thus, two fresh surfaces are exposed for further interaction with polycations. Eventually, all layers are aggregated inter-leaved with polycations, and form thick particles [153,154].

Parazak et al. [155] studied flocculation of dispersed montmorillonite, kaolinite, illite, and silica by three types of polycations: poly(dimethylamine epichlorohydrin), poly(dimethyl diallylammonium chloride), and poly(1,2-dimethyl-5-vinylpyridinium chloride). The results were interpreted in terms of *hydrophobic interactions*. The idea was that adsorbed polycations, their charges neutralized, are considered as hydrophobic moieties. Contact between these hydrophobic patches of two approaching particles will induce flocculation by hydrophobic interaction, even if only a part of the particle charges are neutralized by the polycations.

In papermaking cationic lattices (*polymer microparticles*) were introduced as flocculants to improve retention, drainage, and formation (*retention aids*) [156]. In other two-component microparticle retention systems cationic polyacrylamide and bentonite particles induce the deposition of the calcium carbonate particles on fiber surfaces. The montmorillonite particles form bridges between the fiber and calcium carbonate particles, both covered with the polycations. Therefore, conditions such as pro-longed or intense mixing that cause delamination of thicker mont-morillonite particles can lead to detachment and subsequent flocculation of the polyamide coated carbonate particles [157].

C. Peptization (Deflocculation) of Clays by Macromolecule

As seen in the previous section, optimal flocculation is reached at distinct polymer concentrations (Figure 8.20). At larger polymer dosages restabilization occurs, and the amount of flocculated clay decreases again. Restabilization is accompan-ied by an increased salt stability. A very instructive example was reported by van Olphen [79] (Figure 8.21). Addition of CMC (sodium carboxy methylcellulose) to a sodium bentonite dispersion first decreased the critical NaCl concentration from 20 mmol/l to 10 mmol/l (sensitizing action of CMC), and then increased the colloidal stability strongly up to 3.5 mol/l. Steric stabilization is the main cause of the enhanced salt stability.

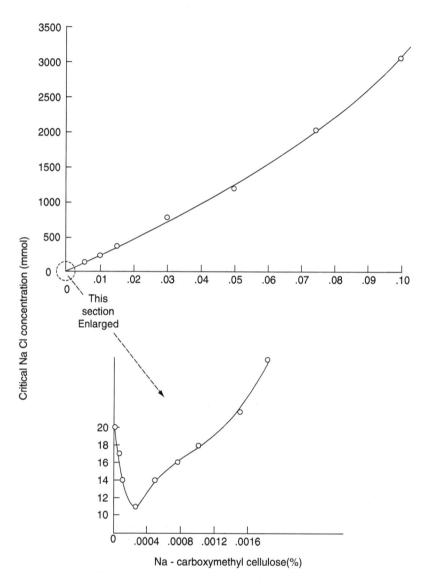

Figure 8.21   Effect of sodium carboxy methylcellulose on the salt stability of sodium bentonite dispersions [79].

The effect of polymer addition seen in Figure 8.21 is basically different from the action of several other polyanions like natural tannates (Figure 8.22(a)). The addition of very small amounts of Quebracho tannate increased the critical salt concentration strongly to a plateau at 270 meq/l NaCl. This polyanion does not exert steric stabilization but simply acts by recharging the edges. Because tannate is added as sodium salt, the total amount of sodium ions in the dispersion increases to 430 meq/l at the highest dosage of tannate (dotted line in Figure 8.22(a)). This range of the critical counterion concentration is typical of face–face aggregation of clay mineral particles (Section V.A).

Polyphosphate ions belong to the most important deflocculants in practical applications (Section VIII.A). In the presence of polyphosphate, the critical coagulation concentration of NaCl increased to a maximum and decreased with further addition of the polyanion (Figure 8.22(b)). However, the total concentration of sodium ions increased only slightly at higher amounts of polyphosphate (≥ 200 meq/l) and is, again, typical of face–face aggregation.

VII. AGGREGATION OF CLAY MINERAL PARTICLES AND GELATION

A. Types of Aggregation

The most popular is the model of house of cards where the clay mineral particles are held together by edge–face contacts [158–160] (Figure 8.23(a)). However, this type of network only forms when the edges are positively charged or in slightly alkaline medium above critical salt concentrations. The formation of edge–face contacts below pH ≈ 6 is a heterocoagulation between the positive edges and the negative faces of the particles or silicate layers. Establishment of card-houses results in non-Newtonian flow of the dispersions and development of yield stresses in the slightly acidic medium (Figure 8.24 and Figure 8.25). Note the yield value increasing with temperature at pH $= 4.5$. The formation of the card-house

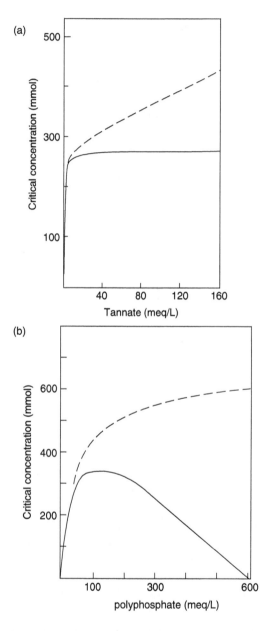

Figure 8.22 Effect of polyanions on the salt stability of sodium bentonite dispersions [50]. —, critical NaCl concentration to attain coagulation; - - -, total Na^+ concentration at the point of coagulation. (a) Quebracho tannate (sodium salt). (b) Sodium polyphosphate [79].

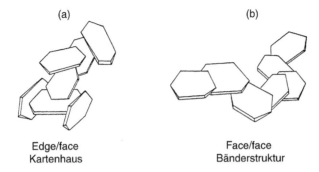

(a)   (b)

Edge/face   Face/face
Kartenhaus   Bänderstruktur

Figure 8.23 Aggregation of clay mineral platelets in (a) card-house and (b) band-type networks by edge–face and face–face contacts.

structure is one of the main causes of plasticity of ceramic masses [158,159].

With increasing pH the network composed of edge–face contacts breaks down and, in the rheological experiments, the shear stress τ (at a given rate of shear, $\dot{\gamma}$) of the sodium montmorillonite dispersion decreases to a sharp minimum (Figure 8.25). The increase in the shear stress at pH > 5 results from the higher degree of delamination and, therefore, from the higher number of particles in the dispersion. Raising the pH above 7 requires increased amounts of NaOH, which reduce the degree of delamination and the electroviscous effect [161] (Section VII.C), and the shear stress decreases again. The maximum of the shear stress at pH ~7 disappears in the presence of calcium ions which impede the delamination of the particles (Figure 8.25) [121].

Weiss and Frank [162,163] first stressed the importance of face–face contacts and the possibility of formation of three-dimensional band-type networks (*Bänderstrukturen*) (Figure 8.23(b)). The band-type network of calcium kaolinite was found to show some elasticity in contrast to the more rigid card-house structure.

The addition of calcium ions has a pronounced effect on the type of aggregation. As discussed in Section III.D, calcium ions held together the silicate layers at maximum distances of

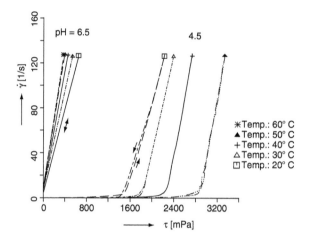

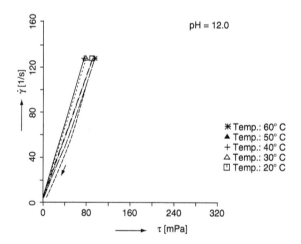

Figure 8.24 Flow curves (shear rate $\dot{\gamma}$ versus shear stress τ) for sodium montmorillonite dispersions (Wyoming, 4% by weight) at various pH values and different temperatures (pH adjusted by addition of HCl or NaOH).

1 nm (basal spacing 2 nm). Even small amounts of calcium ions nucleate face–face contacts and build up band-type networks. The flow behavior of Ca/Na–bentonite dispersions

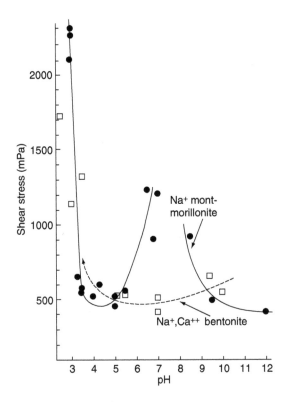

Figure 8.25 Dependence of the shear stress (at $\dot{\gamma} = 94.5$ s^{-1}) on the pH value for the sodium montmorillonite dispersion (from Wyoming, M 40; ● —) and the dispersion of the parent sodium calcium bentonite (□ - - - -). The pH values were determined with indicator sticks just before the rheological measurements. (From T. Permien, G. Lagaly, *Clays Clay Miner.*, 43, 229–236, 1995. With permission.)

is, therefore, complex and depends sensitively on the ratio of Ca/Na in the dispersion [164]. Small amounts of calcium ions added to a sodium montmorillonite dispersion promote face–face contacts and stabilize band-type networks (Figure 8.26(b)). At larger amounts of calcium ions, the bands are contracted to smaller aggregates and, eventually, particle-like assemblages, and the network break asunder (Figure 8.26(c)). Thus, homoionic calcium smectites show only a modest tendency of formation of band-type networks.

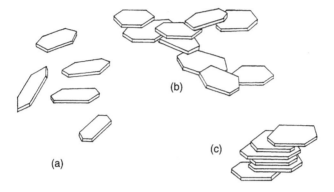

Figure 8.26 Modes of aggregation in the presence of calcium ions: (a) single platelets or silicate layers, (b) band-type aggregates, (c) compact *crystals*.

It may be assumed that calcium ions held between the negative charges at the edges and faces of two approaching particles act in a similar way as in the interlayer space and limit the particle distance to 2 nm. Stable edge$(-)$–Ca^{2+}–face$(-)$ contacts would then be created which would build up a stable card-house network. However, I consider this behavior very unlikely. Calcium ions lying at the edges of particles with very irregular contour lines are in a quite different force field than between the planar silicate layers in the interlayer space. The modest importance of the edge$(-)$–Ca^{2+}–face$(-)$ contacts is expressed by the low shear stresses of the calcium montmorillonite dispersions in alkaline medium.

B.   Sedimentation, Filtration

In many practical applications (Chapter 8) sedimentation and type of the sediments play an important role. Figure 8.27 shows highly dispersed clay mineral plates and two types of aggregates. The sediments or filter cakes formed from these dispersions are substantially different. Highly peptized dispersions of individual platelets (or silicate layers) form virtually impermeable filter cakes and compact, dense sediments that are difficult to stir. The formation of filter cakes

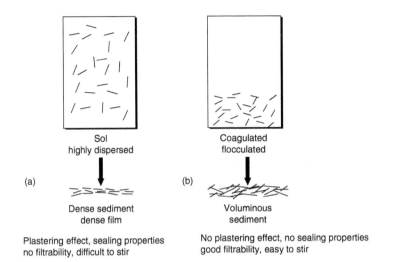

Figure 8.27 Structure and properties of sediments formed from well dispersed (a) and aggregated (b) particles.

finds many applications in sealing operations and causes the plastering effect of drilling muds (Section VIII.B). It is of great importance in producing ceramic casts (Section VIII.A).

The compact sediments or filter cakes form if the interaction between the particles is repulsive. When the particle–particle interaction becomes attractive, the particles aggregate to some extent, and voluminous sediments are formed with large pores between the clay mineral plates. These sediments are easy to redisperse by stirring and show no plastering effect. In the simplest way settling of the particles and properties of the sediments are adjusted by the Na^+/Ca^{2+} ratio (Section VII.A). Dispersions of calcium montmorillonite showed an unusual settling behavior [165].

The influence of bentonite content, pH, and pressure on the permeability of filter cakes was thoroughly studied by Benna et al. [166,167]. The effect of the aggregation of the clay mineral platelets on the filterability was clearly established.

C. Sol–Gel Transitions

The transition from a sol into a gel and vice versa is very
important in many practical applications (Chapter 8). Gels
are usually described as dispersed systems, which show certain stiffness so that, for instance, the vessel containing the
dispersion can be turned without the dispersion flowing out.
Gels show certain elasticity, and we used creeping measurements to distinguish between sol and gel [168].

The experiment is shortly explained in Figure 8.28.
When a constant shear stress τ_e is applied to the dispersion
within time t_e, the strain increases as shown. At $t = t_e$ the
shear stress is set to zero, the sample relaxes and in case of
a viscoelastic behavior, the strain decreases to a plateau. If the
dispersion is a viscous fluid, the strain increases further. The
reversible part of the compliance $(J = \gamma/\tau_0)$ is $J_{rev} = 100$
$(J_0 + J_R)/(J_0 + J_R + J_N)$. The position of the plateau gives the
elastic $(J_0 + J_R)$ and viscous (I_N) contribution to the compliance. A sol shows $J_{rev} = 0$, a gel $J_{rev} > 0$.

Figure 8.29 shows the reversible compliance of a sodium
montmorillonite dispersion as a function of the NaCl concentration. $J_{rev} > 0$ indicates gel formation at low- and high-salt
concentrations. Stiffening at low-salt concentration results

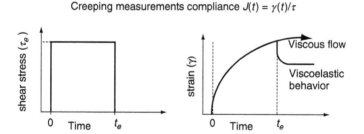

Creeping measurements compliance $J(t) = \gamma(t)/\tau$

Figure 8.28   Creeping experiments: The strain (relative deformation) γ is shown as a function of time t. During the time t_e the
sample is deformed by applying the constant shear stress τ_e. At
$t = t_e$, τ is set to zero and the sample relaxes (viscoelastic behavior,
full line) or flows further (viscous fluid). (From S. Abend and
G. Lagaly, *Appl. Clay Sci.*, 16, 201–227, 2000. With permission.)

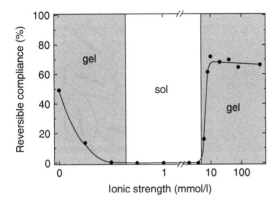

Figure 8.29 Reversible compliance as a function of the NaCl concentration. The 4% (w/w) dispersion of sodium montmorillonite (M 50, Ordu Turkey). (From S. Abend and G. Lagaly, *Appl. Clay Sci.*, 16, 201–227, 2000. With permission.)

from the electroviscous effect (Figure 8.30) [161]. At these conditions the single silicate layers, lamellae, or packets of them are surrounded by the diffuse layers of counterions and are repelled from each other by electrostatic forces [2,79,123,133,169]. When the particle concentration is sufficiently high ($\geq 1\%$ (w/w) for many sodium montmorillonite dispersions). The diffuse ionic layers around the silicate layers, lamellae, or packets restrict the translational and rotational motion of these units. The thin shape of these units is seen as a critical factor for the appearance of the electroviscous effect [170]. The consequence is a certain parallel orientation of the platelets (see Ref. [168]). The dispersion stiffens and becomes gel-like above 3–3.5% (w/w) montmorillonite. This type of gel is called *repulsive gel* (not correctly because the interparticle force, and not the gel, is repulsive). The addition of salt reduces the thickness of the diffuse ionic layer, the particle become again more mobile, and the gel turns into a sol.

The principles of the structure of sodium smectite gels were derived from small-angle scattering experiments of synchrotron beam [171]. A sodium smectite gel (Wyoming montmorillonite, 20 g in 100 g water $\equiv$ 17% (w/w) montmoril-

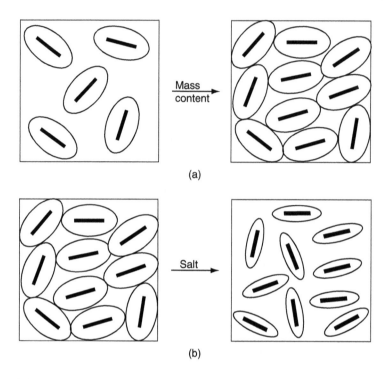

(a)

(b)

Figure 8.30 The electroviscous effect: (a) Sufficiently high-particle concentration reduces the mobility of the particles. (b) When the thickness of the diffuse ionic layers decreases at higher salt concentration, the particles again become more mobile.

lonite) was studied before and after several cooling–heating cycles between $-70°C$ and room temperature. Below $-10°C$ the silicate layers were aggregated to thick particles (500–600 nm thick) which were composed of domains of four to five silicate layers in regular distances with two water layers in between. These domains were separated by zones of one or two silicate layers spaced less regularly by one, three, or four layers of water. During heating to room temperature the interlamellar spaces took up water molecules to form the discrete hydrates with up to four water layers. This process proceeded slowly and could be followed by the changes in the scattering patterns. The insertion of more than four water layers was accompanied by a dramatic expansion of the interlayer space to form a gel in which assemblages of (in an average) four to five almost parallel silicate layers (with d_{001} $\approx 8\,nm$) were still retained. Isolated silicate layers filled the space between these units and the domains of four to five layers, and the single layers were no longer parallel. About 25% of the layers were distributed as single layers. An interesting point is that the domains of four to five layers (in distances of about $8\,nm$) reflected the structure of the particle in the frozen gels. As the changes during cooling–heating cycles were nearly reversible, the deviation of the silicate layers from parallel orientation during heating must be modest, say, by no more than $15°$. Several causes of this peculiar behavior were mentioned [50]. At the same water content the number of layers in the domains was specific of the mineral. An interesting influence of the charge distribution in beidellite gels was noted. The layers constituting the domains remained at spacings of $1.54\,nm$ and did not move in distances of 7–$8\,nm$ as in montmorillonite, saponite, and hectorite [50]. Organization of the layers at different levels was also observed in TEM and small-angle x-ray scattering (SAXS) studies [172,173].

Gel formation at higher salt concentration, above the critical coagulation concentration, is caused by attractive forces between the particles when the van der Waals attraction dominates the electrostatic repulsion (*attractive gel*). At lower salt concentration the interaction is attractive between edges($-$) and faces ($-$), at higher concentrations it becomes also attractive between the faces. Likely, there is a continuous transition from the edge($-$)–face($-$) (card-house) to the face($-$)–face($-$) aggregation (band-type structure). If the forces between the faces are strongly attractive at high-salt concentration, the network is contracted and disrupted (Figure 8.23(b) and (c)). Distinct particles form, the dispersion destabilizes and forms flocs which settle into a sediment.

Attractive gels often show thixotropic behavior. In this case the gel is liquefied by shaking, stirring, or pouring but stiffens again with time. Stiffening and liquefaction are strongly reversible. The cause is a network of weakly adhering particles. When mechanical energy is applied, particle contacts are broken, and the network is disintegrated into many fragments. When resting, the fragments driven by the Brownian motion of the solvent molecules came into contacts again forming an extended network, and the liquefied dispersion becomes gel-like. This reversible process requires attraction, but not too strongly attractive particle–particle interactions. Thixotropy is a very important property in many practical applications of clay, kaolin, and bentonite dispersions (Chapter 8).

The domains of sol, repulsive gel, attractive gel, and flocs are clearly seen in the phase diagrams for two sodium montmorillonites (Figure 8.31). The large domain of sol separates the two types of gels. As expected, the repulsive gel only forms when the particle concentration is >3% (w/w) and 3.5% (w/w), respectively. The salt concentration at which the gel liquefies into the sol increases with the particle concentration because more densely packed particles require thinner diffuse ionic layers to become mobile again. In the case of attraction between the particles, the attractive gel is also built-up at smaller solid contents because band-type aggregates can span a distinctly larger volume. Flocs are only formed at the highest salt concentration and not at too high particle concentrations. When sodium ions are replaced by potassium and cesium ions, the attraction between the particles becomes stronger because these counterions are more strongly adsorbed in the Stern layer. The band-type aggregates are more stable and resist to floc formation. A 2% (w/w) sodium montmorillonite dispersion does not coagulate forming flocs even at the highest KCl and CsCl concentrations, but remains in the state of the gel (Figure 8.32). The liquefying action of phosphate addition [125] is also seen in the sol–gel diagram (Figure 8.32). This effect is also very pronounced for kaolinite dispersions (see Figure 8.34) [125].

D. Hydrogels of Organoclays

Garett and Walker [174,175] first described formation of gels of low-charged vermiculites in water when the inorganic counterions are replaced by butylammonium ions. The gels

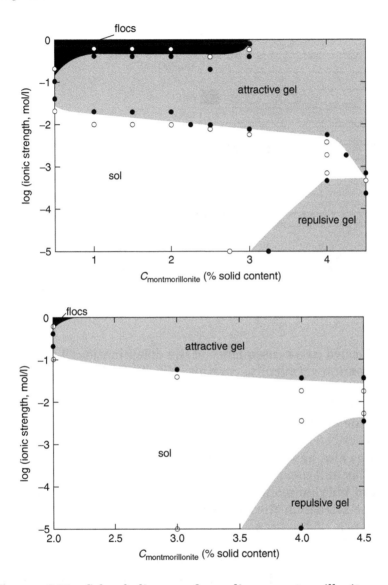

Figure 8.31 Sol–gel diagram for sodium montmorillonite and NaCl. (a) montmorillonite of Ordu, Turkey (M 50); (b) montmorillonite of Wyoming (M 40A). (From S. Abend and G. Lagaly, *Appl. Clay Sci.*, 16, 201–227, 2000. With permission.)

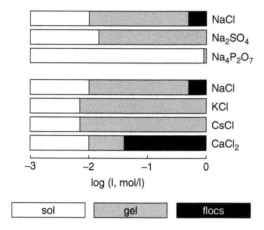

Figure 8.32 Effect of a few salts on the transitions sol–gel and gel–flocs. The 2% (w/w) sodium montmorillonite (Ordu, Turkey, M 50), $I =$ ionic strength. (From S. Abend and G. Lagaly, *Appl. Clay Sci.*, 16, 201–227, 2000. With permission.)

are formed as a consequence of the delamination of the butyl-ammonium vermiculite crystals [176–180]. This very peculiar behavior seems to be related to the organization of the water molecules between the alkyl chains [181]. It is evident that the swelling of the butylammonium vermiculites cannot be described by the DLVO theory because hydrophobic interactions also have to be considered. In a quite different model, Smalley and co-workers discussed the swelling and gel formation of butylammonium vermiculite on the basis of the Coulombic attraction theory (Sogami theory) and postulated the existence of long-range attraction between the vermiculite layers [182,183].

Rausell-Colom and Salvador [184,185] described gel formation of vermiculites in solutions of amino acids like γ-aminobutyric acid, ω-aminocaproic acid, and ornithine. The repulsion between the carboxylate groups accumulated in the interlayer spaces promotes the delamination of the particles. The gels are composed of independently diffracting large packets of several silicate layers (19 layers spaced around $d = 13.5$ nm; Santa Ollala vermiculite in the presence of 2×10^{-2} mol/l ornithine, confined by a pressure of 99.4 g/cm^2). These packets have an average thickness of 260 nm, and they are completely separated from each other by the solution phase. Within the packets, ordered coherent domains (with about six silicate layers in equal distances of 13.1 nm) are separated by layers also in parallel orientation but less regularly spaced [186].

Stable colloidal dispersions of fully delaminated montmorillonites were obtained by exchanging the inorganic interlayer cations by betaines $(CH_3)_3$ $N^+-(CH_2)_n-CO_2^-$ Li^+ (Na^+). These dispersions were more stable against salts than Li^+ and Na^+ montmorillonite (LiCl: $c_K = 60$ and 50 mmol/l for $n = 7$ and 11). Very high LiCl concentrations (>1 mol/l) were needed to coagulate the betaine derivatives ($n > 5$) in the presence of diphosphate [35].

E. Gelation in Organic Solvents

Thickening of organic solvents by hydrophobized bentonites (in a few cases also clays and kaolins) is needed in many practical applications [2,187]. Gelation of these dispersions often requires small amounts of polar additives (water, ethanol) to increase the gel strength [2,188–190]. This effect was explained by the strong orientation of the adsorbed water molecules, which creates giant dipole moments on the particles and hydrogen bonds between the particles [189].

In referring to practical applications the following effect must be noted. In an industrial scale organophilic bentonites usually are prepared by reacting the bentonite with quaternary alkylammonium ions without removing an excess of these cationic surfactants. The presence of these excess amounts can sensitively influence the flow behavior of dispersions in organic solvents (Figure 8.33). The 1% dispersion of a technical dimethyl dioctadecylammonium bentonite in xylene (activated with 0.2% ethanol and 0.02% water) showed low shear stress values ($\tau = 400$ mPa at $\dot{\gamma} = 130\,s^{-1}$). The excess surfactant cations adsorbed by the particles enhanced the electrostatic repulsion, acting like a lubricant between the particles. When these cations were removed by washing, the dispersion

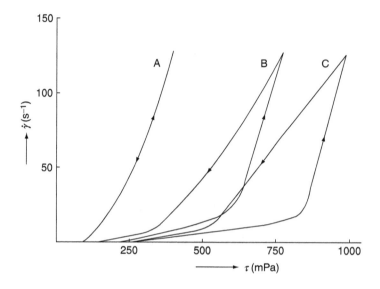

Figure 8.33   Flow behavior (rate of shear $\dot{\gamma}$ against shear stress τ) of a 1% (w/w) dispersion of dimethyl dioctadecylammonium bentonite in xylene containing 0.2% ethanol and 0.02% water. (a) Organo bentonite as obtained with an excess of the alkylammonium salt (1.22 mmol N/g bentonite); (b) after washing out the excess of the alkylammonium salt (1.02 mmol N/g bentonite); (c) again exchanged with alkylammonium ions with the excess of the salt carefully removed (1.33 mmol N/g bentonite). (From K. Jasmund and G. Lagaly, Eds., *Tonminerale und Tone: Struktur, Eigenschaften, Anwendung und Einsatz in Industrie und Umwelt*, Steinkopff Verlag, Darmstadt, 1993, pp. 1–490. With permission.)

stiffened and showed pronounced thixotropy. This example shows that even small changes in the ratio surfactant/bentonite can distinctly change the flow properties and the thixotropic (or antithixotropic) behavior.

VIII.   APPLICATIONS

A.   Common Clays and Kaolins

Ceramic masses contain 25–55% (by mass) of common clay and, for porcelain production, kaolins. Their plastic properties are brought about by the interactions between the clay mineral particles, predominantly by edge(+)–face(−) contacts. Ceramic bodies are produced by manual or mechanical shaping, in the simplest way on a potter's wheel, or by pouring a deflocculated slip (with solid contents up to 72% by mass) into plaster molds. A filter cake builds up on the inside surface of the mold which finally is dried and fired. The final properties of the deposited particle layers depend on the state of aggregation of the dispersion and a delicate balance between coagulation or flocculation and peptization has to be maintained to achieve the optimum properties of the final product.

To make the ceramic mass sufficiently fluid, the card-house-type interactions between the plates must be broken by the addition of deflocculants. Deflocculant addition must be carefully adjusted and should be less than the amount required for optimum deflocculation (cf. Ref. [3]). For instance, a plastic mass containing about 70% by mass kaolin is *liquefied* by 0.3% tetrasodium diphosphate (*pyrophosphate*) at pH = 8. The dispersion can be easily poured when freshly made. Figure 8.34 very nicely demonstrates the effect of diphosphate addition on a kaolin slurry. Increased face–face attraction after addition of 3 g KCl again stiffens the dispersion.

Other deflocculants are sodium salts of inorganic or organic polyanions such as polyphosphates, water glass, and polyacrylates. Dispersions produced by polyphosphate addition become unstable with time, probably because of the formation of complex aluminum phosphates. Some kaolin slurries in the presence of polyphosphate form gels that can give discharge problems. The addition of polyacrylate reduces the risk of gelation [4].

The deflocculating agents are strongly adsorbed at the edges. In acidic medium, they recharge the edges from positive to negative or, at somewhat higher pH, increase the negative charge density. Due to ligand exchange reactions (Figure 8.3) these anions also are bound around the neutral point and in slightly alkaline medium. The edge (+)–face(−) contacts are eliminated, and the card-house structure is no longer stable. Adjusting pH to ≈ 8 supports the effect of re-

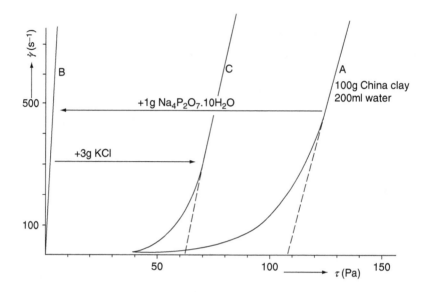

Figure 8.34 Liquefaction of 100 g kaolin in 200 ml water by addition of 1 g tetrasodium diphosphate, $Na_4P_2O_7 \cdot 10\ H_2O$ (A → B) and stiffening of the liquefied kaolin by addition of 3 g potassium chloride (B → C). (Courtesy of Professor Dr. M. Müller–Vonmoos, ETH Zurich; see also Ref. [38]).

charging by the deflocculants. The deflocculating agents also have to keep the calcium and magnesium ions in solution (or, in some cases, to precipitate them) because otherwise these ions would increase the viscosity by initiating face(−)–face(−) and edge(−)–face(−) contacts [38].

The use of kaolins as fillers and coating pigments in the production of paper requires a very strong control of the rheological properties. In paper coating the coating color (a mixture of the deflocculated kaolin and an adhesive) is applied at shear rates as high as $10^6\ s^{-1}$. The important variables that govern the flow properties are not only type and concentration of the deflocculants [149] but also kaolinite particle size, shape, and packing [4].

B. Bentonite Dispersions

Colloidal bentonite dispersions find many important applications. In civil engineering the dispersions are wanted for a variety of uses [2,3]. For slurry walling a thixotropic bentonite dispersion is used as an intermediate supporting material, which obviates the need for any revetting. Only sodium bentonites or soda-activated bentonites can be used which show a significant level of thixotropy. The formation of a thin impermeable filter cake of clay mineral plates on the walls prevents excessive loss of fluid to the surrounding formation (plastering effect).

The second, very important practical application of bentonites is for use in drilling fluids [2,3,79]. About 2.5 million tons of bentonite were required in 1982 for the preparation of drilling muds. The smectites in these dispersions have to act in different ways. During the drilling operation the impermeable filter cake built up by the clay mineral plates on the wall of the hole prevents the loss of fluid into the formation. The drilling fluid is circulated to bring the drill cuttings up and to cool the bit efficiently. When the drilling operation is interrupted before the drill cuttings and sand have been circulated out of the hole, the setting of the cuttings and sand particles is prevented by thixotropic stiffening. In offshore drilling or in drilling through salt domes the mud must be protected against destabilization by salts. A further problem arises from the large increase in viscosity of common drilling muds at elevated temperatures. Hectorite-type minerals were specially formulated to resist high temperatures for drilling deep wells and geothermal areas.

The properties that are most significant in the above-mentioned uses are dispersibility, viscosity, yield value, thixotropy, and good plastering effects. Only natural sodium bentonites or certain sodium-exchanged calcium bentonites have these necessary properties. In practice, a compromise often must be obtained between the different requirements. For instance, development of thixotropic drilling muds requires conditions, which are adverse to an optimum of plastering and pumping properties. Thus, many types of organic and inorganic additives are used to regulate the behavior of the muds.

Stability against salts is obtained by addition of nonionic or anionic polymers (modified starches, sodium carboxymethyl cellulose, tannates) (see also Section VI.C).

Quebracho tannates (Figure 8.22(a)) are the peptizing agents in the *red muds*. The large increase of viscosity in deeper holes can be shifted to higher temperatures by the addition of solid calcium or barium hydroxide. The addition of calcium ions is somewhat contradictory in view of the powerful coagulating action of the divalent metal ions. Complexing of the calcium ions by tannate reduces their coagulating effect. A certain degree of face–face aggregation still occurs, as the particles in the fine red muds are thicker than in the red muds.

The *surfactant muds* provide an interesting example of a dispersion which is in coagulated state but which still shows the properties required for drilling fluids at a reasonable level. Surfactant muds contain polyoxyethylene-type nonionic polymers, polyanions such as sodium carboxymethyl cellulose, and strongly coagulating electrolytes such as calcium sulfate.

In the practical uses described above, natural sodium bentonites (e.g., bentonite from Wyoming) or calcium bentonites converted to the sodium form have to be selected. Natural sodium bentonites generally contain some amounts of calcium ions (sometimes also magnesium ions). Soda activation consists of reacting natural calcium bentonites with soda. The amount of sodium ions added corresponds to about the ion exchange capacity. As the calcium ions are not removed from the system, optimal delamination is not reached (and also not wanted in many practical applications). When about 1 meq soda/g bentonite is added, the interlayer calcium ions are only partially replaced by sodium ions. Increasing amounts of soda displace progressively the interlayer calcium ions. Calcium ions that are not precipitated as $CaCO_3$ promote face–face contacts. The shear stress (at a given shear rate) as a function of the amount of soda added increases to a maximum value at 2–5 meq soda/g bentonite (Figure 8.35) [38]. The increasing shear stress results from the progressive delamination which is accompanied by formation of band-type networks, predominantly with face$(-)$–Ca^{2+}–face$(-)$ contacts. At very high concentrations of soda higher amounts of calcium ions are

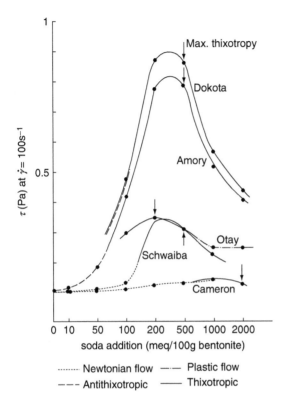

Figure 8.35 Effect of soda addition on the flow behavior. The 2% (w/w) dispersion of several bentonites. The arrows indicate the point of maximum thixotropy (maximum hysteresis in the flow curves.... Newtonian flow, $-\cdot-\cdot-\cdot-$ plastic flow; $----$ antithixotropic behavior; ———— thixotopic behavior. (From G. Lagaly, *Appl. Clay Sci.*, 4, 105–123, 1989. With permission.)

precipitated and are withdrawn from the face–face contact areas. The network breaks down and the shear stress decreases from the maximum to a low value.

IX. FINAL REMARKS

Many scientists reject the study of clay dispersions. They consider clays as dirty materials, mixtures of minerals and consisting of particles of irregular shape. A few scientists and users arrive at a positive attitude and then are fascinated by the variety and variability of the reactions of the different clay minerals. The same is true for colloid scientists: the first contacts with clay dispersions can be disgusting, but further acquaintance produces fascination. The variety of properties discussed above and the variety of practical applications have no parallel compared to other colloidal systems.

REFERENCES

1. G. Lagaly and R. Fahn, in *Ullmann's Encyclopedia of Technical Chemistry*, 4th ed., Vol. 23, Verlag Chemie, Weinhein, 1983, pp. 311–326.
2. K. Jasmund and G. Lagaly, Eds., *Tonminerale und Tone: Struktur, Eigenschaften, Anwendung und Einsatz in Industrie und Umwelt*, Steinkopff Verlag, Darmstadt, 1993, pp. 1–490.
3. I.E. Odom, *Philos. Trans. R. Soc., London A*, 311, 391–409, 1984.
4. W.B. Jepson, *Philos. Trans. R. Soc., London A*, 311, 411–432, 1984.
5. B.S. Neumann and K.G. Sansom, *J. Soc. Cosmet. Chem.*, 21, 237–258, 1970.
6. B.S. Neumann and K.G. Sansom, *Israel J. Chem.*, 8, 315–322, 1970.
7. B.S. Neumann and K.G. Sansom, *Clay Miner.*, 9, 231–243, 1971.
8. J.D.F. Ramsay, *J. Colloid Interf. Sci.*, 109, 441–447, 1986.
9. R.G. Avery and J.D.F. Ramsay, *J. Colloid Interf. Sci.*, 109, 448–454, 1986.
10. J.T. Kloprogge, S. Komarneni, and J.E. Amonette, *Clays Clay. Miner.*, 47, 529–554, 1999.
11. K.A. Carrado, *Appl. Clay Sci.*, 17, 1–23, 2000.
12. T. Kodama, Y. Harade, M. Ueda, K. Shimizu, K. Shuto, S. Komarneni, W. Hoffbauer, and H. Schneider, *J. Mater. Chem.*, 11, 1222–1227, 2001.
13. J.T. Kloprogge, *J. Porous Mater.*, 5, 5–41, 1998.
14. G.W. Brindley and G. Brown, Eds., *Crystal Structures of Clay Minerals and Their X-ray Identification*, Mineralogical Society, London, 1980.

15. R.T. Martin, S.W. Bailey, D.D. Eberl, D.S. Fanning, S. Guggen-
    hein, H. Kodama, D.R. Pevear, J. Środoń, and F. Wicks, *Clays
    Clay Miner.*, 39, 333–335, 1991.
16. S.P. Altaner and R.F. Ylagan, *Clays Clay Miner.*, 45, 517–533,
    1997.
17. D.D. Eberl, R. Nüesch, V. Sucha, and S. Tsipursky, *Clays Clay
    Miner.*, 34, 185–191, 1999.
18. D.M. Moore and R.C. Reynolds, *X-ray Diffraction and the
    Identification and Analysis of Clay Minerals*, 2nd ed., Oxford
    University Press, Oxford, 1997, pp. 1–378.
19. A. Plançon, *Clay Miner.*, 36, 1–14, 2001.
20. H. Vali and H.M. Köster, *Clay Miner.*, 21, 827–859, 1986.
21. C. Chenu and A.M. Jaunet, *C.R. Acad. Sci., Paris 30*, Series 2,
    1990, pp. 975–980.
22. O. Touret, C.H. Pons, D. Tessier, and Y. Tardy, *Clay Miner.*, 25,
    217–233, 1990.
23. G. Lagaly and R. Malberg, *Colloids Surf.*, 49, 11–27, 1990.
24. J. Benbrahin, N. Armagau, G. Besson, and C. Tschoubar, *Clay
    Miner.*, 21, 111–124, 1986.
25. H. Suquet and H. Pézérat, *Clays Clay Miner.*, 35, 353–362,
    1987.
26. C. Tschoubar, A. Plancon, J. Benbrahin, C. Clinard, and C.
    Sow, *Bull. Miner.*, 105, 477–491, 1982.
27. G. Lagaly, *Clays Clay Miner.*, 27, 1–10, 1979
28. J. Cuadres and J. Linares, *Clays Clay Miner.*, 43, 467–473,
    1995.
29. M. Müller-Vonmoos, G. Kahr, and F.T. Madsen, in *Identifizier-
    ung und Nachweis der Tonminerale*, H. Tributh and G. Lagaly,
    Eds., Berichte der DTTG, Giessen, 1990, pp. 131–155.
30. G. Lagaly, *Clay Miner.*, 16, 1–21, 1981.
31. G. Lagaly, *Clays Clay Miner.*, 30, 215–222, 1982.
32. G. Lagaly, in *Layer Charge Characteristics of 2:1 Silicate Clay
    Minerals*, A.R. Mermut, Ed., CMS Workshop Lectures, Vol. 6,
    The Clay Mineral Society, Boulder, CO, USA, 1994, pp. 1–46.
33. D.C. Bain and B.F.L. Smith, in *A Handbook of Determinative
    Methods in Clay Mineralogy*, M.J. Wilson, Ed., Blackie, Glas-
    gow, 1987, pp. 248–274.
34. M. Janek and G. Lagaly, *Appl. Clay Sci.*, 19, 121–130, 2001.
35. C.U. Schmidt and G. Lagaly, *Clay Miner.*, 34, 447–458, 1999.
36. M.R. Böhmer and L.K. Koopal, *Langmuir*, 8, 2649–2665, 1992.
37. U. Brandenburg and G. Lagaly, *Appl. Clay Sci.*, 3, 263–279,
    1988.

38. G. Lagaly, *Appl. Clay Sci.*, 4, 105–123, 1989.
39. R. Keren and D.L. Sparks, *Soil Sci. Soc. Am. J.*, 59, 430–435, 1995.
40. B.K.G. Theng, *The Chemistry of Clay Organic Reactions*, A Hilger, London, 1974.
41. G. Lagaly, *Philos. Trans. R. Soc. A*, 311, 315–332, 1984.
42. G. Lagaly, in *Developments in Ionic Polymers*, Vol. 2, A.D. Wilson and H.J. Prosser, Eds., Elsevier, London, 1986, pp. 77–140.
43. G. Lagaly, Proceedings of the International Clay Conference, L.G. Schultz, H. von Olphen, and F.A. Mumpton, Eds., Clay Mineral Society, Bloomington, IN, 1987, pp. 343–351.
44. P.M. Costanzo, R.F. Giese, and C.V. Clemency, *Clays Clay Miner.*, 32, 29–35, 1984.
45. P.M. Costanzo and R.F. Giese, *Clays Clay Miner.*, 38, 160–170, 1990.
46. Y. Komori, Y. Sugahara, and K. Kuroda, *Chem. Mater.*, 11, 3–6, 1999.
47. P.G. Slade, J.P. Quirk, and K. Norrish, *Clays Clay Miner.*, 39, 234–238, 199.
48. G. Frens and J.Th.G. Overbeek, *J. Colloid Interf. Sci.*, 38, 376–387, 1972.
49. J.Th.G. Overbeek, *J. Colloid Interf. Sci.*, 58, 408–422, 1977.
50. C.H. Pons, E. Rousseaux, and D. Tschoubar, *Clay Miner.*, 17, 327–338, 1982.
51. A. Weiss, A. Häbisch, and A. Weiss, *Ber. Dtsch. Keram. Ges.*, 41, 687–690, 1964.
52. H. Suquet, C. de la Calle, and H. Pézérat, *Clays Clay Miner.*, 23, 1–9, 1975.
53. J. Hougardi, W.E.E. Stone, and J.J. Fripiat, *J. Chem. Phys.*, 64, 3840–3851, 1976.
54. G. Besson, C. Mifsud, D.D. Tschoubar, and J. Mering, *Clays Clay Miner.*, 22, 379–384, 1974.
55. R. Glaeser, I. Mantin, and J. Mering, *Bull. Groupe Fr. Arg.*, 19, 125–130, 1967.
56. J. Mamy and J.P. Gaultier, Proceedings of the International Clay Conference, Mexico, Applied Pub. Ltd., Wilmette, III, 1975, pp. 149–155.
57. C. Poinsignon, J.M. Cases, and J.J. Fripiat, *J. Phys. Chem.*, 82, 1855–1860, 1978.
58. R.F. Giese and J.J. Fripiat, *J. Colloid Interf. Sci.*, 71, 441–450, 1979.

59. J.J. Fripiat, J. Cases, M. Francois, and M. Letellier, *J. Colloid Interf. Sci.*, 89, 378–400, 1982.
60. J.J. Fripiat, M. Letellier, and P. Levitz, *Philos. Trans. R. Soc., London A*, 311, 287–299, 1984.
61. C.T. Johnston, G. Sposito, and C. Erickson, *Clays Clay Miner.*, 40, 722–730, 1992.
62. A. Delville and P. Laszlo, *New J. Chem.*, 13, 481–491, 1989.
63. M.M. Mortland and K.V. Raman, *Clays Clay Miner.*, 16, 393–398, 1968.
64. P. Touillaux, P. Salvador, C. Vandermeersche, and J.J. Fripiat, *Israel J. Chem.*, 6, 337–348, 1968.
65. S.S. Cady and T.J. Pinnavaia, *Inorg. Chem.*, 17, 1501–1507, 1978.
66. I. Grandjean and P. Laszlo, *Clays Clay Miner.*, 37, 403–408, 1989.
67. D.J. Mulla and P.F. Low, *J. Colloid Interf. Sci.*, 95, 51–60, 1983.
68. F.R.C. Chang, N.T. Skipper, and G. Sposito, *Langmuir*, 13, 2074–2082, 1997.
69. F.R.C. Chang, N. Skipper, and G. Sposito, *Langmuir*, 11, 2734–2741, 1995.
70. N.T. Skipper, G. Sposito, and F.R.C. Chang, *Clays Clay Miner.*, 43, 285–293, 294–303, 1995.
71. F.R.C. Chang, N.T. Skipper, and G. Sposito, *Langmuir*, 14, 1201–1207, 1998.
72. J.A. Greathouse, K. Refson, G. Sposito, *J. Am. Chem. Soc.*, 122, 11459–11464, 2000.
73. J. Norris, R.F. Giese, P.M. Costanzo, and C.J. van Oss, *Clay Miner.*, 28, 1–11, 1993.
74. P. Nadeau, *Clay Miner.*, 20, 499–514, 1985.
75. J. Środoń, D.D. Eberl, and V.A. Drits, *Clays Clay Miner.*, 48, 446–458, 2000.
76. L.L. Schramm and J.C.T. Kwak, *Clays Clay Miner.*, 30, 40–48, 1982.
77. K Norrish, *Disc. Faraday Soc.*, 18, 120–134, 1954.
78. J.D. Cebula, R.K. Thomas, and J.W. White, *J. Chem. Soc. Faraday I*, 76, 314–321, 1980.
79. H. van Olphen, *An Introduction to Clay Colloid Chemistry*, John Wiley and Sons, New York, 1977.
80. L.M. Barclay and R.H. Ottewill, *Spec. Disc. Faraday Soc.*, 138–147, 1970.
81. R.H. Ottewill, *J. Colloid Interf. Sci.*, 58, 357–373, 1977.

82. Y. Sun, H. Lin, and P.F. Low, *J. Colloid Interf. Sci.*, 122, 556–564, 1986.
83. P.F. Low, Proceedings of the International Clay Conference, Denver, 1985, L.G. Schultz, H. van Olphen, and F.A. Mumpton, Eds., Clay Minerals Society, Bloomington, IN, 1987, pp 247–256.
84. C.J. van Oss, R.F. Giese, and P.M. Costanzo, *Clays Clay Miner.*, 38, 151–159, 1990.
85. R.D. Fitzsimmons, A.M. Posner, and J.P. Quirk, *Israel J. Chem.*, 8, 301–314, 1970.
86. W.B. Kleijn and J.D. Oster, *Clays Clay Miner.*, 30, 383–390, 1982.
87. R. Kjellander, S. Marčelja, and J.P. Quirk, *J. Colloid Interf. Sci.*, 126, 194–211, 1988.
88. R. Kjellander, *Ber. Bunsenges Phys. Chem.*, 100, 894–904, 1996.
89. O.P. Mehra and M.L. Jackson, *Clays Clay Miner.*, 7, 317–327, 1960.
90. G.G.S. Holmgreen, *Soil Sci. Soc. Am. Proc.*, 31, 210–211, 1967.
91. M.S. Stul and L. van Leemput, *Clay Miner.*, 17, 209–215, 1982.
92. H. Tributh and G. Lagaly, *GIT Fachz. Lab.*, 30, 524–529, 771–776, 1986.
93. V.C. Farmer and B.D. Mitchell, *Soil Sci.*, 96, 221–229, 1963.
94. B. Siffert and P. Espinasse, *Clays Clay Miner.*, 28, 381–387, 1980.
95. J.K. Anderson, *Clays Clay Miner.*, 10, 380–388, 1963.
96. B.D. Mitchell and B.F.L. Smith, *J. Soil Sci.*, 25, 239–241, 1974.
97. W.D. Keller, *Clays Clay Miner.*, 26, 1–20, 1978.
98. W.D. Keller and R.P. Haenni, *Clays Clay Miner.*, 26, 384–396, 1978.
99. W.D. Keller, *Clays Clay Miner.*, 33, 161–172, 1985.
100. A. Weiss, *Angew. Chem.*, 72, 755–762, 1963.
101. N. Lahav, *Clays Clay Miner.*, 38, 219–222, 1990.
102. R. Grim and N. Güven, *Bentonites: Geology, Mineralogy and Uses*, Elsevier, New York, 1978.
103. W.D. Keller, R.C. Reynolds, and A. Inoue, *Clays Clay Miner.*, 34, 187–197, 1986.
104. L.A. Pérez-Maqueda, O.B. Caneo, J. Poyato, and J.L. Pérez-Rodriguez, *Phys. Chem. Miner.*, 28, 61–66, 2001.
105. M. Janek, P. Komadel, and G. Lagaly, *Clay Miner.*, 32, 623–663, 1997.

106. U. Schwertmann, Proceedings of the International Clay Conference, Tokyo, 1969, Israel University Press, Jerusalem, 1969, pp. 683–690.
107. I. Barshad, *Soil Sci.*, 108, 38–42, 1969.
108. H.R. His and D.F. Clifton, *Clays Clay Miner.*, 9, 269–275, 1962.
109. B.G. Williams and S.P. Drover, *Soil Sci.*, 104, 326–331, 1967.
110. L.S. Swartzen-Allen and E. Matijević, *J. Colloid Interf. Sci.*, 56, 159–167, 1976.
111. H. Jenny and R.F. Reitemeier, *J. Phys. Chem.*, 39, 593–604, 1935.
112. A. Kahn, *J. Colloid Interf. Sci.*, 13, 51–60, 1958.
113. A.K. Helmy and E.A. Ferreiro, *Electroanal. Chem. Interf. Electrochem.*, 57, 103–112, 1974.
114. H.S. Arora and N.T. Coleman, *Soil Sci.*, 127, 134–139, 1979.
115. E. Frey and G. Lagaly, *J. Colloid Interf. Sci.*, 70, 46–55, 1979.
116. J.D. Oster, I. Shainberg, and J.D. Wood, *Soil Sci. Soc. Am. J.*, 44, 955–959, 1980.
117. R. Keren, I. Shainberg, and E Klein, *Soil Sci. Soc. Am. J.*, 52, 76–80, 1988.
118. S. Goldberg and H.S. Forster, *Soil Sci. Soc. Am. J.*, 54, 714–718, 1990.
119. R. Perkins, R. Brace, and E. Matijević, *J. Colloid Interf. Sci.*, 48, 417–426, 1974.
120. T. Permien and G. Lagaly, *J. Colloid Polym. Sci.*, 272, 1306–1312, 1994.
121. T. Permien, G. Lagaly, *Clays Clay Miner.*, 43, 229–236, 1995.
122. H. Heller and R. Keren, Clays Clay Miner., 49, 286–291, 2001.
123. G. Lagaly, O. Schulz, and R. Zimehl, *Dispersionen und Emulsionen. Eine Einführung in die Kolloidik feinverteilter Stoffe einschließlich der Tonminerale*, Mit einem historischen Beitrag über Kolloidwissenschaftler von Klaus Beneke, Steinkopff Verlag, Darmstadt, 1997, pp. 1–560.
124. D. Penner and G. Lagaly, *Clays Clay Miner.*, 48, 246–255, 2000.
125. D. Penner and G. Lagaly, *Appl. Clay Sci.*, 19, 131–142, 2001.
126. W. Stumm, C.P. Huang, and S.R. Jenkins, *Croatia Chem. Acta.*, 42, 233–244, 1970.
127. J.N. de Rooy, P.L. de Bruyn, and J.T.G. Overbeek, *J. Colloid Interf. Sci.*, 75, 542–554, 1980.
128. S.D. Lubetkin, S.R. Middleton, and R.H. Ottewill, *Philos. Trans. R. Soc., London, A*, 311, 353–368, 1984.

129. E, Tombácz, I, Abrahám, M, Gilde, and F. Szántó, *Colloids Surf.*, 49, 71–80, 1990.
130. R.O. James and G.A. Parks, in *Surface and Colloid Science*, Vol. 12, E. Matijević, Ed., Plenum Press, New York, 1982, p. 119.
131. J.A. Davies, R.O. James, and J.O. Leckie, *J. Colloid Interf. Sci.*, 63, 480–499, 1978.
132. S.E. Miller and P.F. Low, *Langmuir*, 6, 572–578, 1990.
133. J.P. Quirk and S. Marčelja, *Langmuir*, 13, 6241–6248, 1997.
134. R.M. Pashley and J.P. Quirk, *Colloids Surf.*, 9, 1–17, 1984.
135. E. Frey and G. Lagaly, Proceedings of the International Clay Conference, Oxford, 1978, M. Mortland and V.C. Farmer, Eds., Elsevier, Amsterdam, 1979, pp. 131–140.
136. G. Lagaly, G. Schön, and A.Weiss, *Kolloid Z. Z. Polym.*, 250, 667–674, 1972.
137. B.V. Derjaguin, N.V. Churaev, and V.M. Müller, *Surface Forces*, Consultants Bureau, New York, 1987
138. R.J. Pugh and J.A. Kitchener, *J. Colloid Interf. Sci.*, 35, 636–664, 1971.
139. P. Nadeau, *Appl. Clay Sci.*, 2, 83–93, 1987.
140. G. Lagaly, in Lectures Conferencias, Euroclay '87, J.L. Pérez-Rodriguez and E. Galan, Eds., Sociedad Espanola de Arcillas, Sevilla, 1987, pp. 97–115.
141. T. Stutzmann and B. Siffert, *Clays Clay Miner.*, 25, 392–406, 1977.
142. J.Y. Bottero, M. Bruant, J.M. Cases, D. Canet, and F. Fiessinger, *J. Colloid Interf. Sci.*, 124, 515–527, 1988.
143. A.F. Hollander, P. Somasundaran, and C.C. Gryte, in *Adsorption from Aqueous Solution*, P.H. Tewari, Ed., Plenum Press, New York, 1981, pp. 143–162.
144. E. Pefferkorn, I. Nabzar, and A. Carroy, *J. Colloid Interf. Sci.*, 106, 94–103, 1985.
145. L. Nabzar and E. Pefferkorn, *J. Colloid Interf. Sci.*, 108, 243–248, 1985.
146. E. Pefferkorn, L. Nabzar, and R. Varoqui, *Colloid Polym. Sci.*, 265, 889–896, 1987.
147. P. Stenius, L. Järnström, and M. Rigdahl, *Colloids Surf.*, 51, 219–238, 1990.
148. L. Nabzar, E. Pefferkorn, and R. Varoqui, *J. Colloid Interf. Sci.*, 102, 380–388, 1984.
149. L.T. Lee, R. Rahbari, J. Lecourtier, and G. Chauveteau, *J. Colloid Interf. Sci.*, 147, 351–357, 1991.

150. H.S. Kim, C. Lamarche, and A. Verdier, *Colloid Polym. Sci.*, 261, 64–69, 198.
151. G. Durand-Piana, F. Lafuma, and R. Audebert, *J. Colloid Interf. Sci.*, 119, 474–480, 1987.
152. N. Larsson and B Siffert, *J. Colloid Interf. Sci.*, 93, 424–431, 1983.
153. C. Breen, J.O. Rawson, and B.E. Mann, *J. Mater. Chem.*, 6, 253–260, 1996.
154. J. Billingham, C. Breen, J.O. Rawson, J. Yarwood, and B.E. Mann, *J. Colloid Interf. Sci.*, 193, 183–189, 1997.
155. D.P. Parazak, C.W. Burkhardt, K.J. McCarthy, and M.P. Stehlin, *J. Colloid Interf. Sci.*, 123, 59–72, 1988.
156. H. Xiao, Z. Lui, and N. Wiseman, *J. Colloid Interf. Sci.*, 216, 409–417, 1999.
157. B. Alince, F. Bednar, and T.G.M. van de Ven, *Colloids Surf. A*, 190, 71–81, 2001.
158. U. Hofmann, *Ber. Dtsch. Keram. Ges.*, 38, 201–207, 1961.
159. U. Hofmann, *Keram. Z.*, 14, 14–19, 1962.
160. U. Hofmann, *Ber. Dtsch. Keram. Ges.*, 41, 680–686, 1964.
161. T. Permien and G. Lagaly, *Clay Miner.*, 29, 751–760, 1994.
162. A. Weiss and R. Frank, *Z. Naturforsch.*, 16b, 141–142, 1961.
163. A. Weiss, *Rheologica Acta*, 2, 292–304, 1962.
164. E. Tombácz, I. Balázs, J. Lakatos, and F. Szántó, *Colloid Polym. Sci.*, 267, 1016–1025, 1989.
165. I. Lapides and L. Heller-Kallai, *Colloid Polym. Sci.*, 280, 554–561, 2002.
166. J.M. Benna, N. Kbir-Ariguib, C. Clinard, and F. Bergaya, *Prog. Colloid Polym. Sci.*, 117, 204–210, 2001.
167. J.M. Benna, N. Kbir-Ariguib, C. Clinard, and F. Bergaya, *Appl. Clay Sci.*, 19, 103–120, 2001.
168. S. Abend and G. Lagaly, *Appl. Clay Sci.*, 16, 201–227, 2000.
169. N. Güven and R.M. Pollastro, Eds., *Clay–water Interface and its Rheological Implications*, CMS Workshop Lectures, Vol. 4, The Clay Minerals Society, Boulder, CO, USA, 1992, pp. 1–244.
170. Y. Adachi, K. Nakaishi, and M. Tamak, *J. Colloid Interf. Sci.*, 198, 100–105, 1998.
171. C.H. Pons, F. Rousseaux, and D. Tschoubar, *Clay Miner.*, 16, 23–42, 1981.
172. F. Hetzel, D. Tessier, A.M. Jaunet, and H. Doner, *Clays Clay Miner.*, 42, 242–248, 1994.

173. K. Faisander, C.H. Pons, D. Tchoubar, and F. Thomas, *Clays Clay Miner.*, 46, 636–648, 1998.
174. G.F. Walker, *Nature*, 187, 312, 1960.
175. W.G. Garett and G.F. Walker, *Clays Clay Miner.*, 9, 557–567, 1962.
176. J.A. Rausell-Colom, *Trans. Faraday Soc.*, 60, 190–201, 1964.
177. M.V. Smalley, R.K. Thomas, L.F. Braganza, and T. Matsuo, *Clays Clay Miner.*, 37, 474–478, 1989.
178. L.F. Braganza, R.J. Crawford, M.V. Smalley, and R.K. Thomas, *Clays Clay Miner.*, 38, 90–96, 1990.
179. H.L.M. Hatharasinghe, M.V. Smalley, J. Swenson, A.C. Hannon, and S.M. King, *Langmuir*, 16, 5562–5567, 2000.
180. M.V. Smalley, H.L.M. Hathrasinghe, I. Osborne, J. Swenson, and S. King, *Langmuir*, 17, 3800–3812, 2001.
181. G. Lagaly, in *Interaction of Water in Ionic and Nonionic Hydrates*, H. Kleeberg, Ed., Springer-Verlag, Berlin, 1987, pp. 229–240.
182. M.V. Smalley, *Langmuir*, 10, 2884–2891, 1994.
183. M.V. Smalley, *Prog. Colloid Polym. Sci.*, 97, 59–64, 1994.
184. J.A. Rausell-Colom and P.S. Salvador, *Clay Miner.*, 9, 139–149, 1971.
185. J.A. Rausell-Colom and P.S. Salvador, *Clay Miner.*, 9, 193–208, 1971.
186. J.A. Rausell-Colom, J. Saez-Aunion, and C.H. Pons, *Clay Miner.*, 24, 459–478, 1989.
187. T.R. Jones, *Clay Miner.*, 18, 399–410, 1983.
188. W.T. Granquist and J. McAtee, *J. Colloid Sci.*, 18, 409–420, 1963.
189. V.N. Moraru, *Appl. Clay Sci.*, 19, 11–26, 2001.
190. B. Gherardi, A. Tahani, P. Levitz, and F. Bergaya, *Appl. Clay Sci.*, 11, 163–170, 1996.

9

Stabilization of Aqueous Powder Suspensions in the Processing of Ceramic Materials

Christian Simon

Functional Ceramics, SINTEF, Oslo, Norway

I. INTRODUCTION

Ceramics, together with new polymers and composites, belong
to today's advanced materials. Better performance, resulting
from better control of processing and the ability to tailor their

properties, has placed ceramics in the ranks of emerging technologies.

The market for advanced ceramics is still dominated by (micro) electronic and cutting-tools applications and to a lesser extent by structural applications, but there is an increase in ceramic gas-turbine engine technology. Emerging technologies like gas separation by membranes, and nanotechnology, use mostly ceramics. High-performance ceramic coatings are still an area in expansion. The largest projected growth rate for coatings is attributed to wet-based methods like sol–gel, spraying, and dipping [1]. Ceramic powders have also expanded their share of the market. Silicon carbide, silicon nitride, sialons, alumina, and zirconia (tetragonal and partially stabilized) are useful compounds in structural ceramics. Al_2O_3 is used in many wear and load-bearing applications, while ZrO_2 in its stabilized form is valued for oxygen sensors and as an electrolyte for fuel cells [2,3]. Perovskite materials are used as electrode materials for fuel cells, dense ceramic membranes, and sensors [4,5]. Although most nanostructured materials are still in their development stage, several nanostructured ceramic products are already in the market. Nanosized oxide particles, e.g., SiO_2, Fe_3O_4, CeO_2, Al_2O_3, TiO_2, Sb_2O_3 are found in abrasive polishing slurries, fire-retardant materials, sunscreens, or transparent wood stains [6].

Ceramics are mostly covalently bonded inorganic materials. The chemical structure that confers superior thermal and mechanical properties to ceramics also imparts brittleness. The presence of very small structural defects can, therefore, result in catastrophic failure. In the production of reliable ceramics, processing techniques must be developed to ensure that heterogeneities arising from the powders are eliminated and that they will not be reintroduced in later processing steps.

Advanced ceramics are fabricated in a manner similar to traditional ceramics, i.e., by compacting and then firing powders. Two fundamental research areas in the processing of ceramic materials are: (i) powder synthesis and characterization, and (ii) improving techniques for consolidation of the powder, so as to increase reliability by reducing structural defects. The joint optimization of these two processing steps — powder synthesis and powder consolidation — has been shown to be a very efficient method to form better ceramics.

Many microstructural heterogeneities originate in the powder itself. The presence of hard agglomerates is a major heterogeneity, and is probably the most difficult powder property to control in the whole processing cycle. Excessive impurities affect the densification behavior and the final properties (formation of second phases during processing, voids, pore retention). A fine grain size is important because the strength of a ceramic part depends heavily on the size of the flaws present. New routes to obtain particles with high chemical purity and controlled size, through sol–gel, vapor synthesis, or polymer precursor, are emerging [7–9]. This improvement is central to the development of better ceramics.

The second step in ceramic processing, powder consolidation, has very much to do with colloid chemistry, i.e., powder dispersions and their stabilization. Many efforts and much progress have been made in colloid science during the past 20 years. In particular, the theoretical and experimental studies on aqueous suspensions of silver iodide have led to a better comprehension of interfacial electrochemistry. Considerable progress in the understanding of the behavior of aqueous suspensions of oxide powders has also been made recently [6,10,11], particularly in the description of the solid–liquid interface in the presence of simple electrolytes. Even more recent are adsorption studies of organic polymers and surfactants on oxide powder suspensions and theoretical calculations for such systems [12,13].

The understanding of electrostatic stabilization in colloid science has now reached the stage where it can be utilized to eliminate heterogeneities during the powder processing of ceramics. Certain aspects of colloidal processing require repulsive interparticle forces, whereas others require attractive forces. Powders disperse to form a system of separated particles when repulsive forces dominate, and they flocculate to form a low-density network of touching particles when attractive forces dominate. Repulsive interparticle forces are used to break apart weak agglomerates, to fractionate inclusions greater than a given size, and to mix powders with different fractionations.

The combination of polymer–surfactant chemistry and ceramic surface science should help to predict which dispersing agent–binder is consistent with a specific ceramic powder.

The characteristics of the powder (uniformity, homogeneity, purity, etc.) are essential to the integrity of the final part, which explains why colloid science is so relevant to the development of powder processing. Using the colloidal approach, powders can be fractionated to achieve desired particle-size distributions (which maximize packing density) when one of the colloidally compatible consolidation methods is employed. Such a material can have a very high strength, compared to ceramic made from the as-received powder [14]. Figure 9.1 shows the influence of several processing steps on the modulus of rupture, which characterizes the strength of the material.

More recently, it has been shown that ceramics processed from nanosized particles (1–100 nm) could have drastically improved properties. Especially, the size reduction in nanosized particles compared with micropowders provides materials that shrink less during sintering, sinter at temperatures as low as 600°C, and are much more ductile.

In the first part of this chapter, we will concentrate on some theories and models, used in colloid science to describe stabilization, that can be applied to the processing of ceramics. In the second part, several methods for forming ceramics are discussed, with emphasis on powder consolidation by colloidal techniques. The effect of the improvement in powder packing on further processing steps, e.g., drying and sintering, is studied in the third part. Finally, the effects of the stabilization of powder suspensions on the processing of ceramics are reviewed.

II. DISPERSION OF POWDERS

The field of research in colloid chemistry has expanded steadily for the last 20 years. Progress in understanding colloidal

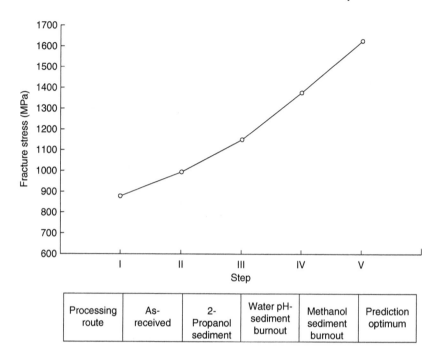

Figure 9.1 Improvement of the strength of ceramic materials through several processing steps. Step I, as-received powder. Steps II to IV, powder sedimented in, 2-propanol (II), water pH 2 (III), and methanol (IV). Step V, prediction realizable by carefully controlled processing [14].

phenomena has been exploited in many industrial applications, such as oil and mineral recovery, water treatment, food and paint production. An effort has been made to apply colloid science in the ceramics industry during the last 10 years.

The word *colloid* includes both sols (particle sizes between 1 and 60 nm) and suspensions (particle sizes between 60 and 1000 nm). Their most typical feature is their high surface-to-volume ratio, which is the reason surface and interface chemistry form a significant part of colloid science.

Before dealing with important parameters which influence the dispersion of oxide powders in aqueous systems, it is necessary to provide a description of the solid–liquid interface of such systems. The possible ways that ions, organic polymers,

and surfactants molecules adsorb at an oxide–solution interface are then described, and finally, the implications of adsorption for the stability of colloidal dispersions are examined.

A. Oxide–Solution Interface in Simple Electrolytes

1. Surface Charge Formation

When an oxide powder such as ZrO_2, Al_2O_3, or TiO_2 is placed
in water, the surface tends to coordinate water molecules, and
further dissociation of these molecules leads to a fully hydroxylated surface [15,16] as shown in Figure 9.2(a) and (b).

The surface hydroxyl groups have an amphoteric character [10,17,18], and the surface may become positively or negatively charged depending on the pH:

$$M_s-OH + H^+ \rightleftharpoons M-OH_2^+, \text{ and} \qquad (1)$$

$$M_s-OH + OH^- \rightleftharpoons M_s-O^- + H_2O \qquad (2)$$

These two equilibria show the direct connection between surface charge and H^+ consumption by the surface. The surface
charge can thus be estimated by pH measurement and

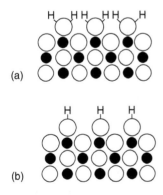

(a)

(b)

Figure 9.2 Cross section of the surface layer of a metal oxide. (l)
metal ions; (O) oxide ions. (a) Coordination of water molecules by the
surface metal ions. (b) Dissociative chemisorption leading to a
hydroxylated surface [16].

evaluation of the amount of H^+ and OH^- ions consumed by the solid surface.

2. Experimental Determination of Surface Charge

Using classical models of aqueous suspensions of silver iodide, colloid science has clearly demonstrated that Ag^+ and I^- ions play a predominant role in surface potential measurements and may be called potential-determining ions [19]. This notion has since been extended to oxide powders and we may also consider H^+ and OH^- ions as potential-determining ions. It follows that surface charge σ_0 may be defined as

$$\sigma_0 = F(\Gamma_{H^+} - \Gamma_{OH^-}), \tag{3}$$

where F is Faraday's constant (96,500 C/mol) and Γ_i is the surface excess of species i (in mol/m^2).

Experimentally, the pH variation of a suspension is measured as a function of the amount of acid or base added [20]. This is generally done by first adding an excess of a basic solution and then, after pH stabilization, measuring the variation of the suspension pH as a function of acid added. In cases where the oxide is partly soluble in water (e.g., SiO_2, Al_2O_3, Y_2O_3-stabilized ZrO_2), the computation of proton consumption by the surface must take the solubility into account.

Using this method, we can determine the variation of the surface charge density $\Delta\sigma_0$ with each acid increment

$$\Delta\sigma_0 = -\frac{F}{AmM}\{(v'_{OH} - v_2) - (v_{OH} - v_1)\}, \tag{4}$$

where A is the surface area of the powder (in m^2/g), m the quantity of solid in the suspension (in g), F the Faraday's constant, M the molarity of the acid or base (in mol/l), v'_{OH} and v_{OH} are the excess volumes of base added to the suspension and the solution, respectively, and v_2 and v_1 are the increment of acid added to the suspension and the solution, respectively.

To assess the absolute value of the surface charge, the intersection point of the $\Delta\sigma_0$–pH curves at several indifferent (simple) electrolyte (e.g., KNO_3) concentrations is chosen as a reference for which the surface charge is taken as equal to zero (the so-called pristine point of zero charge, PPZC [10]). Some typical examples are shown in Figure 9.3. In the case of TiO_2 (rutile), we notice that TiO_2 particles are positively charged for pH values lower than 5.8 and negatively charged for pH value higher than 5.8. The PPZC is 5.8 in this case and is a characteristic of the particular oxide. The PPZC is usually very sensitive to the history of the powder (preparation, impurities, heat treatment, etc.). Table 9.1 shows PPZC values for various oxides used in ceramic manufacturing.

In Figure 9.3, it is also possible to see an increase of surface charge with increasing electrolyte concentration at constant pH. Davis et al. [34] have suggested a complexation model for adsorbed species at the surface. In the case of a suspension of TiO_2 at pH below the PPZC, the surface is positively charged (Equation (1)). This charge is compensated by ions with the opposite charge (anions in this case) and the formation of complex species needs to be considered

$$Ti_s-OH_2{}^+ + NO_3{}^- \rightleftharpoons \{Ti_s-OH_2{}^+\text{---}NO_3{}^-\} \qquad (5)$$

An increase in electrolyte concentration consumes the $Ti_s-OH_2{}^+$ species, and shifts the equilibrium (Equation (1)) to the right and increases the H^+ consumption by the surface. In a similar way, the formation of $Ti_s-O^- \ldots K^+$ species accounts for an increase in surface charge for pH values above 5.8.

3. Specific Adsorption of Ions

Other ions than H^+ and OH^- exhibit chemical affinity for the surface. Heavy metal ions, such as Cd^{2+} or Pb^{2+} for example, shift the common intersection point (which is no longer the PPZC) to lower pH values, whereas specifically adsorbed anions tend to shift the common point to higher pH values. This variation is shown in Figure 9.4 [10,35].

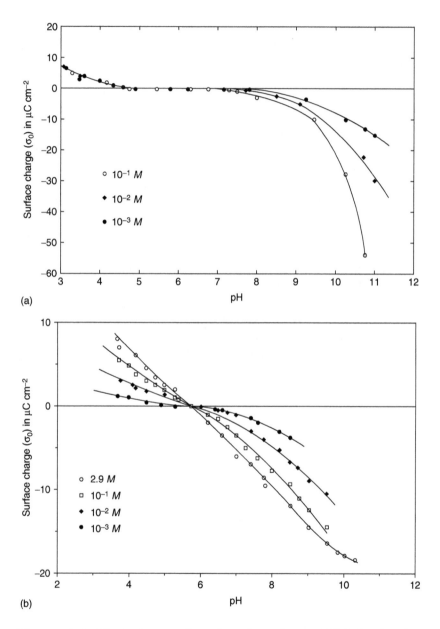

(a)

(b)

Figure 9.3 Typical examples of surface charge/pH behavior for several metal oxides in an inert atmosphere at 25°C. (a) α-Al$_2$O$_3$ in KNO$_3$ solutions [21]. (b) TiO$_2$ rutile in KNO$_3$ solutions [22]. (c) Precipitated SiO$_2$ in KCl solutions [23]. (d) ZrO$_2$ in NaCl solutions [24].

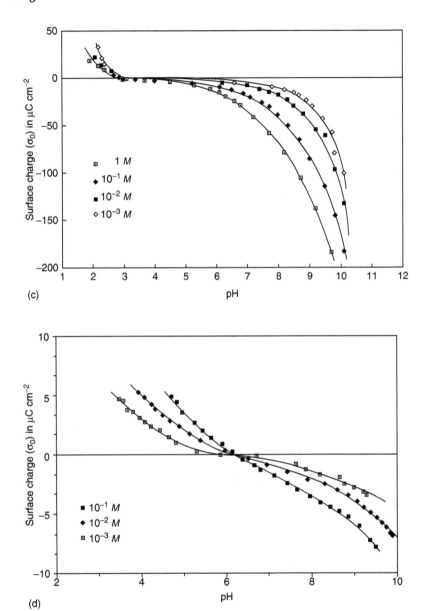

Figure 9.3 *(Continued)*

Table 9.1 Pristine points of zero charge of some ceramic oxides

Materials	ppzc	Ref.
SiO$_2$ quartz	2.0	[25]
Pyrogenic SiO$_2$	3–4	[26]
ZrO$_2$	5.5–6.4	[24, 27, 28]
TiO$_2$ (rutile)	5.8	[29]
TiO$_2$ (anatase)	6.0–6.6	[30–32]
α-Al$_2$O$_3$	9.1	[33]
α-Al$_2$O$_3$	8.5	[21]

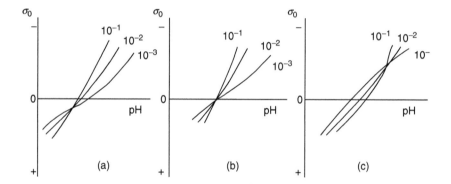

Figure 9.4 Schematic representation of the variation of the common intersection point determined by surface charge titration when specific adsorption occurs. (a) Specific adsorption of cations; (b) No specific adsorption; (c) Specific adsorption of anions [10].

4. Formation of an Electrical Double Layer Around Charged Oxide Particles

Because oxide particles in aqueous solutions are mostly charged, they will tend to be surrounded by ions of opposite charge coming from the electrolyte. Together, they form an electrical double layer. Several models have been developed to describe the electrical double layer in simple electrolytes [11]. In the surface complexation model developed by Davis et al. [34], the oxide–water interface is divided into three planes (Figure 9.5): the surface with a charge σ_0 and a mean

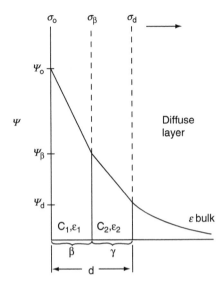

Figure 9.5 Model representing an idealized planar surface and the potential decay away from the surface [34].

potential Ψ_0, an average plane of ions forming surface complexes (charge σ_β and mean potential Ψ_β) and an average plane representing the start of the diffuse layer (charge σ_d and mean potential Ψ_d).

The compact layer of thickness d is divided into an inner part with capacitance C_1 and dielectric constant ε_1, and an outer part with capacitance C_2 and dielectric constant ε_2. The model permits the simultaneous estimation of interface parameters like surface charge, ion adsorbed density, and potential in the diffuse layer.

Electroneutrality at the interface requires that

$$\sigma_0 + \sigma_\beta + \sigma_d = 0. \tag{6}$$

This has been experimentally verified for TiO_2 suspensions in NaCl by using three different techniques: surface titration for surface charge measurements, radiotracers for ion adsorption analysis, and microelectrophoresis for charge determination in the diffuse layer [30].

The technique of microelectrophoresis has been commonly utilized in ceramic science on account of its versatility in the characterization of the stability of suspensions. The principle is to measure the velocity of charged particles (v) in an applied electric field (E). The charged particle surrounded by ions coming from the solution forms an electrokinetic entity. The plane separating the electrokinetic entity from the bulk of the solution is called the slipping plane. The electrophoretic mobility (μ in (μm/s)/(V/cm)), for spherical particles, can be defined as

$$\mu = \frac{v}{E} \tag{7}$$

The zeta potential ζ (in V), defined as the potential at the slipping plane, can be calculated by Equation (8):

$$\xi = \frac{K\mu\eta}{\varepsilon} \tag{8}$$

where η and ε, respectively, are the viscosity and the dielectric constant of the medium, and K is a constant depending on the thickness of the double layer κ^{-1}, and the particle radius a:

$$\text{for } \kappa a < 0.1,$$
$$K = 1/6\pi \text{ (Helmholtz–Smoluchovski equation)} \tag{9}$$

$$\text{and for } \kappa a > 100, \quad K = 1/4\pi \text{ (Huckel equation)} \tag{10}$$

There is an intermediate domain of particle size or electrolyte concentration, where Equation (8) does not apply. However, a set of computed curves permits the calculation of ζ from μ [36].

Figure 9.6 shows the evolution of the electrophoretic mobility for TiO_2 particles at different pH values and electrolytes. The pH at which $\zeta = 0$ is called the isoelectric point (IEP) and is particular to the oxide used. The IEP is often confused with the PPZC, measured by surface titration. These two points can only be identical when the determination is made in the presence of indifferent electrolytes. When specific adsorption occurs, the two points are shifted in opposite directions. This has been discussed in detail by Lyklema [10].

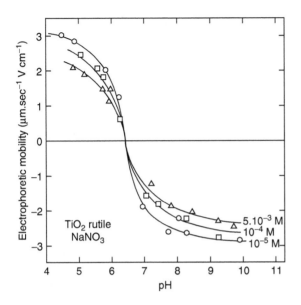

Figure 9.6 Electrophoretic mobility of TiO_2 rutile in the presence of different concentrations of $NaNO_3$ [37].

To evaluate σ_d ($\mu C/cm$) and Ψ_d (mV), it is often assumed that the slipping plane is located at the same place as the outer plane, and ζ can then be identified with Ψ_d. In this case, the charge in the diffuse layer can be computed by (in the case of monovalent electrolyte)

$$\sigma_d = -11.74I^{1/2} \sin h\left(\frac{e\Psi_d}{2kT}\right), \tag{11}$$

where I is the ionic strength, k the Boltzman constant and T the absolute temperature.

Recently, surface titration by internal reflection spectroscopy (STIRS) has been used to determine the IEP of TiO_2 directly on gel films [38]. This method permitted to identify specific adsorption of oxalate ions in the gel. In a similar approach, ζ potential could be determined inside the pores of ceramic membranes by conductivity and streaming potential measurements [39,40].

B. Adsorption at the Oxide–Solution Interface

The fundamental aspects of interfacial phenomena that affect the stabilization of suspensions in ceramic processing have been discussed by Fuerstenau et al. [41]. The adsorption of inorganic electrolytes, polymeric additives, and organic surfactants in oxide–aqueous systems has been especially emphasized. A thorough description of analytical methods and models applied to surface ionization of various silica sols was done by Foissy et al. [42].

1. Ion Adsorption

The distribution of adsorbed ions at the solid–solution interface can be described, as we saw above, by models. According to the type of electrolyte and its concentration, ion adsorption can range from long distance (5–10 nm), of van der Waals-type, to short distance, where electronic clouds overlap.

The extent of adsorption from solutions is controlled by a number of parameters of which pH, ionic strength, and oxide characteristics are the most important [16,43,44]. The relationship between ion adsorption and stability will be examined in Section II.C.

2. Polymer Molecules

Recently, studies of interactions between polymers and oxide surfaces have assumed more importance, due to the numerous technological and industrial applications, including ceramics technology. These studies have been reviewed by several authors [12,13,45,46].

Polymer molecules can adopt several conformations when adsorbed on a surface. Figure 9.7 shows segments, called trains, directly attached to the surface. Others stretch more or less out into the solution and form loops or tails. The structures of adsorbed molecules play a central role in the stabilization of dispersed systems. There exist both theoretical and experimental descriptions of the adsorption of neutral polymers [13], while there is still need for theoretical investigations of polyelectrolytes [12]. Other parameters affecting

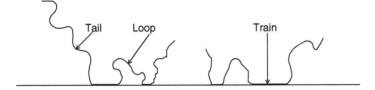

Figure 9.7 Typical conformation of a polymer molecule adsorbed on a surface.

polymer adsorption on oxide particles, include the molecular weight and functional segments of the adsorbate, and surface properties of the powder.

3. Surfactants

The possibility of using surfactants in oil recovery has considerably increased the volume of fundamental studies in this domain. This has had repercussions in several technologies and particularly in the ceramics industry [47].

The adsorption of surfactants on oxides is well reviewed [48–52]. It may involve electrical interactions with charged solid surfaces or specific interactions, e.g., of the chain–chain type. Figure 9.8 shows the effect of surface charge on alumina in suspension on the adsorption of sodium dodecylsulfonate. Around pH 9 (PPZC of $Al_2O_3 = 9.1$), the alumina surface is uncharged and almost no surfactant is adsorbed. When σ_0 increases with decreasing pH, the surfactant adsorption is markedly enhanced [51].

C. Stability of Aqueous Suspensions

1. DLVO theory

Derjaguin and Landau [53], and Verwey and Overbeek [54] in 1941 and 1948, respectively, developed a theory for dilute suspensions based on electrical double layer interactions, e.g., repulsion and van der Waals attraction.

The total free energy of interaction, ΔG_T (in mJ/m²), can be represented as the sum of the variation in the free energy

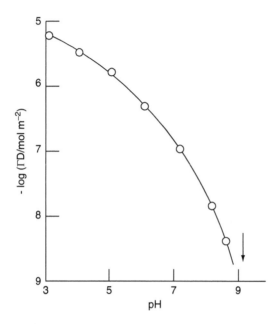

Figure 9.8   Amount of sodium dodecylsulfonate adsorbed on Al_2O_3 as a function of pH. The arrow indicates the ppzc of Al_2O_3 powder [50].

due to double layer overlap, ΔG_{rep}, and the free energy of van der Waals attraction, ΔG_{att}:

$$\Delta G_T = \Delta G_{rep} + \Delta G_{att}. \tag{12}$$

Both ΔG_{rep} and ΔG_{att} can be expressed as a function of H, the distance separating the surfaces of two particles that are assumed to be spherical (of radius a):

$$\Delta G_{rep} \approx 2\pi \varepsilon_r \varepsilon_0 a \left(\frac{4RT}{zF} \gamma \right)^2 \exp(-\kappa H), \tag{13}$$

where σ_r and σ_0 are, respectively, the dielectric constant of the solution and the permittivity of vacuum, R is the gas constant, F the Faraday constant, T the absolute temperature, κ the inverse of the thickness of the double layer, $\gamma = \tanh(zF\sigma_d/ 4RT)$, z the charge number of the counter ions, and ψ_d the potential at the plane d.

$$\Delta G_{att} \approx -\frac{aA}{12H},$$ (14)

where A is the Hamaker constant (J), depending strongly on the nature of the particles and the medium. The evolution of the free energies with the interparticle distance H is represented in Figure 9.9. For the case of a high potential, leading to high repulsion (ΔG_{rep1}), the total interaction energy exhibits one maximum and two minima (curve 1). When the maximum is high enough ($\geq 25\,kT$), particles cannot approach closer than the distance corresponding to the maximum in energy, and the suspension is thermodynamically stable. Otherwise, Brownian motion can surmount a maximum of lower energy, and gives rise to irreversible coagulation. Coagulation also takes place in the secondary minimum, but is often weak and reversible.

At lower potentials (ΔG_{rep2}), the resulting total free energy is attractive at all distances, and coagulation occurs (curve 2).

The thickness of the double layer (characterized by the Debye length, $1/\kappa$, is very sensitive to the electrolyte concentration of the dispersion medium. The effect of the concentration of a 1:1 electrolyte (e.g., KCl) on the thickness of the double layer is shown in Table 9.2. When the thickness of the double layer is larger than 10 nm, i.e., at electrolyte concentrations lower than $10^{-3}\,M$, van der Waals attractions no longer have any action and electrostatic stabilization is expected. On another hand, at high electrolyte concentrations, the van der Waals attraction dominates and suspensions coagulate.

Another approach of suspensions stability concerns the study of coagulation kinetics, i.e., phenomena originating from Brownian motion of dispersed particles, as reported by Stechemesser and Sonntag [55].

2. Stability in the Presence of Polymers

Polymer molecules with high molecular weights can attain large dimensions. Table 9.3 presents typical dimensions for polymer chains of different molecular weights. We can notice

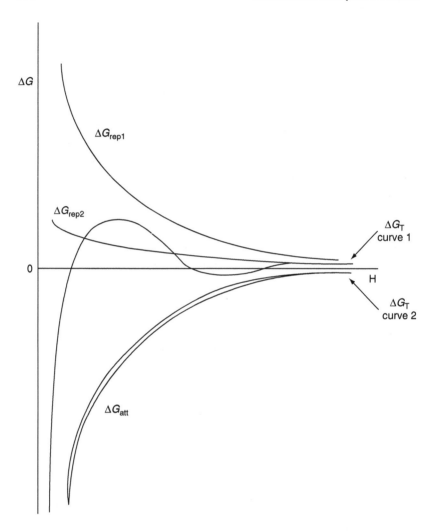

Figure 9.9 Representation of the interaction free energies of repulsion (ΔG_{rep}), attraction (ΔG_{att}), and the total-interaction free energy ($\Delta G_T = \Delta G_{rep} + \Delta G_{att}$, curves 1 and 2) for electrostatic stabilized systems.

that for molecular weights higher than 10 000, the dimensions of the chains become comparable to the range of the van der Waals attraction [12]. Two types of stabilization are encountered: (i) steric stabilization, in which polymer molecules are

Table 9.2 Effect of the concentration of a 1:1 electrolyte on the thickness of the double layer

C (M)	$1/\kappa$ (nm)
10^{-3}	10.4
10^{-2}	3.3
10^{-1}	1.0

Table 9.3 Illustration of the change in dimensions of polymer chains with their molecular weight

Molecular weight	Dimension (nm)
1,000	2
10,000	6
100,000	20
1,000,000	60

adsorbed on the surface of the particles, leading to stabilization, and (ii) depletion stabilization, in which stability is conferred by free polymer molecules in the solution. The magnitude of steric repulsion depends on the chemical nature and configuration of the adsorbed molecules, as well as on the surface coverage and the thickness of the adsorbed layer. Figure 9.10 shows typical forms of the total-interaction free energy for sterically stabilized systems. In case 1, the adsorbed polymer layer is dense, and behaves like a hard sphere. Stabilization is then favored. In cases 2 and 3, the density of the adsorbed polymer layer decreases, and collisions between particles cause interpenetration of the chains. Consequently, the thickness of the adsorbed layer decreases, and stabilization is weakened [56].

3. Interactions at High Solid Concentration

The DLVO theory considers only very dilute systems in which contacts between particles occur occasionally. However, most industrial applications of colloidal dispersions require

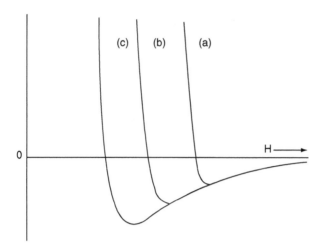

Figure 9.10   Typical forms for the total interaction free energy for sterically stabilized systems. The density of the adsorbed layer decreases from (a) to (c) [56].

concentrated suspensions. In ceramic technology, most preparations involve suspensions containing 10 to 80 wt% of solid. It is therefore very important to consider the stability of concentrated suspensions, and the applicability of the theory of the stability of dilute suspensions to this case.

Few papers can be found on the interrelationship between dilute and concentrated systems. For ζ potential measurements, an electrophoretic mass transport apparatus has been improved to measure the potential in concentrated suspensions (20 wt% of solid) [57]. Values of IEP measured for several oxides by this technique were comparable to those measured in dilute systems, and the evolution of the ζ potential with pH was similar. In TiO_2 suspensions, an increase of the particle size measured by laser light scattering was observed when the solid concentration was increased in the range 0.1–1.5%, in KCl solutions [58].

In dilute systems, particles interact randomly under the influence of Brownian motion. In concentrated suspensions, many more interactions are involved, and highly ordered dispersions appear. Ottewill [59] described the distribution of particles around a central reference particle by defining a pair distribution function $g(r)$ varying with the distance r from the center of the particle

$$g(r) = \frac{\rho(r)}{\rho_0},\tag{15}$$

where ρ_0 is the average density of particles contained in the whole volume, and $\rho(r)$, the radial density function. Figure 9.11 illustrates the variations of $g(r)$ with r, for the cases of dilute and concentrated systems. In the case of concentrated

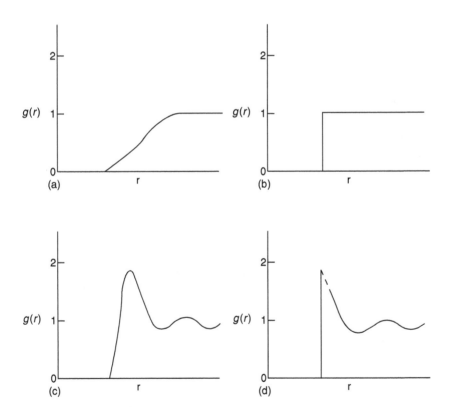

Figure 9.11 Schematic representation of the pair distribution function $g(r)$ as a function of the center-to-center distance r between spherical particles for dilute (a) soft-sphere, (b) hard-sphere, and concentrated (c) soft-sphere, (d) hard-sphere systems [60].

suspensions, $g(r)$ exhibits maxima, indicating that shells of particles are formed around the reference particle. This kind of ordering is encountered in the case of electrostatic stabilization (soft-sphere interaction), and in the case of steric stabilization (hard-sphere interaction).

It follows from this discussion that fundamental studies on dilute systems are still important for a further understanding of the behavior of concentrated suspensions.

III. APPLICATION IN THE PROCESSING OF OXIDE CERAMICS

As shown in Figure 9.12, the processing of ceramic materials in solution includes many steps. At present, many commercial oxide powders are manufactured by a precipitation process (e.g., the Bayer process for alumina production). Starting often from purified salts obtained from ore dissolution, the oxide powder is synthesized by precipitation in alkaline solution. At this stage, it could be relevant to dope the oxide, e.g., by adding YCl_3 to a $ZrOCl_2$ solution and coprecipitating the two salts in order to prepare yttrium-stabilized zirconia

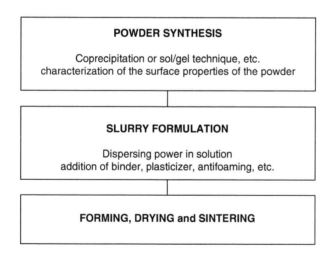

Figure 9.12   Main steps of the processing of ceramic materials.

powder. After successive washings, the powder is dried and often calcined to form aggregates with size adapted to the forming technique.

The step that usually follows powder preparation involves dispersing it in solution, either by pH adjustment or by adding organic polymers or surfactants. The suspensions prepared are very concentrated and called slurries, or slips. Relatively few studies on the adsorption properties of dispersing agents, in the preparation of ceramic materials, are to be found in literature. This stage is, however, important for the sintering properties of the powder and therefore for the performance of the final material. Knowledge of surface properties of the oxide powder is of great help in the understanding of its adsorptive properties. This makes powder characterization an important and complementary step in ceramics processing.

After dispersing, the rheology of the suspension and the strength of the material can be controlled by the addition of organic binders and plasticizers. These additions can also affect a suspension's stability and, consequently, the particle packing in the slurry [59,61]. It is therefore important to know how binders and plasticizers interact with the powder in suspension. The combination of the two steps, powder dispersion and addition of binders, plasticizers, antifoaming, etc., is often called slurry formulation. It is currently recognized that powder packing, which occurs during this stage, determines the microstructures, which develop during sintering.

The final shape of the material is given in the following stage when the slurry is either pored in a form (slip casting or gel casting), spread on to a smooth surface (tape casting or spraying [62]), or coated on a support.

After drying, the material (called *green* before firing) is sintered at a temperature that depends on particle size, composition, sintering aids, etc. The temperature program is chosen to take into account the removal of organic matter. Figure 9.13 summarizes the evolution of the microstructure from the powder state to the sintered material. In the wet processing, the densification occurs in two steps, first during drying, and then during sintering. The green material has an open pore-structure, whereas the sintered material contains

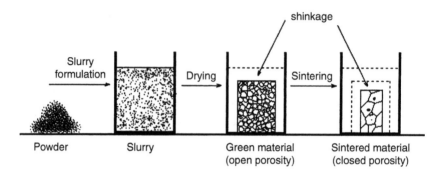

Figure 9.13 Evolution of the microstructure in ceramic materials from powder to the final sintered product.

mostly closed pores. Colloidal stability, powder dispersion, and powder packing have been discussed and applied to the fabrication of advanced ceramic composites [63].

A. Powder Synthesis and Characterization

Powder synthesis is the first stage in the fabrication of ceramics. The microstructure of sintered ceramic materials depends strongly on the characteristics of the initial powder. For example, coprecipitation of zirconia doped with Nb_2O_5 or Ta_2O_5 followed by hydrothermal conversion yielded to oxide precursors with narrow size distribution and small particle size, and permitted to produce ceramics with high relative green density at lower sintering temperatures [64].

A number of other techniques have been developed to yield fine powders with good control of their composition, shape, size, and structure. This control and the transfer of techniques from colloid science have permitted the fabrication of ceramic materials with improved performance. Colloid chemistry is not only valuable for the powder preparation itself, but also for its characterization and for the study of interactions with complexing agents.

Matijevic [65,66] has reviewed methods used in the preparation of uniform particles both those with simple composition and those of composite or coated materials. Zirconia powders coated with various elements have been made for the manufacture of water-resistant ceramics with high fracture strength [67]. Techniques such as chemical reactions in aerosol or homogeneous precipitation have been utilized successfully to form TiO_2, Al_2O_3, or Fe_2O_3 powders with well-defined chemistry and morphology. The ability of polyacrylic acid to form crosslinked chains bridged with metal ions permits the synthesis of fine particle ceramic powders such as yttria, partially stabilized zirconia or alumina. Other processes such as emulsion–precipitation, spray pyrolysis, spray-drying, freeze-drying, or sol–gel have also been used in wet-chemical powder manufacturing [9,68–71]. The latter technique will be discussed in a later paragraph.

Matijevic [66] showed that ions and molecules used in powder manufacture, may interact with the powder itself, and disturb parameters which characterize the oxide–solution interface, e.g., IEP. Both oxalic and citric acids adsorb strongly on colloidal hematite, as shown by the shift of the IEP.

It is well known in the field of colloid science that surface properties of a powder and its characterization are of great importance for the understanding of interactions with inorganic ions or organic molecules [17]. This is also the case in ceramics processing. The history of the powder, its preparation, the degrees of purity and crystallinity, mean that the surface of a given material does not always have the same properties [72]. A clear example is SiO_2, which can be prepared either by precipitation or by flame oxidation of a silicon chloride solution. Measurements of the total number of hydroxyls on the surface show a much larger number for the precipitated silica (over 20 OH/nm^2) than for silica prepared by the flame process (4–6 OH/nm^2) [17]. In the latter case, the high temperature used during the oxidizing step leads to silanol bridges, which are not broken upon rehydration [73]. Consequently these two powders exhibit different surface charge vs. pH behavior as shown in Figure 9.14.

In the characterization of aqueous, yttria-stabilized zirconia suspensions, the high solubility of yttria in acids has to be taken into account in the global consumption of protons by the surface. Some adsorption of Y^{3+} ions from the solution on the surface could also affect the surface properties of the

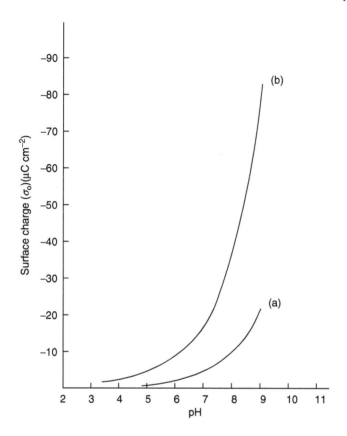

Figure 9.14 Comparison of surface charge/pH behavior for two types of SiO_2 (a) pyrogenic SiO_2 and (b) precipitated SiO_2 [74].

powder [24]. Beside this, the reprecipitation of yttria may change the properties of the final product.

Spray drying is a technique used to produce granulated ceramic powders. It usually starts with a water-based slurry that is sprayed into hot air, generating large spherical droplets of diameters in the range 20–100 μm. In this process, including slurry preparation, the characterization of the powder and the slurry are essential for the production of flowable powders with homogeneous density [70].

Alumina-based ceramic powder containing MgO and La_2O_3 has been made for applications in thermal barrier coatings [75]. Stable aqueous slurries containing various precursors were prepared using additives like dispersants and binders. The stability was followed by ζ potential and pH measurements.

B. Slurry Formulation and Forming Techniques

The next stage in the solution processing of ceramic materials is the stabilization of the concentrated oxide suspension and the forming of well-packed green bodies. For quality, it is important at this stage to prepare stable slurry containing as high solid as possible [76]. This is certainly the most decisive stage with regard to defect control and minimization in the whole fabrication process. Starting from a powder with a certain degree of agglomeration, it is possible, during this stage, to deagglomerate particles, e.g., with help of dispersing agents for weak agglomerates or sedimentation for hard agglomerates. Some precautions should, however, be taken in order to avoid introduction of new impurities, for example, when organic materials are added. If defects are present in the green material, it will be difficult to remove them in a later stage [7].

The strength σ of the green body is a way of measuring the presence of flaws. It can be defined as the sum of the attractive forces between individual particles [77]:

$$\sigma = \frac{1.1\Phi}{1 - \Phi} \frac{F_{att}}{a^2} \tag{16}$$

for a volume fraction Φ occupied by the particles of radius a, $F_{att} = Aa/12l^2$ is the attractive force between two spheres, separated by a distance l. A is the Hamaker constant. Equation (16) shows the importance of both powder packing efficiency and particle size in obtaining strong green bodies. Small solid volume fractions (low-packing density) will generate green bodies with low strength, but very fine powders (small a) will generate tougher materials.

The effect of low molecular weight dispersants on properties of electrostatically stabilized aqueous alumina suspensions has been determined by its adsorption ability on the

particle surface and by the nature and the number of its
dissociable groups [78]. Block copolymers have been used as
well to stabilize alumina suspensions in a wide range of pH
and in the presence of salts [79]. An optimum of stability was
found when the block copolymer had an average molecular
weight of ca. 5000, and consisted of short poly(methacrylic
acid) anchoring block and a longer polyethylene stabilizing
block.

Several forming methods using the solution processing of
ceramics are discussed in the following part. Applications of
these methods in the preparation of ceramic membranes are
also considered.

A continuous coating method of carbon fibers has been
applied, using fluid ceramic precursors as an alternative to
chemical vapor deposition [80].

1. Tape Casting

Table 9.4 shows a typical example of formulation of an aqueous slurry, used in a technique called doctor-blade casting, or
tape casting. This technique permits fabrication of thin layers

Table 9.4 Typical example of formulation of an aqueous slurry for
tape casting of alumina substrates [81]

Process step	Material	Function	Weight (g)
Premix in beaker	Distilled water	Solvent	465.0
	Magnesium oxide	Grain growth inhibitor	3.8
	Polyethylene glycol	Plasticizer	120.0
	Butyl benzyl phthalate	Plasticizer	88.0
	Nonionic octylphenoxy-ethanol	Wetting agent	5.0
	Condensed arylsulfonic acid	Deflocculant	70.0
Add and ball mill 24 h	Alumina powder	Substrate material	1900.0
Add, ball mill 30 min	Acrylic polymer emulsion	Binder	200.0
Add, ball mill 3 min	Waxed based emulsion	Defoamer	2.0

of ceramic materials. It starts with a slurry of inorganic powders dispersed in a liquid containing dissolved organic binders, plasticizers, etc. The properties of these materials have been reviewed and discussed by several authors [81,82]. The slip is spread on a flat support by means of a blade, and can, after drying, be stripped from the support and allowed to form. The technique finds wide applications in electronic industry, fuel cell technology, and oxygen sensors.

Up to now, most commercial tape casting processes have used nonaqueous systems, but the considerable amount of work done on aqueous colloidal systems is leading to the development of aqueous tape casting slurries for new applications.

The thickness of the cast depends very much on its fluid properties. Shear stresses appear when the slurry is spread on the flat support. Two types of rheological behavior are generally encountered in tape casting slurries: plastic and pseudoplastic response. At high shear rates, nonspherical particles are oriented in the direction of motion of the blade, causing nonuniform packing [83]. But binder molecules become oriented at the same time, and keep the structure of the slurry homogeneous [82]. Equation (17) gives the thickness δ of the cast, where it is assumed that the ceramic slurry has a Bingham (plastic) flow behavior [83]:

$$\frac{\delta}{h} = \left(0.5 + \frac{P}{6}\right)\left(1 - \frac{\tau_0^* - 1 + P}{2P}\right), \tag{17}$$

where h is the thickness of the gap between the blade and the support, P the pressure gradient across the blade, and τ_0^* the Bingham yield stress. Both P and τ_0^* depend on slurry viscosity. Consequently, it is important to control parameters characterizing oxide–solution interface in order to improve this technique.

The great number of materials used in the procedure shown in Table 9.4, reflects the complexity of the system. It may be advisable to formulate slurries with less components, for example by developing systems with a dispersing agent that at the same time has the properties of a binder (conferred by a higher molecular weight) and plasticizer. More extensive research is needed in this domain in order to optimize existing formulations or to develop new systems. The knowledge of the interactions between the material, and how to control them is therefore necessary in order to minimize defects both in the green film and in the final product [84].

Until the end of the 1980s, few studies had been undertaken on interactions between dispersing agents and binders, or on their interactions with the particle surface. The formation of complexes between cetyltrimethylammonium bromide (usually used as dispersing agent) and polyvinyl alcohol, denoted PVA (often used as binder) has been emphasized [85]. These interactions enhance polymer adsorption on silica particles at high pH, as shown in Figure 9.15.

An increased adsorption of the polymer could also diminish its *binder* effect, and reduce the flexibility of the green tape. Such adsorption has also been correlated to the shift of PPZC in the system alumina–sodium dodecylsulfate–PVA [86]. The study of interactions between surfactants and polymers has been reviewed by several authors [87,88].

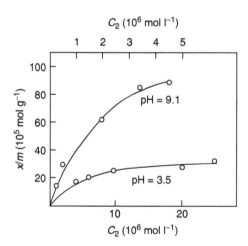

Figure 9.15 Adsorption isotherms of cetyltrimethylammonium bromide on silica at low and high pH. (x is the amount adsorbed by m grams of silica and C_2 the equilibrium concentration) [85].

Analysis of interactions between polymers and oxides find applications in the preparation of slurries. The stabilization of concentrated alumina slurries by polyelectrolytes has been studied as a function of several parameters including pH, solid concentration, and polymer molecular weight [89]. Figure 9.16 shows the amount of Na–poly(methacrylic acid), denoted PMAA–Na required to prepare stable suspensions of 20 vol% α-Al$_2$O$_3$ suspensions at various pH values. We can see that below pH 3.3, the suspension is stabilized by electrostatic interactions between double layers. The amount of polymer adsorbed on the surface is small, with the polymer remaining-mainly uncharged. Above pH 3.3, the amount of PMAA–Na needed to stabilize the slip follows the solid line, and

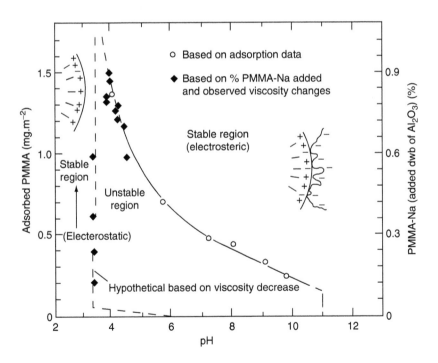

Figure 9.16 Stability of Al$_2$O$_3$ suspensions in the presence of PMAA–Na as a function of pH. The amount of adsorbed PMAA–Na required to form stable suspensions is also shown as a function of pH [89].

decreases with increasing pH. All of the PMAA–Na is adsorbed on the surface in a flat conformation induced by the negatively charged sites on the polymer chains. The polymer adsorption also causes a change in sign of the ζ potential from positive (for pH below IEP in the absence of PMAA–Na) to negative.

The method of agitation has a large effect on the dispersion of slurries for tape casting. Ultrasonic agitation has been studied and compared with the conventional ball milling [90]. In some cases, ultrasonic agitation was clearly more effective in breaking down agglomerates than ball milling. The powder packing has consequently been improved. Figure 9.17 shows, for $BaTiO_3$ slurries, how the two means of dispersing powders in solution affect particle settling.

Alumina slurries suitable for tape casting containing polyacrylate salts as dispersant and latex emulsions as binder, were optimized to obtain disks with high sintered density after sintering at 1600°C [91]. The sintered density could be

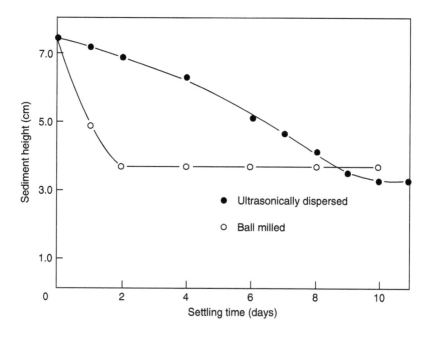

Figure 9.17   Height and sediment as a function of settling time for ultrasonically dispersed and ball-milled $BaTiO_3$ slurries [90].

directly correlated to the volume fraction of powder and to the binder content.

2. Slip Casting

Slip casting is widely used for making both traditional and advanced ceramics. Ceramic materials with intricate shapes can be formed by pouring a slip into a porous mold, and allowing the liquid to be absorbed by the mold. After drying, the green material is removed from the mold, and sintered. In slip casting, colloidal stabilization can also be applied directly.

The slip casting technique is governed by the flow through the porous medium of permeability K, as described by Darcy's law:

$$\frac{\mathrm{d}P}{\mathrm{d}x} = \frac{\eta q}{K},\tag{18}$$

where $\mathrm{d}P/\mathrm{d}x$ is the pressure gradient due to suction pressure P, η the slip viscosity and q, the apparent flow rate of the filtrate. The relationship between the thickness L of a cast layer, on the time t required to obtain that thickness, has been shown to be [92]:

$$L^2 = \frac{2Ct}{\eta}\tag{19}$$

where C is a proportionality constant, depending on mould and cake properties. Equation (19) involves the slip viscosity η, which is important to control in order to optimize the homogeneity of the green material. The viscosity is affected by the pH, as shown in Figure 9.18. Here, the TiO_2 slips attained a maximum in viscosity at the IEP (pH $= 5.6$). Complementary ξ potential and sedimentation rate measurements could be correlated to this evolution and used in optimizing slip stability.

Slip casting of alumina ceramics was performed at temperatures above $40°C$ to increase the casting rate [94]. This increase was related to the reduction of the slip viscosity and to change of the dispersion state of the slip, i.e., flocculation above $40°C$.

The rheology of alumina suspensions and the packing behavior during slip casting have shown to be influenced by

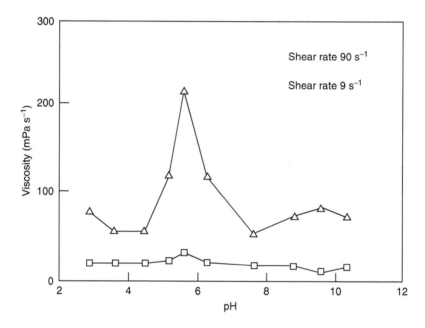

Figure 9.18   Viscosity of 10 wt% TiO_2 suspensions in 10^{-3} KNO_3. Shear rate ($\triangle$) $90\,s^{-1}$ and ($\square$) $9\,s^{-1}$ [93].

the stabilizing performances of two dispersants, i.e., a low Mw sulfonic acid and a polyacrylic acid [95]. It is shown that the polyacrylic acid used was more effective in promoting high green packing densities at high solid loading.

Slip casting and tape casting techniques have much in common with regard to slurry formulation, so several characteristics of the slurry mentioned for tape casting are also relevant here.

3. Colloidal Filtration

Colloidal filtration is based on the same principle as slip casting. But colloidal filtration is viewed rather as a technique that transforms a colloid into a consolidated state by removing the solvent by filtration. Figure 9.19 shows how different the filtration thickening rate can be for two alumina slips: one

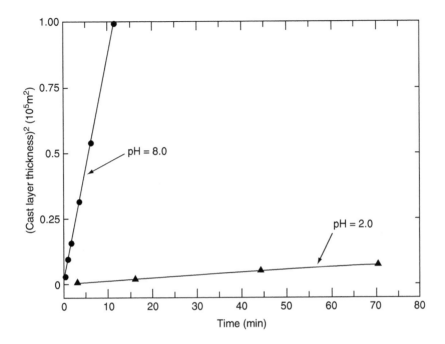

Figure 9.19 Filtration thickening rates for Al_2O_3 suspensions at pH 8.0 ($\Psi_z = 17\,\text{mV}$) and pH 2.0 ($\Psi_z = 55\,\text{mV}$) [96].

flocculating, was prepared at pH 8.0 ($\zeta = 17$ mV), and the other, well dispersed, at pH 2.0 ($\zeta = 55\,\text{mV}$). In the case of the slip prepared at pH 2.0, electrostatic repulsion between particles leads to a better packing, and a slower filtration rate, whereas the weak agglomerates formed at pH 8.0 permit a higher filtration rate, but the microstructures of the consolidated layers and the sintered materials change drastically with agglomeration. Finely dispersed slips at pH 2.0 lead to packed layers with a higher efficiency and a dense microstructure. However, both slips provided homogeneous microstructures, as shown by the linearity of the two curves of Figure 9.19 (in accord with Equation (19)) [96].

Alumina samples could be produced with a high filtration rate and a high green density when a proper size distribution and a proper additive (PVA) were chosen [97].

4. Gel Casting

This method, comparable with slip casting or injection molding, is now well recognized as an alternative route to the fabrication of near-net-shape ceramic parts [98–101]. Gel casting consists in initiating a gel by in situ polymerization of an aqueous dispersion containing a ceramic powder and organic monomers that is previously cast into a mould. The process has been applied to oxides like dense [102] or porous alumina [103] and iron titanates [104].

Recently, boehmite powder has been used as inorganic binder to promote gelation and provide good fluidity of the dispersion [105]. As shown on Figure 9.20, the addition of a concentration of 0.21 mol/l of hexamethylenetetramine (HMTA) to a dispersion of 55 vol% alumina containing 10 wt%

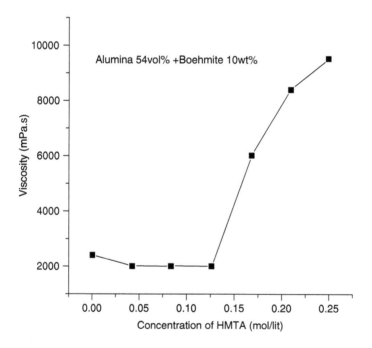

Figure 9.20 Effect of hexamethylenetetramine concentration on viscosity of 54 vol% alumina slurry containing 10 wt% boehmite [105].

boehmite, increased drastically the viscosity of the slurry after 15 min. This was a priori due to rapid bridging between the charged boehmite particles and HMTA molecules. Boehmite helped also alumina to sinter.

Several apparent methods to gel casting, starting from aqueous powder suspensions to form near net-shape ceramics, have been newly developed, like hydrolysis-assisted solidification for alumina components [106] and direct coagulation casting for several oxide ceramics [107].

5. Electrodeposition Methods

There are mainly two processes to prepare ceramic films by electrodeposition, namely electrophoretic deposition starting from a suspension or a sol, and electrolytic deposition starting from ions in solution.

Electrophoretic deposition [77,108] is used to make ceramic coatings on metals, or to prepare ceramic ware with unusual shapes. A powder is deposited on a metallic electrode, by applying an electric field in the suspension of charged particles. Compared to slip casting, this shortens the processing time considerably. Knowledge from colloid chemistry is also relevant to this technique, as the powder must have controlled surface charge, and be well dispersed in order to give homogeneous deposition. Ferrari et al. [109] have studied slip characteristics like viscosity, ξ potential and conductivity, to optimize the formation of coatings and laminate ceramics by electrophoretic deposition. The effect of dispersant concentration on the deposition parameters and the quality of the deposit was emphasized. Especially, Al_2O_3/Y-TZP layered composites with thickness control of each layer could be obtained.

Electrolytic deposition has become a new processing method for the fabrication of nanostructured thin films and powders [110]. Metal ions or complex precursors are hydrolyzed by an electrogenerated base to form ceramic coating on a cathodic substrate. Oxide coatings or powders could be processed by electrolytic deposition, like CeO_2, RuO_2, La_2O_3, Al_2O_3, TiO_2, ZrO_2, but also complex oxide compounds like $ZrTiO_4$, PZT, and $BaTiO_3$. Multilayered ceramics with ultrathin layers (2 nm), exhibiting thickness-dependent quantum optical, electronic, or optoelectronic effects were prepared [111].

6. Plastic-Forming Processes: Injection Molding and Extrusion

Other techniques used for producing small complex ceramic parts, are injection molding and extrusion. Injection molding differs from the other methods in that its processing stages require: (a) mixing the powder with an organic polymer with binder properties, (b) molding the mixture, and (c) removing the binder. The binder used in this process must exhibit thermal gelation characteristics, and is usually a thermoplastic resin, e.g., polyethylene, with some additives. Powder characteristics are important in determining how the solid can be formed in the green state. Aqueous powder processing can be applied to injection molding in order to ameliorate the homogeneity of the mixture by improving the wetting properties of the powder [77]. Recently, new processes have been developed for the aqueous injection molding of oxide ceramics [112–115], using smaller amounts of binder, i.e., 2% to 4%, instead of 10% to 20% for conventional paraffin-based systems [115].

In extrusion, a plastic material, prepared by filtration of a slurry, is forced through a rigid die. Hydroxyethylcellulose, methylcellulose, and polyvinyl alcohol are common binders used in extrusion [115]. Aqueous extrusion of alumina-zirconia (12 mol% ceria) composite with homogeneous constituent phases has been successfully performed by adding boehmite nanoparticles and pH adjustment, and induce flocculation [116].

7. Sol–Gel Techniques

Sol–gel technology was first developed to synthesize homogeneous fine-grained ceramics. It offers simultaneously an optimum control of both purity and crystallinity at a molecular level, and a means of processing ceramics without going through a powder step. The use of the method has lately been extended to catalysis [117], sensors [118], coatings [119], and organic–inorganic hybrid materials [120–122].

Sol–gel involves in most case use of molecular precursors, e.g., metal alkoxides as a starting material [123]. It relies on a high potential in the development of new precursors to adapt the nanostructure of ceramics [123–125]. The processing method can be divided into two different routes [126,127], as shown in Figure 9.21. On one hand, fine colloidal particles or oxide colloids prepared by hydrolysis or precipitation are dispersed in aqueous solution where stabilization occurs under electrostatic repulsions, i.e., control of surface charge. The sol–gel transition occurs by water removal, e.g., evaporation, and is often reversible. The gels obtained can then be redispersed in water, due to the weak nature of the bonds between particles. This method is often called the colloidal route.

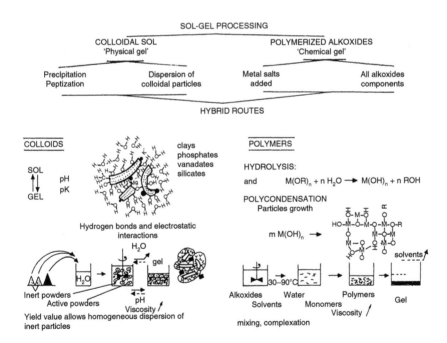

Figure 9.21 Representation of different sol–gel methods. In the colloidal route, gel formation involves electrostatic or steric interactions. In the alkoxide route, chemical reactions and polymerization are the main reactions leading to a gel [126].

On the other hand, alkoxides, alkoxysilanes, or hydroxylated metal ions can be polymerized under controlled hydrolysis, following the reaction

$$MOR + HOH \rightleftharpoons MOH + ROH \text{ (hydrolysis)} \tag{20}$$

$$MOH + HOM \rightleftharpoons MOM + H_2O \text{ (condensation)} \tag{21}$$

$$MOR + HOM \rightleftharpoons MOM + ROH \text{ (condensation)} \tag{22}$$

In this case, the gel transition is more complex, involving chemical reactions and particle growth. A complete hydrolysis of the alkoxide will lead to gel in powder form, whereas high alkoxide–water ratios will generate monolithic samples. In the case of the alkoxide route, the sol–gel transition is often irreversible.

Sol–gel techniques utilize directly most of the principles of colloid science, and have found increasing application in the

Figure 9.22 Influence of the drying conditions on SiO_2 gels fired at 600°C. (a) Thin membrane (3 μm) slowly dried at 30°C; pH = 7. (b) Membrane (7 μm) quickly dried at 30°C, pH = 7. (c) Thick membrane (10 μm) quickly dried at 100°C, pH = 9 [128].

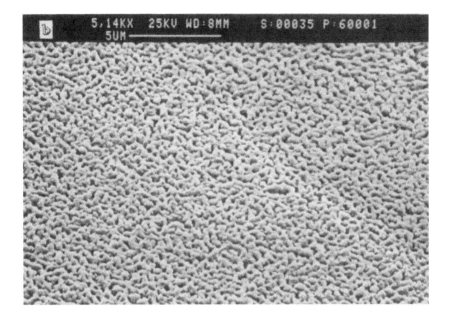

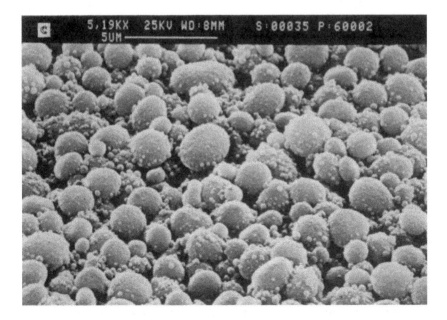

Figure 9.22 (*Continued*)

processing of ceramic materials, i.e., powders, coatings, and thin films (e.g., optical applications), nanostructured materials and membranes. Figure 9.22 shows the effect of pH, temperature, and drying on SiO_2 gels fired at 600°C. The importance of these parameters for the control of the pore size of the ceramic membrane should be noted. Stable zirconia sols coated with yttrium could be synthesized by hydrolysis and peptization of zirconia tetra-*n*-butoxide, with monodisperse particle size of 4 nm. These sols could be applied to produce cubic yttria-stabilized zirconia at temperature as low as 453°C [129].

8. Example of Application: Ceramic Membranes

A novel application of sol–gel technique is the fabrication of ceramic membranes for ultra- and nanofiltration, and gas separation. This new type of membrane presents several advantages over polymeric membranes, e.g., it is high-temperature- and pressure-resistant, chemically and biologically inert, sterilizable, and has a long lifetime. Such a membrane may either be dense or porous, depending on the applications. Dense ceramic membranes are applied as oxygen-selective membranes for the separation of oxygen from air for example [5,130,131]. A ceramic membrane with controlled pore size is typically composed of a porous substrate in tubular, multichannel or flat form [132,133], covered inside by one or more separating layers. The pore size of these layers is chosen according to the size of the species to be separated, and vary from ca. 0.5 nm (e.g., microporous membranes for gas separation [134]) to several micrometers. The possibility of fabricating ceramic membranes with controlled ultrafine pores, opens up avenues for new applications and new markets in ceramic technology, extending from water treatment to gaseous separation and catalysis.

Several techniques described above, like sol–gel or slip casting, are used to prepare and characterize inorganic membranes. Extrusion may be used to fabricate the porous substrates. Slip casting is used to deposit the different separating layers, with control of thickness and homogeneity. Sol–gel is employed to prepare stable sols with monosized particles [135]. Parameters such as pH, temperature, and type of electrolyte control the pore size of the membrane [58]. Surface and colloidal phenomena were applied for a better understanding of transport phenomena in membrane processes [136].

C. Ceramic Materials from Nanosized Powders

Processing of ceramic materials from nanosized particles has been reviewed by several authors [123,137,138]. Preparation of ceramics from nanosized powders is discussed below.

1. Synthesis of Nanosized Ceramic Powders

Beside classical methods for the preparation of micro- or sub-micrometer-sized particles, new technologies have recently been developed to synthesize powders containing nanosized particles. The preparation methods of nanosized particles extend from mechanical processing, plasma- and gas-phase reaction processes, to liquid-state processes like sol–gel, organometallic and spray pyrolysis, and combustion synthesis. Both pure and nanocomposite ceramic powders may be prepared by these methods. Bhaduri et al. [139] have reviewed various types of synthesis methods according to the phase to be processed.

Due to the broad range of applications, the production of nanosized ceramic particles is a rapidly growing field. A limiting parameter is however the production of sufficiently large quantities of these powders, although this problem is overcome for specific systems [137].

2. Compaction and Sintering Methods

Due to their small size, nanosized particles are much more sensitive to agglomeration during powder synthesis step and before processing into ceramics. Careful attention must be paid to control parameters like temperature during the synthesis, and additional stages like surface modification before drying of the powder, are necessary. Modified thermal processes have been applied to produce nanophase alumina, manganite, zirconia, and yttria-stabilized zirconia, at low cost and high rate of synthesis [140].

Beside these precautions to avoid agglomerates formation, more conventional methods may also be applied to eliminate large aggregates during compaction and sintering, e.g., centrifugation, hot pressing, or sinter-forging.

By a combination of control of hydrolysis conditions of alkoxides and peptization, crystallized nanosized TiO_2 could be synthesized in rutile form by hydrothermal process at 250°C under saturated water vapor pressure [141]. Ultrafine TiO_2 powders could be formed from homogeneous precipitation of aqueous $TiOCl_2$ solution. Pure rutile phase was obtained at temperature below 65°C, while anatase formed above that temperature [142].

Alumina and zirconia powders with an average particle size of 100 and 20 nm, respectively, have been processed for slip casting [143]. Slips with lower solid contents than using micropowders were prepared and thixotropy was encountered due to high specific area of the nanosized powders.

III. EFFECT OF STABILIZATION ON DRYING AND SINTERING

Drying is an important operation prior to firing ceramic materials. It is the carefully controlled removal of liquid from the porous material. Interactions between liquid flow and dilatation of the solid induce stresses and strains that may lead to defects in the product, e.g., cracks. The stabilization process also influences this step.

The final main step in the processing of ceramic materials is sintering. The particles are joined together during firing, resulting in strengthening of the product. Two different stages can be identified during firing: elimination of organic matter by decomposition and oxidation, and the actual sintering. A umber of variables controlling the stabilization of colloidal suspensions are of primary importance for the sintering step.

A. Drying

Differential shrinkage during drying can generate deformation and defects in the green material. Several techniques can be used to avoid or to minimize differential shrinkage during drying [115]. Several parameters, such as solid content or type of binder, affect shrinkage conditions greatly. In tape casting for instance, the linear shrinkage $\Delta L/L_0$, is proportional to the mean reduction in interparticle spacing and the mean number of interparticle liquid films per unit length N_l, when particle sliding and rearrangement do not contribute to the shrinkage [115]:

$$\frac{\Delta L}{L_0} = \bar{N}_l \Delta \bar{l} \tag{23}$$

Equation (23) shows that the linear shrinkage can be reduced by increasing the solid content in the slurry (Δl decreases). This is achieved, as seen in Section III, by preparing a well-dispersed suspension, i.e., adjusting pH to have a high surface charge and a low electrolyte concentration, or by adding dispersing agents [144]. These two methods produce suspensions with low viscosity, and therefore allow higher solid contents. How the state of stabilization of the suspension (controlled by pH in case of electrostatic stabilization) influences drying is shown in Figure 9.21.

The addition of organic binders (e.g., polyvinyl alcohol [58]), which have the ability to bind the ceramic particles may also reduce Δl and shrinkage.

The effect of addition of binder to the slurry on the distance between particles is shown in Figure 9.23. Too much binder results in green materials with low density, whereas the flexibility of the material (often desired in tape casting) is poor when the quantity of binder is very low.

B. Firing

1. Removal of Organic Matter

After formation, the green material is fired to densify the powder, and to give a more or less dense material. At first, organic materials added during slurry formulation are burned out, at rather low temperature, usually below 500°C. Depending on the amount of dispersing agent or binder added, and what kind of material it is, e.g., ionic, long chained, a fraction

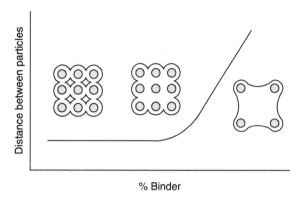

% Binder

Figure 9.23 Effect of binder addition on the distance between particles [145].

of these compounds can become trapped in the material. These impurities, as well as gas from the atmosphere, will limit pore shrinkage in the final stage of sintering. Rhodes and Natansohn [71] showed that residual gases still evolved at temperatures as high as 1000°C during sintering of Y_2O_3, synthesized from yttrium oxalate at 1500°C. It is thus essential to optimize the amount of organic material added during the stabilization procedure. The temperature program must take into account the decomposition and calcination of organic materials. Nitrate compounds are special, as they can produce explosive mixtures with organic materials. Sulfate and chloride salts are therefore preferred to nitrates.

2.   Sintering

The transformation from a green compact to a dense body occurs in three different stages, as shown in Figure 9.24. As the contact area of the particles increases with time, the pores reduce in size, to be isolated in the final stage. The decrease of pore size and volume give the final increase in density of the material.

It is commonly accepted that agglomerates present in the powder affect sintering conditions [146]. The effect of particle size on the final density of yttria-stabilized zirconia is shown

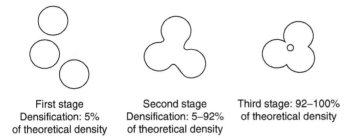

First stage
Densification: 5%
of theoretical density

Second stage
Densification: 5–92%
of theoretical density

Third stage: 92–100%
of theoretical density

Figure 9.24 Schematic representation of the different stages of sintering and densification.

in Figure 9.25. The powder sinters only to 95% of the theoretical density at 1500°C, when cold pressed without pretreatment (curve a). When the powder is conditioned for 72 h in an aqueous solution of pH 1.2 (maximum deflocculation), and concentrated into compacts by centrifugation, the green material has a density close to the theoretical value for close-packing of equal-diameter spheres. A further sintering of the compacts shows that 95% of the theoretical density is already reached at 950°C, and 99.5% after 1 h at 1100°C (curve b). It has also been demonstrated that agglomerates affect the initial stage of sintering [147]. Similar observations have been made on nanocrystaline yttria-stabilized zirconia [137].

Alford et al. [148] have studied the effect of agglomerates in a commercial alumina powder on the strength of ceramics made from it. It is shown that weakness in the ceramics originates from agglomerates. They demonstrated later the effectiveness of colloidal processing in the minimization of flaws, i.e., removal of agglomerates by stabilization control of the colloidal dispersion. The bend strength of alumina could be then improved from 0.37 to 1.04 GPa.

In a study of sintering behavior of compacts obtained from alumina slurries and powders, Zheng and Reed [149] define a critical ratio of pore size to mean particle size for pore shrinkage, above which pores cannot be totally eliminated during sintering. They further showed that an

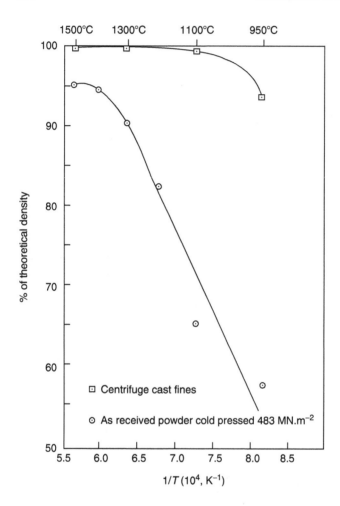

Figure 9.25 Effect of firing temperature (1 h cycle) on the density of sintered of 6.5 mol% Y_2O_3-stabilized ZrO_2 (2.9 μm agglomerate size and agglomerate-free powder) [147].

appropriate processing of the green body, in a solution containing deflocculant, permits the elimination of pores larger than the critical ratio before drying, and gave green and sintered materials with the highest density.

A higher solid content in the slurry, allowed by means of polyelectrolyte-stabilized suspensions, decreases the linear shrinkage during drying, as observed above. The sintering behavior is also affected by the solid content. Cesarano and Aksay [144] obtained 99% of the theoretical density after sintering at 1350°C for 60 vol% α-alumina suspensions, whereas below 90% of the theoretical density the result was obtained with the dried pressed powder at the same temperature.

During solid-state sintering, densification occurs via boundary diffusion or lattice diffusion. In the latter case, the sintering time Δt_{SSL} can be estimated from Equation (24) [77]:

$$\Delta t_{SSL} = \frac{\beta_L d^3 kT}{D_L \gamma_\pi \Omega} \tag{24}$$

and in the case of boundary diffusion:

$$\Delta t_{SSB} = \frac{\beta_B d^4 kT}{D_B \delta_B \gamma_\pi \Omega} \tag{25}$$

where D_L and D_B are the lattice and the boundary diffusion coefficients, respectively, b_L and b_B are constants of proportionality, d_B is the thickness of the boundary, g_p is the specific surface work, W is the atomic volume, and d the particle diameter. Considering these two equations, it is interesting to notice that Δt_{SSL} is proportional to d^3 for lattice diffusion and to d^4 for boundary diffusion. This reinforces the interest to produce powders with smaller particles. The small size is however limited by other diffusion mechanisms, which do not provide densification [77].

Particle-size distribution also has to be considered. α-Alumina powder, with a narrow particle-size distribution obtained by centrifugal classification, sinters more rapidly than a powder having the same mean particle size, but with a broader particle-size distribution [150]. Yeh and Sacks [151] showed that both broad and narrow particle size distributions may produce sintered bodies with high density, in the case of alumina slips stabilized at pH 4. Figure 9.26 shows the size distribution for samples sintered at 1340°C for 48 h. Grain coarsening has not occurred in the case of powders with a

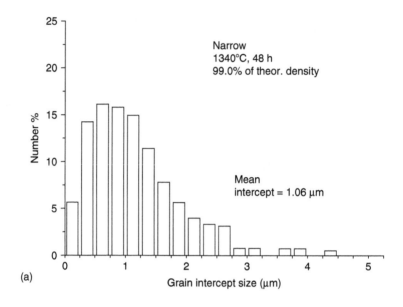

(a)

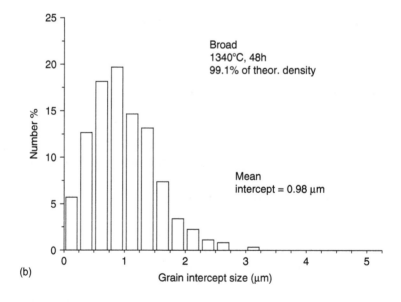

(b)

Figure 9.26   Particle size distribution for sintered Al_2O_3 (1340°C, 48 h). Samples prepared with narrow (a) and broad (b) size distributions [151].

broad size distribution, and the two sintered samples have analogous microstructures.

It is common to add elements to ceramic powders to improve their sintering properties. These elements, called sintering aids, may be added in the form of salts, powders, or suspensions. Clay minerals, magnesium salts, or boehmite are usual sintering aids for alumina. Sintering aids are usually mixed with the original powder, and it is therefore fundamental to prepare a homogeneous and stable mixture before further processing. This renders often the stabilization of the system more complex due to different surface properties of the components. Gosky et al. [152] prepared stable alumina suspensions in the presence of mineral kyanite ($Al_2O_3.SiO_2$) by pH adjustment. An optimum pH of 3–4 was found where the system was stable. In this pH range, it was suggested that kyanite was responsible for the good flow properties of the system.

V. SUMMARY

In spite of the progress made in the fabrication of ceramic materials, still many obstacles remain that have to be overcome. Many impurities and defects are introduced in the numerous manipulations that lead to the final product. The new techniques and routes, e.g., sol–gel, hydrothermal, recently applied to manufacture fine powders of controlled size, purity, and homogeneity, can reduce the addition of impurities in this early phase of ceramic processing.

The wet processing includes the preparation of a stable and concentrated suspension of powder. Most of advanced ceramic materials are, in their early processing stage, in a colloidal state, and therefore methods and developments from colloid science may be applied to control the size of agglomerates in the suspension. Hard agglomerates produced during powder manufacturing, may be eliminated by filtration or centrifugation. The size of soft agglomerates may be adjusted by addition of dispersants (either inorganic or organic materials). Furthermore, studies in colloid science may ameliorate the quality of slurry formulation, e.g., by a better understanding of interactions between dispersants and binders. Slurries of better quality, i.e., well-packed particles, are easier to form, and lead to green materials with fewer defects. Such materials will sinter at a lower temperature and have a higher density. Many of the properties of the final sintered material are already decided before sintering.

REFERENCES

1. T. Abraham, *Am. Ceram. Soc. Bull.*, 78(3), 69–71, 1999.
2. J. Kubel, Jr., *Adv. Mater. Proc.*, 9, 55, 1989.
3. T. Yao, Y. Uchimoto, T. Sugiyama, and Y. Nagai, *Proc. Electrochem. Soc.*, 99(19), 483–492, 1999.
4. C. Simon, R. Glenne, and T. Norby, in B. Bergman, Ed., Proceedings of the First Nordic Symposium on Materials for High Temperature Fuel Cells, RIT, Stockholm, 1991.
5. B. Ma, U. Balachandran, J.-H. Park, and C.U. Segre, *Solid State Ionics*, 83, 65–71, 1996.
6. M.N. Rittner and T. Abraham, *J. of Metals*, 1, 36–37, 1998.
7. F.F. Lange, *J. Am. Ceram. Soc.*, 72, 3, 1989.
8. P.D. Morgan, in *Ceramic Powder Science*, Vol. 1, G.L. Messing, E.R. Fuller, and H. Hausner, Jr., Eds., The American Ceramic Society, Westerville, 1988, pp. 3–7.
9. E.A. Barringer and H.K. Bowen, *J. Am. Ceram. Soc.*, 65, C 199, 1982.
10. J. Lyklema, in *Solid/Liquid Dispersions*, Th.F. Tadros, Ed., Academic Press, London, 1987, pp. 63–90.
11. J.C. Westall, *ACS Symp. Ser.*, 323, 55, 1986.
12. D.H. Napper, *Polymeric Stabilization of Colloidal Dispersions*, 1st ed., Academic Press, London, 1983.
13. G.J. Fleer and J. Lyklema, in *Adsorption from Solution at the Solid/Liquid Interface*, G.D. Parfitt and C.H. Rochester, Eds., Academic Press, London, 1983, pp. 153–220.
14. J. Sung and P.S. Nicholson, *J. Am. Ceram. Soc.*, 71, 788, 1988.
15. H.P. Boehm, *Adv. Catal.*, 16, 179, 1966.
16. P.W. Schindler, in *Adsorption of Inorganics at Solid–Liquid Interfaces*, M.A. Anderson and A.J. Rubin, Eds., Ann Arbor Science, Ann Arbor, 1981, pp. 1–49.

17. R.O. James, in *Advances in Ceramics*, Vol. 21, G.L. Messing, K.S. Mazdiyasni, J.W. McCauley, and R.A. Haber, Eds., The American Ceramic Society, Colombus, 1987, pp. 349–410.
18. G.Y. Onoda, Jr. and J.A. Casey, in *Ultrastructure Processing of Ceramics, Glasses and Composites*, L.L. Hench and D.R. Ulrich, Eds., John Wiley & Sons, New York, 1984, pp. 374–390.
19. B.H. Bijsterbosch and J. Lyklema, in *Advances in Colloid Interface Science*, Vol. 9, A.C. Zettlemoyer and J.T.G. Overbeek, Eds., Elsevier, Amsterdam, 1978, pp. 147–251.
20. Y.G. Berube and P.L. De Bruyn, *J. Colloid Interf. Sci.*, 27, 305, 1968.
21. S.M. Ahmed, *J. Phys. Chem.*, 73, 3556, 1969.
22. D.E. Yates and T.W. Healy, *J. Chem. Soc. Faraday Trans. I*, 76, 9, 1980.
23. Th.F. Tadros and J. Lyklema, *J. Electroanal. Chem.*, 17, 267, 1968.
24. C. Simon, in *Ceramics Today — Tommorrow's Ceramics*, P. Vincenzini, Ed., Elsevier, Amsterdam, 1991, pp. 1043–1051.
25. H.C. Li and P.L. de Bruyn, *Surf. Sci.*, 5, 203, 1966.
26. H. Pilgrimm, *Colloid Polym. Sci.*, 259, 1111, 1981.
27. S.M. Ahmed, *Can. J. Chem.*, 44, 1663, 1966.
28. A.E. Regazzoni, M.A. Blesa, and A.J.G. Maroto, *J. Colloid Interf. Sci.*, 91, 560, 1983.
29. R.O. James, P.J. Stiglich, and T.W. Healy, in *Adsorption from Aqueous Solutions*, P.H. Tewari, Ed., Plenum Press, New York, 1981, p. 19.
30. A. Foissy, A. M'Pandou, J.M. Lamarache, and N. Jaffrezic-Renault, *Colloids Surf.*, 5, 363, 1982.
31. R. Sprycha, *J. Colloid Interf. Sci.*, 102, 173, 1984.
32. M.J.G. Janssen and H.N. Stein, *J. Colloid Interf. Sci.*, 111, 112, 1986.
33. J. Yopps and D.W. Fuerstenau, *J. Colloid Interf. Sci.*, 19, 61, 1964.
34. J.A. Davis, R.O. James, and J.O. Leckie, *J. Colloid Interf. Sci.*, 63, 480, 1978.
35. J. Lyklema, *J. Colloid Interf. Sci.*, 99, 109, 1984.
36. R. O'Brien and L.R. White, *J. Chem. Soc. Faraday Trans. I*, 74, 1607, 1978.
37. D.W. Fuerstenau, D. Manmohan, and S. Raghavan, in *Adsorption from Aqueous Solutions*, P.H. Tewari, Ed., Plenum Press, New York, 1981, p. 93.

38. K.D. Dobson, P.A. Connor, and A.J. McQuillan, *Langmuir*, 13, 2614–2616, 1997.
39. P. Fievet, A. Szymczyk, C. Labbez, B. Aoubiza, C. Simon, A. Foissy, and J. Pagetti, *J. Colloid Interf. Sci.*, 235, 2, 2001.
40. A. Szymczyk, C. Labbez, P. Fievet, B. Aoubiza, and C. Simon, *AIChE J.*, 47, 2349–2358, 2001.
41. D.W. Fuerstenau, R. Herrera-Urbina, J.S. Hanson, *Ceram. Trans.*, 1, 333–351, 1988.
42. A. Foissy and J. Persello. in *The Surface Properties of Silicas*, A.P. Legrand, Ed., John Wiley & Sons, New York, 1998, pp. 365–414.
43. F.J. Hingston, in Adsorption of Inorganics at Solid–Liquid Interfaces, M.A. Anderson and A.J. Rubin, Eds., Ann Arbor Science, Ann Arbor, 1981, pp. 51–89.
44. C.P. Huang, in Adsorption of Inorganics at Solid–Liquid Interfaces, M.A. Anderson and A.J. Rubin, Eds., Ann Arbor Science, Ann Arbor, 1981, pp, 183–217.
45. T. Sato and R. Ruch, *Stabilization of Colloidal Dispersions by Polymer Adsorption*, 1st ed., Marcel Dekker, New York, 1980.
46. G.J. Fleer and J.M.H.M. Scheutjens, in *Coagulation and Flocculation*, 1st ed., Vol. 47, B, Dobias, Ed., Marcel Dekker, New York, 1993, pp. 209–264.
47. E.S. Tormey, in *Surfactants in Emerging Technologies*, Vol. 26, M.J. Rosen, Ed., Marcel Dekker, New York, 1987, pp. 85–93.
48. B.M. Moudgil, P. Somasundaran, and H. Soto, in *Reagents in Mineral Technology*, Vol. 27, P. Somasundaran and B.M. Moudgil, Eds., Marcel Dekker, New York, 1988, pp. 79–104.
49. B. Vincent, in *Solid/Liquid Dispersions*, Th.F. Tadros, Ed., Academic Press, London, 1987, pp. 149–162.
50. L.K. Koopal, in *Coagulation and Flocculation*, 1st ed., Vol. 47, B. Dobias, Ed., Marcel Dekker, New York, 1993, pp. 101–208.
51. R. Aveyard, in *Solid/Liquid Dispersions*, Th.F. Tadros, Ed., Academic Press, London, 1987, pp. 111–129.
52. R.J. Pugh and L. Bergström, *Surface and Colloid Chemistry in Advanced Ceramics Processing*, 1st ed., Marcel Dekker, New York, 1994.
53. B.V. Derjaguin and L. Landau, *Acta. Physicochi. URSS*, 14, 633, 1941.
54. E.J.W. Verwey and J.Th.G. Overbeek, *Theory of Stability of Lyophobic Colloids*, 1st ed., Elsevier, Amsterdam, 1948.
55. H. Stechemesser and H Sonntag, this volume

56. D.H. Everett, *Basic Principles of Colloid Science*, The Royal Society of Chemistry, London, 1988.
57. M. Hashiba, H. Okamoto, Y. Nurishi, and K. Hiramatsu, *J. Mater. Sci.*, 23, 2893, 1988.
58. A. Larbot, J.P. Fabre, C. Guizard, L. Cot, and J. Gillot, *J. Am. Ceram. Soc.*, 72, 257, 1989.
59. R.H. Ottewill, in *Colloidal Dispersions*, J.W. Goodwin, Ed., The Royal Society of Chemistry, London, 1982, pp 197–217.
60. R.H. Ottewill, in *Solid/Liquid Dispersions*, Th.F. Tadros, Ed., Academic Press, London, 1987, pp. 183–198.
61. M.D. Sacks, C.S. Khadilkar, G.W. Scheiffele, A.V. Shenoy, J.H. Dow, and R.S. Shieu, in *Advances in Ceramics*, Vol. 21, G.L. Messing, K.S. Mazdiyasni, J.W. McCauley, and R.A. Haber, Eds., The American Ceramic Society, 1987, pp. 495–515.
62. J.F. Kneller and D.R. Cosper, US Patent No. 5366243, 1993.
63. H.M. Jang, *Mater Eng. (N.Y.)*, 8, 157–196, 1994.
64. A. Pissenberger and G. Gritzner, *J. Mater. Sci. Lett.*, 14(22), 1580–1582, 1995.
65. E. Matijevic, in *Ultrastructure Processing of Ceramics, Glasses and Composites*, L.L. Hench and D.R. Ulrich, Eds., John Wiley & Sons, New York, 1984, pp. 334–352.
66. E. Matijevic, in *Advances in Ceramics*, Vol. 21, G.L. Messing, K.S. Mazdiyasni, J.W. McCauley, and R.A. Haber, Eds., The American Ceramic Society, Colombus, 1987, pp. 423–437.
67. G.P. Dransfield and T.A. Egerton, European Patent Appl. EP 535796, 1993.
68. J.V. Sang, *Can. Br. Ceram. Proc.*, 47, 47–52, 1991.
69. T. Ogihara, T. Ookura, T. Yanagawa, N. Ogata, and K. Yoshida, *J. Mater. Chem.*, 1(5), 789–794, 1991.
70. S. Lukasiewicz, *J. Am. Ceram. Soc.*, 72, 617, 1989.
71. W.H. Rhodes and S. Natansohn, *Ceram. Bull.*, 68, 1804, 1989.
72. G.D. Parfitt. in *Progress in Surface and Membrane Science*, Vol. 11, D.A. Cadenhead and J.F. Danielli, Eds., Academic Press, New York, 1976, pp. 181–226.
73. A.C. Zettlemoyer and E. McCafferty. *Croat. Chem. Acta.*, 45, 173, 1973.
74. J. Lyklema, *Croat. Chem. Acta.*, 43, 249, 1971.
75. G.W. Schäfer and R. Gadow, *Ceram. Eng. Sci. Proc.*, 20(3), 291–297, 1999.
76. A. Bleier, in *Ultrastructure Processing of Ceramics, Glasses and Composites*, L.L. Hench and D.R. Ulrich, Eds., John Wiley & Sons, New York, 1984, pp. 391–403.

77. I.J. McColm and N.J. Clark, *Forming, Shaping and Working of High-performance Ceramics*, 1st ed., Blackie, London, 1988.
78. P.C. Hidber, T.J. Graule, and L.J. Gaukler, *J. Eur. Ceram. Soc.*, 17, 239–249, 1997.
79. J. Orth, W.H. Meyer, C. Bellmann, and G. Wegner, *Acta Polym.*, 48(11), 490–501, 1997.
80. R. Gadow, S. Kneip, and G.W. Schäfer, *Ceram. Eng. Sci. Proc.*, 20(3), 571–577, 1999.
81. J.C. Williams, in *Ceramic Fabrication Processes*, Vol. 9, F.F.Y. Wang, Ed., Academic Press, New York, 1976, pp. 173–198.
82. A. Roosen, in *Ceramic Powder Science*, Vol. 1, G.L. Messing, E.R. Fuller, and H. Hausner, Jr., Eds., The American Ceramic Society, Westerville, 1988, pp. 675–692.
83. T.A. Ring, in *Surfaces and Interfaces in Ceramic Materials*, Vol. 173, L.C. Dufour, C. Monty, and G. Petot-Ervas, Eds., Kluwer Academic Publishers, Dortdrecht, 1989, pp. 459–505.
84. E.S. Tormey, R.L. Pober, H.K. Bowen, and P.D. Calvert, in *Forming of Ceramics*, Vol. 9, J.A. Mangels and L. Messing, Eds., The American Ceramic Society, Colombus, 1984, pp. 140–149.
85. Th.F. Tadros, *J. Colloid Interf. Sci.*, 46, 528, 1974.
86. S. Chibowski, *Mater. Chem. Phys.*, 20, 65, 1988.
87. E.D. Goddard, *Colloids Surf.*, 19, 255, 1986.
88. I.D. Robb, in *Anionic Surfactants*, Vol. 11, E.H. Lucassen-Reynders, Ed., Marcel Dekker, New York, 1981, pp. 109–142.
89. J. Cesarano III and I.A. Aksay, *J. Am. Ceram. Soc.*, 71, 250, 1988.
90. R.J. MacKinnon and J.B. Blum, in *Forming of Ceramics*, Vol. 9, J.A. Mangels and L. Messing, Eds., The American Ceramic Society, Colombus, 1984, pp. 158–163.
91. R. Greenwood, E. Roncari, and C. Galassi, *J. Eur. Ceram. Soc.*, 17, 1393–1401, 1997.
92. F.M. Tiller and C.D. Tsai, *J. Am. Ceram. Soc.*, 69 882, 1986.
93. A.S. Rao, *Ceram. Int.*, 14, 71, 1988; *Ceram. Int.*, 13, 233, 1987.
94. A.M. Murfin and J.G.P. Binner, *Ceram. Int.*, 24, 597–603, 1998.
95. G. Tari, J.M.F. Ferreira, and O. Lyckfeldt, *J. Eur. Ceram. Soc.*, 18, 479–486, 1998.
96. I.A. Aksay and C.H. Schilling, in *Ultrastructure Processing of Ceramics, Glasses and Composites*, John Wiley & Sons, New York, 1984, pp. 439–447.
97. T. Betz, W. Riess, J. Lehmann, and G. Ziegler, *Ceram. Forum Int.*, 74(2), 101–105, 1997.

98. O.O. Omatete, M.A. Janney, and R.A. Strelow, *Am. Ceram. Soc. Bull.*, 70, 1641–1647, 1991.
99. M. Takahashi and H Unuma, *Ceram. Jpn.*, 32, 102–105, 1997.
100. J.J. Nick, D. Newson, B. Draskovich, and O.O. Omatete, *Ceram. Eng. Sci. Proc.*, 20(3), 217–223, 1999.
101. O.O. Omatete and J.J. Nick, *Ceram. Eng. Sci. Proc.*, 20(3), 241–248, 1999.
102. O.O. Omatete, R.A. Strelow, and C.A. Walls, in Proceedings of the 37th Sagamore Army Materials Research Conference, Plymouth, MA, 1990, pp. 201–212.
103. S. Pilar, *Am. Ceram. Soc. Bull.*, 76, 61–65, 1997.
104. M.H. Zimmermann and K.T. Faber, *J. Am. Ceram. Soc.*, 80, 2725–2729, 1997.
105. K. Prabhakaran, S. Ananthakumar, and C. Pravithran, *J. Eur. Ceram. Soc.*, 19, 2875–2881, 1999.
106. T. Kosmač, S. Novak, M. Sajko, and D. Eterovič, EP Patent No. 0-813-508 BI, 1999.
107. T.J. Graule, L.J. Gauckler, and F.H. Baader, *Ind. Ceram.*, 16, 31–35, 1996.
108. P. Sarkar and P.S. Nicholson, *J. Am. Ceram. Soc.*, 79, 1987–2002, 1996.
109. B. Ferrari, A.J. Sanchez-Herencia, and R. Moreno, *Mater. Res. Bull.*, 33(3), 487–499, 1998.
110. I. Zhitomirsky, *Ceram. Bull.*, 9, 57–63, 2000.
111. J.A. Switzer, R.J. Phillips, and R.P. Raffaelle, *ACS Symp. Ser.* 499, 244–253, 1992.
112. G.B. Marsh, A.J. Fanelli, J.V. Burlew, and C.P. Ballard, US Patent No. 5087595, 1992.
113. R.R. Landham, P. Nahass, D.K. Leung, M. Ungureit, W.E. Rhine, H.K. Bowen, and P.D. Calvert, *Am. Ceram. Soc. Bull.*, 66, 1513, 1987.
114. A.J. Fanelli and S.R. Dean, European Patent Publication No. EP 246438 A2, 1987.
115. J.S. Reed, *Introduction to the Principles of Ceramic Processing*, John Wiley & Sons, New York, 1988, pp. 411–425.
116. C.S. Kumar, U.S. Hareesh, B.C. Pai, A.D. Damodaran, and K.G.K. Warrier, *Ceram. Int.*, 24, 583–587, 1998.
117. C. Simon, H. Grøndal, R. Bredesen, A.G. Hustoft, and E. Tangstad, *J. Mater. Sci.*, 30, 5554–5560, 1995.
118. O. Lev, M. Tsionsky, L. Rabinovich, L. Sampath, I. Pankratov, and J. Gun, *Anal. Chem.*, 67(1), 22–30, 1995.

119. C. Simon, M. Seiersten, and P. Caron, *Mater. Sci. Forum*, 251(54), 429–436, 1997.
120. R.M. Ottenbrite and J.S. Wall, *J. Am. Ceram. Soc.*, 83(12), 3214–3215, 2000.
121. F. Ribot, F. Banse, C. Sanchez, M. Lahcini, and B. Jousseaume, *J. Sol–Gel Sci. Technol.*, 8, 529–534, 1997.
122. H.K. Schmidt, *J. Sol–Gel Sci. Technol.*, 8, 557–566, 1997.
123. J. Livage, F. Beteille, C. Roux, M. Chatry, and P. Davidson, *Acta Mater.*, 46(3), 743–750, 1998.
124. S.M.M. Veith, M. Haas, H. Shen, N. Lecerf, V. Huch, S. Hüfner, R. Haberkorn, H. Beck, and M. Jilavi, *J. Am. Ceram. Soc.*, 84, 1921–1928, 2001.
125. M. Veith, A. Altherr, N. Lecerf, S. Mathur, K. Valtchev, and E. Fritscher, *Nanostruct. Mater.*, 12, 191–194, 1999.
126. Ph. Colomban, *Ceram. Int.*, 15, 23, 1989.
127. C.J. Brinker and G.W. Scherer, Eds., *Sol–Gel Science*, Academic Press, San Diego, 1990.
128. A. Larbot, A. Julbe, C. Guizard, and L. Cot, *J. Membr. Sci.*, 44, 289, 1989.
129. T. Okubo and H. Nagamoto, *J. Mater. Sci.*, 30, 749–757, 1995.
130. R. Bredesen, C. Chalvet, H. Ræder, C. Simon, and H. Fjellvåg, in Proceedings of the Fourth Workshop on ESF Network on Catalytic Membrane Reactors: Optimization of Catalytic Membrane Reactor Systems, R. Bredesen, Ed., SINTEF, Oslo, 1997, pp. 181–186.
131. R. Bredesen, L. Beluze, C. Simon, K. Redford, and A. Holt, in Proceedings of the Fifth Workshop on Catalytic Membrane Reactors, Application and Future Possibilities of Catalytic Membrane Reactors, J. Luyten, Ed., VITO, Turnhout, 1997, pp. 23–28.
132. C. Simon, J.F. Glez, and R. Bredesen, in *Récents progrès en génie des procédés*, P. Aimar and P Aptel, Eds., Lavoisier Technique, Paris, 1992, pp. 85–90.
133. C. Simon, A. Solheim, and R. Bredesen, Testing and evaluation of flat ceramic membrane modules, in Proceedings of the ICIM$_3$ International Conference on Inorganic Membranes, Worcester, 1995, pp. 637–640.
134. W.J. Elferink, B.N. Nair, R.M. DeVos, K. Keizer, and H. Verweij, *J. Colloid Interf. Sci.*, 180(1), 127–134, 1996.
135. L. Cot, Sol–gel and inorganic membranes, in Proceeding of the First International Conference on Inorganic Membranes, Montpellier, 1989, pp. 17–24.

136. P. Aimar, C. Causserand, and M. Meireles, *Colloids and Interfaces. A: Physicochemical and Engineering Aspects*, Elsevier, Amsterdam, 1998.
137. M.J. Mayo, *Int. Mater. Rev.*, 41(3), 1996.
138. J.E. Otterstedt and D.A. Brandreth, *Small Particles Technology*, 1st ed., Plenum Press, New York, 1998, pp. 235–323.
139. S. Bhaduri and S.B. Bhaduri, *J. of Metals*, 1, 44–51, 1998.
140. J. Karthikeyan, C.C. Berndt, J. Tikkanen, J.Y. Wang, A.H. King, and H. Herman, *Nanostruct Mater.*, 9(1–8), 137–140, 1997.
141. R.R. Basca and M. Grätzel, *J. Am. Ceram. Soc.*, 79(8), 2185–2188, 1996.
142. S.J. Kim, S.D. Park, and Y.H. Jeong, *J. Am. Ceram. Soc.*, 82(4), 927–932, 1999.
143. Z. Zhang, L. Hu, and M. Fang, *Am. Ceram. Soc. Bull.*, 75(12), 71–74, 1996.
144. J. Cesarano III and I.A. Aksay, *J. Am. Ceram. Soc.*, 71, 1062, 1988.
145. R.A. Gardner and R.W. Nufer, *Solid State Technol.*, 17, 38, 1974.
146. W. Halloran, in *Ultrastructure Processing of Ceramics, Glasses and Composites*, John Wiley & Sons, New York, 1984, pp. 404–417.
147. W.H. Rhodes, *J. Am. Ceram. Soc.*, 64, 19, 1981.
148. N. McN Alford, J.D. Birdall, and K. Kendall, *Nature*, 330, 51, 1987.
149. J. Zheng and J.S. Reed, *J. Am. Ceram. Soc.*, 72, 810, 1989.
150. R.A. Hay, W.C. Moffatt, and H.K. Bowen, *Mater. Sci. Eng. A*, 108, 213, 1989.
151. T.S. Yeh and M.D. Sacks, *J. Am. Ceram. Soc.*, 71, C 484, 1988.
152. D.G. Goski, A.T. Paulson, R.A. Speers, and W.F. Caley, *Can. Br. Ceram. Trans.*, 98(4), 192–195, 1999.

10

Surfactant Adsorption and Dispersion Stability in Mineral Flotation

Bohuslav Dobiáš

Institute of Physical and Macromolecular, Chemistry, University of Regensburg, Regensburg, Germany

I. INTRODUCTION

Mineral flotation is a heterocoagulation process for selective separation of individual components out of polymineral suspensions of ground materials (ores) by using dispersed gas bubbles. This method is based on different attachments of hydrophobized and hydrophilic mineral particles to gas bubbles, mostly in an aqueous medium. Hydrophobized mineral particles attach to the gas bubbles and are carried out as aggregates of lower specific density to the surface of the suspension where they form a froth layer. This froth layer, called concentrate, can be removed mechanically [1].

The mineral particle to be floated is hydrophobized by adsorption of a suitable surfactant (called collector) whereby the nonhydrophobized particles remain dispersed in the mixture (Figure 10.1). The adsorption selectivity which is a necessary prerequisite for the separation of individual mineral components is controlled or modified by adding a depresssant or an activator, and by changing the pH or the ionic strength of the solution, or both. The froth layer must be sufficiently stable until the mineral particles are removed. This is achieved by the addition of, for example, nonionic surfactants, the so-called frothers.

The adsorption of a surface-active substance on the mineral surface leads not only to the attachment of the particle to the gas bubble but also to a destabilization of the mineral dispersion resulting from the aggregation of hydrophobized mineral particles. This phenomenon eventually leading to a spontaneous flocculation of the solids is a characteristic of higher concentrations of the adsorbing surfactant at which higher adsorption densities are attained. Under these conditions a deterioration of flotation process selectivity occurs, especially in the range of small-size particles.

A partial or total destabilization of a mineral dispersion can occur prior to the addition of a collector due to a concurrent presence of fine particles electrically charged with an opposite sign. The fine particles aggregate with each other or deposit on the surface of larger mineral particles. This phenomenon hinders a later collector adsorption or attachment of the mineral

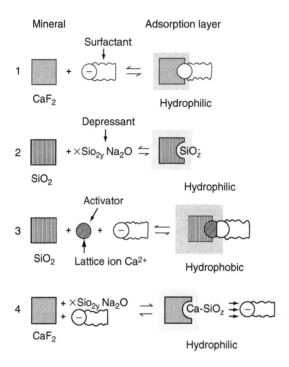

Figure 10.1   Formation of hydrophobic and hydrophilic adsorption layer on solid–water interface in quartz–fluorite system.

particles to gas bubbles. As decreasing particle size the surface energy of the small particles increases relative to the larger ones, the interparticle interaction forces must be considered as process determining in these cases. This is especially the case when fine grinding of the material is necessary.

To describe the effect of the particle size on the interparticular interactions, Warren [2], on the basis of theoretical considerations, characterized the interaction energy as dependent on the particle size (Figure 10.2). The polyvalent lattice ions released from the mineral surface during grinding and mixing of the mineral in water, as well as their hydrolytic products, also affect the stability of a mineral dispersion from a certain concentration onwards [3]. This applies especially for flotation processes in the presence of salt-type minerals.

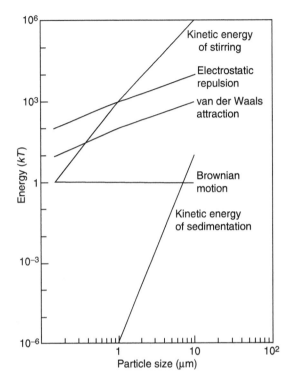

Figure 10.2 Energy change of various interparticular interactions as a function of particle size.

The aim of this chapter is to discuss the processes affecting the hydrophobicity of the mineral surfaces during a selective mineral flotation, whereby the hydrophobic interactions will be considered that lead to interparticular destabilization of mineral dispersions due to the appearance of an adsorption layer.

II. ELECTRIC DOUBLE LAYER: BASIC TERMS

The surface properties of most minerals are generally influenced by the chemical composition and the structure of the mineral–water interface. A knowledge of these characteristics is required to understand the surface chemistry of solids.

From the thermodynamic point of view, adsorption of surfactants on a mineral surface is very complicated. This is caused by the complex structure and ionic composition of the mineral–water interface formed during grinding the mineral in water.

Wetting of a mineral with water generates a surface charge and a region of electrical inhomogeneity at the solid–water interface as a result of the interaction of water molecules with the crystal surface. A surplus of positive or negative charge, which is typical for the broken crystalline structure of a mineral surface, is compensated by a region of counterions (H^+, OH^-, lattice ions, etc.) of the opposite sign from the bulk phase (H_2O). Thus an electrical doublelayer (EDL) results at the mineral–water interface which in many cases plays an important role in the adsorption of a surfactant. The classical model of EDL results from Gouy and Chapman's conception of the diffuse layer that is characterized by the Poisson–Boltzmann equation, and from a layer of specifically adsorbed ions according to Stern and Grahame (GChSG model) [4–6]. This model assumes a firmly bound layer of adsorbed dehydrated (Grahame plane or inner Helmholtz plane) and hydrated (Stern plane or outer Helmholtz plane) ions on the electrically charged mineral surface, and a layer of free mobile hydrated ions which are on the boundary toward the bulk phase in a diffuse state (Gouy's layer). A schematic GCHSG model of the EDL is shown in Figure 10.3, together with appropriate potentials.

For an interpretation of the adsorption process it is important to know the so-called zeta potential (ζ) that can be calculated from electrokinetic measurements. It may be defined as the potential difference at the shear plane (near to the outer Helmholtz plane) and the bulk phase when the solid and liquid phases are moved tangentially to each other. The location of this plane is not exactly known, but it can be assumed that the shear plane is very little further outward from the surface than the outer Helmholtz plane; alternatively, it is located within the Stern layer.

As an example, Hunter and Alexander [7] located the slipping plane 1 nm from the kaolinite surface and Delahay

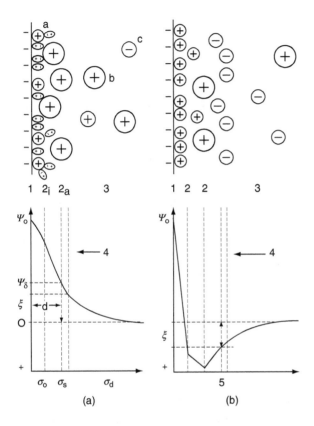

Figure 10.3 Structure of the EDL at the solid–water interface. 1, layer of charged ions 2_i inner and 2_a outer Helmholtz layer (also charged layer as in B); 3, diffuse layer; 4, shear plane; 5, layer of counterions; Ψ_0, phase potential; Ψ_δ, Stern potential; a, H_2O dipoles; b, hydrated counterions; c, negatively charged ions; d, thickness of the GS layer, and σ, charge density.

gives a value of 0.5 nm for the thickness of the Stern layer on clay minerals [8]. Furlong et al. [9] suppose the distance of a few tenths of a nanometer between the shear plane and the outer Helmholtz plane, so that $\zeta = \Psi_\delta$. The identity of ζ and Ψ_δ is valid especially at low potentials ($< 100\,mV$) and at low electrolyte concentrations ($< 10^{-2}\,mol/l$). At higher potentials and electrolyte concentrations the shear plane probably shifts further from the surface and ζ is expected to be less than Ψ_δ.

However, for the right interpretation of a ζ potential it is necessary to consider all reactions in the EDL since different values of Ψ_δ can correspond to one ζ potential value. Changes in the ζ potential will be reflected in the changes in the surface potential only when the ionic strength remains constant and there is no specific adsorption other than by potential-determining ions, i.e., when only these and indifferent ions are present [10].

The formation of a charge on a mineral surface wetted with water is controlled by two processes:

1. Adsorption of water molecules (dipoles) on the solid surface partially followed by chemisorption of water molecules, and by a dissociation of the hydrolytic product formed as dependent on pH, and
2. Preferential release of lattice ion species from the solid phase as a result of different hydration energies, or entropy effect, and readsorption of hydrolytic lattice products.

Mechanism 1 is characteristic for simple and complex metal oxides and a number of silicates whereas mechanism 2 is typical for sparingly soluble salts such as CaF_2, $BaSO_4$, or $CaCO_3$. These mechanisms represent two extreme cases of a surface charge formation since in most minerals both mechanisms proceed simultaneously in proportions depending on the chemical composition and crystalline structure [11,42].

Ions causing electrical charge and a change of the charge sign of a mineral surface are called *potential-determining ions* (PDI). The surface charge density of a mineral according to mechanism 1 can be calculated in simple cases from the change of pH value in an aqueous suspension of the solid by using the following equation:

$$\sigma_s = F(\Gamma_{H^+} - \Gamma_{OH^-}) \tag{1}$$

where Γ_{H^+} and Γ_{OH^-} are the adsorption densities of H^+ and OH^- in mol/cm^2, respectively, as determined from acid–base titrations, σ_S is the surface charge, and F is the Faraday constant. The rate of PDI adsorption is biphasic. The first adsorption step is fast and completed in a few minutes; the second step is slow and the equilibrium is reached in many days [12] (salt-type minerals)].

Two important parameters describing the EDL of a mineral are the *point of zero charge* (PZC) and the *isoelectric point* (IP). Healy et al. [9] define the PZC as the concentration of PDI with the surface charge of mineral $\sigma_s = 0$ and $\Psi_\delta = 0$. In metal oxides, PZC is determined by the concentration of PDI H^+ or OH^- in sparingly soluble salts by the concentration of PDI of the lattice. When both mechanisms of surface charge formation operate simultaneously, both ion species and the reaction products determine the PZC [6,11,13]. The IP is defined as the concentration of PDI at which the electrokinetic potential $\zeta = 0$ [9].

Somasundaran and Goddard [14] cite the current IUPAC definition of these terms. Thus, PZC refers to a particle or surface carrying no fixed charge and IP to a particle showing no electrophoresis or a surface showing no electroosmosis. These quantities are not interchangeable, since the IP characterizes the charge in the diffuse layer ($\sigma_d = 0$) whereas PZC should refer only to the surface charge. In the absence of a specific adorption, PZC and IP should coincide and are called intrinsic. In the presence of unequal specific adsorption the PZC and IP move in opposite directions [15]. Electrolytes in which the IP and PZC are identical and independent of the electrolyte concentration are called *indifferent* and the values of each term (IP, PZC) depend only on the surface properties of minerals and the composition of the aqueous phase. The indifferent electrolytes are adsorbed only by electrostatic attractions.

The value of the surface potential Ψ_0 at any activity of PDI and a constant potential difference due to dipoles, etc., is given by the Equation (2) [16]:

$$\Psi_0 = \frac{RT}{v_+ F} \ln \frac{a_{Me^+}}{(a_{Me^+})_{PZC}} \quad \text{or} \quad \frac{RT}{v_- F} \ln \frac{a_{A^-}}{(a_{A^-})_{PZC}} \tag{2}$$

where a_{Me^+} and a_{A^-} are the activities of the positive and negative PDI in the solution, respectively, $(a_{Me^+})_{PZC}$ and $(a_{A^-})_{PZC}$ are their activities at PZC, and v_+ and v_- are the appropriate valencies. The charge in the diffuse double layer

σ_d given by the Gouy–Chapman relation as modified by Stern for a symmetrical electrolyte is [16]:

$$\sigma_d = -\sigma_s[(v\varepsilon RT/\pi)c]^{1/2} \cdot \sinh(vF\Psi_\delta/RT) \tag{3}$$

where ε is the dielectric constant of the liquid and c is the concentration of the added electrolyte. For theoretical considerations, see Chapters 1 and 2 of this book.

III. POTENTIAL-DETERMINING IONS

The character of the PDI that controls the surface reactions such as the surfactant adsorption is the most important physical chemical parameter for judging adsorption mechanisms [17,18] and the stability of mineral dispersions [19]. According to this, the pH value of the IP of a mineral particle is a characteristic material quantity.

A.   Correlation Between IP and EAP

Because the IP obtained from electrokinetic measurements represents the sum of all interactions at the mineral–water interface, it is not possible to interpret the surface reactions using only IP data without further information on the contribution of the PDI and other ion species at the mineral–water interface in relation to the final value of the IP. This was confirmed by our study on a series of minerals using potentiometric and microcalorimetric measurements for the determination of points of equal adsorption density (EAP) of H^+/OH^- ions as well as comparisons with IP values [17,20–22].

Calorimetric measurements with oxides, sparingly soluble salt-type minerals, and silicates showed a relatively good agreement between EAP_{cal} values and values obtained from adsorption measurements (EAP_{ads}), as outlined in Table 10.1 and Table 10.2 [17].

In oxides a relatively good correlation was found for Al_2O_3 and Fe_2O_3 at all values. In these minerals, H^+ and OH^- determine the surface potential, and at the IP surface charges are compensated by the adsorption of the same amount of H^+

Table 10.1 EAP and IP of Oxides and Silicates

Mineral	SiO_2	SnO_2	Al_2O_3	Fe_2O_3	Kaolinite	Feldspar
EAP_{cal}	5.4 ± 0.4	5.5 ± 0.2	8.1 ± 0.3	6.8 ± 0.1	8.4 ± 0.2	8.2 ± 0.1
EAP_{ads}	5.2 ± 0.2	5.6 ± 0.1	8.3 ± 0.1	7.0 ± 0.1	8.3 ± 0.1	8.3 ± 0.1
IP	2.0	3.0	8.0	6.8	2	2
Electric charge in EAP	Neg	Neg	Near zero	Near zero	Neg	Neg

Table 10.2 EAP and IP of Salt-Type Minerals

Mineral	$CaCO_3$	CaF_2	$Ca_5F(PO_4)_3$	$CaWO_4$	$BaSO_4$
EAP_{cal}	10.0 ± 0.1	6.5 ± 1.0	7.2 ± 0.3	7.6 ± 0.2	—
EAP_{ads}	10.3 ± 0.1	6.95 ± 0.05	7.85 ± 0.05	8.2 ± 0.1	8.6 ± 0.1
IP	10.45	10.2	4.05	2	5.3
Electric charge in EAP	Near zero	Pos	Neg	Neg	Neg

and OH^-. In the case of SnO_2 and SiO_2, each IP differs substantially from both EAP values. This means that the surface of both minerals is negatively charged in the region of the EAP. The reason for this negative charge can be seen in the presence of nondissociable and therefore pH-independent adsorption sites on the surface of the minerals studied.

When compared to oxides, the H^+ exchange heats of the salt-type minerals, except those of barite and fluorite, attain very high values although the H^+/OH^- adsorption densities of salt-type minerals are 100–1000 times lower than those of the oxides (Figure 10.4). The heats are therefore not only caused by H^+ adsorption but are also due to a strong dissolution of the salt lattice and to the hydration of the free lattice ions. Schulz and Dobias [20] found a more pronounced dissolution of lattice ions of calcite, apatite, and fluorite below pH 10, 6.5, and 4, respectively, during their measurements of the mineral solubility dependence on the pH of the solution. The same is true for scheelite below pH 6.5.

On the other hand, sparingly soluble barite shows neither a dissolution heat nor a significant H^+ adsorption heat, so that the EAP_{cal} could not be determined.

Fluorite differs from all these minerals in the acidic range by its endothermic heat. This difference is caused by the behavior of the fluoride ion toward the proton. While no measurable association heat of dissolved ions with H^+/OH^-

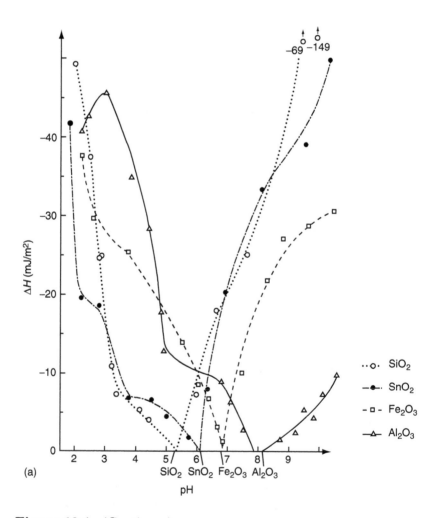

(a)

Figure 10.4 (*Continues*)

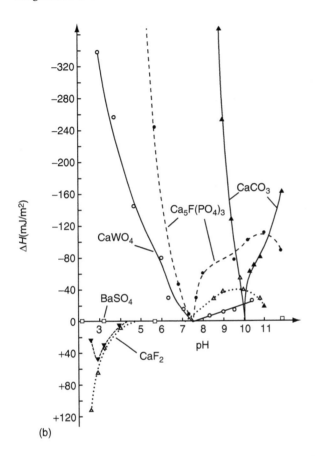

(b)

Figure 10.4 Exchange enthalpy (ΔH) of H^+/OH^- on oxides (a) and salt-type minerals (b) as a function of pH at constant ionic strength (1×10^{-3} mol/l NaCl). $\Delta H_{\text{fluorite}} = \Delta H_{\text{ass}} + \Delta H_{\text{ads}} + \Delta H_{\text{diss}}$; $\Delta H_{\text{fluorite}}$ (Δ) $= \Delta H_{\text{ass}}$.

occurred in the case of the above-mentioned minerals, H^+ ions reacted with dissolved F^- ions endothermically. Beside the exchange heats, Figure 10.4 also shows the heat of association of the following reactions for fluorite:

$$H^+ + F^- \rightarrow HF \qquad k = 0.0015 \qquad (4)$$

$$F^- + HF \rightarrow HF_2^- \qquad k = 4.3 \qquad (5)$$

Because of the endothermic association reaction of H^+ with dissolved F^-, it can be expected that the H^+ adsorption on the fluorite surface also proceeds endothermically like an H^+/F^- association and that the overall exchange enthalpy is positive.

Figure 10.5 shows that H^+/OH^- exchange enthalpy of kaolinite and feldspar. In feldspar, the exchange heat of H^+ increased strongly with increasing H^+ concentration and reached more than tenfold higher values than the reaction heats of kaolinite. Thus, the behavior of feldspar is comparable to that of salt-type minerals (release of Al^{3+}, K^+, and Na^+ into the solution). The pH-dependent dissolution of Al^{3+} from

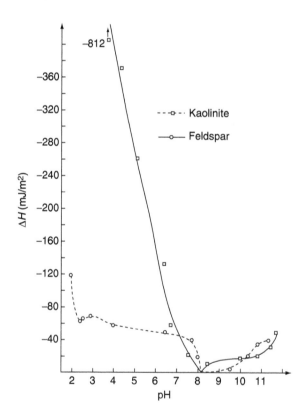

Figure 10.5   Exchange enthalpy (ΔH) of H^+/OH^- on kaolinite and feldspar as a function of pH at constant ionic strength (1×10^{-3} mol/l NaCl).

kaolinite was investigated by Siracusa and Somasundaran [23]. Since in feldspar the adsorbed amount of H^+ is approximately ten times higher than that of kaolinite, it can be concluded that in the latter mineral that H^+ ions adsorb mostly at the edges where the lattice structure is broken and similar to the feldspar surface.

The EAPs are summarized together with the IPs in Table 10.1, with the EAP_{ads} values from a former paper [17]. As these data show, the EAP values agree well with each other. The difference between the pH values of IP and EAP can be explained first by a greater release of lattice ions of both minerals which is characterized by a higher value of the dissolution heat rather than by H^+ and OH^- adsorption, as in salt-type minerals, and second on isomorphic substitution of Si^{4+} by Al^{3+} and of Al^{3+} by Mg^{2+} on the flat sides of kaolinite, resulting in a pH-independent negative charge [24].

B. Stability of Mineral Dispersions

The effect of PDIs on the stability of mineral dispersions is evident from the plot in Figure 10.6. The stability of Al_2O_3, SiO_2, and CaF_2 suspensions, as characterized by the changes in turbidity (TU), is plotted here against the equilibrium pH value [19]. Since these minerals have an origin other than those investigated in the previous paper, one finds only small pH differences of IP and EAP in comparison to values shown in Table 10.1 and Table 10.2.

As the plots for corundum and quartz show, the appropriate critical coagulation concentrations (CCC) of pH 6.5 and 4.4 agree better with the EAPs ($\Gamma_{H^+} = \Gamma_{OH^-}$) of 7.3 and 4.9, respectively, than with the IPs of 9.0 and 2.0.

The stability of the fluorite dispersion is only slightly dependent on pH in spite of an enormous H^+ adsorption density increase at pH under 5.8. When the lattice ions Ca^{2+} and F^- are considered as PDIs, the increase in the F^- concentration at a positively charged CaF_2 surface (Ca^{2+}) should have led to a reduction of the potential and destabilization of the dispersion. This was also confirmed experimentally (Figure 10.7, curve 1). The same experiment with NaCl instead of NaF (curve 2)

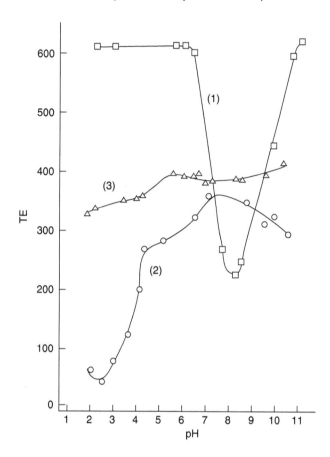

Figure 10.6 Turbidity (TE) or corundum (1), quartz (2), and fluorite (3) as a function of pH.

confirmed the specific role of F^- in the Stern layer, which is to bring about a destabilization of the fluorite dispersion due to an increase in NaCl concentration at very high concentrations, where the thickness of the electric double layer is somewhat compressed due to the adsorption of Cl^- (curve 2).

In contrast to it is the behavior of the negatively charged quartz surface (Figure 10.8, curve 1), where the Ca^{2+} ions are specifically bound, whereas the fluorite dispersion gets destabilized only at high Ca^{2+} concentrations (curve 2) that correspond roughly to a concentration needed for fluorite

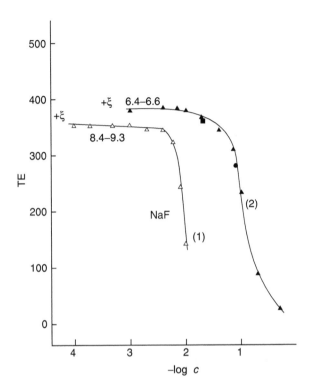

Figure 10.7 Turbidity (TE) of fluorite dispersion as a function of the concentration of NaF (1) and NaCl (2) (mol/l). The pH range and ζ potential signs are given at the curves.

destabilization by NaCl (Figure 10.7, curve 2). From this point of view it is possible to say that in both cases Cl$^-$ ions are responsible for compressing the EDL, leading to the destabilization of fluorite suspension.

IV. MECHANISMS OF SURFACTANT ADSORPTION

Various hypotheses have been conceived concerning the elementary processes during the adsorption of surfactants on minerals under conditions of flotation (see, e.g., Refs. [1,25–30]). In the following section, basic data are

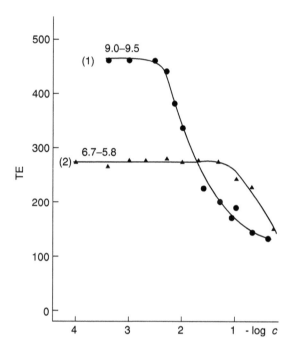

Figure 10.8   Turbidity (TE) of quartz (1) and fluorite (2) dispersion as a function of the Ca^{2+} concentration (mol).

summarized concerning adsorption isotherms that are used to interpret surfactant adsorption on a mineral surface. For theoretical consideration, see Chapters 1 and 4 of this book.

A.   Basic Adsorption Isotherms and Terms

Adsorption isotherms represent a relationship between the adsorption density Γ (mol/m^2) at an interface and the equilibrium activity of an adsorbed substance in the bulk phase at a constant temperature. Between the degree of coverage θ and the adsorption density Γ a relationship exists: $\theta = \Gamma/\Gamma_{mono}$, where Γ_{mono} is the saturation value for the adsorption of a substance in a single layer. The analysis of adsorption isotherms can yield thermodynamic data for the given adsorption system. Theoretical adsorption isotherms derived from statistical and kinetic data and, using the described assumptions, are known only for the gas–solid interface or for dilute solutions of surfactants (Gibbs). Those for the system gas–solid are a few basic types that can be thermodynamically predicted. From temperature relations it is possible to calculate adsorption and activation energies or rate constants for individual isotherms. Since there are no theoretically founded equations of adsorption isotherms for dissolved surfactant on solids, the adsorption of gases on solids can be used as a starting point for an interpretation.

The Langmuir adsorption isotherm is the most frequently used relationship for such interpretations [1]. According to the Langmuir conception, there are certain preferential adsorption sites on a mineral surface called *active centers*, but they are not exactly specified. Once adsorbed, the particles remain firmly bound.

The active centers are at such distances that the adsorbed particles cannot affect each other. The adsorption eventually leads to a monomolecular cover and all sites occupied are energetically equal $[\Delta H^{ads} \neq f(\Gamma)]$. This hypothesis agrees well with conceptions of crystal formation where steps, edges, corners, shifts, and atomic defects can act as active centers. The lattice geometry of a crystal surface has no influence on adsorption. The presence of *defects on the surface* and preferential adsorption sites was demonstrated, for example, by Boudriot et al. [31,32]. Although the importance of structural defects, especially of shifts, for adsorption was clearly shown, it has not yet been possible to develop a quantitative relationship between structural defects and adsorption.

The Langmuir isotherm is valid under the following conditions:

$$\theta = \frac{c}{c + K}$$

or, with $1/K = B$:

$$\frac{\theta}{1 - \theta} = Bc \tag{6}$$

The factor K represents the so-called half-saturation concentration ($c = K \Rightarrow \theta = 0.5$) and is dependent on temperature.

For K the following relationship applies: $K = K^0 \exp(\Delta H^{ads}/RT)$ where $\Delta H^{ads} \neq f(T)$. An isotherm analog to the Henry–Dalton solubility law for gases in fluids is valid only for the adsorption from very dilute solutions:

$$\theta = Bc \qquad (7)$$

where B is the adsorption coefficient.

According to the Henry isotherm, a linear function $\theta = \theta(c)$ results. For very large values of c, $\theta \to 1$ is true. The space required by the adsorbed molecule is characterized by the term $1 - \theta$ (concentration of the free adsorption sites).

An empirically found Freundlich isotherm for the adsorption from a liquid phase can be expressed as follows:

$$\theta = B(c/c^0)^{1/m} \qquad (8)$$

where $m \geq 1$ represents an undefined system-specific constant and for $m = 1$ the Henry equation results again.

If Equation (8) is solved for c, under the consideration that both B and m are temperature-dependent and using the calculation of ΔH^{ads}, it can be shown that the adsorption enthalpy of the Freundlich isotherm is a logarithmic function of θ, meaning that the adsorption energy is distributed exponentially.

If the ratio of covered and free surface sites increases exponentially, an extension of the Langmuir isotherm can be used according to Volmer [33] to express the heat motion of adsorbed molecules:

$$\frac{\theta}{1 - \theta} \exp\left(\frac{\theta}{1 - \theta}\right) = Bc \qquad (9)$$

One can consider the adsorption from a solution as an exchange reaction according to the following scheme:

$$A^V + nX^A \Leftrightarrow A^A + nX^V \qquad (10)$$

where A represents the adsorbed substance and X the desorbed substance (solvent molecules or counterions). To take this exchange into consideration, Zuchovickij [34] substituted the surface $1-\theta$ that is available for the adsorbent according to Langmuir by the term $n(1-\theta)^n$:

$$\frac{\theta}{n(1-\theta)^n} = Bc \tag{11}$$

For $n = 1$ Equation (11) becomes the Langmuir isotherm (Equation (6)).

A consideration of interactions of the mobile adsorbent in the adsorption layer under the assumption of a linear dependence between the adsorption enthalpy ΔH^{ads} and θ leads to the isotherms of the form described in Equation (12) that are associated with different names (Frumkin, Temkin, Fowler-Guggenheim, Bragg-Williams):

$$\frac{\theta}{1-\theta}\exp(a\theta) = Bc \qquad \text{(Frumkin isotherm)} \tag{12}$$

where $a = e/RT$, e is the lateral interaction parameter of the dimension energy–mass amount, and θe represents a surplus adsorption enthalpy due to this interaction. If $e < 0$ ("positive cooperativity"), the adsorption density — and thus θ — increases steeply at higher concentrations due to attraction forces than in the Langmuir isotherm and the course of the curve has an S-shape. If a exceeds a critical value ($a = -4$), a phase transformation occurs in the adsorption layer according to this isotherm.

Equation (13) cannot be solved for θ. In the literature expressions of the form2

$$\theta = \frac{\ln(B'c)}{a} \tag{13}$$

can be found that are called Temkin or Frumkin isotherms [34]. The transformation

$$\exp(a\theta) = B'c \tag{14}$$

shows that the term $\theta/(1-\theta)$ disappears. Neglect of this term is allowed only at intermediate coverage degrees when $\theta/(1-\theta) \approx 1$.

As in the case of the Frumkin isotherm that has been derived from the Langmuir isotherm introducing the lateral interaction, the Hill–de Boer equation results from the Volmer isotherm:

$$\frac{\theta}{1-\theta} \exp\left(\frac{\theta}{1-\theta}\right) \exp\left(a\theta\right) = Bc \qquad (15)$$

Equation (15) represents an isotherm for nonlocalized adsorbates and shows a course qualitatively similar to that of the Frumkin isotherm [35].

The course of all adsorption isotherms mentioned in this section is shown according to Orthgiess [36] in Figure 10.9. The classification of the isotherms was taken over from Volke [37]. We published a number of papers [25,26,38–42] on the interpretation of adsorption isotherms of surfactants on minerals. In most cases these isotherms have an S or double-S form and can be described by the Langmuir or Temkin equation. Accordingly, the adsorption of surfactants conforms to the Langmuir equation in the concentration range dose to the monomolecular coverage ($\theta \cong 1$) on minerals with PDI H^+ or OH^-. For adsorptions on mineral surfaces with lattice ions as PDI the mentioned equation is less suitable because it is valid only for lower surfactant concentrations. Here the Temkin's equation fits well within the range of intermediate coverage values ($\theta = 0.2$–0.8). Freundlich's adsorption equation can be used in both cases only for very low surfactant concentrations.

B. Effect of Surfactant Concentration

Regarding the basic mechanism of surfactant adsorption on mineral surfaces different conceptions have been developed using indirect investigation methods. Only a few methods, such as surface spectroscopy, can provide direct evidence of interactions in the adsorption layer. They are of limited use since the measurements must be performed in high vacuum. Besides, these methods can be employed only for sulfides. A promising method seems to be Fourier transform infrared (FTIR) spectroscopy, which allows for example, an in situ investigation of the mineral–aqueous solution interface by using the diminished total reflexion (DTR) technique [43].

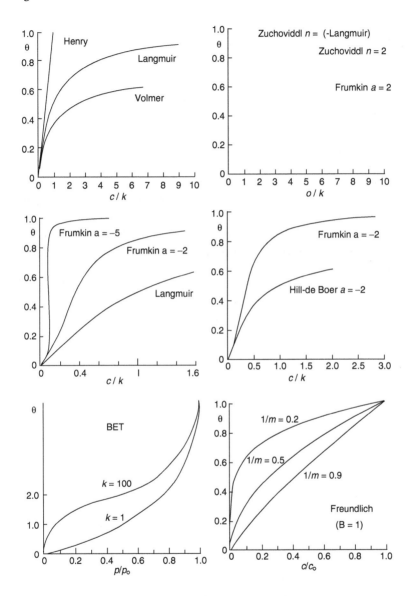

Figure 10.9 Various types of adsorption isotherms.

The adsorption of an ionogenic surfactant at the polar mineral–water interface generally depends on physical–chemical properties of its molecule enabling the substance to bind and orientate there. Possible mechanisms will be listed in the following text; however, they often act together successively or even simultaneously in the course of the adsorption process [1,25,26,44–46]:

Ion exchange: exchange of counterions in the electric double layer by adsorbing surfactant ions (Figure 10.10).

Ion pair formation: adsorption of ions with an opposite sign to the surface on surface sites not occupied by counterions (Figure 10.10).

Adsorption through activating bridges: adsorption of ions of the same sign as the surface, mediated by polyvalent counterions in the electric double layer (Figure 10.11).

Covalent binding: formation of a covalent bond between the adsorbent and an atom or ion of the mineral surface (Figure 10.11).

Surface complexation: by complexing mineral surface cations, the adsorbent forms a surface complex, also of the chelate type (see Chapter 4, Section 1)

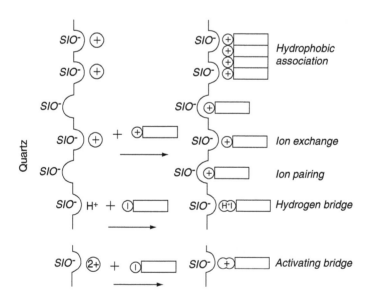

Figure 10.10   Possible adsorption mechanisms of surfactants on minerals with H^+/OH^- potential determining ions.

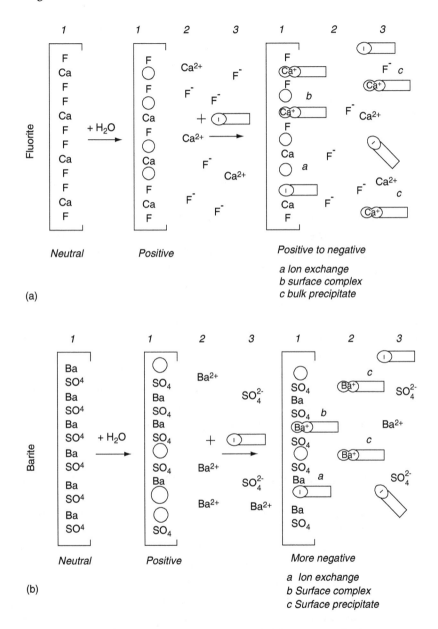

Figure 10.11 Possible adsorption mechanisms of surfactants on minerals with lattice ions as potential determining ions. (a) Charge of mineral surface and surfactant are opposite to each other. (b) Charge of the surfactant head and mineral surface are the same.

Surface precipitation: the adsorbent forms a slightly soluble product with the double-layer ions or also with the bordering solution that precipitates (first) specifically on the mineral surface, not in the solution (Figure 10.11)

Binding of hydrogen bridges: adsorption due to the formation of H bridges between the adsorbent and the adsorbate (e.g., in the adsorption of nonionic detergents and polyelectrolytes)

Adsorption due to van der Waals forces: an adsorbent–adsorbate interaction due to dipole–dipole, induction, or dispersion forces (Figure 10.10).

During adsorption on a solid, surfactants differ from other adsorbates because of the polar–apolar structure of their molecules. Their adsorption from an aqueous medium on a surface of a distorted crystalline lattice is the result of a reaction equilibrium between binding forces of all species participating in this process. Because of the heterogeneity of such forces acting on the mineral–water interface it is necessary to consider several simultaneous, subsequent reactions. Binding forces in the process of surfactant adsorption on the surface of ionic crystals were studied in detail by Richter and Schneider [47] who analyzed the adsorption conditions for low θ values and values of θ in the range of a monomolecular layer. For low θ values the following energetic contributions may play a role:

E^1: energy needed for dehydration of polar groups of the surfactant

E^2: energy needed for dehydration of the appropriate ions of the crystal surface

E^3: energy obtained by association of water dipoles derived from dehydration according to E^1 and E^2

E^4: energy obtained by binding polar groups of the surfactant with ions of the heteropolar crystal surface

E^5: energy needed for repulsion of polar groups of the surfactant having the same charge sign

E^6: energy obtained at the parallel (or close to parallel) arrangement of apolar groups of surfactant molecules on the crystal surface due to their interaction with ions of the crystalline lattice. It is supposed that at the perpendicular arrangement of isolated surfactant molecules, their apolar groups remain hydrated

E^7: energy needed for a partial dehydration of apolar chains of the surfactant molecule according to E^6

E^8: energy obtained from the association of water dipoles after dehydration of apolar chains according to E^6.

In terms of the structure of the adsorbed layer, it is usually assumed that the adsorbate molecules form island-like structures (patches) at an already very low adsorption density due to energetical heterogeneity of the solid surfaces. Whether a closed monolayer develops during the adsorption of surfactants at higher coverage degrees before a double layer forms on preferred adsorption sites has not been decided for most of the systems. This is among other factors dependent on the extent of lateral interactions in the adsorption film. In connection with considerations regarding the structure of the adsorption layer, some ideas should be presented.

1. Hemimicelle Model According to Fuerstenau

Fuerstenau subdivides the course of adsorption into three regions (Figure 10.12(a)) [30]: In the range of low concentrations (region 1), the surfactant adsorption takes place through ion exchange with ions in the electric double layer. During this exchange the ζ potential remains almost constant. It is a matter of purely electrostatic interactions.

With an increasing concentration of the surfactant (region II), the effect of mutual hydrophobic interactions of hydrocarbon chains becomes more important: two-dimensional associates, called hemimicelles, are formed [30,48,49]. This phenomenon is analogous to the three-dimensional micellization in the bulk phase and at the solid–water interface it is called hemimicellization. The mechanism of hemimicelle formation is shown in Figure 10.13. This two-dimensional aggregation is dependent on the chemical composition of the surfactant's polar head (adsorption reactivity), structure and length of the hydrocarbon chain, ionic strength, surface potential of the substrate, pH, and temperature. Hemimicelles

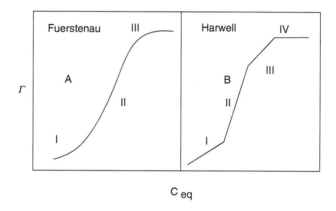

Figure 10.12 Characteristic regions of surfactant adsorption on solids after Fuerstenau and Harwell, respectively. Γ = adsorption density.

obviously contain more water in their environment than micelles since smaller fractions of the CH_2 groups are possibly removed from the aqueous phase. The hemimicellization at the mineral–solution interface causes characteristic changes in the adsorption density, flotation recovery, ζ potential, and contact angle.

At the concentration C^{HM} when hemimicelles start to form it is possible to use the Stern–Graham equation for calculating the energy φ, needed to transfer a CH_2 group from the aqueous phase to the hemimicelle [50]:

$$C^{HM} = \frac{\Gamma'}{2r} \exp\left[\frac{W'e - n\varphi}{kT}\right] \text{ or } \ln C^{HM} = \ln\left(\frac{\Gamma'}{2r}\right) + \frac{W'}{kT} - \frac{n\varphi}{kT}$$

$$(16)$$

where $W'e$ is the electrostatic interaction and n is the number of CH_2 groups. The plot of $\ln C^{HM}$ against n should be a straight line with a slope of $(-\varphi/kT)$ which can be used to calculate φ.

After reaching the concentration at which the sign of the ζ potential reverses, the electrostatic interaction and specific adsorption are opposed to each other and the slope of the

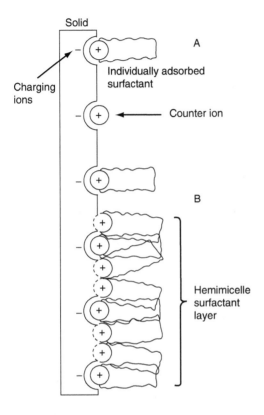

Figure 10.13 Adsorbed long-chain surfactant on negatively charged solids. (A) Without lateral interactions ($c \ll c_{\text{monolayer}}$). (B) With lateral interaction between hydrocarbon chain at higher concentrations.

adsorption isotherm gets reduced (up to ≈ 0; region III). The reaction ends in an upper range at a constant adsorption density.

Scamehorn et al. [28] divide the course of adsorption isotherms into four regions similarly to Fuerstenau, with a general configuration of the adsorbed surfactant molecules. In the first region only unassociated first layer molecules are present. The adsorption data conform with Henry's law at sufficiently low surfactant concentrations. In the second region the adsorption density increases rapidly due to lateral interactions between molecules of the adsorbed surfactant; this occurs first on the most energetic surface patches (hemimicelles). The standard hemimicelle is defined as a patch having an almost complete first-layer coverage ($\theta \cong 1$) and an arbitrary second-layer coverage ($\theta_2 < \theta_1$), where the coverage of a given layer is identical for different isomeric surfactants. In the regions III and IV (plateau), the surface is largely covered with hemimicelles, with a substantial second-layer adsorption. Above the critical micelle concentration (CMC) the adsorption of the surfactant is independent of the concentration. The hydrocarbon group environment in the hemimicelle should be about the same for the first and second layers and is more favorable than that in the micelle, but a low second-layer parameter sensitivity precludes such a comparison.

2.   Admicelle Model of Harwell

Harwell [51] starts his considerations on the surfactant adsorption on mineral surfaces at the $\log \Gamma$–$\log c$ plot and divides the course of an adsorption isotherm into four regions (12b), similarly to Fuerstenau and Scamehorn (see above):

1. The region of a low surface coverage (region 1) is without signs of aggregates formation.
2. The beginning of aggregate formation (transition region I/II) is characterized by a large slope increase in the adsorption isotherm. These aggregates are formed by a patch surfactant adsorption and called hemimicelles or admicelles, in order to illustrate their structure and behavior, which are similar to that of micelles. As opposed to Fuerstenau, according to Harwell two adverse oriented layers are formed.
3. The start of region III is characterized by a distinctly smaller increase in the adsorption isotherm slope. This is caused by an electrostatic repulsion of surfactant ions on the surface that is caused by a sign change of the ζ potential.
4. The attainment of CMC concentration in the solution or a complete bilayer formation on the surface occurs in the transition regions III and IV.

In his experiments, Harwell considers the fact that different (cationic) counterions and surface defects can permanently influence the adsorption isotherm course.

3. Model of Gu

Gu [52] starts his characterization of the adsorption process at the Γ'/c plot and divides it into four regions (Figure 10.14). The surfactants adsorb in a first layer on the mineral surface. Because of the small surfactant concentration and large distances of the molecules to each other, the intermolecular interactions can be neglected (region a). The adsorption in this region is due chiefly to electrostatic interactions and therefore is controlled by the surface charge of the mineral. The individually adsorbed surfactant ions serve as *anchor* for the deposition of further ions from the solution.

In the first adsorption plateau (region b) the adsorption results from both electrostatic attraction and specific interactions between the surfactant and the mineral surface. Gu found in his investigations with pyridinium ions on silica gel that in this plateau region a sign change of ζ potential values occurs, i.e., the IP lies in this range.

In the third region (c) aggregates occur and the formation of a second adsorption layer permanently affects the adsorption course. These associates are, similarly to Fuerstenau, also called hemimicelles. The extreme increase in ions in the adsorption isotherm is described by the hemimicelle concentration.

The second adsorption plateau (region d) is reached at concentrations higher than the CMC. The adsorption densities remain constant because above the CMC the monomer concentration is supposed to be constant and thus — supposing that only monomers adsorb — no *relevant* concentration increases occur any more.

In this region the surface is vastly covered with hemimicelles whereby the upper limit of the surface aggregation is reached. A second *discrete* layer will be formed that again has a hydrophibic character.

According to Gu, adsorption is determined both by the surface charge of the mineral and by the number of CH_2

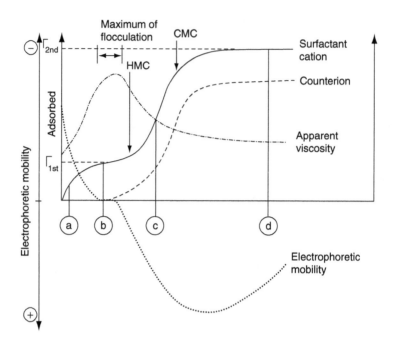

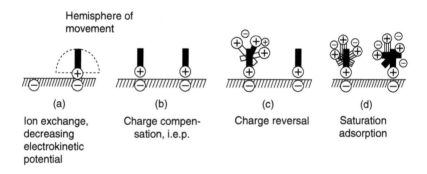

Figure 10.14 Characteristic regions of surfactant adsorption on solids after Gu with corresponding state of adsorbed layer at different concentration regions. $\Gamma =$ adsorption density.

groups in the hydrocarbon chain and therefore by the association of chains.

The adsorption model of Fuerstenau is well suited for the interpretation of the flotation behavior of minerals in solutions of surfactants regarding the hydrophobicity of the adsorption layer. On the other hand, in the Harwell and Gu models one finds substantial differences if one wishes to relate the individual regions of the adsorption isotherm to the structure of the adsorption layer.

While the steep increase in the adsorption density in the Fuerstenau isotherm (region II) is characterized by a hydrophobic layer of the associated surfactant molecules (hemimicelles), the same region in Harwell's model means that two reversely oriented layers are formed because of the adsorption of admicelles and thus they are hydrophilic. According to this conception, flotation would be possible only in region I.

According to the Gu's adsorption model, the mineral surface has hydrophobic properties (regions a and b) without aggregating surfactant molecules. Also, the IP is reached in region b (plateau). It can be assumed that mineral particles attach to gas bubbles in both these regions.

None of the models explains the structure and behavior of a surfactant adsorption layer beyond the CMC as, for example, the adsorption of anionic surfactants on a negatively charged surface of a mineral such as barite. In the Fuerstenau model it would not be possible at all since he explains the adsorption isotherm only up to the first plateau.

Starting from a number of adsorption isotherms measured with surfactants on minerals with a different character of PDIs, we like to present the following adsorption model.

Adsorption isotherms can be schematically characterized by two extreme curves that are either S-shaped or double-S-shaped (Figure 10.15, curve a–c) [1]. Most adsorption isotherms have a form fitting between these two extremes. It depends on the adsorption density of the adsorption sites that can be occupied as to what extent the first plateau is expanded. The adsorption mechanism has the decisive importance.

In the first region surfactant molecules adsorb on the mineral surface sparsely from one another, without lateral interactions. This adsorption cannot be based only on electrostatic forces since surfactants also adsorb which have the same charge sign as the mineral surface (chemisorption). The adsorption of anionic surfactants on negatively charged

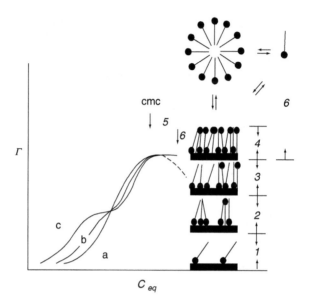

Figure 10.15 Characteristic regions of surfactant adsorption on solids for the transition from a double S-shaped to a single S-shaped isotherm. Γ = adsorption density.

$BaSO_4$ surface serves as an example (Figure 10.11). The individually adsorbed molecules that can be viewed as anchor molecules are very probably not oriented perpendicular to the mineral surface.

The second region corresponds to a rise of adsorption density of individually adsorbing molecules and to the formation of adsorbate patches or hemimicellization of the mineral surface in the course of which the association energy would depend on the coverage of the surfactant [25]. A partial adsorption of reversely oriented molecules is supposed to take place here. In this region of the surfactant concentration one finds an increased floatability of the mineral as well.

A first plateau formation characterizes region 3 during which the building up of a hydrophobic layer gets completed. This range is often incorrectly called *monomolecular*. In minerals where H^+ and OH^- are the chief PDIs we observed very often the attainment of the IP and a change of the charge sign.

Presumably the transition from a double-S curve to a single-S curve c → a depends on the surfactant density in region 2 at which the aggregation can start. This transition is considered to be a result of all interactions in the phase interface, which are very complicated, especially in salt-type minerals. Due to an overlap of more interactions the first plateau may become indistinct or extinct (the isotherm gets an S-form).

In region 4 a second adsorption layer is formed by the surfactant, whose polar heads are oriented toward the bulk phase together with the appropriate counterions. Therefore the adsorption layer becomes more and more hydrophilic. The adsorption is terminated with a plateau (region 5).

How the floatability of mineral particles alters in the region of second-layer formation is shown in Figure 10.16. A maximum is very often found in the plateau region after the CMC had been reached (region 6, Figure 10.15). This phenomenon can be explained as follows: In the presence of

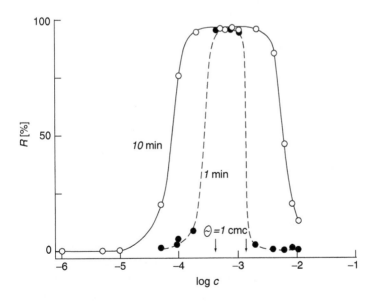

Figure 10.16 Flotation recovery (R in %) of barite as a function of the concentration (mol/l) of Na-dodecylbenzenesulfonate.

micelles the mineral surface with its adsorption layer must be viewed as one giant micelle that takes part in the overall adsorption equilibrium of the system. Beside this it can be assumed that the equilibrium is shifted in favor of a three-dimensional micelle instead of a two-dimensional, laminar one. In systems where polyvalent lattice ions cause a precipitation reaction with the surfactant used, solubility effects of precipitation products play an important role.

The ζ potential in relation to the surfactant concentration reached its maximum in the region of micelle formation and thus confirmed the similar shape of the adsorption isotherm.

Maxima, often followed by minima, were observed on adsorption isotherms of surfactants adsorbed on solids by a number of other authors [39,40,53–55]. However, it is not possible to say that one and the same surfactant causes maxima above the CMC on all kinds of solid substances. The surface properties of minerals are one of the chief factors determining the behavior of surfactants at concentrations greater than CMC.

Somasundaran and Hanna [56] mention two possible reasons for the presence of the adsorption maxima:

(1) exclusion of micelles from the near-surface region of the solid due to higher electrostatic repulsion between the particles and the micelles than that between the former and singly charged surfactant monomers, and
(2) alterations in solid properties such as the effective surface area, owing to change in particle morphology upon excessive surfactant adsorption that can cause repulsion between various parts of the particle.

Experiments of Ebworthy and Mysels [54] brought evidence against the consideration of micelles as a separate phase due to a decrease in the surface tension of a surfactant at concentrations above the CMC, which is not compatible with the phase separation hypothesis. To explain the formation of adsorption maxima, Sexsmith and White [55] used the law of mass action and calculated the monomer concentration in the bulk phase for the micelles with different ratios of surfactant ion number to counterion number. Appropriate calculations showed that variations in ion and counterion numbers have an influence on the concentration of monomers and that theoretical adsorption isotherms feature maxima.

The occurrence of maxima on adsorption isotherms is further explained by the presence of impurities [57] due to adsorption of laminar micelles, or by formation of hydrophobic clusters [58,59,72]. Trogus et al. [48] proposed a hypothesis that minima and maxima in total adsorption isotherms for mixtures of surfactants are predictable from considerations of monomer–micelle equilibrium relationship.

According to Scamehorn et al. [28], maxima adsorption isotherms result from isomeric inhomogeneity of the surfactant and at a high solution-to-solid ratio.

C. Exchange Enthalpies of Surfactant Adsorption

Based on the exchange enthalpies of surfactants it is possible to describe the interaction and mode of attachment of surfactant molecules at the mineral–water interface [47,60–63].

As shown in Chapter 3, PDIs have a decisive significance for the adsorption mechanism. The following section discusses the exchange enthalpies of surfactant adsorption on minerals with a different character of PDIs [64].

1. PDI H^+ and OH^-: Oxides

The adsorption of an ionogenic surfactant on an oxide–water interface is usually determined by electrostatic interactions [25]. According to this mechanism, the electric charges of the surfactant's polar group and the mineral surface must have opposite charges. Based on this assumption results of adsorption measurements will be shown for cetylpyridinium chloride (CPCl) adsorbed on negatively charged quartz and positively charged corundum. The data an IP and EAP are given in Chapter 3.

The adsorption of cationic surfactants is favored in quartz and made harder in corundum due to repulsion of identically charged surfactant heads and the mineral surface of the latter. The measurements and calculation of the adsorption enthalpy was difficult in both minerals because the surfactant's dilution heat was strongly endothermic and superimposed the exothermic adsorption heat. Table 10.3 shows the numerical values of the CPCl adsorption on quartz and corundum, and in Figure 10.17a the adsorption enthalpies and adsorption densities are plotted against the log of the surfactant's equilibrium concentration. It can be seen that the adsorption density of CPCl in corundum is very low and is accompanied by a low adsorption heat. In quartz the adsorbed densities as well as the adsorption heats attain higher values but show a certain dispersion beyond the CMC. The molar adsorption heat amounts to $-3.7\,\text{kJ/mol}$ ($1.49kT$), as shown in Table 10.3, and it is surprising that it is lower than in corundum ($\sim -8.5\,\text{kJ/mol}$).

Assuming that CPCl adsorption on quartz takes place mainly by electrostatic interactions, we can judge the free adsorption energy ΔG^{el} using the Stern potential Ψ_δ of the

Table 10.3 Calorimetric Data on the Adsorption of CPCl on Quartz ($13.5\,\text{m}^2/30\,\text{ml}$ of 10^{-3} mol/liter NaCl, pH 3.8–4.0) and α-Corundum ($10.7\,\text{m}^2/30\,\text{ml}$ of 10^{-3} mol/liter NaCl, pH 7–8)

CPCl final conc. (mol/liter)	Total heat[a] (mJ)	Dilution heat[a] (mJ)	Adsorption heat (mJ/m^2)	Adsorbed amount (μmol/m^2)	Molar adsorption (kJ/mol)
Quartz					
1.75	+104	+206	−11.6	2.7	−4.2
1.8	+96	+233	−10.1	2.2	−4.5
1.4	+79	+222	−10.6	2.2	−4.8
1.1	+97	+213	−8.60	2.2	−3.9
0.55	+91	+208	−7.93	2.5	−3.1
0.23	+101	+153	−3.85	1.3	−3.0
0.09	+80	+65	−1.07	1.1	−0.95
0.41	+101	+197	−7.11	1.3	−5.4
α-Corundum					
2.88	+302	+340	−3.7	0.42	−8.8
1.90	+261	+291	−2.8	0.42	−6.7
0.87	+157	+200	−4.0[b]	0.36[c]	−11.1
0.85	+157	+200	−4.0[b]	0.42[c]	−9.5

[a] $\pm 15\,\text{mJ}$.
[b] ± 2.
[c] ± 0.1.

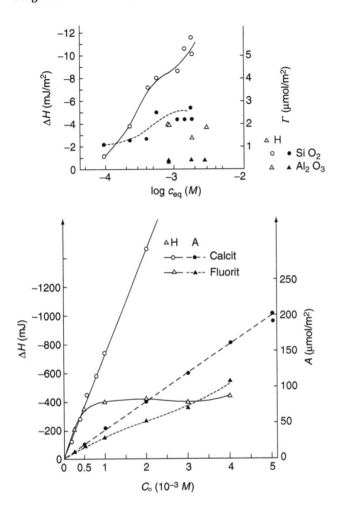

Figure 10.17 (a) Exchange enthalpies (ΔH) and adsorption densities (Γ) of cetylpyridinium chloride on quartz (pH 3.95 $\pm$ 0.05) and α-corundum (pH ~7.5) at a constant ionic strength of NaCl (1×10^{-3} mol/l). (b) Exchange enthalpies (ΔH) and abstraction densities (A) of Na-aleate an calcite and fluorite as a function of the initial concentration (pH 10 $\pm$ 0.05) at a constant ionic strength of NaCl (1×10^{-3} mol/l).

mineral surface, the charge number z of the surfactant ion, and the Faraday constant F according to Equation (17):

$$\Delta G^{el} = zF\Psi_\delta \qquad (17)$$

At a low ionic strength ($< 10^{-1}\,M$) the ζ potential can be used instead of Ψ_δ. Then we obtain for quartz at pH 4 a potential of $\sim -100\,mV$ and it results for the CPCl adsorption:

$$\Delta G^{el} = +1(-0.1V)F = -9.64\,kJ/mol = -3.96RT \qquad (18)$$

A comparison of this theoretical value with the measured adsorption enthalpy ($-3.7\,kJ/mol$) shows a difference of $\sim -6\,kJ/mol$. This can be explained according to the Gibbs–Helmholtz equation ($\Delta G = \Delta H - T\Delta S$) by a marked entropy increase (due to a release of H_2O molecules on the mineral surface and on surfactant heads) or the Stern equation is not sufficiently exact for the complex adsorption process involving exchange reactions of ions and dipoles.

2.   PDI Lattice Ions: Salt-Type Minerals

The adsorption of ionogenic surfactant on the surface of salt-type minerals is characterized by a manifold adsorption mechanism and is often combined with a parallel occurring precipitation reaction. A typical example is the adsorption of Na oleate on the fluorite or calcite surface (Figure 10.17b). The experimental conditions were that the surface of both minerals offered for adsorption and the pH value (~ 10) were the same. Thereby we obtained an almost identical abstraction at low concentrations; at intermediate and high oleate concentrations the abstraction isotherm has a linear course with calcite, while it tends to saturation with fluorite. A linear increase in the abstraction enthalpy of calcite that stretches far beyond a concentration of $2 \times 10^{-3}\,M$ indicates that beside an oleate adsorption on the surface, Ca^{2+} ions are released from the mineral surface and react with the added oleate under formation of a Ca oleate (OL) precipitate:

$$Ca^+ + 2(OL^{-1}) \rightarrow Ca(OL)_2 \quad H^m = -28\,kJ/mol$$

The precipitate remains partly in the bulk phase, and partly it binds on the adsorption layer. The calculation of the molar abstraction enthalpy results in a mean value of $H^m = -29.4\,kJ/mol$. This value agrees very well with the Ca oleate formation enthalpy ($-28\,kJ/mol$) found by von Rybinski at 40°C [65].

A division of the abstraction in adsorption and precipitation in calorimetric measurements is not possible experimentally. Experiments carried out with the filtrate to determine the precipitated portion cannot be compared with experiments performed with mineral dispersions because the amount of dissolved Ca^{2+} ions is different. While the Ca^{2+} ion amount in the filtrate is firmly given, in the dispersion new Ca^{2+} ions are supplied steadily on oleate addition from the calcite surface so that more Ca oleate particles precipitate.

A different situation exists in fluorite. Here the dissolution of Ca^{2+} is reduced by the adsorption of oleate on the mineral surface and the abstraction tends to a maximum. The oleate adsorption combined with a coadsorption of the Ca oleate precipitate results in a firmer and more hydrophobic adsorption layer than in the case of calcite [66] and enables a very good floatability even at very low oleate concentration. In fluorite the molar abstraction enthalpy ΔH^m decreases with increasing oleate concentration (Table 10.4). This means that the adsorption which took place through Ca oleate complex formation at the beginning ($\Delta H^m = 26.6\,kJ/mol$) is controlled by weaker forces (such as hydrophobic interactions) at higher concentrations.

In the literature on flotation with oleate as collector [67] one can find hints that the reactivity of double bonds in the oleate contributes to its good collector effectivity. Cross-linkages between alkyl chains by epoxide bridges should lead

Table 10.4 Molar Abstraction Enthalpy ΔH_m of Oleate on Fluorite at Different Initial Oleate Concentrations c_0

c_0 ($\times 10^{-3}$ mol/l)	0.5	1	2	3	4
$-\Delta H^m$ (kJ/mol)	26.5	18.8	11.7	8.2	8.0

to a polymerized adsorption layer with a larger hydrophobicity. To judge the effect of double bonds, flotation experiments and abstraction measurements were carried out with salts of polyunsaturated C_{18} fatty acids, Na linolate (OL 2, with two double bonds), and Na linolenate (OL 3, with three double bonds). The double bonds of OL 2 and OL 3 are conjugated and more reactive than those of OL 1. The solubility products of Ca salts of OL 1, OL 2, and OL 3 are almost equal ($pK = 12.4$, 12.4, and 12.2, respectively) [68] and therefore a similar abstraction of the three soaps on calcite can be expected. This has been also confirmed (Figure 10.18). However, the released enthalpy is lower with OL 2 and OL 3 than with OL 1. This means that OL 2 and OL 3 do not bind as firm complexes with Ca^{2+} ions as OL 1. Possible polymerization reactions are not provable by measuring the enthalpies.

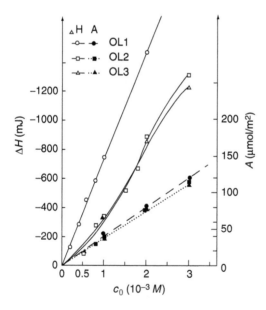

Figure 10.18 Abstraction isotherm (A) and enthalpy (ΔH) of oleate (OL 1), linolate (OL 2), and linolenate (OL 3) on calcite as a function of the initial surfactant concentration at pH ~10.

3. PDI H^+, OH^- and Lattice Ions: Silicates

Kaolinite and feldspar are good examples for an examination of the simultaneous influence of both PDI types together because these minerals are composed of the same elements (K, Na, Al, Si, and O). Kaolinite forms bilayer silicate sheets consisting of SiO_4 tetrahedral silicate layer and an $Al(OH,O)_6$ octahedral aluminum oxide layer. The flat sides of these sheets are provided with $=O >$ and $-OH$ groups while the edges are covered with broken valencies of $Al-O$ and $Si-O$. Feldspar creates a continuous three-dimensional network of SiO_4 and AlO_4 tetrahedra with positively charged mono- and divalent cations in the interstices of this negatively charged network.

On the basis of the negative ζ potential of both minerals (IP $\cong$ pH2) the adsorption of the cationic surfactant CPCl was investigated at approximately pH 8.5 and 25°C. In Figure 10.19a the adsorption densities and enthalpies are plotted against equilibrium concentrations of CPCl. While in feldspar a marked adsorption does not start unless a high surfactant concentration is used and no reaction heat is measurable, in kaolinite the adsorption increases steadily and reaches its maximum at the CMC. The adsorption enthalpy remains almost constant at the beginning and then it increases quickly with elevating adsorption. This increase is caused by a strongly exothermic reaction at higher adsorption densities as two exothermic peaks were found on a calorimetric sheet (Figure 10.19b). Thus the kaolinite surface possesses reactive adsorption sites that get occupied by CPCl at a low concentration and at an exothermal reaction enthalpy of -10 kJ/mol. On the other hand, in feldspar the surface is almost inactive toward CPCl. In kaolinite specific interactions with CPCl must operate in addition to electrostatic attraction forces because in this case the adsorption enthalpies are 3–7 kJ/mol more negative than in quartz, the surface of which shows a negative ζ potential and thus has higher electrostatic attraction forces for CPCl. If only electrostatic interactions were operating between CPCl and the kaolinite surface the free adsorption energy ΔG_{ads} would reach a value of 5.79 kJ/mol

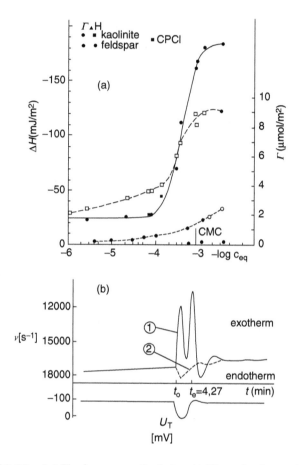

Figure 10.19 (a) Exchange enthalpies (ΔH) and adsorption densities (Γ) of cetylpyridinium chloride (CPCl) on kaolinite and feldspar
at a constant NaCl ionic concentration (1×10^{-3} mol/l) and pH 8.5.
(b) Calorimetric record of the CPCl adsorption as in (a). ν, number of
the heater impulses; U_T, thermal voltage; t_0, t_e, starting time and
ending time of the titrations; 1, heat of reaction; 2, heat af dilution.

(according to Equation (17) for $z = +1$ and $\Psi_\delta \equiv \zeta = -60\,\text{mV}$)
and thus it would be smaller than the enthalpy measured.

The occurrence of a second exotherm peak at a higher
CPCl concentration is surprising. It can be speculated that it
indicates the formation of a second adsorption layer. If this

build-up takes place due to hydrophobic van der Waals inter-actions, the enthalpies of the second peak would be too high (see below).

D. Binding Energies of Hydrophobic Chains

1. From Adsorption Isotherms

The adsorption of surfactant on a mineral surface is generally an exothermic process and at a constant surfactant concentration in the solution its adsorbed amount decreases with increasing temperature. Using the partial derivation $\delta \ln c / \delta T$ for a constant number of CH_2 groups n of a hydrocarbon chain of a surfactant and constant values of the surface coverage θ, we arrive at an equation for the calculation of the total adsorption energy E. From the relation $E = E_p + E_\theta$, Richter and Schneider [47] derived an equation which is valid for all isotherms:

$$\left(\frac{\delta \ln c}{\delta T} \right)_{n,\theta} = -\frac{E}{kT} \qquad (19)$$

This equation is verified by plotting the experimental data against $1/T$; the slope of the line determines the value $-E/k$.

If the experimental isotherms agree with adsorption isotherms according to Langmuir or Temkin, it is possible to calculate the adsorption energy E using the constant β or A, which were obtained from isotherms at different temperatures. E is estimated from the slope of the regression lines $\ln \beta$ against $1/T$ (Langmuir or $\ln A$ against $1/T$ (Temkin), at a constant number of CH_2 groups of the hydrocarbon chain.

Schubert et al. [49,69], Schneider [70], and Matthé [71] analyzed the adsorption isotherms of N-alkylammonium chloride on SiO_2 in order to estimate the binding energy of the hydrophobic groups. They used the constants β and A from the appropriate equations of the Langmuir and Temkin isotherms, respectively. Their results are summarized in Table 10.5. Dobias [25] calculated the association energy for adsorption af C_{10}–C_{18} N-alkyl ammonium chlorides in the range of surface coverage 0.2–0.8 an biotite according to

Table 10.5 Bonding Energy E_A of Hydrocarbon Chains for One CH_2 Group

Surface coverage θ Author and conditions	Equation used: $\Gamma = 2rc$ $\exp(-E/kT)$	From Langmuir isotherm	From Temkin isotherm
Schneider: $\theta = 0.0275$–0.275, low conc. range of the surfactant	0.44–0.64kT $n = 14$ and 16	0.31kT $n = 12$ and 16	0.41kT $n = 12$ and 16
Schubert: $\theta = 0.2$–0.8 higher conc. range of the surfactant	0.54–0.90kT $n = 12, 14, 16$	0.88kT $n = 12, 14, 16$	1.05kT $n = 12, 14, 16$
Schubert same conditions but a constant ionic strength (10 g KCl/liter)	0.74–0.94kT $n = 12, 14, 16$	1.03kT $n = 12, 14, 16$	1.22kT $n = 12, 14, 16$
Matthé low conc. range of surfactant	$kT\, k_{12}\, n_1 - n_2$		$n = 8$ and 18: 0.23kT $n = 12$ and 18: 0.20kT $n = 8$ and 18: 0.27kT

Schubert [49] and found values of E_A ranging from 0.401 to 0.459 kT. From these results it follows that with increasing surface coverage by surfactant molecules the association energy also increases slightly. This binding energy value is of the order of van der Waals energy (0.008–25 kJ/mol, whereby 1 $kT = 2.5$ kJ/mol). At higher concentrations of indifferent electrolytes, such as KCl or NaCl, E_A increases since the surfactant adsorption also rises as a result of binding water molecules by KCl.

2. From Calorimetric Measurements

When the calculated association energies are supplemented by calorimetric measurements it is possible to obtain further information on the buildup and structure of the adsorption layer. This can be demonstrated, for example, by alkylpyridinium salt adsorption on kaolinite. In this connection the effect of the chain length of C_{10}–C_{16} pyridinium chlorides (PCl) on changes of exchange enthalpies and adsorption density as dependent on equilibrium concentration was studied (Figure 10.20).

It can be assumed that when a plateau is reached in the range of $\Gamma = 3.6\text{–}4.5\,\mu\text{mol/m}^2$ on the adsorption isotherm of C_{14} and C_{16} PCl, and the C_{10} and C_{12} PCl display a maximum density, a monolayer gets formed. The surface area of $0.45\,\text{nm}^2$ required per one molecule agrees well with the literature [72].

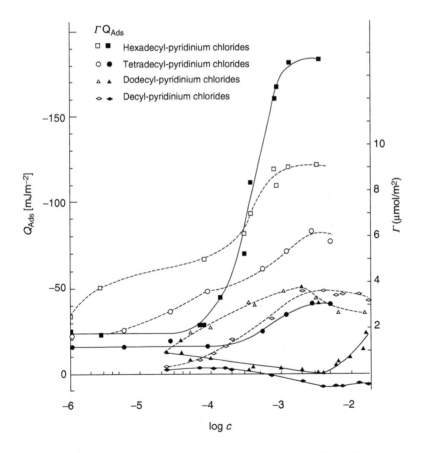

Figure 10.20 Exchange enthalpies (Q_{Ads}) and adsorption densities (Γ) of various alkylpyridinium chlorides on kaolinite at a constant concentration of NaCl ($1\times10^{-3}\,\text{mol/l}$) and pH ~ 8.5.

The natural logarithms of occupancy densities of PCl derivatives obtained from adsorption isotherms at the same concentrations can be plotted against the chain length n. Then the line slope indicates the free association energy $s \cdot \Phi$ in RT units. This calculation leads to the following values of $s \cdot \Phi$: $C_{eq} = 3 \times 10^{-4}$ M for $s \cdot \Phi = \sim -2.0RT$ (5 kJ/mol) and $C_{eq} = 5 \times 10^{-5}$ M for $s \cdot \Phi = \sim -0.5RT$ (1.25 kJ/mol).

If it now holds, according to Mukerjee [73], that $\Phi = 1.39RT$, we obtain an association degree $s = 0.35$ at a low surfactant concentration. The association degree represents a mean value of different surfactant occupancies and is somewhat lower when compared with literature data [74] for n-alkylammonium ions on quartz $(s = 0.39–0.65$ at 20–80% occupancy). At higher surfactant concentrations $(C_{eq} = 3 \times 10^{-4}$ $M)$ the association energy amounts to $-2RT$, thus more than the maximum value of -1.39 assumed by Mukerjee. The question arises as to whether further chain length-dependent energy terms appear here besides the association energy. Since in addition, the surfactant adsorption an agglomeration of kaolinite sheets also occurs, additional terms are possible.

An evaluation of adsorption isotherms in the range of very low concentrations shows that all four enthalpy isotherms are exothermic. For the C_{14} and C_{16} surfactants they remain constant with increasing concentration and then rise steeply, accompanied by a second exothermic peak. On the other hand, for the C_{10} and C_{12} surfactants they sink continuously (in the former one even in the endothermic range) and only in the latter one finds a small increase at very high concentrations. These results allow a conclusion that the adsorption occurs in three steps:

1. *At a low surfactant concentration.* On the kaolinite surface there are reactive negatively charged adsorption sites that get occupied first under a simultaneous heat release. The alkyl chain length plays an inferior role. The energy gain is mainly due to electrostatic and chemical interactions between the surfactant head and the surface.

2. *At intermediate surfactant concentrations.* Surfactant molecules deposit on less reactive sites or an anchor molecules, namely, as long as a monolayer gets formed. The adsorbed surfactant heads of the same sign as the surface repulse one another. Water molecules in spaces between alkyl chains are driven away, and the chains start to associate due to van der Waals forces. The reaction enthalpy of these steps amounts to approximately zero in the C_{14} and C_{16} surfactant; in the C_{10} and C_{12} surfactant it is endothermic. The reaction proceeds here only on the basis of an entropy increase that originates from the release of water molecules previously arranged around the alkyl chains into the bulk phase.

3. *At higher surfactant concentrations* (C_{14}: $c > 3 \times 10^{-4}$ *M*; C_{16}: $c > 1 \times 10^{-5}$ *M*). The third step is observable only with the C_{14} and C_{16} surfactants. At these concentrations a second layer gets formed due to van der Waals forces between the chains with surfactant heads directed toward the bulk phase. This manifests itself both in a stronger increase in the adsorption isotherm and in the appearance of a second exothermic peak. Further, the surface hydrophobicity and thus the floatability decrease, which was confirmed experimentally.

E. Stability of Mineral Dispersion as Dependent on Surfactant Concentration

The effect of surfactant concentration on the stability of a mineral dispersion can be demonstrated, for example, in an aqueous dispersion of Al_2O_3 or CaF_2, where these minerals differ from each other by the character of the PDIs. Na-dodecylbenzenesulfonate (NaDBS) and CPCl were used as the anionic and cationic surfactants, pH employed was under the IP of both substances (9.0 and 10.75, respectively); this means that the potential of both minerals was positive.

In agreement with our earlier observations [6,75] and those of others [77,78] on the adsorption of anionic and cationic surfactants on corundum, we demonstrated [19] that the stability of the Al_2O_3 suspension is dependent on the IP as well. In Figure 10.21 the suspension turbidity and the potential are plotted against the NaDBS and CPCl concentrations. According to the curve shape we find that the coagulation of Al_2O_3 does not occur when the polar surfactant group and the mineral surface have the same sign of charge. The NaDBS concentration of the minimum of stability corresponds to the first plateau of the adsorption isotherm as described above

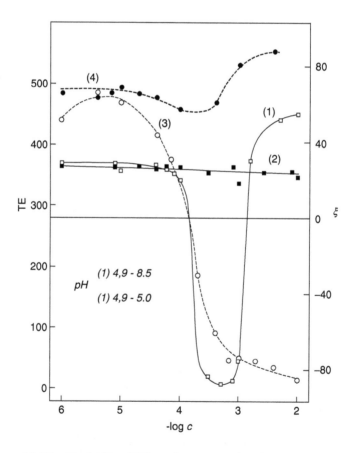

Figure 10.21   Turbidity (TE) and ζ potential (mV) of pyrolytic Al_2O_3 as a function of the Na-dodecylbenzenesulfonate (1) and cetylpyridinium chloride (2) concentration (mol/l). ζ potential, dashed curves.

[6,75]. This plateau represents the hydrophobicity optimum of the adsorbed layer. Here also the flotation optimum of a-corundum is reached (Figure 10.22). The critical flocculation concentration (CFC) of Al_2O_3 (Figure 10.21, curve 1) agrees reasonably well with the appropriate IP according to electrophoretic measurements (curve 3). Corresponding to the course of the ζ potential, the stability of the Al_2O_3 suspension does not change in CPCl solutions (curve 2). Similarly, neither corundum flotates in solutions of CPCl when the potential is positive in spite of a certain increase in the positive value (curve 4), which can be explained as a sign of surfactant adsorption.

The CFC of fluorite in NaDBS solutions (Figure 10.23) has the value of 5×10^{-5}, thus over the IP (2.7×10^{-5}), and is not reached until the ζ potential gets negative ($-25.6\,\mathrm{mV}$). The reason can be a precipitation reaction between the surfactant and Ca^{2+} ions, since the adsorption density of the surfactant is in fact lower (see Chapter 5).

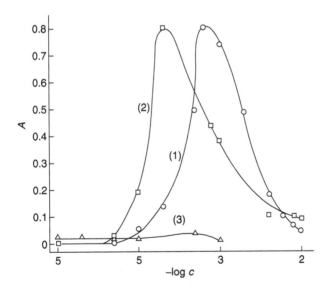

Figure 10.22 Floatability (A, arbitrary units) of Al_2O_3 (1) and fluorite (2, 3) as a function of NaDBS or CPCl (3) concentration (c-mol/l).

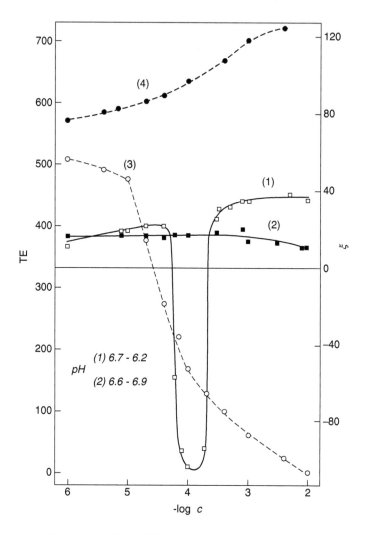

Figure 10.23   Turbidity (TE) and ζ potential (mV) of fluorite as a function of the Na-dodecylbenzenesulfonate (1) or cetylpyridinium chloride (2) concentration (c-mol/l). ζ potential, dashed curves.

The NaDBS concentration at which the fluorite dispersion reaches the stability minimum corresponds, similarly to Al_2O_3, to the concentration range of the adsorption isotherm with a plateau [40] and the maximum floatability of fluorite (Figure 10.22, curve 2). An increase in CPCl concentration leads to an increase in the positive ζ potential without any alteration of the dispersion stability (CPCl adsorption at negative F^- sites). In addition, the flotation of fluorite with CPCl does not occur under these conditions (curve 3).

A simplified presentation of the mineral dispersion stability in relation to coverage degree of the surfactant is shown in Figure 10.24. The mode of binding of the surfactant's polar group on the mineral surface is not considered here.

F. Effect of pH Value

The influence of pH on the surfactant adsorption on a mineral surface is closely coupled with the character of the PDI.

1. PDIH$^-$ and OH$^-$

The mechanism of interaction of an ionogenic surfactant with a mineral surface occupied by PDI H^+ and OH^- is generally

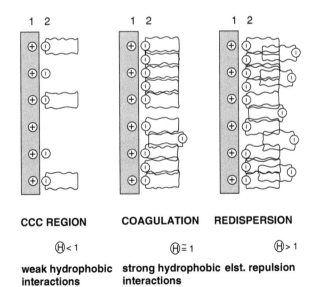

Figure 10.24 Stability of a mineral dispersion as a function of the character of the adsorbed surfactant layer.

confirmed by the position of IP in relation to the optimum of mineral floatability. The limit of efficiency of an anionic or cationic surfactant is IP, as follows from the correlation of adsorption density, potential, and floatability, which are all dependent on pH [41,56,75,80–82]. The course of such an adsorption is schematically shown in Figure 10.25. As an example, the adsorption isotherms of sodium dodecyl sulfate at different constant concentrations on α-corundum (IP $= 8.0$) are displayed in Figure 10.26 as dependent on the pH value [79].

If H^+ or OH^- ions react with the surface of one mineral the released ions or their hydrolytic products can adsorb on the unequally charged surface of the other mineral and cause an activated adsorption of the surfactant, or they can inhibit the adsorption, as shown in the following schemes:

$$Fe(OH)^{2+} + [SiO_2]^- \rightarrow [SiO_2]\text{-}Fe(OH)^+ \qquad \text{activated complex}$$

$$[SiO_2]\text{-}Fe(OH)^+ + - \rightarrow [SiO_2]\text{-}Fe^+ - \qquad \text{flocculation}$$
$$\text{surfactant} \qquad\qquad | \qquad \text{(adsorption)}$$
$$OH$$

$$[SiO2]\text{-}Fe(OH)^+ + \oplus - \rightarrow [SiO_2]\text{-}Fe^+ \rightleftarrows \oplus - \qquad \text{dispersion}$$
$$\text{surfactant} \qquad\qquad\quad | \qquad\qquad \text{(repulsion)}$$
$$OH$$

In surfactants that dissociate in a limited way especially at higher H^+ or OH^- concentrations, these ions are built in undissociated molecules due to coadsorption in the adsorption layer formed by electrostatic forces.

A number of models have been proposed concerning the way of adsorption of ionogenic surfactants that are of use especially for oxides. As an example, the model of Lai and Fuerstenau [83] should be mentioned here. These authors, using the mass action law, derived an equation to describe the distribution of charged and neutral surface sites in relation to the H^+ concentration in a solution and to the PZC of oxides. On the basis of flotation experiments they characterized the adsorption of anion- and cation-active surfactants presuming an exact correlation between the adsorption and

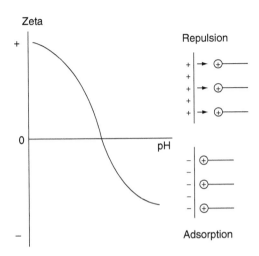

Figure 10.25   Adsorption range of cationic surfactants on mineral surface with H^+/OH^- as PDIs as a function of pH. Using anionic surfactants, adsorption occurs on a positively charged surface.

floatability. For this purpose, Na dodecyl sulfonate and dodecylammonium chloride were used to examine the charged sites of rutile, and Na oleate for examining the neutral surface sites. Using the charging reactions (Equations (20) and (21)):

$$MeOH_{surf} \Leftrightarrow MeO^-_{surf} + H^+_{aq} \tag{20}$$

$$MeOH_{surf} + H^+_{aq} \Leftrightarrow MeOH^+_{2surf} \tag{21}$$

and the appropriate equilibrium constants

$$\begin{aligned} K_1 &= a_{MeO^-} \times a_{H^+}/a_{MeOH} \quad \text{and} \\ K_2 &= a_{MeOH^+_{2surf}}/a_{MeOH_{2surf}} \times a_{H^+_{aq}} \end{aligned} \tag{22}$$

they developed equations for individual fractions θ of the surface sites of the solids:

$$\theta_0 = \frac{K_1(a_{H^+})}{(a_{H^+})^2 + K_1(a_{H^+}) + K_1 K_2} \qquad \text{neutral sites MeOH}$$

$$\tag{22a}$$

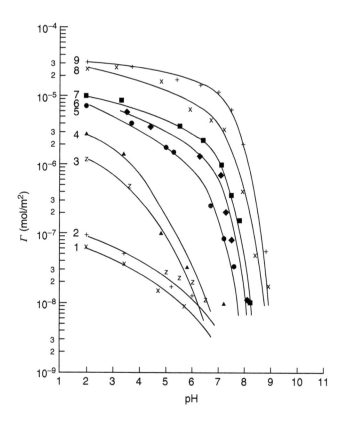

Figure 10.26 Adsorption isotherms (Γ in mol/m^2) of Na-dodecyl sulfate at different constant initial concentration (1–9) on α-corundum as a function of pH at constant ionic strength of NaCl (2×10^{-3} mol/l). 1, 5×10^{-7}; 2, 1×10^{-6}; 3, 5×10^{-6}; 4, 1×10^{-5}; 5, 5×10^{-5}; 6, 1×10^{-4}; 7, 5×10^{-4}; 8, 2.5×10^{-3} and 9, 9×10^{-3} (mol/liter). IP of α-A$_2$O$_3$ = 8.0; EAP = 8.1.

$$\theta_+ = \frac{(a_{H^+})^2}{(a_{H^+})^2 + K_1(a_{H^+}) + K_1 K_2} \qquad \text{positive sites MeOH}_2^+$$

$$(22b)$$

$$\theta_- = \frac{K_1 K_2}{(a_{H^+}) + K_1(a_{H^+}) + K_1 K_2} \qquad \text{negative sites MeO}^-$$

$$(22c)$$

The concentration of the surfactants used was below the concentration of hemimicelle formation, so that they interacted in the Stern layer only with the appropriate surface site, since a physically adsorbing surfactant should adsorb over an opposite charged surface site. A study of the effect of pH on the relationship between floatability and the surface charge of rutile showed that the adsorption of a surfactant on a solid is well correlated with the distribution of the positive, negative, and neutral sites at the interface. The character of ion surfactant adsorption is of a physical nature whereas Na oleate chemisorbs with an optimum in the region of the PZC.

An interpretation of Iwasaki et al. [84] results of flotation measurements on goethite with a model of charged surface sites confirms a general validity of the proposed equations. The flotation limit for Na oleate related to the calculated neutral MeOH site distribution agrees well with the Wadsworth postulate that surface hydroxyls are responsible for the chemisorption of carboxylate surfactants on solids [85].

The presence of polyvalent lattice ions in the system containing minerals with PDI H^+ and OH^- leads to their specific adsorption in the EDL and is often accompanied by a change in IP and PZC provided the surface charge has the opposite sign of that of the adsorbing lattice ion. This leads to an inhibition or activation of the mineral surface [86]. The same is true for a hydrolytic product of lattice ions exhibiting a stronger surface activity than nonhydrolyzed ions, as a result of a combined electrostatic and chemisorptive effect.

2. PDI Lattice Ions

The adsorption of a surfactant on the mineral surface with polyvalent lattice ions as dependent on pH is always coupled with a simultaneous H^+ and OH^- adsorption and with an increased output of individual lattice ions from the mineral surface crystalline lattice due to their solubility. In this case the pH value at which IP is attained is usually not the criterion for the adsorption ability of a cationic or anionic surfactant, but polyvalent lattice ions are the decisive factor. This is especially the case where the surfactant has the same sign of the electric charge as the adsorbent. A typical example for it is the adsorption of Na dodecyl sulfate on barite at different pH values (Figure 10.27). It follows from the relation that the $BaSO_4$ surface has a negative ζ potential from pH 3.3 upward and in this region the anionic surfactant adsorption takes place [6].

Since the effect of pH on surfactant adsorption on minerals with PDI lattice ions is very complicated, it is extremely difficult to process adsorption isotherm data to more general conclusions or model conceptions.

G. Stability of Mineral Dispersions as Dependent on pH

To study the effect of H^+ and OH^- concentrations on the mineral dispersion stability in solutions of surfactants, measurements were carried out at a constant surfactant concentration and varying pH [19,87].

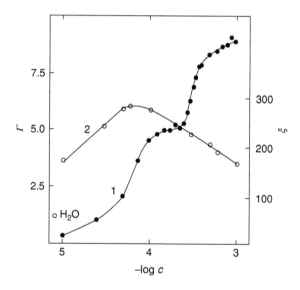

Figure 10.27 Adsorption density (Γ in 10^{14} surfactant molecules/cm^2) (1) and ζ potential (mV) (2) of barite as a function of Na-dodecyl sulfate concentration (mol/l).

Figure 10.28 shows the rest turbidity of an Al_2O_3 dispersion in dependence on pH for three different constant NaDBS
concentrations. These concentrations are either under the
CFC (8×10^{-5} M, curve 1), or at the stability minimum
(5×10^{-4} M, curve 2), or above the critical redispersion concentration (CRC), thus above the CMC (3×10^{-3} M, curve 3).
The concentrations were chosen according to curve 1 of Figure
10.21.

The NaDBS concentration of 8×10^{-5} M is not sufficient
at pH < 9 to form an adequate hydrophobic layer (curve 1,

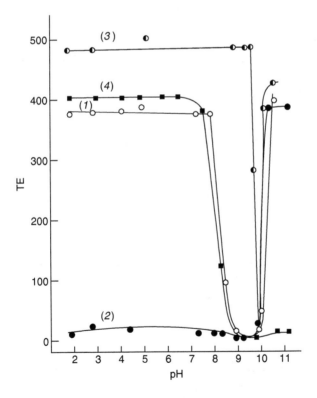

Figure 10.28 Turbidity (TE) of α-Al_2O_3 dispersion at a constant
Na-dodecyl sulfate concentration or cetylpyridinium chloride concentration as a function of pH. NaDBS: 1, 8×10^{-5}; 2, 5×10^{-5}, and 3,
3×10^{-3} mol/l, CPCl: 4, 1×10^{-4}.

Figure 10.28). The destabilization occurs only within the IP of Al_2O_3.

At 5×10^{-4} M NaDBS the dispersion is unstable in a wide range of pH < 9, under a positive ζ potential. This means that the adsorption density of anionic surfactants on the positively charged Al_2O_3 surface at pH < 9 is sufficient to form a sufficiently hydrophobic adsorption layer (curve 2, hydrophobic flocculation).

At concentrations above the IP of the surfactant (1×10^{-3} M) the Al_2O_3 dispersion remains stable (a second, hydrophilic layer forms) and gets destabilized only at a narrow pH range dose to IP (curve 3).

Opposite to the behavior in NaDBS solutions the Al_2O_3 suspensions get unstable in solutions of CPCl only at pH > 9, as expected. From this value onwards the adsorption density of HP^+ ions increases very sharply. Because of hydrophobic interactions of the adsorbed layers particle aggregation occurs and the dispersion remains unstable at higher OH^- concentrations (curve 4).

Because H^+ and OH^- determine the adsorption of ionogenic surfactants on the phase interface in general, it can be concluded that the destabilization of Al_2O_3 suspensions in the pH range of IP is caused by a loss of electrical charge rather than by hydrophobic interaction of the adsorbed layers.

The effect of pH on the stability of fluorite dispersions (Figure 10.29) was studied at a NaDBS concentration of 1×10^{-4} M at which the fluorite suspension reaches a coagulation optimum (Figure 10.23). As can be seen (Figure 10.29, curve 1), the fluorite dispersion is unstable in the whole pH range. The explanation may be that an increasing adsorption density of H^+ has no distinct effect on the alteration of fluorite dispersion stability. The NaDBS adsorption proceeds only over the binding Ca^{2+}-lattice ion-surfactant; the aggregation is caused by hydrophobic interparticle interactions.

The pH value alterations at a constant CPCl concentration has no effect on the dispersion stability (curve 2), as in experiments with NaDBS. In spite of an adsorption of the cationic surfactant, the hydrophobic forces are not sufficient at the given concentrations to overcome the electrostatic

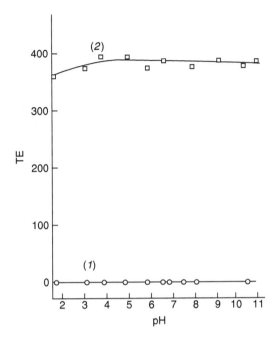

Figure 10.29 Turbidity (TE) of the fluoritce dispersion at the constant concentration of Na-dodecylbenzenesulfonate $(1, 1\times10^{-4})$ or cetylpyridinium chloride $(2, 1\times10^{-4})$ as a function of pH.

forces resulting from the high positive ζ potential (Figure 10.23, curve 4). Under these conditions the fluorite dispersion remains stable.

In the stability minimum of a mineral dispersion an optimum floatability is attained. This was demonstrated in appropriate experiments with fluorite, barite, and pyrite as dependent on pH [87].

H. Effect of Counterions

In general, counterions of a surfactant molecule affect the structure of their environment. This effect can be demonstrated on the example of CMC. Many authors integrate the CMC sequences within the known Hofmeister series and explain the results obtained with the aid of Luck's theory [88] of water structure alterations that should be induced by the ions added.

In this connection adsorption measurements were carried out by Petzenhauser [89] with five different cetylpyridinium (CP) salts on quartz.

The results are displayed in Figure 10.30. All curves show a similar sigmoidal course with a plateau region that corresponds reasonably well with the appropriate CMC values in CPCl, CPBr, and $CPNO_3$. CP acetate and benzoate exhibit

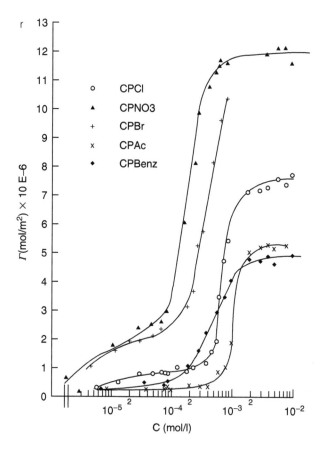

Figure 10.30 Adsorption densities (Γ in $1\times10^{-6}\,mol/m^2$) of different cetylpyridinium salts on kaolinite as a function of their concentration (mol/l).

this almost constant adsorption behavior at concentrations that lie far beyond the CMC. The CMC values are marked with perpendicular arrows in the figure. Only when the maximum adsorption densities Γ_{max} are considered can the following rank order be obtained:

Γ_{max}: $CPNO_3 > CPBr > CPCl > CPAcet > CPBenz$.

The series of the first three substances agrees well again with the Hofmeister series.

It can therefore be concluded that the higher the adsorption densities, the smaller are the CMC values. In other words, the more the anions favor the association the more cations adsorb at the interface. Basing on the performed verification of the adsorption isotherms it can generally be deduced that the behavior of surfactants examined here is not sufficiently described by the Langmuir equation. Other quantitative models described in the literature, for example, by Temkin, Freundlich, and Matthé [71,90] are not in satisfactory accord with the isotherms measured either. Similarly, if one tries to reconcile the results of the adsorption measurements with the different qualitative models such as that of Fuerstenau, Harwell, Gu, etc., substantial differences soon arise in some cases.

In spite of a low occupancy of the mineral surface with surfactant molecules, a maximum hydrophobicity must be achieved in the plateau region of the adsorption isotherm since flotation maxima were also attained here. From Figure 10.31 it follows that the flotation recovery increases markedly with all five surfactants at very low concentrations ($\sim 5 \times 10^{-7}$ mol/l). After a surfactant concentration has been reached that corresponds roughly to the IP, the recovery remains almost constant over a certain range of concentrations. After the CMC is exceeded the flotation recovery decreases again up to zero, with an exception of CPBenz. The flotation experiment again confirmed the earlier assertion that the complete monomolecular coverage of the adsorption layer should be assessed according to the maximum hydrophobicity and not by a theoretically calculated area requirement. The maximum hydrophobicity would then be related to different coverage

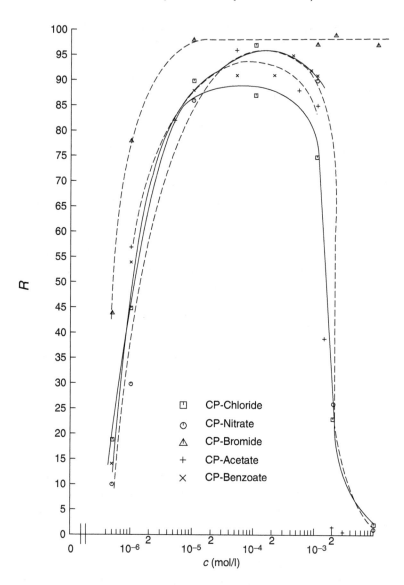

Figure 10.31   Recovery (R in %) of kaolinite as a function of concentration (mol/l) of different cetylpyridinium salts.

values θ, depending on the energetic conditions in the mineral–water phase interface.

To investigate the effect of ionic strength on the adsorption of the CP cation, measurements were performed at constant concentrations (1×10^{-2} mol/l of the appropriate sodium salt of the surfactant counterion. As a rule, a rise in the electrolyte concentration results in an elevation of the adsorption densities. This connection is explained by the effect of electrolytes on the thickness of the electric double layer of surfactant heads. If the double-layer thickness gets reduced, the electrostatic repulsion of head groups to one another decreases and the formation of associates both in the bulk phase and on the mineral surface should get easier [72].

The adsorption isotherms that were obtained at a constant ionic strength show a course similar to that without electrolyte addition (Figure 10.32). The rank order of maximum adsorption densities is the same as in experiments without a constant ionic strength. At a low equilibrium concentration the adsorption densities differ only weakly from one another. The curves diverge in a range of $C = 2.5$–6.0×10^{-5} mol/l. The differences thus occur in a concentration range when the IP of all five surfactants has been passed. It can therefore be concluded that under IP the adsorption process is independent of the electrolyte concentration.

Similar to experiments without a constant ionic strength, the adsorption densities of CPCl, CPBr, and CPNO$_3$ become nearly constant after the CMC has been reached. As in experiments without a constant ionic strength, CPAcet and CPBenz behave differently. Their passing the CMC is not connected with constant adsorption densities either. Their saturation range is reached at concentrations that lie markedly beyond the CMC.

It is striking for all surfactants investigated that they achieve substantially lower adsorption densities at an electrolyte concentration of 1×10^{-2} mol/l than in experiments without a constant ionic strength. Since an increase in ionic strength leads to a decrease in the CMC (accompanied by a plateau shift), an occurrence of micelles can be the reason why the adsorption density cannot rise further.

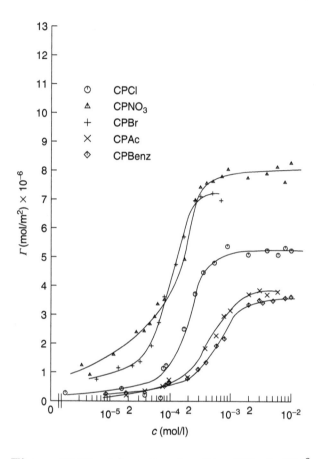

Figure 10.32 Adsorption densities (Γ in 1×10^{-6} mol/m^2) of different cetylpyridinium salts on kaolinite depending on their concentrations at constant ionic strengths (1×10^{-2} mol/l) of the Na salts of the same surfactant anion.

The effect of surfactant counterions was investigated calorimetrically also by Wierer [64,91] in the system CP salt–kaolinite. As counterions of CP$^+$ he used Cl$^-$, H$_2$PO$_4^-$, HSO$_4^-$, Br$^-$, and NO$_3^-$. Exchange enthalpy measurements show (Figure 10.33) a second exothermic peak at higher concentrations in all surfactant salts, the start of which is marked with a perpendicular dash in the figure and which

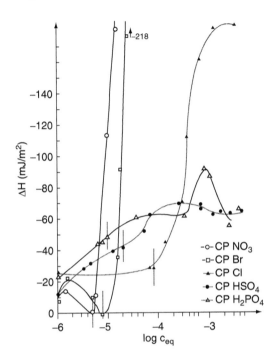

Figure 10.33 Exchange enthalpies (ΔH) of different cetylpyridinium salts on kaolinite as a function of their concentrations (mol/l) at constant ionic strengths (1×10^{-2} mol/l) of the Na salts of the same surfactant anion.

causes a strong enthalpy increase. In $CPNO_3$, a third exothermic peak occurs beyond $C_{eq} = 1.1\times10^{-5}$ mol/l.

To clarify the molar adsorption enthalpies, the enthalpies were plotted against the occupation density (Figure 10.34). The slopes of the lines $\Delta H/\Delta \Gamma$ directly represent the appropriate molar adsorption enthalpies. Except for $CPHSO_4$ and CPH_2PO_4 (where an H^+ adsorption takes place as well), a mean adsorption enthalpy of -22 kJ/mol thus results for very low occupancies (>1 μmol/m^2) in the case of the other salts. At greater occupancies, the course and the slope of the lines are very inhomogeneous. It is remarkable that the molar adsorption enthalpy of $CPNO_3$ and $CPBr$ is endothermic up to $\Gamma = 2.5$ and 3.8 μmol/m^2, respectively. At higher occupancies no molar

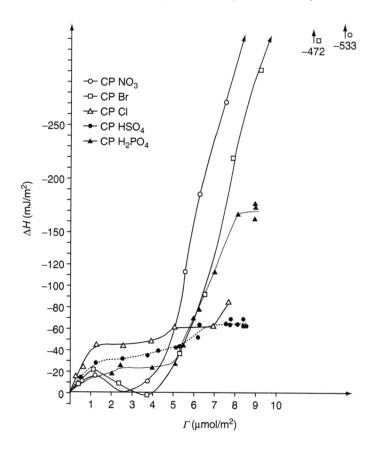

Figure 10.34 Exchange enthalpies (ΔH) of different cetylpyridinium salts as a function of adsorption density (Γ).

heats can be read out because of the adsorption of counterions. Thus, the effect of counterions on CP^+ is considerable. The anions do not perform via adsorption in the electric double layer (the ζ potential does not change) but rather by influencing the structure of the aqueous bulk phase and by their capacity to bind in the double or multiple layer of the adsorbed surfactant. The effect is thus similar to that in the formation of micelles. If we compare the CMCs of salts with one another, i.e.:

$$\text{CMC}(CP^+ - \text{salt}): \ Cl^- > Br^- > NO_3^- > H_2PO_4^- > HSO_4^-$$

we observe that the rank order of the monovalent anions agrees well with the lower adsorption density:

$$(CP^+ - \text{kaolinite}): \ H_2PO_4^-, \ HSO_4^-, \ Cl^- < Br^- < NO_3^-$$

In other words, the better the micelle formation, the more the surfactants adsorb on the mineral surface. This is true not only for $H_2PO_4^-$ and HSO_4^- counterions that are handicapped for the coadsorption because of their large radius. The rank order agrees well with that of the sequence mentioned by Underwood and Anacker [92] for the CMCs of decyltrimethylammonium salts.

I. Effect of Complexing Agents

Complexing and chelating agents are used in flotation to control surfactant adsorption at the mineral interface in connection with the formation of a hydrophobic layer. In terms of the adsorption mechanism, it is assumed that a complexing agent that forms a chelate with cations of the crystalline lattice can also adsorb similarly on the mineral surface. In this way the mineral surface becomes hydrophilic and is thus excluded from the attachment process [93–97].

It is a question, however, how comparable the steric conditions are for chelate formation in the bulk phase and on a completely or partially dehydrated mineral–water phase interface. In this connection it should be mentioned that these substances can be used in flotation not only as collectors but also as depressants (in most cases).

To reinvestigate the above-mentioned hypothesis of the depressant effect of complexing and chelating agents, extensive measurements have been performed by Orthgiess [36] on salt-type minerals (Table 10.6 and Table 10.7). Complexing agents were chosen according to the greatest possible binding variability of these substances with metal ions, both in solution and at the mineral–water interface. The substances examined are listed in Table 10.8 together with their chemical formulas.

When the summarized flotation data are examined, the following results are conspicuous:

Table 10.6   Effect of Complexing Agents on Flotation with NaDBS

Complexing agent	Calcite	Barite	Fluorite
NTA	−	−	−−
EDTA	−	−−	−−
CDTA	0	−	++
AGTA	−	−	0
SO	n.d.	n.d.	0
HBO	0	0	−
AHS	−	+	−
8-HQ	0	+	++
NaDBS, mol/liter	2×10^{-4}	5×10^{-5}	2×10^{-6}
pH	9	8	8

0 almost no effect, +light activation, ++strong activation, −light depression,
−−strong depression, n.d. = not determined. For definitions acronyms, see Table 10.8.

From Table 10.6:

The effect of the chelating agents on floatabibity is not uniform: besides depressing effects, activating ones are present as well.

The effects of complexing reagents on calcite flotation are in general negligible. Strong effects are observable mainly in chelating agents of the complexone type (on barite and in the first line on fluorite).

Table 10.7   Effect of Complexing Agents on Flotation with Na Oleate

Complexing agent	Calcite	Barite	Fluorite	Hornblende
NTA	0	−−	−	−−
EDTA	0	−−	−	−−
CDTA	0	−−	++	n.d.
AGTA	0	−	+	n.d.
SO	+	+	0	n.d.
HBO	+	0	0	n.d.
AHS	+	0	0	n.d.
8-HQ	+	+	++	n.d.
NaOl, mol/liter	5×10^{-4}	4×10^{-6}	8×10^{-7}	5×10^{-4}

0 almost no effect, +light activation, ++strong activation, −light depression, −−strong depression, n.d. = not determined. For definitions of acronyms, see Table 10.8.

The complexones show, in spite of their structural relations, different effects, especially on fluorite.

From Table 10.7:

Similar to the flotation with NaDBS, both depressing and activating effects of complexing agents are present.

The effects are often similar with both anionic surfactants examined.

The strong chelating agents do not affect the calcite floatability at all, whereas the weaker complexing agents are slightly activating.

CDTA and 8-HQ are effective stimulants of the fluorite flotation.

The most important examples of the effectivity of complexing agents on flotation will be now presented.

1. Hydrophilization of the Mineral Surface by the Adsorption of Complexing Agents

Experiments carried out on fluorite with complexing agents HBO and SO in combination with an anionic surfactant as a collector (NaDBS) showed that the two molecules that differ from each other only by the position of the hydroxy group on the benzene ring have a significantly different effect on the floatability of the mineral. While SO is almost ineffective, HBO acts as a depressant in a concentration range of 10^{-5}–10^{-3} mol/l.

Table 10.8 Complexing Agents Used

Complexing agent	Abbreviation
Nitrile triacetic acid	NTA
Ethylenediaminetetraacetic acid	EDTA
Trans-1,2-diamine cyclohexanetetraacetic acid	CDTA
Bis(2-amineethylether)ethylene glycol tetraacetic acid	AGTA
8-Hydroxyquinoline	8-HQ
Acetohydroxam acid	AHS
Salicylaldoxime	SO
p-Hydroxybenzaldoxime	HBO

In agreement with these observations it was found that HBO, but not SO, adsorbs on fluorite (Figure 10.35).

Due to this adsorption the mineral surface gets hydrophilized and this is responsible for the depressing effect observed. It can be assumed that the HBO molecule interacts with the mineral surface through its oxime group while the OH group, as a second hydrophilic group, points to the bulk phase. This can mean that in this case the mechanism, that has been postulated as the standard mechanism of complexing agents acting as depressants operates (see scheme in Figure 10.36).

SO is very probably able to chelate Ca^{2+} ions in the bulk phase, but not those of the mineral surface. Therefore no SO adsorption takes place because of a steric hindrance. The

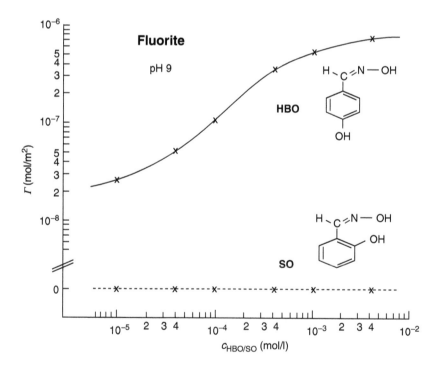

Figure 10.35 Adsorption density (Γ) of HBO or SO on fluorite as a function of the complexing agent concentrations.

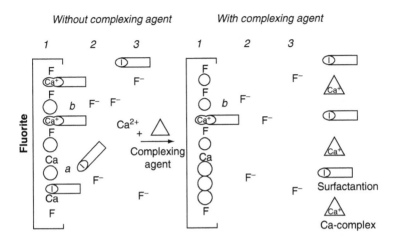

Figure 10.36 Proposed complexing mechanism from adsorption isotherms in Figure 10.35; (a) ion exchange; (b) surface complex.

weaker complexing agent HBO, however, that forms spatially simpler complexes (no chelates) is able to adsorb on fluorite. In this connection it should be mentioned that chelating agents showed a very small propensity to adsorb on minerals in further experiments as well.

On fluorite flotation with Na oleate almost no depressing effect of HBO was evident, in contrast to flotations with NaDBS. This can be linked to the special adsorption mechanism of oleate on salt-type minerals. Contrary to NaDBS, an adsorption takes place here in the sense of a surface precipitation that can hinder the adsorption of HBO or at least make it ineffective as far as the floatability of the mineral is concerned.

2. Modification of Collector Adsorption by Complexing Agents

On barite flotation with NaDBS as collector it has been shown that EDTA acts as a strong depressant in a wide concentration range whereas AHS exhibits almost no effect. However, an experimental verification of this observation showed that a significant complexing agent adsorption on the mineral occurs neither with AHS nor with EDTA. If the effect of both complexing agents is examined in the presence of NaDBS, a marked reduction of the surfactant density is observed with increasing EDTA concentration (Figure 10.37).

To explain this phenomenon, electrophoretic measurements were carried out. They showed that the ζ potential of barite turns more negative in the presence of EDTA whereas the effect of AHS is small. This can be explained according to our findings to date only by a removal of Ba^{2+} from the mineral surface and from immediately bordering layers by the strong chelating agent EDTA.

The described effect must be accompanied by an increase in the total Ba^{2+} concentration (free, complexed, and associated) in the bulk phase. This could be confirmed experimentally. Figure 10.38 demonstrates distinctly different effects of EDTA and AHS on the composition of the solution, which is generally not altered by an increasing surfactant concentration.

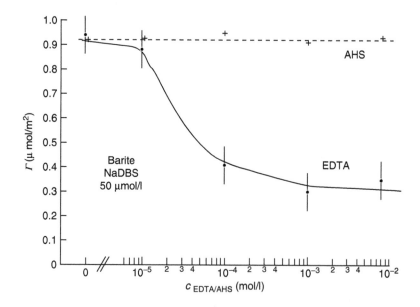

Figure 10.37 Adsorption density (Γ in $1 \times 10^{-6}\,mol/m^2$) of Na-dodecylbenzenesulfonate on barite as a function of AHS or HBO concentration (mol/l).

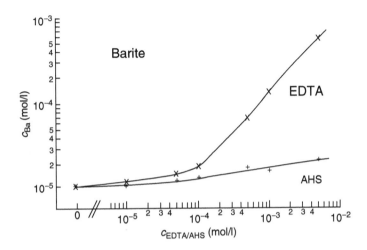

Figure 10.38 The total Ba^{2+} concentration (mole/l) in a barite dispersion as a function of EDTA or AHS concentration (mol/l).

From the enhanced ζ potential negativity observed it follows that an adsorption of the anionic surfactant is made more difficult (energetically less favorable) by the electrostatic repulsion or that a surfactant ion binding with Ba^{2+} cannot occur in the compact part of the electric double layer.

To confirm the conclusions up to now, a "reverse" experiment was performed. Instead of the addition of a complexing agent (reduction of free cation concentration) the addition of Ba^{2+} in terms of its effect on the ζ potential, collector adsorption, and floatability was examined. The barite surface was rendered more positive, the surfactant adsorption increased, and the floatability of barite improved.

The mechanism of action of complexing agents described in this section is presented schematically in Figure 10.39 and can be summarized as follows: A complexing agent removes Ba^{2+} from the mineral–water interface. This renders the ζ potential more negative, hindering the collector adsorption, with a depressing effect. This mechanism of action is well confirmed experimentally and occurs very often especially in chelating agents that are not well fitted for an adsorption in the sense of a complexation of cationic surface sites because of

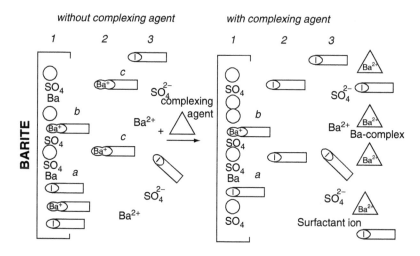

Figure 10.39 Proposed complexing mechanism in connection with removing lattice cations from the mineral surface: (a) ion exchange; (b) surface complex; (c) surface precipitate.

steric hindrances. In the systems examined, with NaDBS as collector, the more or less pronounced depressing effect of complexing agents can be caused by the described process in the following cases: barite–NTA, barite–EDTA, barite–CDTA, barite–AGTA; fluorite–NTA, fluorite–EDTA; calcite–NTA, and calcite–EDTA.

Complicated adsorption mechanisms of complexing agents occur in solutions where an anionic surfactant forms sparingly soluble compounds with polyvalent lattice ions. For example, oleate adsorption, at least at higher concentrations, must be understood as precipitation of low-solubility metal ions on the mineral surface. This surface precipitation (compact part of the electric double layer) starts at much lower oleate concentrations than the precipitation of, say, $CaOl_2$ or $BaOl_2$ in the bulk phase.

If the composition of solutions of a system such as calcite–EDTA or barite–EDTA is considered, it is clear that free Ca^{2+} ions are available in the case of calcite also in the absence of EDTA (or other chelating agents) and that they form insoluble complexes with oleate — in the first instance limited to the phase interface — which later adsorb on calcite particles as a surface precipitate. On the contrary, in barite and fluorite the cation concentration soon decreases to zero with increasing amount of chelating agent (in the mineral dispersion there are even free complexing agents present). As a result, no $BaOl_2$ (or $CaOl_2$) could be formed and the oleate adsorption density is reduced.

The reason for this different availability of free cations could be a relatively fast dissolution kinetics of calcite. In this way a rapid substitution is possible of bound Ca^{2+} ions that had been consumed to form a complex.

In this way it is understandable why no flotative separation of, say, calcite and barite is possible. As soon as calcite is present in a bimineral system, sufficient free Ca^{2+} ions are available to provide for adequate adsorption density of the collector on both minerals. A selective flotation is then not possible.

In summary, the mechanism of action of chelating agents on flotation with oleate can be described as follows: Complexing agent binds free cations in the mineral–solution interface, resulting in inhibition of formation of metal–oleate associates on slow cation delivery from the mineral surface, diminishing oleate adsorption due to surface precipitation, and a resultant depressing effect.

Since Na oleate is an anionic surfactant like NaDBS it can be also assumed that the mineral surface gets more negative as well, resulting in a diminishing collector adsorption in addition to the mechanism discussed in the above section. The results obtained with the calcite–oleate system are, however, not in favor of this hypothesis. It is therefore more credible that a probably uncharged metal–oleate complex adsorbs on the mineral relatively independently of the surface charge.

The negativeness of the ζ potential caused by the strong chelating agents NTA, EDTA, and CDTA causes a reduction of the adsorption density of the anionic collector NaDBS on the minerals examined. This results in a marked decrease in floatability. On the other hand, the use of a cationic surfactant should promote the collector adsorption and activate flotation. This was confirmed in flotation experiments with calcite and in particular with fluorite. Compatiable with the flotation results is also an increase in the collector adsorption density.

3. Cationic Surfactant–Complexing Agent Interaction

If cationic surfactants are present in the systems examined along with anionic complexing agents, a surfactant–complexing agent interaction is possible for electrostatic reasons. Such an interaction between CTAB and EDTA or CDTA was demonstrated by surface tension measurements.

The complexing agents alone are not surface-active at the pH 9 used here, but they reduce surface tension when CTAB is present in constant amount. This can be explained by the formation of an associate consisting of a cationic surfactant and an anionic complexing agent. This probably uncharged associate must be more interface-active than CTAB.

If relatively concentrated solutions of CTAB and EDTA (both at 1×10^{-3} mol/l) are mixed together, turbidity is seen first and after some time a colorless precipitate forms. This precipitate was examined after filtration and drying spectroscopically by IR spectroscopy and analytically. The IR spectrum shows only a superimposition of the CTAB and EDTA spectra without any new bands. The elemental analysis (Table 10.9) is compatible with a mixture of CTAB and EDTA associates in the ratios of 2:1 and 3:1. A surplus of hydrogen can be explained by traces of water while the nitrogen loss is not immediately visible.

Surfactant–chelating agent complexes, which could be demonstrated with all complexes investigated and not only

Table 10.9 Elemental Analysis (%) of the CTAB–EDTA Precipitation Product

Element	Found	$Na_2H(EDTA)(CTA)$ calculated	$NaH(EDTA)(CTA)_2$ calculated	$H(EDTA)(CTA)_3$ calculated
C	67.5 ± 0.2	56.2	65.42	70.42
H	12.5 ± 0.2	8.95	11.20	12.26
N	5.6 ± 0.1	6.78	6.35	6.13

with EDTA and CDTA, that are strong interface active components should also act as collectors. In this way it can be explained why chelating agents activate CTAB adsorption on hornblende and the mineral floatability simultaneously. An additional collector adsorption results from an adsorption of a CTAB–chelating agent complex on the mineral surface, not of CTAB alone.

This proposed mechanism of action is compatible with the results of electrophoretic measurements. It can be seen in Figure 10.40 that in the presence of EDTA a more negative effect of the chelating agent occurs but not until the CTAB concentration increases and this happens at a simultaneously markedly increased CTAB adsorption. This can only be explained by a preferential adsorption of the uncharged (or even anionic) CTAB–EDTA associates instead of cationic

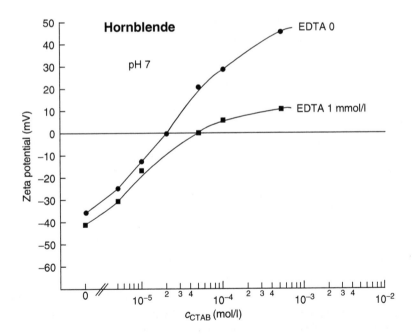

Figure 10.40 Change of ζ potential (mV) with increasing the CTAB concentration (mol/l) in the presence of EDTA.

CTAB. Two mechanisms of action of complexing agents are therefore possible in flotation with cationic surfactants:

Mechanism A: The complexing agent binds cations on the mineral surface, causing negativeness of the ζ potential, promotion of collector adsorption, and activation of floatability.

Mechanism B: The complex forms from a cationic surfactant and an anionic chelating agent, leading to adsorption of this surface-active complex on minerals, a hydrophobicity increase of the mineral surface, and activation of floatability.

Interactions with chelating agents can occur also in the case of anionic collectors. This was demonstrated, for example, with NaDBS or Na oleate and EDTA, NTA, CDTA, and AGTA. On the other hand, complexing agents such as AHS, 8-HQ, SO, and HBO do not interact with anionic surfactants.

Since the above-mentioned interaction probably does not play a role in mineral floatability with NaDBS, no further experiments have been performed in this direction. Parenthetically, a turbidity increase or even precipitation reactions as with CTAB do not happen in these systems.

4.   Complexing Agents Acting as Collectors

If the action of EDTA or CDTA is studied on the adsorption density of NaDBS in fluorite, it can be concluded from the observed reduction of Γ that both complexing agents will have a depressing effect (Figure 10.41). Flotation experiments confirmed this assumption only in the case of EDTA whereas the fluorite flotation was even activated by CDTA. The collector effect of CDTA without any addition of a surfactant is shown in Figure 10.42. In further flotation experiments it was investigated whether CDTA can serve as collector for other minerals. However, it has been shown that its effect is very selectively limited to fluorite. This selectivity remained preserved on bimineral flotation also, although the fluorite yield was markedly reduced in this case. This is certainly caused by a low ability of the mineral to release cations. In the presence of calcite, which supplies Ca^{2+}, the especially lively fluorite nearly fails to flotate with CDTA.

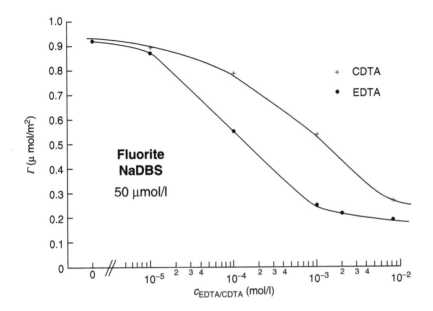

Figure 10.41 Adsorption density (Γ in 1×10^{-6} mol/m^2) of Na-dode-cylsulfonate on fluorite as a function of the EDTA or CDTA concentration (mol/l) at the constant pH 9.

A possible mechanism of the collector action of CDTA is certainly a selective adsorption of this chelating agent on fluorite. The hydrophilic molecule parts interact with the mineral surface whereas the hydrophobic cyclohexane ring of CDTA is directed toward the solution. In this way the already relatively hydrophobic fluorite surface may get sufficiently modified for a mineral flotation.

Like CDTA, 8-hydroxyquinoline acts as activator on fluorite flotation with all three surfactants tested without an observable increase in adsorption of the surfactants. As flotation experiments without a surfactant addition showed, the observed activation may be related to a selective collector effect of the complexing agent for fluorite.

The molecular structure of 8-HQ is similar to that of CDTA in the sense that it contains hydrophobic ring structures as well as hydrophilic parts.

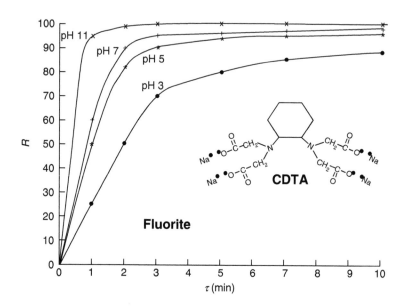

Figure 10.42   Floatability (R in %) of fluorite at a constant concentration of CDTA (1×10^{-4} mol/l) as dependent on time (τ).

5.   Effect of Complexing Agents on the
     Aggregation State of Minerals

The presented adsorption mechanisms of surfactants on the mineral surface have shown that complexing agents have a decisive influence on the formation of a hydrophobic adsorption layer necessary for the attachment process. The extent of this influence is dependent on the chemical and structural character of a particular complexing agent.

A comparison between the aggregation behavior of the system surfactant–mineral and the corresponding floatabilities is shown in Figure 10.43 and Figure 10.44 (CTAB and Na oleate, respectively) depending on the complexing agent concentration. Whereas a good agreement between the two parameters compared can be seen in all cases with Na oleate, the trend runs in opposite directions in the fluorite–CDTA system: the aggregation decreases, but the floatability gets activated.

The behavior of the system CTAB–hornblende is similar. In the system CATB–fluorite the mineral dispersion gets de-

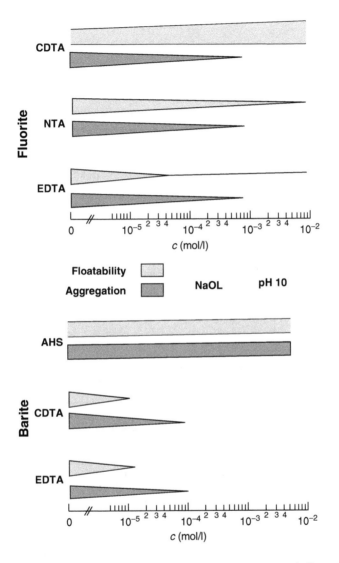

Figure 10.43 Relationship between aggregation and floatability of fluorite, or barite dispersion in the presence of Na-oleate as a function of the concentrations of the complexing agents (mol/l).

stabilized with the increasing complexing agent concentration. Almost no aggregation phenomena occurred in the presence of complexing agents in combination with NaDBS.

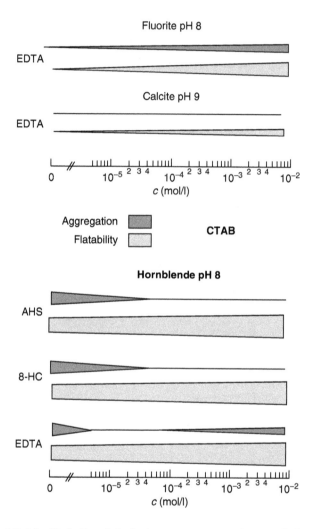

Figure 10.44 Relationship between aggregation and floatability of fluorite, calcite or hornblende dispersion in the presence of CTAB as a function of the concentrations of the complexing agents (mol/l).

V. SURFACTANT–LATTICE ION INTERACTIONS

It is known that polyvalent lattice cations influence the stability of solid dispersions and thus play an important role in surfactant adsorption during flotation in the presence of sah-type minerals.

A. Minerals with PDI H$^+$ and OH$^-$

Most papers on the relationship between the adsorption dens-ity of surfactant and polyvalent ions concentrate on flotation and electrokinetic evaluations [98–101]. A direct determin-ation of the adsorption density of surfactants as dependent on Na$^+$, Ca^{2+}, and Fe^{3+} concentrations was performed by Dobias [86]. His results are shown in Figure 10.45. Under the assumption that N$^+$, Cl$^-$, Br$^-$, and cetyltrimethylammo-nium ion (the surfactant used) are not the PDI, he modified the Stern–Graham equation [102] to the form:

$$\frac{\Gamma_{St}}{\Gamma_{Me}} = \frac{r_{St}c_{St}}{r_{Me}c_{Me}} \exp\left(\frac{(F\zeta - \phi_{St} + \phi_{Me})}{RT}\right) \tag{23}$$

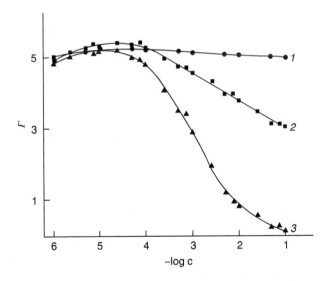

Figure 10.45 Adsorption density (Γ) of CTAB (Γ in 1×10^{-6} mol/g) on quartz as a function of the concentrations (mol/l) of Na$^+$ (1), Ca^{2+} (2), and Fe^{3+} (3).

where Γ_{St} and Γ_{Me} are the adsorption densities of the surfactant and the polyvalent ion, respectively, and ϕ_{St} and ϕ_{Me} are their specific adsorption potentials. For verification of the equation using experimental data, the following simplifications were made: r_{St}/r_{Me} and F/RT are constant, $\ln(c_{St}/c_{Me})$ varies linearly with the change of $\ln(\Gamma_{St}/\Gamma_{Me})$ (this was proved), ϕ_{St} can be neglected since $\theta \ll 1$ (association of hydrocarbon chains cannot be assumed), and ϕ_{Me} remains unchanged within the range of concentrations of $CaCl_2$ and $FeCl_3$ during adsorption, as confirmed by the linearized form of the adsorption isotherms of Ca^{2+} and Fe^{3+}. Because of low coverage of the SiO_2 surface by Ca^{2+} and Fe^{3+} ions, a homogeneous surface of the solid is assumed. A linear relationship between $\ln(\Gamma_{St}/\Gamma_{Me})$ and $\zeta(\zeta \equiv \psi_{\delta})$ confirmed the exchange-mechanism of the adsorption in the EDL without any lateral interaction for $\theta \ll 1$.

Measurements using Na tridecylate showed that the validity of Equation (23) for SiO_2 activation by polyvalent cations is limited by the concentration at which insoluble Ca salts of the surfactant start to occur in the system (Figure 10.46, curve 1). When the SiO_2 surface gets activated in a solution containing Ca^{2+} at first, and the solution is then replaced by a surfactant solution, Ca salts do not precipitate and the adsorption isotherm has its expected form (Figure 10.46, curve 2). The activation of quartz by Ca^{2+}, Sr^{2+}, and Ba^{2+} ions in oleate solutions was examined by Estefan and Malati [104]. As follows from the temperature dependence (at 10°C, 30°C, and 45°C) at pH 10.5 on the SiO_2 surface in the absence and presence of activated Ca^{2+} ions, the adsorption of Na oleate is an exothermic process and is of a physical nature. Assuming a blow surface coverage, the adsorption heat from the Langmuir adsorption isotherm was found to be 9 kJ for the activation reaction. At 30°C the increase in the oleate adsorption density at a constant Ca^{2+} concentration of 2×10^{-4} mol/l with increasing Na oleate concentration ($1-4 \times 10^{-5}$ mol/l) suggests lateral interactions between the adsorbed molecules. The adsorption density of oleate increases in the order $Ca^{2+} < Sr^{2+} < Ba^{2+}$.

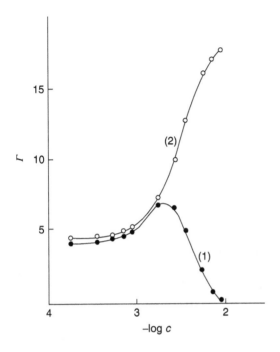

Figure 10.46 Adsorption density (Γ) of Na-tridecylate (NaTD) (Γ in 1×10^{-8} mol/g) on quartz as a function of the concentration of Ca^{2+} (mol/l). 1, Ca^{2+} and NaTD together; 2, Ca^{2+} and NaTD separate.

B. Minerals with PDI Lattice Ions

The solubility of salt-type minerals in water has an influence on the stability of their dispersions. The presence of lattice ions of one component affects the solubility product of another component and this, as well as the formation of hydrolytic products of special ion species, has a substantial influence on the transition states between three systems:

System 1: with PDI H^{+} and OH^{-} and the same mineral group (e.g., α-Al_2O_3– SiO_2)

System 2: with PDI H^{+} and OH^{-} and lattice ions of different mineral groups (e.g., CaF_2–SiO_2)

System 3: with potential-determining lattice ions of the same mineral group (e.g., $CaCO_3$–$Ca_{10}(PO_4)_6F_2$)

Mixtures of salt-type minerals are at present difficult to separate by flotation. Since selective surfactant adsorption at a mineral is the decisive step for selective flotation, many investigations of adsorption at salt-type minerals have been made [20,105–108]. Owing to the limited solubility of the Ca salts used, at a certain surfactant concentration precipitation starts, along with adsorption. For this reason, most published isotherms give the abstraction (adsorption plus precipitation).

1. Anionic Surfactant Precipitation

Flotation of sparingly soluble salts of divalent cations is usually carried out at neutral to alkaline pH. As collectors, along with carboxylates, also alkyl sulfates, alkanesulfonates, alkylbenzenesulfonate, and alkylammonium salts have been used [107]. Many anionic surfactants form sparingly soluble compounds with polyvalent cations. Mostly Ca salts have been investigated since they play an important role not only in flotation but especially in oil extraction and washing processes. The degree of Ca mineral solubility as related to pH is demonstrated in Figure 10.47.

In the system mineral-anionic surfactant the adsorption on the mineral is accompanied by the precipitation of the surfactant with Ca ions [105,106]. From the reactions with fluorite, e.g.,

$$CaF_2 \rightarrow Ca^{2+} + 2F^- \quad \text{(mineral dissolution)} \tag{24}$$

$$Ca^{2+} + 2T^- \rightarrow CaT_2 \quad \text{(precipitation)} \tag{25}$$

Figure 10.48a shows the turbidity of the sol and the ζ potential value of the precipitate depending on the Na dodecylsulfate (NaDS) equilibrium concentration after 2 h.

The total Ca concentration was 10^{-4} mol/l, the pH 10. The turbidity increased with rising surfactant concentration owing to the formation of increasing precipitation. At about the CMC of the surfactant, the turbidity reached a maximum and started to decrease. The ζ potential was strongly negative, with its charge maximum at the CMC.

After filtration of the sols through a 0.22-μm Millipore filter, we measured the Ca and the surfactant concentrations

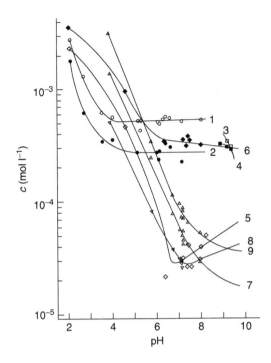

Figure 10.47 Lattice ions concentration in the saturated solutions of mineral–water dispersions. Fluorite: 1, F^-; 2, Ca^{2+}. Calcite: 3, total carbonate; 4, Ca^{2+}. Scheelite: 5, WO_4^{2-}; 6, Ca^{2+}. Fluorapatite: 7, total phosphate; 8, F^-; 9, Ca^{2+}.

in the clear filtrate. The results (Figure 10.48b) show the precipitation boundary.

It has been shown [109,110] that at surfactant concentrations below the CMC, the beginning of precipitation can be described by a solubility product $L = C_{Ca} C_T^2$. Above the CMC, the slope of the precipitation boundary reverses.

In Figure 10.49, the Ca^{2+} and F^- concentrations and the abstracted NaDS concentration are given in relation to the NaDS equilibrium concentration in the system fluorite–NaDS at pH 10 and 20°C after a 24-h conditioning. The Ca^{2+} and F^- concentrations are constant at low surfactant concentrations. The excess of F^- in the solution relative to the crystal stoichiometry shows an excess of Ca^{2+} on the mineral surface.

The excess Ca^{2+} concentration amounts to about 1.03×10^{-5} mol/m^2.

When the surfactant concentration crosses the precipitation boundary the Ca concentration decreases and that of F^- increases, indicating precipitation.

From the reactions in the system (Equations (24) and (25)) it follows that

$$c_{Ca} - c_{Ca}^0 = \Delta Ca = -\frac{1}{2}\Delta c_{Tpre} + \Delta c_{Ca-dis} \qquad (26)$$

$$\Delta c_{Ca} = -\frac{1}{2}\Delta c_{Tpre} + \frac{1}{2}\Delta c_F \qquad (27)$$

$$\Delta c_{Tpre} = -2\Delta c_{Ca} + \Delta c_F \qquad (28)$$

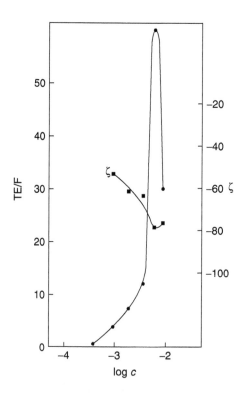

Figure 10.48a   *(Continued)*

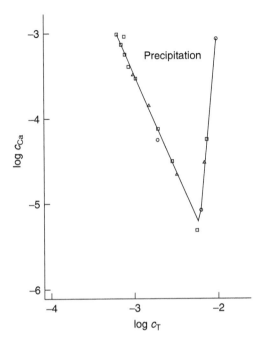

Figure 10.48b (a) Turbidity (TE/F) and ζ potential (mV) of Ca-dodecyl sulfate as a function of the equilibrium surfactant concentration (mol/l) at a constant initial concentration of Ca^{2+} (1×10^{-4} mol/l).
(b) The $\log c_{Ca}$ ($CaCl_2$)$-\log c_T$(NaDS) precipitation domain, c in mol/l.

Using the last equation it is possible to calculate the adsorption isotherm from the abstraction data.

In Figure 10.50 the abstraction and the adsorption densities are given as a function of the NaDS equilibrium concentration. Most of the surfactant abstraction results from precipitation. The adsorption density decreases above the CMC.

Similar results were obtained in the system fluorite–Na dodecylsulfonate (NaDSO) at concentrations below the CMC.

All reactions are similar in both systems, but they occur at a lower surfactant concentration with NaDSO. Even the adsorption density at the beginning of precipitation and at the maximum are equal in both systems.

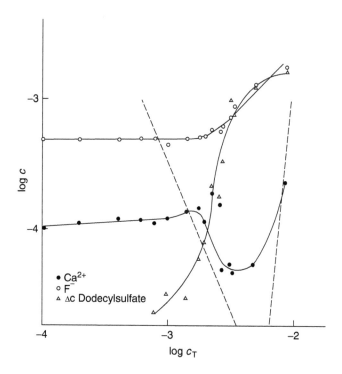

Figure 10.49 The lattice ion concentrations and the decrease of surfactant concentration (mol/l) in a fluorite dispersion as a function of the Na-dodecyl sulfate (c_T in mol/l). Dashed curve: Ca-dodecyl sulfate precipitation boundary.

2. Adsorption of Anionic Surfactant

The adsorption mechanism of surfactants at an equilibrium concentration reduced due to precipitation can be described using the example of NaDS and fluorite.

At low surfactant concentrations no precipitation occurs since the Ca concentration is too low, according to the precipitation boundary shown as a dashed line (Figure 10.49). With increasing surfactant concentration the lattice ion concentrations remain constant until the Ca concentration curve reaches the precipitation boundary. In this concentration range the surfactant adsorption results from ion pair formation, ion exchange with OH^- ions, and hydrophobic inter-

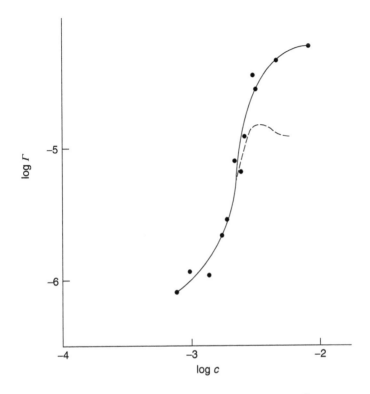

Figure 10.50 Abstraction density (Γ in mol/m^2) of Na-dodecyl sulfate on fluorite as a function of the equilibrium concentration (mol/l) of the surfactant. Dashed curve, calculated adsorption density.

actions. A significant exchange of the surfactant and F$^-$ ions does not happen since in that case the F$^-$ concentration would have to increase and that of Ca^{2+} should decrease because of the mineral solubility product.

On a further increase in NaDS concentration the surfactant starts to precipitate. The Ca concentration is reduced owing to the precipitation up to a concentration of the precipitation boundary, that of F$^-$ elevates. Differences from the precipitation diagram at the highest NaDS concentrations result from the influence of a high ionic strength that is reached due to Na$^+$ and F$^-$ concentrations.

The equation of all possible reactions under the CMC are as follows where T and T_{sol} is monovalent anionic surfactant adsorbed and surfactant in the solution, respectively:

$$T^-_{Sol} \rightarrow fluorite - T^- \quad \text{adsorption 1} \tag{29}$$

$$T^-_{Sol} + fluorite - F^- \rightarrow F^- + fluorite - T^- \quad \text{adsorption 2} \tag{30}$$

$$T^-_{Sol} + fluorite - OH^- \rightarrow OH^- + fluorite - T^- \quad \text{adsorption 3} \tag{31}$$

For solutions with a constant volume, the following concentration balances result from the equations:

$$\Delta c_{Ca} = c_{Ca} - c^0_{Ca} = \Delta c_{Ca-ppt} + \Delta c_{Ca-sol} \tag{32}$$

$$\Delta c_F = c_F - c^0_F = \Delta c_{F-sol} + \Delta c_{F-ads2} \tag{33}$$

$$\Delta c_T = c^0_T - c_T = \Delta c_{T-ppt} + \Delta c_{T-ads1} + \Delta c_{T-ads2} + \Delta c_{T-ads3} \tag{34}$$

as well as the following equations:

$$2\Delta c_{Ca-ppt} = \Delta c_{T-ppt} \tag{35}$$

$$2\Delta c_{Ca-sol} = \Delta c_{F-sol} \tag{36}$$

$$\Delta c_{T-ads2} = -\Delta c_{F-ads2} \tag{37}$$

$$\Delta c_{T-ads3} = -\Delta c_{OH-ads3} \tag{38}$$

where c is the concentration in the suspension, $c_{Ca}{}^0$, $c_F{}^0$ are the concentrations of Ca and F, respectively, in the suspension without the surfactant. $c_T{}^0$ is the starting surfactant concentration, and Δc_{Ca-ppt} is the Ca concentration change due to surfactant precipitation.

From Equations (32), (35), and (36) it follows that

$$\Delta c_{Ca} = -\frac{1}{2}\Delta c_{T-ppt} + \frac{1}{2}\Delta c_{F-sol} \tag{39}$$

Using Equation (33)

$$\Delta c_{Ca} = -\frac{1}{2}\Delta c_{T-ppt} + \frac{1}{2}\Delta c_F - \frac{1}{2}\Delta c_{F-ads2} \tag{40}$$

$$\Delta c_{T-ppt} = -2\Delta c_{Ca} + \Delta c_F - \Delta c_{F-ads2} \tag{41}$$

Suppose $\Delta c_{F-ads2} = 0$, i.e., there is no adsorption owing to surfactant ion–F^- ion exchange like in the range of the surfactant concentration under the precipitation boundary, then it follows that

$$\Delta c_{T-ppt} = -2\Delta c_{Ca} + \Delta c_F \tag{42}$$

Using Equation (42) and the experimentally obtained value Δc_T, the adsorbed surfactant amount Δc_{T-ads} can be calculated:

$$\Delta c_{T-ads} = \Delta c_{T-ads1} + \Delta c_{T-ads2} + \Delta c_{T-ads3} \tag{43}$$

$$\Delta c_{T-ads} = \Delta c_T - \Delta c_{T-ppt} \tag{44}$$

$$\Delta c_{T-ads} = \Delta c_T + 2\Delta c_{Ca} - \Delta c_F \tag{45}$$

The abstraction density Γ_{abs} can be calculated from the diminution of surfactant concentration c_T:

$$\Gamma_{abs} = \Delta c_T V/Am \tag{46}$$

where V is the suspension volume, A the specific mineral surface, and m the mineral amount weighed.

For the adsorption density of nonhydratized ions in the Stern layer the following equation holds (according to Grahame) [103,111]:

$$\Gamma = 2rc \exp\left(-\Delta G_s/RT\right)$$

where r is the radius of ions adsorbed in the Stern layer, c the ion concentration in the solution, and ΔG_s the adsorption energy of one ion in the Stern layer. This equation can be also used for a rough estimate of the surfactant adsorption energy.

Table 10.10 shows the surfactant concentrations of the systems surfactant–mineral investigated and the calculated adsorption energies at an adsorption density of $1 \, \mu mol/m^2$.

Table 10.10 Concentration in the Solution and Adsorption Energy of Some Surfactants at an Adsorption Density of 1 mol/m^2

Mineral	Surfactant	c_T/M	G_s (kJ mol^{-1})
Fluorite	Na-DSO	3×10^{-4}	-21.4
Fluorite	Na-DS	1×10^{-3}	-18.5
Fluorite	Na-TDS	4×10^{-5}	-26.3
Fluorite	Na-Laurate	6×10^{-5}	-25.4
Calcite	Na-DSO	2×10^{-4}	-22.4
Calcite	Na-DS	7.5×10^{-4}	-19.2
Scheelite	Na-DSO	3.5×10^{-4}	-21.0
Fluoroapatite	Na-DSO	2×10^{-3}	-16.8

As ion radius a polar group radius of 2.5×10^{-10} m was employed; the temperature was 20°C.

At the same surfactant chain length the rank order of adsorption energy diminution is carboxylate > sulfonate > sulfate. This corresponds to the order of solubility increase of the Ca–surfactant salts.

The energy difference of -7.8 kJ/mol between the adsorption of NaDS and NaTDS (tetradecyl sulfate) results from the association energy of two additional CH$_2$ groups of NaTDS. A maximum association energy value of one CH$_2$ group $-1.39RT$ is given [112], i.e., 3.39 kJ/mol at 20°C. Since the experimental value is somewhat higher, an almost complete association can be presumed. This can also be expected in other surfactants used because of a similar structure.

The whole association energy of hydrocarbon chains is higher for all systems than the adsorption energy. The adsorption starts on energetically favorable sites. Higher adsorption densities are achieved only owing to the association of hydrocarbon chains of the surfactants. The interaction energies of polar groups with the surface are not high; this might be the reason that no measurable exchange was found for lattice ions for surfactant ions in any system. The main operating adsorption mechanism is thus the exchange of OH$^-$ ions for surfactant ions and a further adsorption due to an association accompanied by a surface negativity. In fluorite, many exchangeable OH^- ions are present on the surface; therefore a double layer forms after a monolayer is complete. In other minerals, a higher negative charge of the mineral surface limits the maximum adsorption density.

The start of a precipitation can be shifted toward higher adsorption densities values by choosing surfactants with a better solubility of their Ca salts. The differences are, however, small and it can be assumed that the same interactions are responsible for adsorption and for precipitation: the interactions of a polar group with Ca^{2+} ions and the surfactant association.

3. Flotation Experiments

With regard to flotation of suspensions that correspond to the adsorption experiments concerning both the lattice ion and the surfactant concentrations, a good agreement usually exists between the mineral floatability and the surfactant adsorption isotherms.

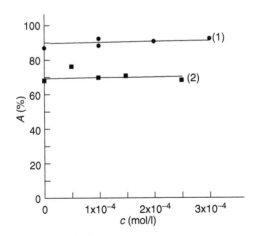

Figure 10.51   Floatability (*A*) of fluorite (1) and calcite (2) as a function of the concentration of Ca-dodecylsulfonate precipitate (mol/l).

High lattice ion concentrations, especially of polyvalent ions, reduce the surfactant adsorption because of their own precipitation in the solution. In addition, phosphates form soluble complexes with Ca^{2+} ions and reduce surfactant precipitation in this way. To study the influence of precipitation on the mineral surface, flotation experiments were carried out with fluorite and calcite. As shown in Figure 10.51, the mineral floatability does not change in either system with increasing precipitation product concentration. The deposition of precipitation products at the mineral–water interface was not demonstrated in the system studied. This probably happens as a result of the strongly negative ζ potentials of the precipitate and the mineral after the surfactant adsorption.

REFERENCES

1. B. Dobias, *Surfactant Adsorption on Minerals Related to Flotation, in Structure and Bonding Series*, Vol. 56, Berlin: Springer-Verlag, 1984, p. 92.
2. L. Warren, *Chem. Aust.*, 12(44), 316, 1977.
3. M. Beccari and R. Passino, *L'ingegneure*, 12, 1, 1970.
4. O. Stern, *Z. Electrochem*, 30, 508, 1924.
5. DC G"bibrahame, *Chem. Rev.*, 41, 441, 1947.
6. B. Dobias, *Erzmetall* 26, 173, 1973.
7. R.J. Hunter and A.E. Alexander, *J. Colloid Sci.* 18, 820, 1963.
8. P. Delahay, *Double Layer and Electrode Kinetics*, New York: John Wiley & Sons, 1966, p. 33.
9. D.N. Furlong, D.E. Yates, and T.W. Healy, *Electrodes of Conductive Metallic Oxides, Part B*, Amsterdam: Elsevier, 1981, p. 367.
10. J.M.W. Mackenzie. *Min, Sci. Eng.*, 25, 1971.
11. B. Dobias, *Tenside Deterg.* 13, 131, 1976.
12. B. Dobias, Uber die Adsorption der Tenside an Mineralien bei der Flotation, in Symp. über Adsorption aus Lösungen, Arbeitstagung der Kolloidgesell. Bochum, 1980.
13. J. Lyklema and JTG Overbeek, *J. Colloid Sci.* 16, 501, 1961.
14. P. Somasundaran and E.D. Goddard, In: *Modern Aspects of Electrochemistry*, Vol. 13, (BE Conway and JOM. Bockris, eds.), New York: Plenum Press, 1979, p. 207.

15. M.A.F. Payman, J.W. Bowden, and A.M. Posner, *Aust. J. Soil Res.*, 17, 191, 1979.
16. D.W. Fuerstenau and S. Raghavan, In: *Flotation, AM Gaudin Memorial*, Vol. 1 (MC Fuerstenau ed.), American Institute of Mining, Metallurgical and Petroleum Engineers: New York, 1976, p. 21.
17. K.A. Wierer and B. Dobias, *J. Colloid Interf. Sci.*, 122, 171, 1987.
18. S.G. Bussetti, M Tschapek, and AK Helmy, *J Electroanal Chem* 36, 507, 1972.
19. P. Schulz and B. Dobias, In: *Proceedings of the World Surfactant Congress*. Munich, München: Carl Hanser, 1984, p. 385.
20. P. Schulz and B. Dobias, In: *Proceedings of the Fifteenth International Mineral Process Congress*, Vol. 2, Flotation and Hydrometallurgy, Cannes, 1985, p. 16.
21. D.A. Griffiths and D.W. Fuerstenau, *J. Colloid Interf. Sci.*, 80, 271, 1981.
22. T. Wakamatsu and Sh. Mukai, *AIChE. Symp. Ser.*, 71(150), 81, 1975.
23. P.A. Siracusa and P. Somasundaran, *J. Colloid Interf. Sci.*, 114, 184, 1986.
24. D.J.A. Williams and K.P. Williams, *J. Colloid Interf. Sci.*, 65, 79, 1978.
25. B. Dobias, *Tenside Deterg*, 9, 322, 1972.
26. B. Dobias, *Tenside Deterg*, 13, 131, 1976.
27. J.J. Prédali and J.M. Cases, Thermodynamics of the adsorption of collectors. In: Proceedings of the Tenth International Mineral Processing Congress, Institute of Mining and Metallurgy: London, Paper 1973, p. 33.
28. J.E. Scamehorn, R.S. Schechter, and W.H. Wade, *J. Colloid Interf. Sci.*, 85, 463, 1982.
29. P. Somasundaran, T.W. Healy, and D.W. Fuerstenau, *J. Phys. Chem.*, 68, 3562, 1964.
30. D.W. Fuerstenau, *Pure. Appl. Chem.*, 24, 135, 1970.
31. H. Boudriot, P. Matthé, and H.A. Schneider, *Freib. Forsch.-H. Reihe A*, 546, 25, 1971.
32. H. Boudriot, H.A. Schneider, and P. Matthé. *Neue Bergbautechnik* 1, 946, 1971.
33. M. Volmer, *Z. Phys. Chem.*, 115, 253, 1925.
34. A.A. Zukhovickij, *J. Phys. Chem.*, 18, 214, 1944.
35. T.L. Hill, *J. Chem. Phys.*, 20, 141, 1952.

36. E. Orthgiess, Complexing Agents as Reagent in the Mineral Flotation, Ph.D. thesis, University of Regensburg, 1991.
37. K. Volke, *Tenside Deterg*, 28, 52, 1991.
38. B. Dobias, *Freib. Forsch-H. Reihe A.* 401, 123, 1966.
39. B. Dobias, *Colloid Polym. Sci.*, 255, 682, 1977.
40. B. Dobias, *Colloid Polym. Sci.*, 256, 465, 1978.
41. B. Dobias. In: *Proceedings of the Fourth International Congress on Surface Active Substances*, Brussels 1964, London: Gordon & Breach, 1967, p. 681.
42. B. Dobias, *Freib. Forsch-H. Reihe. A.* 335, 7, 1965.
43. J. Kellar, W.M. Cross, and J.D. Miller, *Appl. Spectrosc.*, 43, 1456, 1989.
44. M.J. Rosen, *Surfactants and Interfacial Phenomena*, New York: John Wiley, 1989, Chapter 2.
45. K.H. Rao, J.M. Cases, and K.S.E. Forssberg, *J. Colloid Interf. Sci.*, 145, 330, 1991.
46. B.M. Mondgil, H. Soto, and P. Somasundaran, *Reagents in Mineral Technology*, Vol. 27, New York: Marcel Dekker, 1988, p. 79.
47. E. Richter and H.A. Schneider, *Freib. Forsch-H. Reihe. A* 564, 31, 1977.
48. F.J. Trogus, R.S. Schechter, and W.H. Wade, *J. Colloid Interf. Sci.*, 70, 293, 1979.
49. H. Schubert and H. Baldauf, *Tenside Deterg*, 4, 172, 1967.
50. P. Somasundaran and K.P. Ananthapadmanabhan. In: *Solution Chemistry of Surfactants*, Vol. 2 (K.L. Mittal, ed.), New York: Plenum Press, 1979, p. 777.
51. J.H. Harwell, J.C. Heskins, R.S. Schechter, and W.H. Wade, *Langmuir*, 1, 251, 1985.
52. T. Gu, J. Du, and Y. Gao, *J. Chem. Soc. Faraday Trans.*, 183, 267, 1987.
53. P. Mukerjee and A. Anavil, In: *Adsorption on Interfaces*, ACS Symposium Series 8 (KL Mittal, ed.), 1975, p. 107.
54. P.H. Elworthy and K. Mysels, *J. Colloid Interf. Sci.*, 121, 331, 1966.
55. F.H. Sexsmith and H.J. White, *J. Colloid Interf. Sci.*, 14, 630, 1959.
56. P. Somasundaran and H.S. Hanna, In: *Improved Oil Recovery by Surfactant and Polymer Flooding*, New York: Academic Press, 1977, p. 205.
57. H.S. Hanna, Contribution to the Flotation of Phosphate Ores, Ph.D. thesis, American Shams University, Cairo, 1968.

58. A.M. Gaudin and D.W. Fuerstenau, *Trans. AIME*, 202, 958, 1955.
59. M.A. Cook, *J. Colloid Sci.*, 28, 463, 1982.
60. P. Roy and D.W. Fuerstenau, *J. Colloid Interf. Sci.*, 26, 102, 1968.
61. O. Meigren, R.J. Gochin, H.L. Shergold, and J.A. Kitchener, In: Proceedings of the Tenth International Mineral Processing Congress, London, 1973, p. 541.
62. E. Richter, H.A. Schneider, and G Wolf, *Freib Forsch-H. Reihe A*, 593, 1, 1977.
63. B. Ball and D.W. Fuerstenau, *Disc. Faraday. Soc.*, 52, 361, 1971.
64. K.A. Wierer, Adsorption of H^+ and OH^- Ions and Surfactants on Minerals, Ph.D. thesis, University of Regensburg, 1987.
65. W. v Rybinski and M. Schwuger, *Ber. Bunsenges. Phys. Chem.*, 88, 1148, 1984.
66. K.I. Minakis and H.L. Shergold, *Int. J. Min. Process*, 14, 161, 1985.
67. J.D. Miller, ME Wadsworth, M. Misra, and J.S. Hu, Flotation chemistry of the fluorite–oleate system. In: *Principles of Min Flotation* (MH Jones and JT Woodcock, eds.), *Trans. Austr. Inst. Min. Metall* 40, 31, 1984.
68. C. du Rietz, In: *Proceedings of the Eleventh International Mineral Processing Congress*, Cagliari, 1975, p. 29.
69. H. Schubert and W. Schneider, In: *Proceedings of the Eighth International Mineral Processing Congress,* Leningrad, S-9, 1, 1968.
70. W. Schneider, Ph.D. thesis, Bergakademie Freiberg, 1968.
71. P. Matthé, Ph.D. thesis, Bergakademie Freiberg, 1971.
72. H. Rupprecht, *Kolloid Z.Z. Polymere* 249, 1127, 1971.
73. P. Mukerjee, *Adv. Colloid Interf. Sci.*, 1, 241, 1967.
74. H. Schubert, *Freib Forsch-H. Reihe A* 504, 7, 1971.
75. B. Dobias, J Spurny, and J Cibulka, *Colln. Czech. Chem. Commun*, 31, 166, 1966.
76. O. Ozcan, A.N. Bulutcu, O. Recepoglu, and R. Tolun. In: *Proceedings of the Third International Mineral Processing Symposium*, Instanbul (G Onal, ed.), Instanbul Technical University, 1990, p. 388.
77. P. Somasundaran and D.W. Fuerstenau, *J. Phys. Chem.*, 79, 90, 1966.
78. D.W. Fuerstenau and H.J. Modi, *J. Electrochem. Soc.*, 106, 336, 1959.

79. T. Gu and H. Rupprecht, *Colloid Polym. Sci.*, 269, 506, 1991.
80. Γ. Raab, Flotation and adsorption of surfactants on oxides as dependent on the pH value, Diploma thesis, University of Regensburg, 1991.
81. B. Dobias, J. Spurny, and E. Freudlova, *Coll. Czech. Chem. Commun.*, 24, 3668, 1959.
82. D.W. Fuerstenau, T.W. Healy, and P. Somasundaran, *Trans AIME*, 229, 321, 1964.
83. R.W.M. Lai and D.W. Fuerstenau, *Trans. AIME*, 260, 104, 1976.
84. I. Iwasaki, S.R. Cooke, and A.F. Colombo, Flotation characteristics of goethite, Rep Invest 5593, US Bureau of Mines, 1960.
85. A.S. Peck, L.H. Raby, and M.E. Wadsworth, *Trans. SME / AIME*, 253, 301, 1966.
86. B. Dobias. Berichte VI. Int. Kongr. grenzflächenaktiver Stoffe, Zürich 1972, Carl Hanser, Munich, 1973, p. 563.
87. J. Spurny, B. Dobias, and V. Hejl, *Chemicke Listy* 51, 215, 1957.
88. W.A.P. Luck, *Progr. Coll. Polym. Sci.*, 65, 6, 1978.
89. R. Petzenhauser. Effect of counter ions on the adsorption behaviour of cetylpyridinium cation on quartz, Diploma thesis, University of Regensburg, 1989.
90. P. Matthé and H.A. Schneider, *Freib Forsch-H. Reihe A.*, 564, 9, 1977.
91. K.A. Wierer and B. Dobias, *Progr. Colloid Polym. Sci.*, 76, 283, 1988.
92. A.L. Underwood and E.W. Anacker, *J. Colloid Interf. Sci.*, 117, 242, 1987.
93. D.R. Nagaraj, The chemistry and application of chelating or complexing agents in mineral separation. In: *Surfactant Science Series 27*, New York: Marcel Dekker, 1987.
94. Pradip, *Min. Metall. Process*, 1980, 80.
95. X. Wang and E. Forssberg, *J. Colloid Interf. Sci.*, 140, 217, 1990.
96. H. Baldauf, *Freib Forsch-H. Reihe A.*, 619, 1, 1980.
97. T.L. Wei and RW Smith, *Int. J. Min. Process*, 21, 93, 1987.
98. D.M. Hopstock and G.E. Agar, *Trans. AIME*, 241, 466, 1968.
99. G.Y. Onoda and D.W. Fuerstenau, In: Proceedings of the Seventh International Mineral Processing Congress, New York, 1964, p. 301.
100. H.J. Modi and D.W. Fuerstenau, *Trans. AIME*, 217, 381, 1960.
101. M Murata, *J. Chem. Soc. Japan, Chem. and Ind. Chem.*, 1979, 984.

102. J.M.W. Mackenzie, *Min. Sci. Eng.*, 1971, 25.
103. S.F. Estefan and M.A. Malati, *Trans. Sect. C. Int. Min. Metal*, 82, 237, 1973.
104. H. Oberndorfer. Adsorption of Anionic Surfactants on Calcium Salts, Ph.D. thesis, University of Regensburg, 1989.
105. H. Oberndorfer and B. Dobias, *Colloid Surf.*, 41, 69, 1989; *Progr. Coll. Polym. Sci.*, 76, 286, 1988.
106. K. Ananthapadmanabhan and P. Somasundaran, *Colloids Surf* 13, 151, 1985.
107. K. Hannamantha Rao, B-M Antti, J.M. Cases, and E. Forssberg, In: Sixteenth International Mineral Processing Congress, Stockholm, 1988, (E Forssberg, ed.), Amsterdam: Elsevier, Part A, p. 819.
108. M. Baviere, B. Bazin, and R. Aude, *J. Colloid Interf. Sci.*, 92, 580, 1983.
109. J.M. Peacock and J. Matijevic, *J. Colloid Interf. Sci.*, 77, 548, 1980.
110. H. Schubert. Aufbereitung fester mineralischer Rohstoffe, Vol 2, VEB Dt Verlag Grundstoffindustrie, Leipzig, 1977.
111. S.I. Chou and J.H. Bae, *J. Colloid Interf. Sci.*, 96, 192, 1983.

11

Flocculation and Dispersion of Colloidal Suspensions by Polymers and Surfactants: Experimental and Modeling Studies

P. Somasundaran[a], Venkataramana Runkana[a,b], and P.C. Kapur[b]

[a]Langmuir Center for Colloids and Interfaces, Columbia University, New York, 10027, USA
[b]Tata Research Development and Design Centre, 54-B, Hadapsar Industrial Estate, Pune, 411013, India

I. INTRODUCTION

Flocculation and dispersion of colloidal suspensions are important unit operations in many industries such as pulp and papermaking [1], mineral and ceramics processing [2,3], and water treatment [4], to name a few. Flocculation plays a major role also in the fate and transport of contaminants in aquatic environments [5]. Polymers and polyelectrolytes, inorganics, and surfactants are commonly used either to flocculate or to disperse a suspension. Flocculation is a complex phenomenon that involves several steps or subprocesses occurring sequentially or concurrently. These include: (i) mixing of particles and polymers or surfactants in solution, (ii) adsorption of the polymer or the surfactant molecules on particle surfaces, (iii) reconformation of adsorbed chains on the surface, (iv) formation of aggregates due to salt, polymer, or surfactant, (v) breakage of flocs by shear, (vi) restructuring of flocs, (vii) reflocculation of broken flocs, (viii) desorption of polymer under high shear, and (ix) subsidence or sedimentation or creaming of flocs.

The process variables that are commonly measured to evaluate the effectiveness of flocculation are settling rate of flocs, percent solids settled, sediment volume, viscosity, turbidity or supernatant clarity, depending on the industrial application. All these output variables are actually manifestations of the floc or aggregate size distribution and the shape and structure of flocs produced during the flocculation process. The floc size distribution depends strongly on the subprocesses mentioned above and is a function of the characteristics of solids and polymer, solvent, solution chemistry, temperature, and equipment geometry. As a result, the number of variables affecting the floc size distribution is quite high.

Population balance approach is perhaps the most suitable for modeling aggregation phenomena in colloidal suspensions. Starting from the classical work of Smoluchowski [6], it has been employed extensively for the modeling of various agglomeration and aggregation processes, including coagulation and flocculation. As the complete description of polymer-induced flocculation is quite complex, attempts are made to model individual aspects of flocculation separately. Mathematical models for pure aggregation, floc fragmentation, polymer adsorption, and suspension hydrodynamics have been developed and validated with varying degrees of success. However, in general, there is an absence of an integrated model for the flocculation process. Moreover, the models developed thus far have been only for salt-induced coagulation and have not incorporated the most important component of the flocculation process, namely, adsorption of polymers or surfactants and their conformation or orientation at the solid–liquid interface.

In this chapter, we will first discuss the results of experimental studies of the effect of polymers and surfactants on flocculation and dispersion processes and then present an integrated approach for modeling flocculation of colloidal suspensions in the presence of polymers using the population balance approach.

II. STABILITY OF SUSPENSIONS IN THE PRESENCE OF SURFACTANTS, POLYMERS, AND THEIR MIXTURES

Charge generation of particles in aqueous media is primarily a consequence of hydrolysis of pH-dependent surface species. If M represents the metal, charge generation as a function of pH can be written as follows:

$$MOH = H^+ + MO^-$$

$$MOH + H^+ = (MOH)_2^+$$

It is clear that under acidic pH conditions, surfaces are positively charged and under basic pH conditions negatively charged. The pH at which net charge of the surface is zero is referred to as the point of zero charge (PZC). A general rule of thumb is that aggregation occurs in systems where the zeta potential of the particles is less than 14 mV in magnitude.

The electrochemical nature of surfaces can be modified suitably by the adsorption of surface-active agents such as polymers and surfactants. These molecules change properties of the solid–liquid interface by adsorbing onto particle surfaces and affect the behavior of suspensions. Macroscopic properties such as solids settling rate and suspension turbidity are strongly dependent on the structure of the adsorbed layer. In this section, the role of polymers and surfactants and their mixtures on the stability of suspensions is illustrated with experimental results obtained for colloidal alumina suspensions using different polymers and surfactants.

A. Effect of Surfactants

The behavior of colloidal alumina suspensions in the presence of anionic sodium dodecyl sulfonate (SDSO3) is used here to demonstrate the importance of adsorption density and the structure of adsorbed layer in surfactant-induced stabilization [7]. Figure 11.1 and Figure 11.2 show the adsorption density of SDSO3, electrophoretic mobility of alumina particles, and the suspension stability ratio at pH 7.2 and 6.9. The PZC of alumina used in this study was pH 9.1. Microscopic information about the adsorbed layer can be obtained using fluorescence and electron spin resonance (ESR) spectroscopies. Fluorescence spectroscopy, with pyrene as the photosensitive probe, provides information about the micropolarity and size of the surfactant aggregates while ESR spectroscopy, with a paramagnetic probe such as doxyl acid, provides information about microviscosity of the adsorbed layer. A schematic representation of the microstructure of adsorbed layer, deduced from fluorescence and ESR spectroscopic studies, for a similar anionic surfactant, sodium dodecyl sulfate (SDS) at the alumina–water interface [8,9] is shown in Figure 11.3 to explain the adsorption mechanism.

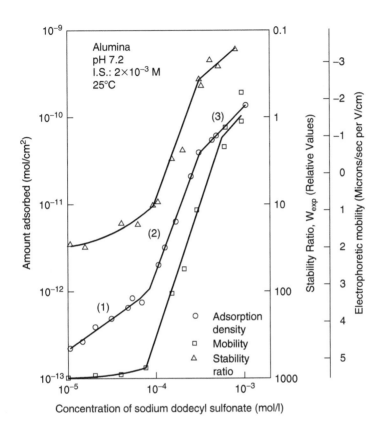

Figure 11.1 Adsorption, electrophoretic mobility, and colloidal stability isotherms for α-alumina at pH 7.2, $0.002\,M$ ionic strength and 25°C as a function of the equilibrium concentration of sodium dodecyl sulfonate.

The adsorption isotherm is characterized by four regions, attributed to three different mechanisms dominant in each region. In region I, adsorption occurs through electrostatic interaction between oppositely charged alumina and SDSO3. At low concentrations, surfactant ions adsorb at the alumina–water interface as individual ions and adsorption occurs by ion exchange of counter ions. This region has a slope of unity under constant ionic strength conditions. Region II is marked by the onset of surfactant aggregation at the interface through

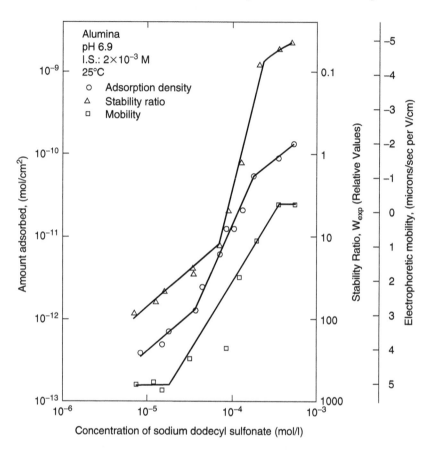

Figure 11.2 Adsorption, electrophoretic mobility and colloidal stability isotherms for α-alumina at pH 6.9, 0.002 M ionic strength and 25°C as a function of the equilibrium concentration of sodium dodecyl sulfonate.

lateral interaction between hydrocarbon chains [10]. These surfactant aggregates were later termed as *solloids* (surface colloids) [11] and include aggregates such as hemimicelles, admicelles, and self-assemblies [12]. In this region, the surface is not completely covered and enough positive sites on the surface are available for further adsorption. The transition from regions II to III corresponds to the isoelectric point (IEP) of the oxide. The slope of the isotherm decreases in

REGION I

No aggregation

REGION II

Number of aggregates increases
~ 120-130 Molecules per aggregate

REGION III

Size of aggregates increases
> 160 Molecules per aggregate

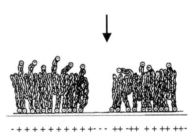

Figure 11.3 Schematic representation of the growth of aggregates of sodium dodecyl sulfate at the alumina–water interface.

region III because of the increasing electrostatic hindrance to surfactant adsorption. This is due to the surface charge reversal caused by surfactant adsorption. Adsorption in region III occurs through the growth of existing aggregates rather than through the formation of new aggregates due to lack of positive adsorption sites. Though it is not shown here, at the end of region III, the adsorption isotherm reaches a plateau level (region IV), corresponding to maximum surface coverage, at higher surfactant concentrations due to the formation of micelles in the bulk or monolayer coverage, whichever is attained at the lowest surfactant concentration [10].

It can be noticed from Figure 11.1 and Figure 11.2 that there is a close similarity between adsorption, electrokinetic and stability isotherms, indicating that flocculation occurs mainly by reduction of diffuse layer potential by surfactant adsorption. But this is applicable only up to the point of zero mobility. Beyond this, the particle surface becomes negative and the alumina suspension should have redispersed. However, the stability ratio continues to decrease, indicating further coagulation, rather extensively. This could probably be due to mutual coagulation and reorientation of adsorbed surfactant layers. It was proposed by the authors that particles with high surface coverage would undergo flocculation in order to reduce their surface energies, with the surfactant molecules acting as bridging agents between different particles or flocs.

Adsorption of ionic surfactants and its effect on suspension stability was discussed above. Though it is not discussed here, the behavior of colloidal suspensions is quite different in the presence of nonionic surfactants and mixtures of surfactants from that in the presence of ionic surfactants. Besides surfactants, polymers are also extensively employed for flocculation and dispersion of suspensions. The effect of polymers on suspension behavior is discussed next.

B. Effect of Polymers

It is well established that polymer or polyelectrolyte adsorption and conformation at the solid–liquid interface are the critical parameters that have a strong bearing on suspension stability. The adsorption and conformation of polymers in solution and at the solid–liquid interface depend on pH, ionic strength, polymer concentration, solids concentration, and the nature of the solvent and the substrate. It has been described in terms of trains, loops, and tails [13] or flat, coiled and stretching or dangling conformations [14] or more artistically as, pancakes, mushrooms, and brushes [15]. Process variables such as solids settling rate, viscosity, and turbidity are strongly affected by polymer adsorption and conformation [14], as the mechanism of flocculation is governed by such polymer behavior at the solid–liquid interface. Flocculation occurs by bridging if the polymer is in dangling form while it takes place by charge or patch neutralization if the polymer adsorbs in the form of coils at the interface.

Polymer conformation can now be determined using fluorescence and electron spin resonance spectroscopies. This involves measuring monomer (I_m) and excimer (I_e) emission peaks, for example, of a pyrene-labeled polymer. The coiling index, defined as the ratio of the excimer to monomer peak, I_e/I_m, gives a measure of the conformation. A high ratio indicates a coiled conformation while a low ratio indicates a dangling or flat conformation. Polymer conformation can be controlled by manipulating pH, ionic strength, solvent and polymer concentrations. Results obtained for polymer conformation in the alumina–polyacrylic acid (PAA) system [16] at different polymer concentrations under varying solution pH conditions and the resultant effect on the suspension stability are discussed here to illustrate the role of polymer conformation in flocculation and dispersion.

Figure 11.4 shows the coiling index of the PAA adsorbed at the alumina–water interface as a function of pH and polymer concentration. At pH 4, PAA has a coiled conformation and as the pH is increased, the coiling ratio decreases, resulting in stretching of the polymer into the solution. On the other hand, as the pH is increased, the coiling index decreases to a lower value at 5 ppm than that at 100 ppm concentration. The corresponding flocculation results and the zeta potential measurements are shown in Figure 11.5 and Figure 11.6, respectively. The pK_a of PAA is around 4.5 and below this pH, the polymer is in the coiled form. As the pH is increased above the pK_a, polymer attains the stretched conformation due to the electrostatic repulsion between segments. Under low pH conditions, flocculation is relatively better with higher polymer concentration possibly due to the reduced electrostatic repulsion between the particles. However, above the pK_a of PAA, flocculation is superior at lower polymer concentrations mainly because of the bridging effects. This could

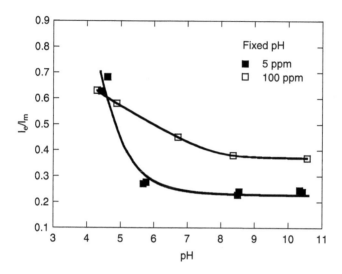

Figure 11.4 Coiling index (excimer-to-monomer ratio, I_e/I_m) of polyacrylic acid at the alumina–water interface as a function of pH (fixed pH conditions) and polymer concentration (ionic strength = $0.03\,M$ NaCl, S/L = 10 g/200 ml).

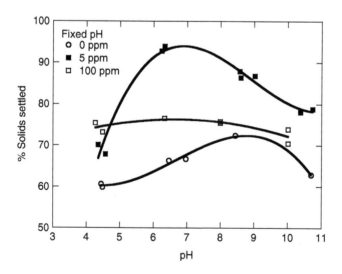

Figure 11.5 Percent alumina settled as a function of pH (fixed pH conditions) with and without polyacrylic acid (ionic strength = $0.03\,M$ NaCl, S/L = 10 g/200 ml).

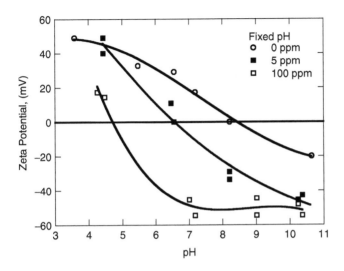

Figure 11.6 Zeta potential of alumina suspension as a function of pH (fixed pH conditions) with and without polyacrylic acid (ionic strength $= 0.03\,M$ NaCl, $S/L = 10\,$g/200 ml).

happen due to two reasons: first, as the polymer stretches into the solution it has a higher collision or capture radius. Second, the number of sites available for bridging on particle surfaces is higher at lower polymer concentration than that at higher concentration. As the pH is increased to higher levels, flocculation deteriorates because of the increased electrostatic repulsion. At higher polymer concentrations, pH does not affect flocculation mainly because of the higher surface coverage and the strong electrostatic repulsion between similarly charged polymer-coated particles.

The above results were obtained under fixed pH conditions. Polymer adsorption is history dependent and one can take advantage of this fact for controlling flocculation or dispersion at the microscopic level through polymer conformation by manipulating the solution pH [14,16]. Figure 11.7–Figure 11.9 show percent of alumina settled, zeta potential of alumina particles, and polymer coiling index obtained at two different polymer concentrations under variable pH conditions, along with percent settled

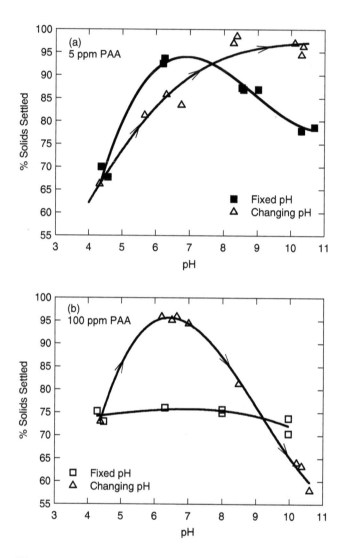

Figure 11.7 (a) Percent alumina settled with 5 ppm polyacrylic acid as a function of pH (fixed and changing pH conditions) (ionic strength $= 0.03\,M$ NaCl, $S/L = 10\,\text{g}/200\,\text{ml}$). (b) Percent alumina settled with 100 ppm polyacrylic acid as a function of pH (fixed and changing pH conditions) (ionic strength $= 0.03\,M$ NaCl, $S/L = 10\,\text{g}/200\,\text{ml}$).

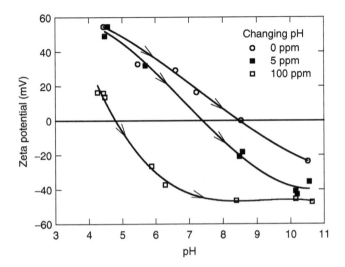

Figure 11.8 Zeta potential of alumina suspension as a function of pH (changing pH conditions) with and without polyacrylic acid (ionic strength $= 0.03\,M$ NaCl, $S/L = 10\,g/200\,ml$).

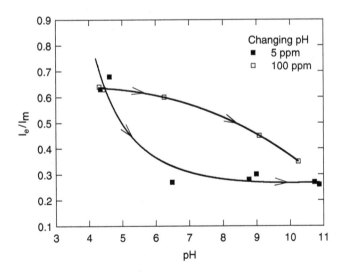

Figure 11.9 Coiling index (excimer-to-monomer ratio, I_e/I_m) of polyacrylic acid at the alumina–liquid interface as a function of pH (changing pH conditions) and polymer concentration (ionic strength $= 0.03\,M$ NaCl, $S/L = 10\,g/200\,ml$).

under fixed pH conditions for comparison purposes. For pH shift experiments, PAA was first adsorbed on alumina at pH 4 and then pH increased incrementally. This did not have any effect on the zeta potential of both bare and polymer-coated particles. But, PAA conformation at the alumina–water interface changed from coiled to stretching conformation as the pH increased from 4 to 10. The sensitivity to pH shifting was more at the lower PAA concentration than at the higher concentration. This could be because the space available at the interface for polymer stretching is less at higher PAA concentration than that at a lower concentration. At 5 ppm PAA, percent solids settled increases as pH is increased. At lower pH values, flocculation was slightly superior under fixed pH conditions than under variable pH conditions because of the relatively lower polymer-coiling index under the latter conditions. At 100 ppm PAA, percent solids settled was always higher under variable pH conditions than that obtained under fixed pH, except at the very high pH.

Flocculation results obtained under variable pH conditions can be explained as follows: At 5 ppm PAA, flocculation appears to be occurring in stages. At low pH, PAA adsorbs in the coiled form and the surface coverage is low. Flocculation occurs by charge neutralization and results in the formation of microflocs. As the pH is increased, PAA stretches into the solution and results in what can be termed as macroflocculation through bridging. As the pH is increased further, PAA conformation does not change as it is already in the stretched form. But because the surface coverage is low, flocculation progresses further by bridging of macroflocs and leads to the formation of superflocs. At 100 ppm, there appears to be an optimum pH for flocculation. Under acidic pH conditions, flocculation occurs by charge neutralization. As the pH is increased, PAA molecules stretch out slightly as indicated by the marginal decrease in coiling index. These stretched out molecules improve flocculation by bridging. So in the intermediate pH range, flocculation appears to occur by both charge neutralization and bridging. As the pH is increased further, the suspension restabilizes, rather surprisingly. It is not clear why the suspension which was completely destabilized by 100 ppm PAA at about pH 6.5 becomes stable again as the pH was increased. One possibility is that PAA molecules that were adsorbed during the second stage (at about say pH 5–6) get desorbed as the pH is increased. This cannot happen if PAA adsorption is reversible. Otherwise, it may be possible that the weak polymer bridges formed during the second stage are broken because PAA is completely dissociated at higher pH values. Since the polymer-coated particles are highly negatively charged, the suspension becomes stable again to some extent.

These results indicate the complexity of polymer-induced flocculation and highlight the importance of understanding the role of all the variables associated with polymer, solvent, solids and at large scale, equipment mixing characteristics and procedure, and their interactions on the process behavior.

C. Multipolymer Flocculation

Adsorption of polyelectrolytes on charged surfaces occurs mainly through the electrostatic interaction. They adsorb strongly on oppositely charged surfaces. The polymer conformation at the solid–liquid interface is same as that in solution and has a direct influence on flocculation. However, conformation of a particular polymer can be changed in the presence of another polymer due to complexation. Depending on the nature of complexation and application, this could be detrimental or beneficial. It was found that a combination of a low molecular weight cationic polymer and a high molecular weight anionic polymer improved flocculation of paper pulp [17]. Recent studies showed that it is possible to enhance flocculation of oxide suspensions using a binary system of oppositely charged polymers [18–20].

In one of the studies on flocculation of alumina [19], a maximum settling rate of 0.8 mm/sec was observed close to the IEP of alumina in the absence of any polymer. The settling rate did not increase in the presence of cationic polydiallyl dimethyl ammonium chloride (PDADMAC), but increased to about 2 mm/sec in the presence of 20 ppm PAA, again close to the IEP. However, sequential addition of PDADMAC followed by anionic PAA resulted in drastic improvement in flocculation. Figure 11.10 shows settling rate of alumina flocs obtained with PDADMAC alone and in the presence of PAA at pH 10.5. It can be seen that the maximum settling rate improved from about 0.9 mm/sec to about 2.4 mm/sec at low concentrations of both polymers. Since the maximum settling rate was observed at pH 10.5, away from the IEP of alumina, it was inferred that PAA molecules acted as bridges between PDADMAC-coated alumina particles, resulting in higher settling rate. This was confirmed by fluorescence spectroscopic studies using pyrene-labeled PAA as well as higher adsorption density of PAA on PDADMAC-coated alumina particles than on bare particles under all pH conditions [19].

Dual or multipolymer systems provide another avenue to optimize flocculation. As seen above in the case of alumina flocculation, PAA alone does not provide the best flocculation as it is in the coiled form at pH below its pK_a and flocculation occurs mainly by charge neutralization. At higher pH values, PAA is in stretched conformation, suitable for bridging floccu-

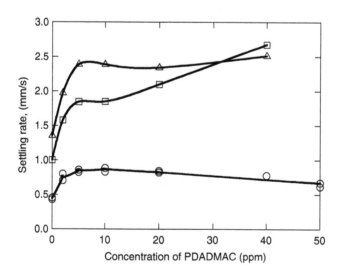

Figure 11.10 Flocculation of alumina suspension with PDAD-MAC and PAA (O PDADMAC alone, □, PDADMAC + 2 ppm PAA, △, PDADMAC + 5 ppm PAA).

lation. But alumina is negatively charged under these conditions and there will be electrostatic repulsion between alumina and PAA. The introduction of cationic PDADMAC resulted in the modification of the surface charge of alumina particles in a manner favorable for the adsorption of PAA in the stretched conformation that subsequently resulted in improved settled rate.

D. Mixed Polymer–Surfactant Systems

Polymers and surfactants are used in combination in many industrial applications. Consumer products like detergents contain polymers such as carboxylates and polyethers and anionic surfactants. They are also employed in paint formulations to stabilize latex particles and fillers and to modify rheological properties. Flotation of minerals like hematite and quartz is enhanced by a combination of polymers and surfactants. This is mainly because the interfacial and bulk properties of polymer–surfactant mixtures are different from those of polymers or surfactants alone. It is important to understand their interaction both in the solution and at the interface in order to control the properties of suspensions.

Stability of alumina suspensions in the presence of polyacrylic acid (PAA) and SDS is discussed here to illustrate their interactions. Both PAA and SDS are anionic in nature and their adsorption on positively charged alumina occurs mainly through electrostatic attraction. Fluorescence spectroscopic studies [21] indicated that PAA and SDS interact with each other in the solution under acidic pH and low PAA concentration conditions. They do not interact directly when PAA is ionized because both bear similar charges. However, if PAA concentration is very high and if it is ionized, the resulting increase in solution ionic strength causes aggregation of SDS at concentrations much below its CMC in the absence of any polymer.

When polymers and surfactants are used together, the sequence of their addition has a significant effect on the stability of suspensions [22]. Figure 11.11 shows the effect of SDS and PAA addition sequence and the concentration of SDS on

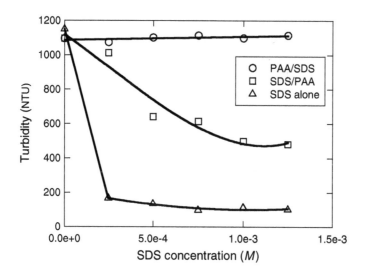

Figure 11.11   Turbidity of alumina suspension for sodium dodecyl sulfate (SDS)–polyacrylic acid (PAA) addition mode, PAA–SDS addition mode, and addition of SDS only ([PAA] = 10 ppm, [NaCl] = 0.03 M).

the turbidity of alumina suspension. When PAA was added first (PAA–SDS mode), SDS concentration did not have any effect on the turbidity. But, when SDS was added first (SDS–PAA mode), turbidity decreased as the SDS concentration is increased. This appears to be mainly because of the change in PAA conformation at the alumina–water interface under different addition sequences (Figure 11.12). In the SDS–PAA addition mode, PAA coiling index at the interface increased with SDS concentration even though its conformation in solution did not change. This is attributed to a decrease in the sites available for PAA adsorption as SDS concentration is increased and hence PAA adopting a coiled conformation. On the other hand, when PAA was added first, subsequent addition of SDS did not cause any change in PAA conformation even though adsorption density measurements showed that preadsorption of PAA did not have any effect on SDS adsorption. This is possibly due to the masking of SDS molecules by

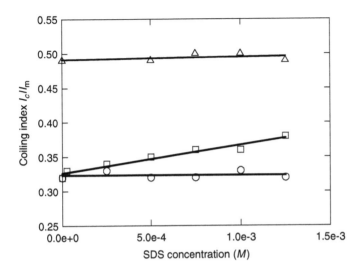

Figure 11.12 Polyacrylic acid (PAA) coiling index at the alumina–water interface under the SDS–PAA ($\square$) and PAA–SDS ($\bigcirc$) addition modes as well as in solution ($\triangle$) in the presence of SDS ([PAA] = 10 ppm, [NaCl] = 0.03 M, 0.5 g alumina/40 ml TDW).

the dangling polymer chains. However, there appears to be a critical PAA concentration below or above which SDS assumes an important role in the system. The results discussed so far were obtained at a PAA concentration of 10 ppm. Under the experimental conditions studied, PAA concentration of 0.5 ppm was found to be optimum for flocculation (Figure 11.13). Below this concentration and especially at very low concentrations, addition of even a small quantity of SDS causes a large decrease in the suspension turbidity. Above 0.5 ppm, SDS does not have any effect because PAA masks SDS at medium PAA concentrations (say 10 ppm) and at high PAA concentrations, adsorption density of SDS decreases, as PAA concentration is increased [22].

Adsorption of PAA on alumina is irreversible while that of SDS is reversible. Because of this, the sequence of addition should not have any effect on the suspension stability, as PAA conformation at the interface would ultimately attain the

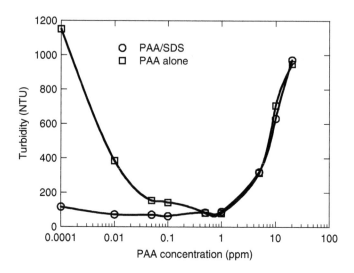

Figure 11.13 Turbidity of alumina suspension as a function of PAA concentration in the absence and presence of sodium dodecyl sulfate (PAA–SDS mode).

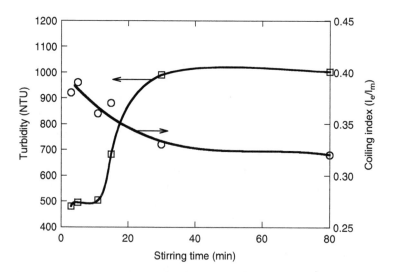

Figure 11.14 Changes in turbidity and PAA coiling index as a function of stirring time under SDS–PAA addition mode.

equilibrium conformation obtained in the PAA–SDS mode. This is evident from Figure 11.14. Suspension turbidity increases as time progresses and it eventually attains a steady-state value that was obtained with the PAA–SDS addition sequence. Though these results cannot be generalized, they emphasize the importance of addition sequence, especially for short duration processes.

We have dealt with the stability of single component (alumina) suspensions above. Large scale manufacturing processes involve multicomponent systems or a mixture of solids and one has to take the interactions between different solids species in solution into account besides their interactions with the solvent and polymers or surfactants. The situation becomes more complex if the substrate is sparingly soluble in the solvent. In that case, the solubility of the substrate is perhaps the most important parameter that determines the efficacy of the external reagent in achieving flocculation or dispersion. For example, in the selective flocculation of minerals, dissolved inorganic species from one mineral can adsorb on surfaces of other minerals and modify their electrochemical nature. In some cases, the dissolved species form complexes with polymers also and change their adsorption behavior. Hence it is necessary to understand these interactions before attempting to flocculate or stabilize a multicomponent suspension with a particular polymer or surfactant.

The flocculation–dispersion behavior of suspensions has so far been discussed with reference to the commonly measured variables such as solids settling rate, percent solids settled, and the suspension turbidity. These variables are actually functions of the size distribution, structure, and morphology of flocs that develop during the process. The size distribution and the fractal dimension of flocs are measured by employing light-scattering techniques, which are quite involved. Mathematical models based on fundamental principles of aggregation, surface chemistry, and interaction forces between particles can be useful in this regard and the population balance approach is probably the most suitable for predicting the floc size distribution.

III. POPULATION BALANCE MODEL FOR FLOCCULATION

The complete description of the flocculation process in the population balance framework results in a set of partial inte-gro-differential equations. Flocculation is essentially an aggre-gation process in which primary feed particles collide with each other to form small clusters or aggregates, which in turn interact with other clusters as well as the primary par-ticles to form larger aggregates. Shear is applied to improve mixing and the frequency of collisions between particles and to uniformly disperse the primary particles and the polymer mol-ecules in the solution. Concurrently, shear leads to fragmen-tation of the aggregates. The population balance equation (PBE) for the processes involving simultaneous aggregation and fragmentation is given by

$$
\frac{\partial n(v,t)}{\partial t} = - \int_0^\infty \alpha(v,u)\beta(v,u)n(v,t)n(u,t)\mathrm{d}u
$$

$$
+ \frac{1}{2} \int_0^v \alpha(v-u,u)\beta(v-u,u)n(v-u,t)n(u,t)\mathrm{d}u
$$

$$
- S(v)n(v,t) + \int_v^\infty S(u)\gamma(v,u)n(u,t)\mathrm{d}u. \tag{1}
$$

The first two terms on the right-hand side account for aggre-gation while the last two represent fragmentation. S is spe-cific rate constant of floc fragmentation, n is number concentration of particles or aggregates, v and u are particle or aggregate volumes, t is flocculation time, α is collision efficiency factor, β is collision frequency factor, and γ is break-age distribution function. This equation is not amenable to analytical solutions except with highly simplifying assump-tions. Therefore, numerical techniques are commonly applied to solve this equation after discretizing it into a set of ordinary differential equations. The numerical solution of the resulting system of equations is not trivial in view of the large number of equations in the system, which is also highly stiff. The number of equations is dependent on the primary particle size as well as the maximum stable floc size besides the floc fractal dimension. For example, if aggregates of 50 μm are to be considered and the primary particle size is 1 μm, the number of equations to be solved will be 2500 if the fractal dimension is 2. In order to reduce the computational burden, the different size classes are lumped into a smaller number of *sections* and solved using a suitable numerical technique for the set of ordinary differential equations [23]. A geometric grid is commonly employed to group the discretized set of equations into a smaller number of *sections*:

$$V_{i+1} = fV_i, \quad f > 1, \tag{2}$$

where V_i and V_{i+1} are section boundaries and f is geometric spacing factor. The discretization method proposed by Hounslow et al. [24] using a sectional spacing factor of 2 has been found to be suitable for flocculation [25,26]. The rate of change of the particle number concentration during simultaneous aggregation and fragmentation is now given by the following discretized population balance equation:

$$\frac{dN_i}{dt} = N_{i-1} \sum_{j=1}^{i-2} 2^{j-i+1} \alpha_{i-1,j} \beta_{i-1,j} N_j + \frac{1}{2} \alpha_{i-1,i-1} \beta_{i-1,i-1} N_{i-1}^2$$

$$- N_i \sum_{j=1}^{i-1} 2^{j-i} \alpha_{i,j} \beta_{i,j} N_j$$

$$- N_i \sum_{j=i}^{\max_1} \alpha_{i,j} \beta_{i,j} N_j - S_i N_i + \sum_{j=i}^{\max_2} \gamma_{i,j} S_j N_j, \tag{3}$$

where N_i is number concentration of particles in section i. The following sections cover the kinetics of aggregation and fragmentation and the methods to evaluate the collision efficiency factor.

A. Kinetics of Aggregation

Aggregation occurs because of particle–particle, particle–floc, and floc–floc collisions due to Brownian motion, differential settling, and applied shear. As aggregation proceeds, flocs develop as porous and irregular fractal objects from perhaps spherical or close to spherical primary particles. The size, shape, and structure are important properties of a floc. These properties not only influence the rate of collisions but also the rate of sedimentation. The floc structure is represented in terms of the mass fractal dimension. The power law that correlates the size of an aggregate with the number of primary particles in the aggregate is given by [27]:

$$r_n = n_{\text{p}}^{1/d_{\text{F}}} r_1, \tag{4}$$

where, r_n and r_1 are aggregate size and primary particle size respectively, n_{p} is number of primary particles in the aggregate, and d_{F} is mass fractal dimension. Evidently, a densely packed floc has a higher fractal dimension than a porous floc.

Depending on the process conditions, different collision kernels are invoked to represent the appropriate aggregation mechanism. The collision frequency factor for the Brownian motion (perikinetic aggregation) is given by

$$\beta_{i,j} = \frac{2k_{\text{B}}T}{3\mu} \left(\frac{1}{d_i} + \frac{1}{d_j} \right) (d_i + d_j), \tag{5}$$

where, k_{B} is Boltzman constant, d_i and d_j are collision diameters of aggregates belonging to sections i and j, T is suspension temperature, and μ is viscosity of the suspension.

The collision frequency factor for the orthokinetic aggregation is given by

$$\beta_{i,j} = \frac{G}{6} (d_i + d_j)^3, \tag{6}$$

where G is average velocity gradient or shear rate.

The collision frequency factor under the differential sedimentation is given by

$$\beta_{i,j} = \frac{\pi g}{72\mu} (d_i + d_j)^2 |d_i^2(\rho_i + \rho_l) - d^2; (\rho_j - \rho_l), \tag{7}$$

where g is acceleration due to gravity, and ρ_i and ρ_l densities of aggregate and liquid, respectively. In general, aggregate density decreases as the size increases and it can be estimated using the following equation [28]:

$$\rho_a = \rho_p \left(\frac{d_a}{d_o}\right)^{d_F - 3},$$ (8)

where d_a and d_o are diameters of the aggregate and the primary particle, respectively, and ρ_a and ρ_p are their corresponding densities. It is evident that aggregate density depends on fractal dimension, which is a function of shear rate, primary particle size, solids and polymer concentration, solution ionic strength, and temperature.

Figure 11.15 shows our simulations for the evolution of mean hydrodynamic diameter of hematite aggregates obtained by solving Equation (3) for perikinetic aggregation, without including the terms for fragmentation. The simulation results are compared with experimental data [29] for aggregation of hematite at different temperatures. The rate

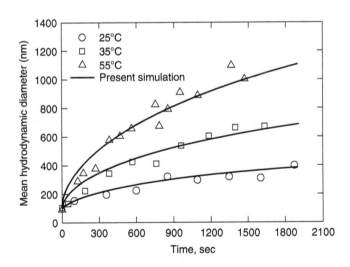

Figure 11.15 Experimental time evolution of the hydrodynamic mean diameter of hematite aggregates in Brownian motion at different temperatures, compared with our simulation results obtained by solving Equation (3) (Data from Ref. [29]) (solids concentration = 2.25×10^{10} particles/ml; pH = 3.0; [KCl] = 50 mM).

of aggregation increases with temperature and as a result the floc fractal dimension decreases as temperature increases. The experimentally determined fractal dimensions of 2.81, 2.62, and 2.37 at 25°C, 35°C, and 55°C, respectively, in 50 mM KCl solution were used as inputs while the collision efficiency was the only fitting parameter employed to match the experimental data. In order to simplify the computations, the collision efficiency was assumed to be independent of size and found to be approximately 0.09, 0.3, and 0.48 at 25°C, 35°C, and 55°C, respectively.

B. Kinetics of Fragmentation

Floc fragmentation can take place when shear is applied for improving the mixing. During shear flocculation, both aggregation and fragmentation occur simultaneously. Aggregation is dominant in the initial stages resulting in a steep increase of aggregate size. However, as the aggregates grow in size, they become susceptible to fragmentation by fluid shear. Eventually the process attains a dynamic equilibrium when the rate of fragmentation is comparable to that of aggregation and the aggregate mean size does not increase any further. It is postulated that fragmentation occurs by splitting or surface erosion [30]. The mode of breakage is influenced by the aggregate structure. The fragmentation of flocs with low fractal dimensions can occur by splitting as they have an open structure, while surface erosion could be the possible mode of breakage for flocs with high fractal dimensions.

Several attempts have been made to describe the fragmentation of flocs mathematically [30–36]. Because of its highly stochastic nature, fragmentation is more difficult to model than aggregation. Kramer and Clark [36] summarized the floc breakup theories and classified them into five basic categories of models based on criteria applied to limiting strength, strain rate, limiting size, reaction rate, and stochastic dependence.

The fragmentation of flocs has been assumed to be identical to either droplet breakage in liquid–liquid dispersions or size reduction of particulate solids, though the actual physical situation of floc breakage may not be similar to either phenomenon. Droplets are supposed to deform before breakage while floc deformation is not essential before fragmentation. Similarly flocs cannot possibly be classified as brittle materials while particulate solids are generally brittle. Moreover, flocs are highly porous and have a branched structure, which is not the case with droplets and solids. In spite of these differences, it turns out that the fragmentation theories have been fairly successful in representing floc breakage.

The specific rate of fragmentation, assuming the floc breakage is similar to the droplet breakage in liquid–liquid dispersions, is given by [37]:

$$S_i = \left(\frac{4}{15\pi}\right)^{1/2} \left(\frac{\varepsilon_\mathrm{E}}{\nu}\right)^{1/2} \exp\left(\frac{-\varepsilon_{b,i}}{\varepsilon_\mathrm{E}}\right), \tag{9}$$

where $\varepsilon_{b,i}$ and ε_E are respectively the critical and average turbulent energy dissipation rates at which floc breakage takes place and ν is kinematic viscosity of the suspending fluid. It is well known that the mean aggregate size decreases as fluid shear increases and it has been assumed that $\varepsilon_{b,i}$ is inversely proportional to the aggregate collision radius:

$$\varepsilon_{b,i} = \frac{B}{R_{c,i}}, \tag{10}$$

where $R_{c,i}$ is collision radius of the aggregate belonging to section i and B is proportionality constant. The specific rate of fragmentation, assuming floc breakage is similar to the size reduction of solid particles, is given by [34]:

$$S_i = S_l \left(\frac{d_i}{d_l}\right)^m, \tag{11}$$

where d_l is diameter of the largest floc size present, S_l is specific rate of breakage for aggregates of size d_i and m is the model fitting parameter.

In addition to the rate of breakage, it is also necessary to have a function which describes the daughter fragments. One could employ binary, ternary, or normal distribution functions for this purpose. For example, the normal distribution function is given by:

$$\gamma_{i,j} = \frac{V_j}{V_i} \int_{b_{i-1}}^{b_i} \frac{1}{\sqrt{2\pi}\sigma_f} \exp\left[-\frac{(V - V_f)}{2\sigma_f^2}\right] dV, \qquad (12)$$

where V_f is mean volume of daughter fragments, σ_f is standard deviation of the volume of daughter fragments, and b is the section boundary volume.

Figure 11.16 shows our simulations for coagulation of polystyrene latex involving simultaneous aggregation and fragmentation. The fragmentation kinetics was assumed to be identical to droplet breakage and a binary breakage distribution function was employed. It can be observed that the rate of aggregation increases and the mean aggregate diameter at steady state decreases as shear rate increases. The model was fitted to experimental data at different shear rates [37] by tuning parameter B in Equation (10).

Model fitting to experimental breakage data suggests that flocs do not break into more than two or three daughter fragments [34]. Hence, the simple binary breakage distribution function should suffice. However, it is often necessary to employ a normal distribution [34,37] in order to fit the experimental data accurately, especially the floc size distributions.

C. Collision Efficiency

One of the most important parameters in the population balance model is the collision efficiency factor. It has been treated as a fitting parameter in most models although it is possible in principle to compute it. There are a few cases where the collision efficiency factor was computed as a function of either surface forces or hydrodynamic forces and included in the population balance equation for salt-induced coagulation [37–39].

One of the main problems in not computing the collision efficiency is the difficulty in understanding the surface-chemical phenomena, especially in the presence of polymers and surfactants. In order to overcome this, simplified models have been developed based on the surface coverage of particles

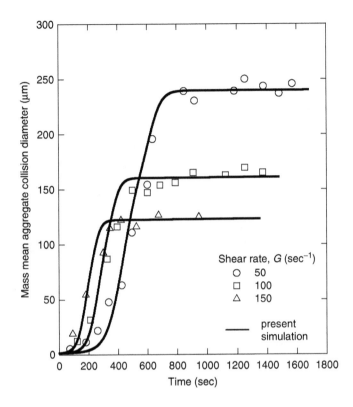

Figure 11.16 Experimental time evolution of the mass mean diameter of latex aggregates in turbulent shear compared with our simulation results obtained by solving Equation (3). (Data from Ref. [37]) (solids volume fraction $= 1.4 \times 10^{-5}$; $[Al_2(SO_4)_3 16H_2O] = 10 \, mg/l$; $[NaHCO_3] = 1 \, mM$; $pH = 7.2 \pm 0.05$).

with polymer molecules. La Mer and coworkers [40,41] were the first to introduce this approach. They assumed that polymer bridging takes place by the adhesion of a polymer covered particle and a bare particle and predicted that optimum flocculation takes place at half the surface coverage. However, later experimental results showed that optimum flocculation occurs at very low surface coverages [42]. Hogg [43] proposed that the collision efficiency depends not only on the interaction between a polymer-covered particle and a bare particle but also on the interaction for the opposite case. Later,

Moudgil and coworkers [44,45] introduced the concept of site blocking agent in order to explain selective flocculation of minerals. Recently, this model was extended for multicomponent systems encountered in papermaking [46]. One of the main drawbacks of the simplified models is that they are derived using the surface coverage alone and do not take interparticle forces into account. On the other hand, theoretical models based on fundamental principles of surface and colloid chemistry and interparticle forces are potentially much more powerful and versatile.

It is well known that interparticle forces play an important role in flocculation and dispersion as the stability of a suspension depends strongly on the forces of attraction and repulsion. The collision efficiency factor α is defined as the inverse of the stability ratio, W [47]:

$$W_{i,j} = (r_i + r_j) \int_{r_i+r_j}^{\infty} \frac{\exp(V_T/k_B T)}{s^2} ds, \tag{13}$$

where V_T is total potential energy of interaction between particles i and j, r_i and r_j are radii of primary particles i and j, respectively, and s is center-to-center separation distance of the particles.

1. Salt-induced Coagulation

The well-known DLVO theory [48,49] is commonly used to compute the interparticle forces that arise during salt-induced coagulation. According to this theory, the total energy of interaction, V_T is a sum of the London–van der Waals energy of attraction, V_A and the energy of electrostatic repulsion, V_R.

$$V_T = V_A + V_R. \tag{14}$$

The energy of attraction between two spherical particles was derived as [50]:

$$V_A = -\frac{A}{6} \left\{ \frac{2r_i r_j}{s^2 - (r_i + r_j)^2} + \frac{2r_i r_j}{s^2 - (r_i - r_j)^2} + \ln\left[\frac{s^2 - (r_i + r_j)^2}{s^2 - (r_i - r_j)^2} \right] \right\}, \tag{15}$$

where A is Hamaker constant. A highly useful but approximate expression for the electrostatic energy of repulsion or attraction was derived as [51]:

$$V_R = \frac{\varepsilon r_i r_j (\psi_{0_i}^2 + \psi_{0_j}^2)}{4(r_i + r_j)} \left\{ \frac{2\psi_{0_i}\psi_{0_j}}{(\psi_{0_i}^2 + \psi_{0_j}^2)} \ln \left[\frac{1 + \exp(-\kappa H_0)}{1 - \exp(-\kappa H_0)} \right] \right.$$

$$\left. + \ln[1 - \exp(-2\kappa H_0)] \right\}, \tag{16}$$

where H_0 is distance of minimum separation between particle surfaces, κ is Debye–Huckel parameter and ψ_0 is particle surface potential. The dielectric constant of the suspending medium, ε is given by

$$\varepsilon = \varepsilon_0 \varepsilon_r, \tag{17}$$

where ε_0 is dielectric constant of free space, ε_r is dielectric constant of water or solvent.

2. Systems involving Polymers

It has been well established that polymers affect the interparticle forces significantly and polymer conformation at the solid–liquid interface is the key variable that governs the suspension stability. Of the number of theories that describe the polymer adsorption, the two most popular models are the self-consistent field model of Scheutjens and Fleer [52–54] and the scaling model of de Gennes [15,55,56]. These models have been used to derive expressions for the energy of interaction between polymer-coated surfaces. For example, the energy of interaction between two flat polymer-coated surfaces, $P(H_0)$, using the scaling theory is given by [15]:

$$P(H_0) = \frac{k_B T}{p^3} \left[\left(\frac{D^*}{H_0} \right)^{9/4} - \left(\frac{H_0}{D^*} \right)^{3/4} \right], \quad H_0 < D^*, \tag{18}$$

where D^* is twice the length of the polymer tail and p is linear distance between the anchored polymer chains on the surface. The above expression was used to derive the forces between a sphere and a flat plate [57] and between two crossed cylinders [58] to fit the interaction force data obtained by the atomic force microscope and the surface forces apparatus, respectively. The Scheutjens and Fleer model was later extended to polyelectrolytes [59] and tested with experimental data [60]. Either of these theories can be employed to compute the interaction energy, which can be subsequently used to evaluate the collision efficiency. Besides this, the frequency of collisions also needs to be modified as the collision radius of a particle does increase due to the adsorbed polymer layer.

D.   Dynamics of Polymer Adsorption

The effect of polymers on flocculation can be incorporated in two ways: one, by assuming that the adsorption attains equilibrium very fast so that adsorption kinetics can be neglected, and two, by treating the adsorption as a rate process. The approach assuming fast equilibrium was presented in Section III.C.2.

Recent studies on polymer conformation [61] and interparticle forces [62] indicated that adsorption does not reach equilibrium within the time scales of flocculation–dispersion. There are also other studies, which indicate that kinetics of polymer adsorption affect the process and result in nonequilibrium flocculation [63,64]. One of the methods to deal with this situation is to treat polymer molecules as *particles* and apply the collision kernels similar to those used for particle–particle interactions [65,66]. As in the case of solids, diffusional characteristics of polymers in the solvent will assume significance in the case of perikinetic aggregation while it may not be that important in the case of orthokinetic aggregation. In order to take the polymer adsorption dynamics into account, an additional equation for polymer concentration at the interface or in solution needs to be solved along with the set of population balance equations for aggregates. This would enable estimation of time-dependent characteristics of polymer conformation and the true dynamics of polymer-induced flocculation.

IV. CONCLUDING REMARKS

Flocculation and dispersion of colloidal suspensions are complex processes and their behavior is critically dependent on characteristics of polymers and surfactants and nature of the solvent and the substrate. Reagent and solids concentrations are also important variables in addition to mixing time and equipment geometry. Bulk properties of the suspension such as percent solids settled, solids settling rate, and suspension turbidity are functions of polymer or surfactant adsorption density and structure of the absorbed layer. Conformation and orientation of polymer and surfactant molecules have a strong bearing on suspension stability. Fluorescence and electron spin resonance spectroscopies are very useful for determining these microscopic properties, which can be controlled by manipulating variables such as pH, ionic strength, and mixing time.

Solution pH is one of the important variables in flocculation. It affects the electrochemical nature of both the substrate and ionic polymers and surfactants and as a result the microstructure of adsorbed layer. Flocculation results obtained under variable pH conditions can be significantly different from those obtained under fixed pH conditions because of the polymer reconformation at the interface and in some cases, detachment of the polymer molecules from the particle surfaces. Polymers can be used in combination to obtain improved flocculation performance than that obtained with single polymers. Interaction between polymers and surfactants is dependent on solution conditions and reagent concentrations. Complexation occurs both in the solution and at the interface, resulting in the modification of the interfacial and bulk properties of the suspension.

During flocculation, aggregation of particles and flocs, fragmentation of flocs, and adsorption of polymers or surfactants take place simultaneously. The commonly measured process variables are functions of the floc size distribution and the shape and structure of flocs. Population balance approach is perhaps the most suitable for predicting the floc size distribution. Since complete description of polymer-induced flocculation is quite complex, models for individual aspects of flocculation are developed separately and in general there is an absence of an integrated model for the flocculation process. An integrated approach is presented here to combine theories related to aggregation, fragmentation, and polymer adsorption so that the effect of variables such as solution pH, ionic strength, polymer and solids concentration, and shear rate on flocculation and dispersion can be predicted.

ACKNOWLEDGMENTS

The authors acknowledge the National Science Foundation for support of this work and the management of Tata Research Development and Design Centre for permission to publish this work.

REFERENCES

1. R.H. Pelton, in *Colloid–Polymer Interactions: From Fundamentals to Practice*, R.S Farinato and P.L. Dubin, Eds., John Wiley & Sons, New York, 1999, pp. 51–82.
2. P. Somasundaran, K.K. Das, and X. Yu, *Curr. Opin. Colloid Interf. Sci.*, 1, 530–534, 1996.
3. B.M. Moudgil, S. Mathur, and T.S. Prakash, *KONA*, 15, 5–19, 1997.
4. D.N. Thomas, S.J. Judd, and N. Fawcett, *Water Res.*, 33, 1579–1592, 1999.
5. G.A. Jackson and A.B. Burd, *Environ. Sci. Technol.*, 32, 2805–2814, 1998.
6. M. Smoluchowski, *Z. Phys. Chem.*, 92, 129, 1917.
7. P. Somasundaran, T. Healy, and D.W. Fuerstenau, *J. Colloid. Interf. Sci.*, 22, 599–605, 1966
8. P. Somasundaran, P. Chandar, and N.J. Turro, *Colloids Surf. A*, 20, 145, 1986.
9. P. Chandar, P. Somasundaran, and N.J. Turro, *J. Colloid. Interf. Sci.*, 117, 31, 1987.
10. P. Somasundaran and D.W. Fuerstenau, *J. Phys. Chem.*, 70, 90, 1966.
11. P. Somasundaran and J.T. Kunjappu, *Colloids Surf. A*, 37, 245–268, 1989.

12. J.H. Harwell and D. Bitting, *Langmuir*, 3, 500, 1987.
13. G.J. Fleer, M.A. Cohen Stuart, J.M.H.M. Scheutjens, T. Cosgrove, and B. Vincent, *Polymers at Interfaces*, Chapman & Hall, New York, 1993.
14. K.F. Tjipangandjara, Y.B. Huang, P. Somasundaran, and N.J. Turro, *Colloids Surf. A*, 44, 229–236, 1990.
15. P.G. de Gennes, *Adv. Colloid. Interf. Sci.*, 27, 189–209, 1987.
16. K.F. Tjipangandjara and P. Somasundaran, *Colloids Surf. A*, 55, 245–255, 1991.
17. K.W. Britt, A.G. Dillon, and L.A. Evans, *Tappi*, 60, 102–104, 1977.
18. X. Yu and P. Somasundaran, *Colloids Surf. A*, 81, 17–23, 1993.
19. X. Yu and P. Somasundaran, *J. Colloid. Interf. Sci.*, 177, 283–287, 1996.
20. A. Fan, N.J. Turro, and P. Somasundaran, *Colloids Surf. A*, 162, 141–148, 2000.
21. C. Maltesh and P. Somasundaran, *Colloids Surf. A*, 69, 167–172, 1992.
22. A. Fan, P. Somasundaran, and N.J. Turro, *Colloids Surf. A*, 146, 397–403, 1999.
23. D. Ramkrishna, *Population Balances: Theory and Applications to Particulate Systems in Engineering*, Academic Press, New York, 2000.
24. M.J. Hounslow, R.L. Ryall, and V.R. Marshall, *AIChE J.*, 34, 1821–1832, 1988.
25. P.T. Spicer and S.E. Pratsinis, *AIChE J.*, 42, 1612–1620, 1996.
26. T. Serra and X. Casamitjana, *AIChE J.*, 44, 1724–1730, 1998.
27. J. Feder, *Fractals*, Plenum Press, New York, 1988.
28. Q. Jiang and B.E. Logan, *Environ. Sci. Technol.*, 25, 2031–2038, 1991.
29. R. Amal, J.A. Raper, and T.D. Waite, *J. Colloid. Interf. Sci.*, 140, 158–168, 1990.
30. J.D. Pandya and L.A. Spielman, *J. Colloid. Interf. Sci.*, 90, 517–531, 1982.
31. C.F. Lu and L.A. Spielman, *J. Colloid. Interf. Sci.*, 103, 95–105, 1985.
32. R.C. Sonntag and W.B. Russel, *J. Colloid. Interf. Sci.*, 113, 399–413, 1986.
33. R.C. Sonntag and W.B. Russel, *J. Colloid. Interf. Sci.*, 113, 378–389, 1987.
34. D.T. Ray and R. Hogg, *J. Colloid. Interf. Sci.*, 116, 256–268, 1987.

35. S.J. Peng and R.A. Williams, *J. Colloid. Interf. Sci.*, 166, 321–332, 1994.
36. T.A. Kramer and M.M. Clark, *J. Colloid. Interf. Sci.*, 216, 116–126, 1999.
37. J.C. Flesch, P.T. Spicer, and S.E. Pratsinis, *AIChE J.*, 45, 1114–1124, 1999.
38. R. Amal, J.R. Coury, J.A. Raper, W.P. Walsh, and T.D. Waite, *Colloids and Surf. A*, 46, 1–19, 1990.
39. K.A. Kusters, J.G. Wijers, and D. Thoenes, *Chem. Eng. Sci.*, 52, 107–121, 1997.
40. R.H. Smellie, Jr. and V.K. La Mer, *J. Colloid. Sci.*, 13, 589–599, 1958.
41. T.W. Healy and V.K. La Mer, *J. Phys. Chem.*, 66, 1835–1838, 1962.
42. P. Somasundaran, Y.H. Chia, and R. Gorelik, in *Polymer Adsorption and Dispersion Stability*, E.D. Goddard and B. Vincent, Eds., ACS Symposium Series 240, American Chemical Society, 1984, pp. 393–410.
43. R. Hogg, *J. Colloid. Interf. Sci.*, 102, 232–236, 1984.
44. B.M. Moudgil and S. Behl, *J. Colloid. Interf. Sci.*, 146, 1–8, 1991.
45. S. Behl, B.M. Moudgil, and T.S. Prakash, *J. Colloid. Interf. Sci.*, 161, 414–421, 1993.
46. A. Swerin, L. Odberg, and L. Wagberg, *Colloids and Surf. A*, 113, 25–38, 1996.
47. N. Fuchs, *Z. Phys.*, 89, 736, 1934.
48. B.V. Derjaguin and L.D. Landau, *Acta. Physiochim. USSR*, 14, 633, 1941.
49. E.J.W. Verwey and J.Th.G. Overbeek, *Theory of the Stability of Lyophilic Colloids*, Elsevier, Amsterdam, 1948.
50. H.C. Hamaker, *Physica IV*: 1058–1072, 1937.
51. R. Hogg, T.W. Healy, and D.W. Fuerstenau, *Trans. Faraday Soc.*, 62, 1638–1651, 1966.
52. J.M.H.M. Scheutjens and G.J. Fleer, *J. Phys. Chem.*, 83, 1619–1635, 1979.
53. J.M.H.M. Scheutjens and G.J. Fleer, *J. Phys. Chem.*, 84, 178–190, 1980.
54. J.M.H.M. Scheutjens and G.J. Fleer, *Macromolecules* 18, 1882–1900, 1985.
55. P.G. de Dennes, *Macromolecules*, 14, 1637–1644, 1981.
56. P.G. de Dennes, *Macromolecules*, 15, 492–500, 1982.
57. H.G. Pedersen and L. Bergstrom, *J. Am. Ceram. Soc.*, 82, 1137–1145, 1999.

58. P.M. Claesson, H.K. Christenson, J.M. Berg, and R.D. Neuman, *J. Colloid. Interf. Sci.*, 172, 415–424, 1995.
59. M.R. Bohmer, O.A. Evers, and J.M.H.M. Scheutjens, *Macromolecules*, 23, 2288–2301, 1990.
60. J. Blaakmeer, M.R. Bohmer, M.A. Cohen Stuart, and G.J. Fleer, *Macromolecules*, 23, 2301–2309, 1990.
61. X. Yu and P. Somasundaran, *J. Colloid. Interf. Sci.*, 178, 770–774, 1996.
62. H.G. Pedersen and L. Bergstrom, *Acta. Mater.*, 48, 4563–4570, 2000.
63. T.G.M. Van de Ven, *Adv. Colloid. Interf. Sci.*, 48, 121–140, 1994.
64. M.A. Cohen Stuart and G.J. Fleer, *Annu. Rev. Mater. Sci.*, 26, 463–500, 1996.
65. Y. Adachi, *Adv. Colloid. Interf. Sci.*, 56, 1–31, 1995.
66. R. Hogg, *Colloids Surf. A*, 146, 253–263, 1999.

12

Flocculation and Dewatering of Fine-Particle Suspensions

R. HOGG

Department of Energy and Geo-Environmental Engineering, The Pennsylvania
State University, University Park, PA 16802, USA

I. INTRODUCTION

Flocculation is a process of aggregation of fine suspended particles that is commonly carried out in order to facilitate solid–liquid separation by sedimentation or filtration. The size of the aggregates (flocs) produced largely determines their settling behavior; floc structure, especially floc density, also affects settling characteristics and is particularly important in controlling filter cake permeability.

Fine particles suspended in a liquid may tend to aggregate spontaneously due to the presence of attractive forces or they may remain dispersed as discrete entities if there are repulsive forces that oppose aggregation. Intermolecular forces, particularly the London–van der Waals type, lead to an attraction between any pair of molecules. Additivity of the forces between individual atoms or molecules gives rise to a significant attraction between solid particles [1]. In the case of particles suspended in a liquid, the liquid itself is also subject to the same kind of interactions. Analysis of these interactions indicates that, for particles of the same composition, suspended in any liquid, the net force is always attractive. Dissimilar particles, on the other hand, may experience either an attraction or repulsion, depending on the nature of the solid–solid, liquid–liquid, and solid–liquid interactions.

The repulsive forces responsible for the stability of colloidal dispersions are usually of electrical origin and result from the development of electrical charges on particle surfaces. These charges may be inherent to the solid — due to imperfections in the solid crystal, as in many clay minerals. Alternatively, they may be a consequence of preferential adsorption–desorption of ionic constituents of the solid — *potential determining ions* — or of specific adsorption of other ionic species from solution. Partial hydrolysis of solid surfaces in aqueous suspensions typically leads to a pH-dependence of the surface charge and, in consequence, provides a means for *chemical* control of the charge. Electrostatic attraction between a charged solid surface and oppositely charged ions — *counter ions* — from solution produces what is known as an *electrical double layer* surrounding each dispersed particle. The interaction of the double layers around adjacent particles gives rise to forces between them. Identical particles in a common solution bear the same surface charge and tend to repel one another. Dissimilar particles may be oppositely charged leading to a double-layer attraction [2].

The presence of surface films of adsorbed species, especially macromolecules, can also affect the interaction forces between particles. Complete surface coverage by a layer of soluble polymer molecules leads to a condition where the coated particles interact more favorably with the liquid medium than with each other causing interparticle repulsion. Incomplete coverage by a high molecular weight polymer can promote aggregation by *molecular bridging*. The use of polymers to prevent aggregation is generally known as *steric stabilization* [3].

Recently observed, strong attractive forces between hydrophobic surfaces, usually referred to as *hydrophobic forces*, can obviously play an important role in dispersion stability [4]. These forces may correspond to a combination

of relatively strong van der Waals forces with capillary interactions due to the presence of microscopic gas bubbles at
hydrophobic surfaces.

The stability of a colloidal dispersion is determined by
the net interaction force between particles. The classical
Derjaguin–Landau–Verwey–Overbeek (DLVO) *theory* is based
on the net effects of the van der Waals and electrical double
layer forces on the aggregation of particles subject to Brownian
motion [5–7]. The sum of the attractive and repulsive forces can
result in attraction at all separations if the repulsive component is relatively weak, or in a region of net repulsion over some
range of particle separations. Aggregation (*coagulation*) occurs
if the forces are attractive at all separations or if the thermal
energy of the particles — due to Brownian motion — is sufficient to drive particles through a repulsive region.

In general, flocculation can be regarded as a two-stage
process involving particle *destabilization*, permitting flocs to
form, followed by a period of *floc growth*. A third stage — *floc
degradation* — is typically inevitable and is often considered
to be detrimental overall. However, it may also provide a
useful contribution in regulating the final floc size distribution and modifying the floc structure. Destabilization is normally accomplished by control of the chemistry of the liquid
medium so as to eliminate repulsive barriers that otherwise
prevent particles from coming into contact. In some cases,
flocculation can be promoted by applying vigorous agitation,
in effect forcing particles to collide by hydrodynamic effects, in
spite of existing repulsive forces. This approach has been
termed *shear flocculation*. Floc growth is a dynamic process
in which destabilized particles and existing flocs collide and
adhere to form larger units. The interparticle collisions may
be the result of Brownian motion or, more commonly in practice, velocity gradients (shear) caused by agitation of the suspension. In the absence of agitation, collisions can also result
from differential settling of particles or flocs of different sizes.
Agitation due to shear also leads to floc degradation and
generally limits the extent to which growth can occur.

This chapter will be mainly concerned with the factors
that control the development of flocs, evaluating the perform-

ance of flocculation processes and a discussion of their role in dewatering of suspensions by sedimentation or filtration.

II. PROCESS EVALUATION AND FLOC CHARACTERIZATION

Traditionally, the performance of flocculation processes has generally been evaluated by experimental observation of various aspects of settling behavior. While these tests give only indirect measures of flocculation, they are simple to perform, provide useful information, and are directly relevant to important objectives such as dewatering by sedimentation. The actual tests used often depend on the type of settling behavior observed.

In relatively dilute suspensions, particles and flocs typically settle as individuals with a range of settling velocities. The determination of a single, characteristic settling rate is not practical for such systems. Common approaches include determination of supernatant turbidity after a fixed settling period (preferably from a fixed location), or measurement of the mass of sediment, again after a fixed settling period.

When the solids concentration in suspension is high, particularly combined with extensive flocculation, so-called *zone settling* occurs. Flocs settle en masse producing a sharp interface — *mud-line* — between the settling solids and a relatively clear supernatant. The rate of fall of the mud-line represents a characteristic *settling rate* for the system and can often be regarded as an indicator of floc size. However, because zone settling occurs under hindered-settling conditions, relationships between settling rate and floc size depend on solids concentration and comparisons should only be made at fixed concentration. In the use of settling rate to evaluate flocculation behavior, in situ measurements in the flocculation vessel are preferable. Transfer to another container can significantly change the characteristics of the flocculated material. Supernatant turbidity measurements provide information on the extent to which primary particles and small flocs are incorporated into the settling mass. Final sediment volumes give an indication of floc density and compressibility.

A. Floc Size Measurement

More complete information on the performance of flocculation processes can be obtained by direct measurement of floc size distribution. Laser light scattering and diffraction methods are well suited to such measurements, which can be carried out using standard, commercial equipment. Generally, samples of the flocculated suspension must be diluted to the appropriate concentration and care must be exercised to avoid breakage of the flocs during sampling and measurement.

B. Floc Density Measurement

The density of individual flocs and its variation with floc size are important characteristics of the process. Estimates of floc density can be obtained from direct observations of the settling of individual flocs in dilute suspensions. The settling velocity v_s of a small particle in a fluid is related to its size x and density ρ_p through Stokes law

$$v_s = \frac{(\rho_p - \rho_l)gx^2}{18\mu},\tag{1}$$

where ρ_l is the fluid density, μ is the fluid viscosity and g is the acceleration due to gravity. Alternative forms are available for larger particles [8]. Direct measurements of floc size and free settling velocity can therefore be used to estimate floc density.

C. Sediment Compressibility

The compressibility of sediments or filter cakes formed from flocculated material is an important factor in determining the extent to which mechanical dewatering is possible. The compression process involves the collapse of structures formed by the interacting particles and the subsequent transport of liquid out of the fine pores between the particles. Compressibility can be measured either by direct force application using some kind of piston arrangement or through the weight of the sediment itself in the presence of gravity or centrifugal forces. The former approach is directly applicable to filtration processes while compression under self-weight is equivalent to the processes responsible for the consolidation of sediments.

III. DESTABILIZATION OF FINE-PARTICLE SUSPENSIONS

The initial step in a flocculation process is to destabilize the dispersed particles allowing them to collide and form aggregates. This step is sometimes referred to as *coagulation* and chemical reagents used are called *coagulants*. However, usage of this terminology tends to be somewhat inconsistent and it will not be adopted here.

In most cases, destabilization involves the reduction of repulsive forces between particles thereby permitting the London–van der Waals attractions to lead to aggregation. For insoluble particles such as many minerals, interparticle repulsion is usually due to electrical double layer interaction. The latter can be controlled by changing either the magnitude of the surface potential or the extension of the double layer out from the particle surface. As noted above, the surface potential — the electrical potential difference between the particle surface and the bulk solution — depends on the concentrations of potential-determining ions, which are normally defined as ionic constituents of the solid particles. Other kinds of ionic species that are specifically adsorbed at the surface can also affect the potential at the solid–liquid interface. These can include multivalent cations and anions, ionic surfactants and, particularly in aqueous systems, hydrogen and hydroxyl ions. Such species are tightly bound to the surface becoming, in effect, part of the solid particle. The magnitude of the double-layer interaction force is determined by the potential drop across the mobile part of the layer, which consists of *indifferent* ions that are attracted to the solid by electrical forces only. This potential difference can reasonably be represented by the so-called *zeta-potential* ζ, which can be determined by electrokinetics measurements such as electrophoresis.

The double-layer thickness is controlled by the overall ionic strength of the solution medium. At high concentrations,

ions are close together and the surface charge can be balanced over a relatively short distance from the surface — the double layer is said to be compressed. Conversely, at low ionic strength, ions are far apart and the double layer extends a considerable distance from the surface.

The interaction between the double layers on adjacent particles can be expressed as changes in free energy with separation. A widely used approximation for spherical particles with the same zeta potential is [7]:

$$\Delta G_{dl} = \frac{\varepsilon_m \bar{x}}{4} \zeta^2 \ln\left(1 + e^{-\kappa H}\right), \tag{2}$$

where ε_m is the dielectric constant of the medium, $\bar{x}$ is the harmonic mean particle size, H is surface-to-surface separation of the particles, and κ is the Debye–Hückel (inverse) thickness parameter. The latter is defined by

$$\kappa^2 = \frac{4\pi e^2}{\varepsilon_m k T} \sum_i c_i z_i^2, \tag{3}$$

where e is the electronic charge; k is Boltzmann's constant; T is absolute temperature; c_i and z_i are, respectively, the concentration and valence of ionic species i in solution.

The London–van der Waals attraction between particles can also be expressed as a change in free energy with separation, which can usually be estimated from [7]:

$$\Delta G_{vdw} = -\frac{A\bar{x}}{24H}, \tag{4}$$

where A is the Hamaker constant for the solid material in the suspending medium. The values of the Hamaker constant for many solids are available in the literature — see, for example, Lyklema [9] and Visser [10]. Procedures for estimating the values have been described by Hunter [11] and Gregory [12]. The effect of the suspending medium can be taken into account using

$$A = \left(\sqrt{A_s} - \sqrt{A_m}\right)^2, \tag{5}$$

where A_s is the value for the solid and A_m is that for the medium. For interaction between solid particles of the same material, it is clear from Equation (5) that the overall Hamaker constant A is always positive, representing an attraction between the particles. However, for particles of different composition in a fluid medium, it is possible for the net van der Waals force to be repulsion.

In the absence of other interaction forces (e.g., magnetic), the overall interaction between particles can be determined from the sum of the double layer and van der Waals interactions as given by Equations (2) and (4). Some examples are given in Figure 12.1. When the double-layer interaction energy is large, the curves generally indicate attraction at very small separations and repulsion at somewhat greater distances (of the order of the double-layer thickness as given by Equation (3)). The region of net repulsion is commonly referred to as an "energy barrier" and represents the repulsive energy that must be overcome for particles to come into

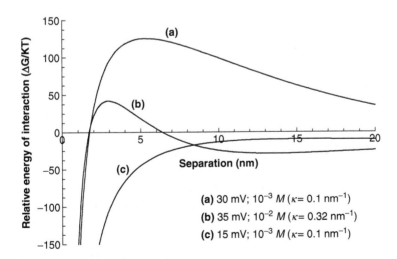

Figure 12.1 Examples of relative interaction energies for $1\,\mu m$ particles dispersed in water (net Hamaker constant A; 5×10^{-20} J) showing the effects of double layer potential and ionic strength of the solution medium.

contact, i.e., to coagulate. The destabilization of a dispersion normally requires reduction of the energy barrier to the point where it is less than the energy (thermal or due to mechanical agitation etc.,) of the individual particles.

It is clear from Equation (2) that destabilization (lowering of the energy barrier) can be achieved by reducing the zeta-potential or by decreasing the double-layer thickness (increasing κ). Either of these can be accomplished by changing the composition of the solution medium. Increasing the ionic strength of the solution, especially with multivalent species, compresses the double layer and often reduces the magnitude of the zeta potential. For aqueous dispersions of solid particles, particularly minerals, solution pH has a major effect on the zeta potential including, typically, a reversal in sign at some value of pH — known as the *point of zero charge* (PZC). The adjustment of pH, towards the PZC normally results in a reduction of dispersion stability. Because the Hamaker constant is a material property, the attractive, van der Waals interaction is not usually amenable to control.

Destabilization can also be accomplished by the addition of oppositely charged polyelectrolytes or fine particles, including colloidal precipitates, leading to heterocoagulation effects. Typically, this approach also has important effects on subsequent floc growth and will be discussed in greater detail below.

IV. FLOC FORMATION AND GROWTH

Destabilization of a dispersion permits aggregation of particles and the formation of flocs. In most applications of the process, the primary objective is to produce flocs that are large enough and dense enough to enhance dewatering by sedimentation or filtration. While destabilization depends mainly on the chemistry of the particles and the solution medium, floc growth is controlled by physical factors such as solids concentration and mechanical agitation of the suspension.

A. Collision Frequency

The formation and growth of flocs results from particle–particle, particle–floc, and floc–floc collisions. Such collisions may be a consequence of the inherent Brownian motion of particles or small flocs, velocity gradients in agitated suspensions or differential settling of particles or flocs varying in size and density. Obviously, the rate of growth is determined by the frequency of these collisions. Based on the work of Smoluchowski [13,14], it can be shown that the frequency f_{ij} of binary collisions between particles of type i and j is given by

$$f_{ij} = K_{ij} n_i n_j \qquad (6)$$

where K_{ij} is a collision frequency factor that depends on the specific mechanism responsible for the collisions; n_i and n_j are the respective number concentrations of the two types. Expressions for the frequency factors corresponding to the above mechanisms are given in Table 12.1.

It can be seen that the factor for Brownian motion depends on the size ratio but not on the absolute size of the particles. In contrast, the factors for shear and sedimentation are strongly size-dependent, varying with absolute size to the third power for shear and the fourth power for sedimentation. Collision frequencies due to shear or sedimentation are negligible for ultrafine particles but increase rapidly with increasing particle size. For typical mineral particles of similar size at ambient temperature, the rates for shear and Brownian motion are equal at sizes between 0.2 and 0.5 μm depending on shear rate (in the 100 to 1000 sec^{-1} range). Thus, while Brownian motion may be the dominant factor in the aging of

Table 12.1 Binary collision frequency factors for suspended particles (sizes x_i x_j).

Collision Mechanism	Frequency Factor (K_{ij})
Brownian motion	$\frac{2kT}{3\mu}\left(2 + \frac{x_i}{x_j} + \frac{x_j}{x_i}\right)$
Hydrodynamic shear (mean shear rate $\bar{G}$)	$\frac{\bar{G}}{6}(x_i + x_j)^3$
Sedimentation (density difference, $\Delta\rho$)	$\frac{\pi\Delta\rho g}{72\mu}(x_i + x_j)^3 |(x_i^2 - x_j^2)|$

dispersions of submicron particles, growth in practical flocculation systems is generally dominated by shear introduced by
mixing with reagents, etc. Collisions due to sedimentation
may be important in systems involving large flocs in the
presence of residual primary particles.

The estimates of collision frequencies in agitated suspensions require a knowledge of the effective mean shear rate, $\bar{G}$.
Reasonable estimates can be obtained from the power input
per unit volume (energy dissipation per unit mass) [15]. Thus,

$$\bar{G} = \left(\frac{P}{\mu V}\right)^{1/2} \tag{7}$$

For stirred tanks, the power input P can be estimated using

$$P = N_p \rho_f N_I^3 D_I^5, \tag{8}$$

where ρ_f is the fluid density, N_I and D_I are the rotational speed
and diameter of the impeller and N_p is the power number for
the particular tank configuration used. Extensive tabulations
and graphical presentations of power numbers are available
(see, for example, Tilton [16]).

If collisions lead to adhesion between particles, the result
is floc formation and growth. Smoluchowski [13] applied population-balance concepts to the growth process and showed
that, in the absence of size dependence of the frequency factors (constant frequency factor K), the rate of change of the
overall number concentration N of particles or flocs is given by

$$\frac{dN}{dt} = -\frac{KN^2}{2}. \tag{9}$$

Experimental and theoretical evidence indicates that floc size
distributions approach a self-preserving form as growth
proceeds, i.e., the form of the distribution becomes time-
independent; only the characteristic size (e.g. the mean size)
increases [17–19]. It follows that the effects of relative size can
be included in the rate factor K. For aggregation due to
Brownian collisions, K is approximately constant and Equation (9) represents a simple second-order process. In the case of
shear-induced collisions, on the other hand, K is proportional

to size cubed, i.e., using the definition of the shear collision factor as given in Table 12.1,

$$\frac{dN}{dt} = -f_p \bar{G} \bar{x}^3 N^2, \tag{10}$$

where $\bar{x}$ is the mean floc size at time t and f_p is a polydispersity factor that depends on the width of the self-preserving floc size distribution.

For a fixed volumetric concentration ϕ, the number concentration is inversely proportional to the mean particle volume, i.e.,

$$N = \frac{\phi}{k_v \bar{x}^3}, \tag{11}$$

where k_v is the volume shape factor ($= \pi/6$ for spheres). With this constraint, Equation (10) reduces to a pseudo-first-order form [17]:

$$\frac{dN}{dt} = -\frac{f_p}{k_v} \bar{G} \phi N, \tag{12}$$

Integration of Equations (9) and (12) leads, respectively, to

$$N = \frac{N_0}{1 + t/\tau_{Br}}, \tag{13}$$

for Brownian collisions and to

$$N = N_0 e^{-t/\tau_{sh}}, \tag{14}$$

in the case of shear. N_0 is the initial number concentration and τ_{Br} and τ_{sh} are characteristic times defined by

$$\tau_{Br} = \frac{\mu}{f_p N_0 kT} = \frac{\mu k_v x_0^3}{f_p \phi kT}, \tag{15}$$

and

$$\tau_{sh} = \frac{k_v}{f_p \bar{G} \phi}, \tag{16}$$

respectively.

Using the relationship between mean floc size and number concentration as given by Equation (11), the kinetics of floc growth can be expressed by [20]:

$$\bar{x} = \bar{x}_0(1 + t/\tau_{Br})^{1/3}, \tag{17}$$

for Brownian collisions, and by

$$\bar{x} = \bar{x}_0 e^{t/3\tau_{sh}}, \tag{18}$$

for shear.

It can be seen from the above that the introduction of shear to a particle dispersion has a twofold effect on floc growth. In addition to increasing the overall rate, the form of the growth curve changes drastically as illustrated in Figure 12.2. Growth in the case of Brownian coagulation slows as the suspended particles are depleted by aggregation. In the presence of shear, however, this effect is offset by a corresponding increase in the collision frequency factor. It is clear from the figure that even a brief period of quite gentle agitation can dramatically increase growth rates.

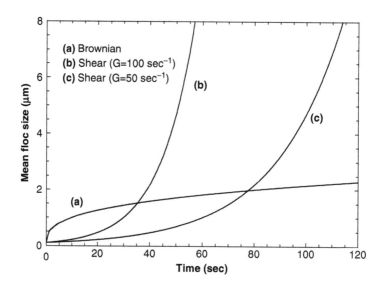

Figure 12.2 Effects of collision mechanism on floc growth kinetics for destabilized 0.1 μm particles in water at 25°C.

The half times given by Equations (15) and (16) vary inversely with solids concentration and indicate that floc growth can be a very slow process at low concentrations. For example, the half time for Brownian coagulation of $1\,\mu m$ particles at a concentration of $10\,ppm$ is almost $3\,h$. Fairly gentle agitation (at $100\,sec^{-1}$) would only reduce this to about $25\,min$. More vigorous agitation, however, may lead to more breakage than growth. A common approach to enhancing floc growth in dilute suspensions, widely practiced in water treatment, is to increase the overall solids content. This can be accomplished by adding fine particles such as colloidal silica, by precipitation of iron–aluminum hydroxy-species using alum or, in some cases, by recycling sediment from a subsequent dewatering operation.

B. Collision Efficiency

The above analysis of floc growth is predicated on the assumption that each collision leads to adhesion of the colliding particles. Repulsive forces due to electrical double layer interactions or hydrodynamic effects can prevent particles from actually coming into contact so that only a fraction of "collisions" lead to adhesion. Alternatively, attractive forces can turn "near misses" into actual collisions, thereby enhancing growth rates. The inclusion of the van der Waals and electrical double layer interaction forces in the analysis of the Brownian collision process is the basis of the well-known DLVO theory of colloid stability. Extensions to shear-induced collisions have also been described [21,22]. In general, the analyses involve the inclusion of an "efficiency factor" in the kinetics equations. For example, Equation (6) is replaced by

$$f_{ij} = E_{ij}K_{ij}n_in_j \tag{19}$$

where E_{ij} is the efficiency factor for collisions between i- and j-type particles or flocs. Based on an analysis by Fuchs [23], the DLVO theory provides explicit relationships between the efficiency factor (expressed as a "stability ratio" W, equivalent to $1/E$) and the particle interaction forces. Because of the complexity of these relationships, their direct incorporation into floc growth models is far from simple. However, the form of the relationships is such that the stability ratio changes abruptly from about one (100% efficiency) to a very large value (0% efficiency) over relatively narrow ranges of important variables such as ionic strength, pH, or particle size. Consequently, it is often reasonable to assume, as a working approximation, that the efficiency is 100% over a certain range (e.g., pH around the PZC) and zero elsewhere. Extensions to include other forces such as solvation [3] and hydrodynamic interactions [24] have also been considered.

C. Floc Degradation

The presence of velocity gradients (shear) in a suspension of fine particles promotes collisions between particles, leading to floc formation and growth. At the same time, however, shear also causes breakage and degradation of the flocs that are formed. Breakage is important both in limiting floc growth and in the degradation of flocs during subsequent handling. Floc degradation is particularly important in applications involving polymeric reagents, for which quite vigorous agitation, in the turbulent regime, is necessary to ensure adequate mixing of the reagents with the dispersed particles.

It is reasonable to assume that floc breakage occurs when the pressure difference across a floc entrained in a turbulent eddy exceeds its mechanical strength. Based on accepted theories for turbulence [25] and agglomerate strength [26], it can be shown that both the hydrodynamic force and the binding force increase with floc size in a given turbulent environment [27,28]. Depending on the floc size relative to the turbulence microscale the breakage force increases with floc size to the 8/3 power for large flocs and to the 4th power for smaller ones [28]. In the case of flocs bound primarily by van der Waals forces, floc strength increases linearly with size. Consequently, flocs become increasingly susceptible to breakage as size increases. Parker et al. [27] and Tambo and Hozumi [28] assumed, in effect, that flocs whose strength is less than the applied force break catastrophically, so that the size for which the forces are equal represents the maximum size in the system. Relationships were presented between the maximum floc size and power input (energy dissipation rate) to the agitated suspension. Tambo and Hozumi [28] showed, for example, that maximum size decreased linearly with agitation speed for alum-flocculated kaolin particles.

As an alternative to the maximum size hypothesis, some authors [29–31] treated floc breakage as a rate process, analogous to emulsification of liquids or comminution of solids. Population-balance methods were used to describe the changes in floc size distribution due to breakage. Recognizing that floc growth and breakage generally occur simultaneously — growing flocs are subject to breakage while fragments of broken flocs undergo reagglomeration — Hogg et al. [32] presented a simplified model to describe the change in median floc size x_{50} as growth and breakage proceeds. Thus,

$$\frac{dx_{50}}{dt} = K_a x_{50} - K_b x_{50}^{1+\alpha}, \tag{20}$$

where K_a is a growth rate constant and K_b is a breakage rate constant. α is a factor that describes the size dependence of the breakage rate. Typically, values of α are close to 1.0 for the comminution of brittle solids. However, larger values might be expected for large, fragile flocs. Unfortunately, in fitting experimental data to this model, it is found that there are compensating effects between the values of K_b and α such that several combinations appear to fit data equally well. Consequently, the use of the model to evaluate the effects of process variables has met with only limited success [33].

Floc breakage is usually considered to be detrimental to the process. However, beneficial effects are also possible. For example, breakage can help prevent the runaway growth of large flocs thereby facilitating the incorporation of fine, primary particles by increasing the likelihood of their colliding with existing flocs.

V. FLOC STRUCTURE

When solid particles aggregate to form flocs, they do not, of course, coalesce into a new solid particle. Rather, they form relatively open structures with considerable internal voidage. Simulation studies of the formation of flocs by random addition of primary particles or other flocs predict the development of fractal structures whose density decreases progressively with increasing size [34–37]. The predicted forms have been confirmed by experimental studies of a variety of systems [38–45]. In general, the variation in internal porosity ε with relative floc size can be expressed by

$$1 - \varepsilon = \left(\frac{x}{x_c}\right)^{-\gamma}, \tag{21}$$

where x_c is a characteristic size, usually related to but not necessarily equal to the primary particle size. The exponent γ is related to the fractal dimension δ through $\delta = 3 - \gamma$. An example of a typical floc size–density relationship is given in Figure 12.3.

Flocculation mechanisms (Brownian motion or shear) and conditions (agitation intensity, the presence or absence of polymers) seem to have relatively minor effects on the

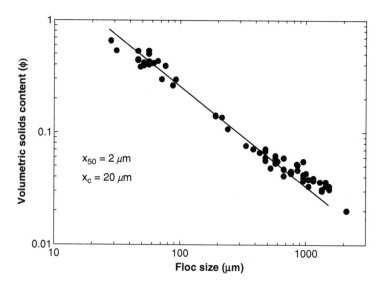

Figure 12.3 Floc size–density relationship for 2-μm quartz particles flocculated with a nonionic polymer (after Klimpel [41]).

structure relationship [43]. The size of the primary particles does, however, appear to play a role, especially for flocs formed in a turbulent environment. Flocs formed from relatively coarse particles (1–10 μm) appear to undergo some degree of compaction in the early stages of their development, leading to a value of the characteristic size x_c that is significantly larger than the primary particle size. For example, the characteristic size for the data shown in Figure 12.3 for 2-μm primary particles is about 20 μm. Such compaction is not generally observed for very fine (submicron) particles, probably because the external forces acting at that scale are small relative to the binding forces between particles.

VI. POLYMER-INDUCED FLOCCULATION

A. Reagents

Soluble polymers are widely used in flocculation processes. They can serve as destabilization agents and can also promote floc growth. Types commonly used in industrial applications include nonionic polyacrylamide, $[CH_2CHCONH_2]_n$; anionic acryamide–acrylate copolymers, $[CH_2CHCONH_2]_n$ $[CH_2CHCOO^-]_m$, or partially hydrolyzed polyacrylamides, and cationic dimethyldiallylammonium chlorides (DMDAAC), $[(CH_3)_2N(CH_2CHCH_2)_2{}^+Cl^-]_n$; or copolymers of the dimethyldiallylammonium ion with acrylamide. The nonionic and anionic polymers used are typically of high molecular weight (10–20 million), while the cationic reagents usually have substantially lower molecular weight — 50,000 to 1.5 million for the DMDAAC types and up to about 5 million for the DMDAAC–acrylamide copolymers.

B. Flocculation Mechanisms

High molecular weight polymers are generally agreed to function through a molecular "bridging" mechanism, whereby individual molecules adsorb onto more than one particle, providing a physical linkage between them [46–48]. For the lower molecular weight polyelectrolytes, flocculation is considered to occur through a "charge-patch" mechanism. The highly charged polymer molecule attaches to the surface of an oppositely charged particle, creating a charged patch on the surface. This patch can then attach itself to a region of bare surface on another particle, leading to aggregation [49]. In many respects, the charge-patch mechanism is simply a special case of molecular bridging. The distinction is useful, however, because of significant differences in the applicability of the two mechanisms to flocculation practice.

C. Polymer Adsorption

The action of polymeric flocculants generally involves adsorption at the surfaces of solid particles. Polymer adsorption differs significantly from that of small molecules. Polymer molecules of the type used as flocculants are essentially linear chains containing many segments, each of which can, potentially, attach to the surface. A 15 million molecular weight polyacrylamide consists of over 200,000 individual acrylamide segments. Since the attachment of any segment constitutes adsorption of the molecule, even a very weak segment-surface attraction is likely to be sufficient. Consequently, polymers tend to adsorb on any available surface.

The polyacrylamides generally used as flocculants in aqueous systems are believed to attach to solid particles primarily through hydrogen bonding with hydroxylated surface sites. For low molecular weight, polyelectrolytes such as the cationic DMDAAC types, electrical interaction with negatively charged particles obviously plays a significant role and favors the formation of the surface patches responsible for charge-patch aggregation. Similarly, anionic polymers will be attracted to positively charged particles. In this case, however, the molecules are not usually completely anionic — typically less than 50% — and other binding mechanisms such as hydrogen bonding may be equally or even more important. It is widely observed that anionic flocculants can adsorb on and flocculate negatively charged particles. Presumably, nonionic molecular segments are hydrogen-bonded to the surface while the anionic segments are repelled out into solution.

Individual segment adsorption by any of these mechanisms should be a reversible process so that molecules can change their configuration and location on the surface. However, desorption of a complete molecule requires that all adsorbed segments desorb simultaneously and are not replaced by segments from the same molecule — an event that seems highly unlikely. Consequently, polymer adsorption acts like an irreversible process with the polymer showing an apparent high affinity for the surface, despite the relatively weak interactions involved. Detailed analyses of the adsorption of soluble polymers at solid–liquid interfaces and their role in flocculation and dispersion stabilization are presented in chapters 6 (Fleer and Leermakers) and 11 (Somasundaran et al.) of this volume as well as elsewhere [50].

The conformation of adsorbed polymer molecules is considered to be important in their function as flocculants. Segments of an adsorbed molecule are considered to belong to one or other of three sequences: loops, tails, and trains, as illustrated in Figure 12.4. Trains consist of groups of adjacent segments that are actually attached to the surface. Tails are the segments at the ends of the linear molecule that are not directly attached to the surface and extend out into solution. Loops are intermediate sequences of segments, between trains and also extend into the solution phase. Bridging flocculation is favored by extended loops and tails.

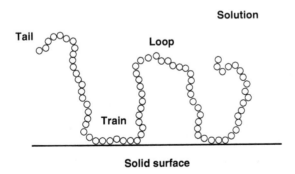

Figure 12.4 Schematic representation of an adsorbed polymer molecule showing the different kinds of segment sequences.

Conformation at the surface is affected by the strength of segment–surface interaction, stronger bonding favoring the formation of trains rather than loops or tails. Segment–segment interactions, repulsion between charged groups in polyelectrolytes, for example, cause expansion of randomly coiled polymer chains and tend to favor the development of extended loops and tails. The amount adsorbed — adsorption density — can also affect conformation. At low densities, individual molecules can spread over the surface with extended train sections. Increasing adsorption density causes crowding effects, reducing the area available for each adsorbed molecule and causing train sections to be partially displaced from the surface.

When polymers are used to promote flocculation, the resulting aggregation affects further adsorption of the polymer by reducing the available surface area.

D. Performance of Polymeric Flocculants

1. Destabilization

It has been demonstrated that low molecular weight cationic polymers are highly effective for destabilizing dispersions of negatively charged particles, but not very effective for promoting extensive floc growth [51,52]. On the other hand, high molecular weight polymers are relatively poor destabilizers but can facilitate the growth of very large flocs. Some illustrative examples are provided in Figure 12.5, which shows the evolution of the floc size distribution following polymer addition to an initially stable dispersion of (negatively charged) fine alumina particles in water at pH 11. A concentration of 1.6 mg/l of a cationic polymer (molecular weight about 50,000) was sufficient to produce a narrow floc size distribution centered at about 10 μm. The same quantity of a nonionic polymer (molecular weight about 15 million) produced a bimodal size distribution with about 50% of the material remaining as dispersed primary particles. More than four times as much of the nonionic polymer was needed to eliminate the fine particle mode. It is worthy of note, however, that increased dosage of the cationic reagent led to little or no further growth while

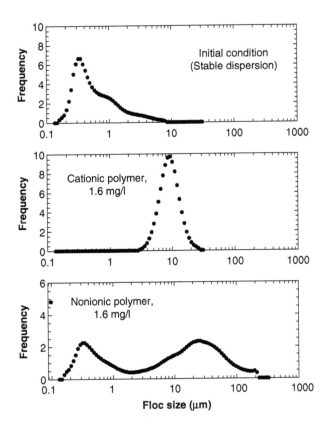

Figure 12.5 Evolution of floc size distributions in suspensions of negatively charged, fine alumina particles dispersed in water at pH 11, following the addition of a cationic (DMDAAC) polymer (molecular weight about 50,000) and a nonionic polyacrylamide (molecular weight about 15 million) (after Rattanakawin [54]).

high nonionic polymer dosage increased the median floc size to 50 μm or more.

It appears that the destabilizing capability of the cationic polymers is related to both charge and molecular weight. The high positive charge on the polymer ensures adsorption on the negatively charged particles, while the relatively low molecular weight provides for the formation of reasonably small surface patches and ensures a sufficiently high ratio of molecules to particles. The molecular weight effect is illustrated in Figure 12.6. The addition of 4.3 mg/l of the low molecular weight cationic polymer at pH 11 destabilizes the suspension, eliminating residual, dispersed primary particles. The same

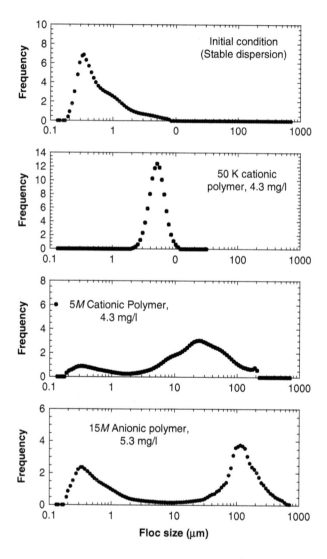

Figure 12.6 Effect of polymer molecular weight on the destabilization of aqueous dispersions of fine alumina particles, by cationic polymers at pH 11 and an anionic polymer at pH 5 (after Rattanakawin [54]).

quantity of a higher molecular weight (about 5 million) cationic polymer leaves a significant amount of dispersed primary particles. Addition of 5.35 mg/l of a high molecular weight (about 15 million) anionic polymer to a suspension of positively charged particles at pH 5 produces large flocs but leaves about half of the solids as dispersed primary particles.

2. Floc Growth

Following destabilization of a suspension, it is usually desirable to grow large flocs that are amenable to solid–liquid separation by sedimentation or filtration. However, the low molecular weight polymers used for destabilization do not generally promote extensive floc growth. The effects of polymer addition to an unstable dispersion of alumina particles near the point of zero charge (pH 8.6) are shown in Figure 12.7. The addition of more than 10 mg/l of the cationic polymer had essentially no effect on the floc size distribution, while as little as 4 mg/l of the higher molecular weight, nonionic flocculant were sufficient to increase the median floc size to about 100 μm.

It is clear that the efficient use of bridging (high molecular weight) polymers to promote the growth of large flocs requires pre-destabilization of the suspension, prior to flocculant addition. The particular destabilization procedure used appears to be relatively unimportant. The elimination of double-layer repulsion (e.g., by pH adjustment or double-layer compression) or charge-patch aggregation seems to be equally effective [52]. Simple measures such as pH control or salt addition have obvious economic advantages but may not always be practical — if extremes of pH are required or salt addition is unacceptable, for example. Furthermore, some systems are not easily destabilized by these simple approaches, especially if steric stabilization is involved. In such cases, the use of polymer "coagulants" such as the low molecular weight cationics may be required. These reagents may also provide greater flexibility and minimize deleterious effects due to changes in system chemistry etc.

In addition to the requirement for predestabilization of the suspension, the performance of polymeric flocculants

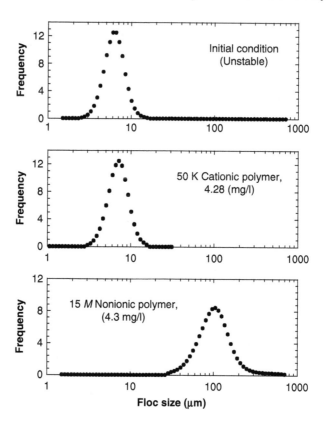

Figure 12.7 Effect of polymer molecular weight on floc growth in unstable suspensions of fine alumina particles (pH 8.6) (after Rattanakawin [52]).

depends on a number of system variables. Chemical factors such as solution pH and ionic strength can affect polymer conformation in solution and on the surface, and influence the development of molecular bridges. Polymer molecular weight and ionicity also affect bridge formation. It is to be expected that bridge formation by adsorbed polymers depends on effective molecular size. The latter is determined by molecular weight and ionicity. Yu and Somasundaran [53] and Rattanakawin [54] have shown that increasing molecular size either through increased molecular weight or due to higher ionicity leads to enhanced floc growth. In addition, the use of partially ionic flocculants in suspensions of particles with similar but small surface charge may promote bridging by favoring surface conformations with extended loops and tails. Indeed, Rattanakawin [54] has shown that a partially anionic polymer added to a suspension of slightly negatively charged alumina particles produces larger flocs than a non-ionic polymer of the same molecular weight.

Flocculant performance is significantly affected by agitation and mixing during and after flocculant addition [55]. Agitation affects several aspects of the overall process, viz.

- mixing of flocculants etc., with the suspension
- flocculant adsorption
- floc growth
- floc breakage

Adsorption of polymers on particle–floc surfaces can be assumed to be a collision process analogous to floc growth and described by similar relationships. For instantaneous addition of concentration C_{po} of polymer to an agitated suspension, it can be shown [56] that the concentration C_p remaining in solution after time t decays exponentially according to

$$C_p = C_{p^0} e^{-(\bar{G}\phi/\pi)t}, \tag{22}$$

Equation (22) indicates that the time for 95% of the polymer to be removed (adsorbed) would be about 10 sec for a suspension at 1% solids by volume under gentle agitation (shear rate $100 \, \text{sec}^{-1}$) and about 1 second with vigorous agitation ($1000 \, \text{sec}^{-1}$). In other words, adsorption occurs very rapidly.

Mixing times τ_m in stirred tanks can be estimated from [57]:

$$\tau_m \approx 36/N_I, \tag{23}$$

For the experimental conditions corresponding to Figure 12.6 and Figure 12.7 ($N_I = 1000 \, \text{rpm}$), the mixing time would be about 2 sec. On an industrial scale, impeller speeds giving the same shear rate and, therefore, similar adsorption rate, would generally be considerably lower with correspondingly longer mixing times. In order to ensure uniform adsorption throughout the system, it is important that mixing occurs more quickly than adsorption. This constraint is especially critical for polymer adsorption in view of the effective irreversibility of the process. Excessive local adsorption close to points of polymer addition cannot be eliminated by subsequent desorption and transfer to regions deficient in polymer.

For a given mixing system, the rates of mixing, polymer adsorption and, ultimately, floc growth and breakage are all determined by agitation speed and cannot be controlled independently. Adsorption rates can be controlled, however, through the rate of polymer addition to the system. The problem of nonuniform polymer distribution can be minimized by ensuring that the rate of addition is significantly less than the prevailing mixing rate. Extensive laboratory testing has demonstrated that continuous polymer addition consistently provides superior flocculation performance as compared to instantaneous addition followed by a mixing period [55,58,59]. The problems of inadequate mixing in concentrated suspensions can also be overcome by increased overall polymer dosage, leading of course to increased cost.

By controlling the adsorption rate, the rate of polymer addition determines the rate of floc growth [52,58,59]. Limiting conditions occur at very high addition rates, when adsorption times again become shorter than the inherent mixing time. The conditions necessary to maintain the appropriate balance between adsorption and mixing rates are illustrated in Figure 12.8 which shows the effects of agitation speed and solids concentration on the mixing and adsorption times on a laboratory-scale (300 ml mixing tank) batch system. It is clear that adsorption occurs faster than mixing at moderate to high concentrations (greater than about 0.1% by volume) while the reverse is true in very dilute systems. Thus, continuous addition is required for efficient flocculation in concentrated suspensions but unnecessary at low concentrations. Some degree of control should be possible at intermediate concentrations by appropriate selection of agitation speed.

During continuous polymer addition to a concentrated suspension, floc growth proceeds until some limiting size is approached [58,59]. Some examples are shown in Figure 12.9.

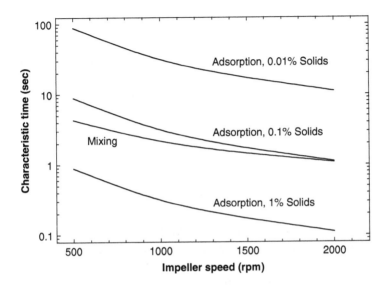

Figure 12.8 Predicted effects of agitation intensity and solids concentration on mixing and polymer adsorption times in a stirred tank.

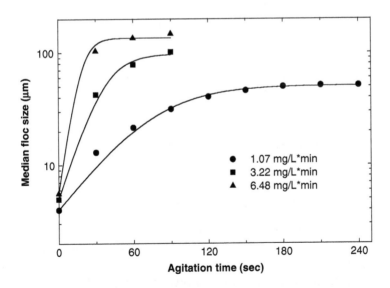

Figure 12.9 Effects on floc growth of the rate of continuous addition of a nonionic polyacrylamide to unstable suspensions of fine alumina particles at pH 9 (after Rattanakawin [52]).

It appears that this limiting size reflects a condition of balance between growth and breakage in the turbulent environment. Since the growth rate is proportional to the rate of polymer addition, increasing the latter generally produces an increase in the limiting floc size. Similarly, reducing the rate of polymer addition has the effect of reducing the limiting floc size. It follows that continued mixing without continued polymer addition generally leads to net breakage of the flocs and should be avoided.

Floc growth rates in dilute suspensions are determined primarily by agitation speed during a period of continued mixing following polymer addition. Again, the growth will be limited by floc breakage. As flocs grow, the external surface area available for polymer adsorption decreases and a point may be reached where a small quantity of additional polymer is sufficient to saturate the surface, leading to steric stabilization. While this condition is unlikely to be reached in concentrated suspensions, it can be quite close to the optimum dosage level for very dilute systems. Indeed, Rattanakawin [52] has observed that increasing the rate of continuous polymer addition to a very dilute suspension can reduce the limiting floc size presumably due to steric stabilization caused by excessive polymer dosage. Because of the risk of overdosage, instantaneous polymer addition, followed by an extended mixing period is the preferred approach to flocculation of dilute suspensions. Furthermore, since mixing is not a limiting factor here, the use of a brief, rapid-mixing step followed by a more extended period of fairly gentle agitation — to minimize breakage — is appropriate in such systems.

E.   Continuous Flocculation Processes

The mixing and polymer dosing procedures described above refer to laboratory-scale batch testing and are appropriate for the evaluation of flocculant performance and reagent selection. Industrial applications, however, typically involve continuous processes in which the flocculant is mixed continuously with a fresh suspension feed. The semibatch tests with continuous polymer addition in a well-mixed tank are not quite analogous to continuous systems where both suspension and flocculant are added continuously. In the former case, the polymer concentration increases progressively with time while it is constant in the continuous system at steady state. Similarly, the floc size increases progressively with time in the semibatch system while the continuous system includes the complete range of sizes from the feed size to the final, limiting size. On the other hand, because flocculant and slurry are both added continuously, the mixing problems associated with instantaneous polymer addition in a batch process are less critical.

Experimental results indicate that continuous flocculation in a stirred tank generally produces smaller flocs than semibatch tests in a similar tank [60,61]. The author has indicated that, in the presence of a range of floc sizes, preferential adsorption on the larger sizes is likely [56]. Since growth depends on adsorption on the smaller flocs, such preferential adsorption could lead to reduced overall growth rates. As in the semibatch system, reducing the residence time, in this case by increasing the flow rates, leads to enhanced floc growth, probably by reducing the time available for floc breakage.

Industrially, continuous flocculation is often carried out by direct injection into a pipe. Actually, this approach is roughly analogous to the batch procedures used in the laboratory. Injection at a single point corresponds to instantaneous addition to a stirred tank, while addition at multiple locations along the pipe is equivalent to continuous polymer addition in the semibatch system. Experimental results, at the laboratory scale, show considerable similarity to the batch tests showing, for example, that multiple injection leads to improved performance [62]. Again, floc sizes were generally smaller than those obtained in the standard batch tests. It should be noted, however, that in this case flocculation conditions in the batch and continuous systems were quite different. Polymer dosage and estimated shear rates covered similar ranges but residence times in the continuous, in-line systems were substantially shorter. A comparison of flocculation performance, as indicated by settling rate, in batch and continuous systems is given in Figure 12.10.

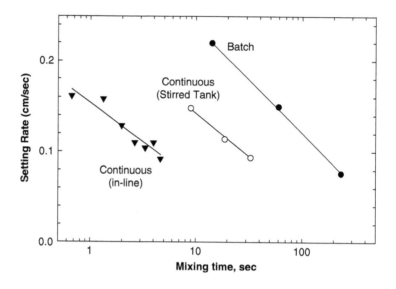

Figure 12.10 Comparison of batch and continuous flocculation of kaolin at 3% solids by weight by 4 mg/l of a nonionic polyacrylamide (after Suharyono [60]).

F. Scale-Up

The above discussions clearly demonstrate the significant effects of mixing and dosing conditions on laboratory-scale flocculation by polymers. Unfortunately, only limited information is available on how these effects can be scaled up from laboratory tests to industrial applications. Maffei [63] investigated the batch flocculation of kaolin in a series of geometrically similar stirred tanks ranging in volume from 0.3 to 52 l. The performance criterion used was settling rate, determined directly in the mixing tanks, and correlations were made of the initial growth rate (increase in settling rate with time) and the limiting settling rate. In all cases, the initial growth rate was found to increase in proportion to the rate of continuous polymer addition, as noted previously, but to be relatively insensitive to tank size and agitation speed. The limiting settling rate was found to vary with the rate of polymer addition, tank size, and agitation speed. For the larger tanks, 0.9–52 l, the limiting settling rates appear to correlate

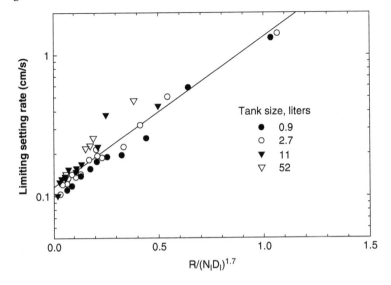

Figure 12.11 Approximate scale-up relationship for batch floccu-lation in stirred tanks (R = rate of polymer addition (mg/l/min); N_I = impeller speed (rpm); D_I = impeller diameter (cm)) (after Maffei [63]).

with the quantity: $R/(N_i D_i^{1.67})$. An example of this correlation is given in Figure 12.11. The results for the 0.3-l tank followed similar trends with respect to polymer addition and agitator speed but appear to scale differently with tank size (impeller diameter). Little or no information on scale-up of continuous flocculation processes, either in stirred tanks or in-line in pipes or static mixers has been found in the literature.

VII. APPLICATIONS TO DEWATERING BY SEDIMENTATION AND FILTRATION

The primary objective of many flocculation processes is to facilitate subsequent dewatering operations. Floc size distri-bution and floc structure are important in dewatering by either sedimentation or filtration. In sedimentation, large flocs settle rapidly but the existence of a broad, especially bimodal distribution can lead to high residual turbidity.

Similarly, residual fine particles can cause clogging of filter media and the development of impermeable filter cakes. Floc structure affects sediment density and consolidation behavior and controls filter cake permeability.

A. Sedimentation

Solid–liquid separations by sedimentation generally fall into one or other of the two broad categories: clarification or thickening. Both have similar objectives but differ in emphasis. In clarification, the primary product is clear liquid containing as little residual solid as possible. The solid is typically a by-product to be disposed of. Thickening processes, on the other hand have the solid phase as the desired product, usually with minimum liquid content. Liquid clarity is of secondary importance. Clarification is normally applied to dilute suspensions while thickening systems typically receive a more concentrated feed.

Flocculation is beneficial to both processes by enhancing settling rates. Indeed, it is a requirement for suspensions of very fine particles. Because of the differences in emphasis, destabilization is probably the major concern in clarification while floc growth takes greater priority in thickening. Solids concentration and the extent of flocculation affect the type of settling behavior which, in turn, determines settling rates and the capacity of sedimentation systems.

1. Settling Characteristics

It is generally recognized that settling behavior in flocculated suspensions involves three fairly distinct modes [64]:

- Free settling of individual particles or flocs
- Zone settling of a mass of material
- Compression of the sediment at the bottom of a container.

Free settling occurs in dilute suspensions when individual particles or flocs are far enough apart that their motion is essentially independent. For well dispersed solid particles, it is usually considered that free settling conditions prevail for solids concentrations up to about 1% by volume [65], although the effects are not very significant for concentrations as high as 5%. However, it should be recognized that, due to the porous structure of flocs, the effective concentration may be considerably higher. For flocs with internal porosity ε in a suspension with solids concentration ϕ, the effective concentration of the settling units is

$$\phi_{\text{eff}} = \frac{\phi}{1 - \varepsilon}. \tag{24}$$

Referring to Figure 12.3, floc porosities as high as 95% are quite common. Then, at a solids content of 2.5% by volume, the effective concentration would be 50%, which corresponds to a relatively close-packed condition. In other words, hindered settling conditions may apply to flocculated suspensions at surprisingly low concentrations or may be established after only a limited amount of thickening by sedimentation.

Zone settling is a special case of hindered settling in which all particles settle at the same rate so that a clear interface — "mud-line" — forms between the settling solids and the supernatant liquid. It can occur for particles or flocs of identical size or, more commonly, when interactions between flocs minimize relative motion. While interactions lead to structure development, settling rates under these conditions are determined by hydrodynamic drag forces due to fluid flow between flocs. For a given system with a fixed extent of flocculation, zone-settling rates depend only on solids concentration.

The compression regime applies when the structure developed by particles or flocs in contact with one another acquires sufficient mechanical strength to support an applied force such as the weight of material above. The strength of the structure can be characterized by a *compressive yield stress*, which is a function of the solids concentration [66]. An example of the concentration-dependence of the compressive yield stress for flocculated clay is given in Figure 12.12. Provided the applied force exceeds the compressive yield stress, the sedimentation rate is determined by the fluid drag due to flow out of the pores and by the rate of deformation of the

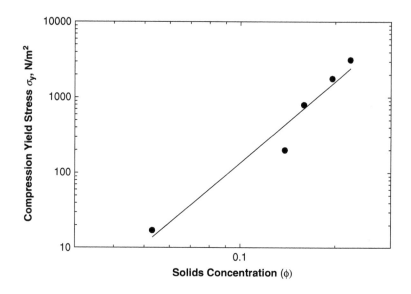

Figure 12.12   Effect of solids concentration on the compressive yield stress for flocculated kaolin.

structure. Sediment compression only occurs for flocculated systems — fully dispersed particles settle so as to give an incompressible, close-packed structure. It has been suggested that compression behavior is dominated by the small flocs produced by dispersion destabilization and is largely independent of further flocculation [67].

2.   Clarification

For the most part, clarification processes are applied to dilute suspensions such that free-settling conditions prevail. Process design is fairly straightforward: the principal requirement is that the solids flux due to settling always exceed the flux of fresh feed into the system. Thus, so long as free-settling conditions apply, it is sufficient that:

$$v_s > Q_f/A_T, \tag{25}$$

where v_s is the settling velocity, Q_f is the volumetric feed rate of slurry, and A_T is the clarifier area. If the feed contains very fine particles (typically smaller than about $1\,\mu m$), settling rates for primary particles are effectively zero and complete destabilization is vital to eliminate residual turbidity. Subsequent growth of flocs to larger sizes serves to increase clarifier capacity.

3. Thickening

a. Required Area

Processes for thickening suspensions are considerably more complex than clarification, especially when highly compressible sediments are formed. If the sediment is essentially incompressible or the required thickened product concentration can be achieved without compression, the above flux condition can be applied, but modified for hindered-settling conditions. Thus, it is necessary that at any location in the thickener

$$\phi v_s(\phi) > Q_f \phi_f / A_T, \tag{26}$$

where $v_s(\phi)$ is the hindered-settling velocity at volumetric concentration ϕ and ϕ_f is the feed concentration. In a continuous thickener, the flux due to removal of thickened product must also be taken into account. A simple mass balance around the thickener (assuming negligible solids content in the liquid overflow) leads to the following expression for the required thickener area [68]:

$$A_T = \frac{Q_f \phi_f}{v_s(\phi)} \left(\frac{1}{\phi} - \frac{1}{\phi_u} \right), \tag{27}$$

where Q_f is the volumetric feed rate of slurry at concentration ϕ_f, $v_s(\phi)$ is the settling velocity at concentration ϕ and ϕ_u is the desired product solids concentration.

The application of Equation (27) requires a knowledge of the concentration dependence of the settling rate over the range from the feed to the underflow. Whichever concentration gives the largest required area is potentially limiting. That area represents the minimum for successful operation. Alternatively, for a thickener with area A_T, Equation (27) specifies the maximum feed rate Q_f that can be used. Further increase in the feed rate would lead to the formation of a layer with the limiting concentration that builds up with time, eventually overflowing the thickener.

The settling velocity–concentration relationship, $v(\phi)$, can be determined by direct measurement on slurries with a range of concentrations, as suggested by Coe and Clevenger [68]. An alternative procedure, based on the Kynch [69] analysis of batch sedimentation processes, was suggested by Talmadge and Fitch [70]. Their approach permits the complete relationship to be estimated from a single batch test in which the feed slurry is allowed to settle until the final concentration exceeds the desired underflow concentration. Ideally, both methods should yield the same result but, in practice, significant discrepancies are commonly observed, especially for highly flocculated systems. This lack of agreement can usually be attributed to variations in the extent of flocculation. The direct (Coe and Clevenger) method requires homogeneous slurries at different solids concentrations. These can be prepared either separately or by thickening of the original feed slurry followed by remixing. Flocculation at different solids concentrations and agitation of flocculated suspensions both have significant affects on floc size and settling rate. For this reason, the Talmadge and Fitch procedure is more appropriate for determining settling behavior in flocculated slurries.

b. Required Depth

The procedures and analysis outlined above specify area only and make no reference to depth. Ideally, if settling rates do depend only on concentration, it is necessary to provide only sufficient depth to allow for underflow removal and to avoid bypassing of the feed directly to the overflow. However, if the specified underflow concentration falls in the compression regime, a minimum sediment depth must be available. In particular, the self-weight of the sediment must be equal to its compressive yield stress σ_y. The self-weight W_s, corrected for buoyancy, is given by:

$$W_s = \Delta\rho g\bar{\phi}H_s, \tag{28}$$

where $\bar{\phi}$ is the appropriate average concentration in the sediment and H_s is the sediment depth. Equating the self-weight to the compressive yield stress, and using the underflow concentration as a conservative approximation for the average, the required depth can be estimated from

$$H_s \cong \frac{\sigma_y}{\Delta \rho g \phi_u}. \tag{29}$$

Referring to Figure 12.12, the compressive yield stress for the flocculated clay at 20% solids by volume ($\phi = 0.2$) is about $2000\,\text{N/m}^2$. The buoyant density of the clay is about $1650\,\text{kg/m}^3$, leading to a required depth of about 0.6 m. To achieve 25% solids would require a depth of about 1.7 m. It should be emphasized that the above calculations do not consider the consolidation *rate*.

The dependence on depth invalidates the simple analyses described above. Models for sedimentation in the compression regime are available [66,71,72]. However, because of their mathematical complexity and the need for empirical permeability–compressibility data, their application in practice has been quite limited. Unfortunately, the required data cannot be obtained from the simple batch settling tests normally used, since the rates also depend on self-weight, i.e., sediment depth. Typical laboratory-scale tests, in columns up to 1 m or so high, involves final sediment depths of a few centimeters only, and the higher final concentrations cannot be achieved. A test to obtain settling rate data for thickening the clay feed slurry at 1% solids up to 25% would require a column over 30 m high. By increasing the effective gravitational force, centrifugal testing can be used to reduce both the time and height requirements. Such tests are, however, considerably more difficult to perform and still provide only macroscopic data on the overall sediment depth.

B. Filtration

1. Deep-Bed Filtration

Both clarification and thickening can also be accomplished by filtration. Deep-bed filtration is used extensively for clarifying dilute suspensions — removal of turbidity from potable water, for example. In this case, the filter medium consists of a porous block of material or a bed of particles such as sand. The principal mechanisms involved are simple straining whereby relatively large particles or flocs are trapped in the pores of the medium, and deposition of individual fine particles on internal surfaces of the medium. The destabilization of the suspension aids in the capture of particles in pores and may help reduce clogging of fine pores. Extensive floc growth is probably not necessary and may, in fact, be detrimental to the process. The deposition of fine particles depends on the same interaction forces responsible for flocculation, but between dispersed particles and the surfaces of the porous medium. The deposition mechanism offers particular advantages for the clarification of extremely dilute suspensions of very fine particles for which flocculation is very slow and individual particles are too small for effective capture by straining. Particle deposition can be encouraged by appropriate selection of filter media and the use of chemical additives — to eliminate repulsive forces or permit heterocoagulation of oppositely charged media and particles.

2. Cake Filtration

Dewatering of suspended solids is commonly carried out by cake filtration in which the principal filter medium is a "cake" built up from the particles themselves. A filter cloth is used primarily to initiate cake formation and support the cake once it has formed. Destabilization and flocculation of the suspension can be used to enhance filter performance and are often mandatory for very fine particles. Thorough destabilization helps prevent fine particles from passing through the filter and also reduces problems due to clogging of the supporting medium. The formation of large, uniformly sized flocs can enhance filter capacity by favoring the development of open, permeable cake structures.

Industrial applications typically involve continuous operations in which the overall cycle: filter–(wash)–dewater–discharge cake, is repeated sequentially. Rotary vacuum drum or disk filters are among the most commonly used types. Both types consist, in effect, of a cylindrical vessel with the filter medium placed either on the ends of a very short cylinder (disk) or around the circumference of a longer unit (drum). The disk or drum is partially immersed in a tank containing the feed slurry. Vacuum is applied to the inside of the disk or drum. As the unit rotates, cake formation occurs where the medium is immersed in the slurry; liquid sprays can be applied after emergence from the slurry tank for washing purposes. Dewatering takes place in the final part of the cycle prior to cake discharge and re-immersion of the medium into the slurry tank. Removal of the cake from the medium is commonly by means of a mechanical scraper for drum filters. In the case of disk filters, the cake is often "blown" off by the application of a positive air pressure from the inside of the disk.

Flocculation plays a role in each of the stages. Filter capacity is determined by the flow rate of filtrate through the cake, which, in turn, depends on cake permeability. A well-flocculated slurry, with large flocs, will generally produce a highly permeable filter cake. It is also necessary for the cake to have sufficient strength to maintain the open structure and resist collapsing under the applied pressure. In this respect, the requirements for filtration differ from those applying to sedimentation. In the latter, compaction of the sediment is usually desirable, to enhance dewatering. Consequently, higher flocculant dosage is often used in filtration applications. When a washing stage is included in the cycle, cake strength and integrity may be even more important, to ensure that the cake can withstand the action of sprays etc., without tending to be re-slurried.

The dewatering stage involves displacement of liquid in the cake by air. The displacement is opposed by capillary forces whose magnitude varies inversely with pore size. Large, highly porous flocs forming an open cake structure generally favor dewatering by gas displacement. Cracking of the cake during the dewatering stage can be a serious problem by creating open channels for airflow, bypassing the cake itself. The binding action of high molecular weight polymeric flocculants can help reduce cracking. Again, relatively high polymer dosing may be appropriate.

3.   Pressure Dewatering

The cake formation stage produces a cake that is saturated with liquid. The removal of liquid by gas displacement can, ideally, reduce the liquid content substantially below the saturation level. In the case of slurries containing extremely fine particles, chemical precipitates and some wastewater sludges, for example, low-density, highly compressible cakes are produced. Filtration rates under the pressures (less than one atmosphere) available in conventional vacuum filters tend to be low. In addition, very fine pores remain in the cake, even after flocculation, so that capillary forces seriously limit the effectiveness of dewatering by gas displacement. Pressure filtration systems such as the plate-and-frame filter or, more recently, the belt filter press are commonly used for these materials. The product of these types of pressure filters is a fully saturated, but dense cake, i.e., residual liquid content is lowered by reduction of cake porosity rather than by displacement from the pores.

In the batch, plate-and-frame filter, slurry is forced through a filter membrane in an enclosed space under pressure. Flow continues until the space is filled, after which the system is disassembled and the cake is removed. Some degree of flocculation is useful to minimize the passage of fine particles though the membrane, but extensive floc growth is probably unnecessary. Continuous pressure filtration is possible using belt-press filters [73]. Typically, in these systems, flocculated slurry is fed onto a filter membrane, which takes the form of a continuous moving belt. Free liquid is first removed by simple gravity drainage through the belt. Following the drainage zone, the belt passes into a low-pressure zone where a porous, upper belt converges on the lower one, squeezing the cake with gradually increasing pressure. The belts, together with the compressed sludge, then pass over sets of rollers, often arranged in a zigzag pattern, that exert increasing pressure on the cake. Slight differences in the belt speeds are often used to cause shearing of the cake to aid in liquid removal. Flocculation is critical to the successful operation of belt presses and is often built into the system, directly ahead of the drainage zone. Proper destabilization of the slurry followed by extensive floc growth using high molecular weight polymers is advisable to ensure complete incorporation of fine particles into the flocs and allow sufficient drainage to form a compact cake. Inadequate drainage can lead to problems of the slurry retaining too much fluidity, allowing it to be squeezed out from between the belts.

In all of the applications discussed above, it is important to recognize that flocculation, especially when accomplished by polymer addition, involves more than simply the appropriate reagent selection and dosage. Proper reagent addition, mixing, and subsequent handling of the slurry may be as important, or even more so.

REFERENCES

1. H.C. Hamaker, *Physica*, 4, 1058, 1937.
2. R. Hogg, T.W. Healy, and D.W. Fuerstenau, *Trans. Faraday Soc.*, 62, 1638–1651, 1966.
3. D.A. Napper, *Polymeric Stabilization of Colloidal Dispersions*, Academic Press, New York, 1983.
4. J.N. Israelachvilli and R.M. Pashley, *Nature*, 300, 341–342, 1982.
5. B.V. Derjaguin and L.D. Landau, *Acta Physiochim.*, 14, 633, 1941
6. E.J.W. Verwey and J.Th.G. Overbeek, *Theory of the Stability of Lyophobic Colloids*, Elsevier, Amsterdam, 1948.
7. J.Th.G. Overbeek, in *Colloid Science* I, H.R. Kruyt, Ed., Elsevier, Amsterdam, 1952.
8. F. Concha and E.R. Almendra, *Int. J. Miner. Process.*, 5, 349, 1979.
9. J. Lyklema, *Adv. Colloid Interf. Sci.*, 2, 65, 1968.
10. J. Visser, *Adv. Colloid Interf. Sci.*, 3, 331, 1972.
11. R.J. Hunter, *Foundations of Colloid Science*, Oxford University Press, London, 1986, p. 222.
12. J. Gregory, *Adv. Colloid Interf. Sci.*, 2, 396, 1968.
13. M. Smoluchowski, *Physik. Z.*, 17, 557, 1916.

14. M. Smoluchowski, *Z. Physik. Chem.*, 92, 59, 1917.
15. T.R. Camp and P.C. Stein, *J. Boston Soc. Civ. Engrs.*, 30, 219, 1943.
16. J.N. Tilton, in *Perry's Chemical Engineers' Handbook*, 7th ed., R.H. Perry and D.W. Green, Eds., McGraw-Hill, New York, 1997, pp. 6-34–6-36.
17. D.L. Swift and S.K. Friedlander, *J. Colloid Interf. Sci.*, 19, 621–647, 1964.
18. C.S. Wang and S.K. Friedlander, *J. Colloid Interf. Sci.*, 24, 170, 1967.
19. K.V.S. Sastry, *Int. J. Miner. Process.*, 2, 187–203, 1975.
20. R. Hogg, *Powder Technol.*, 69, 69–76, 1992.
21. T.G.M. van de Ven and S.G. Mason, *J. Colloid Interf. Sci.*, 57, 505, 1976.
22. H.S. Chung and R. Hogg, *Colloids Surf.*, 15, 119–135, 1985.
23. N. Fuchs, *Z. Physik.*, 89, 736, 1934.
24. L.A. Spielman, *J. Colloid Interf. Sci.*, 33, 562, 1970.
25. V.G. Levich, *Physicochemical Hydrodynamics*, Prentice-Hall, New York, 1962.
26. H. Rumpf, in *Agglomeration*, W.A. Knepper, Ed., Interscience, New York, 1961.
27. D.S. Parker, J. Kaufman, and D. Jenkins, *J. San. Eng. Div. ASCE.*, 98, 79–99, 1972.
28. N. Tambo and H. Hozumi, *Water Res.*, 13, 421–427, 1979.
29. L.A. Glasgow and J.P. Hsu, *AIChE J.*, 28, 779–785, 1982.
30. J.D. Pandya and L.A. Spielman, *J. Colloid, Interf. Sci.*, 90, 517–531, 1970.
31. D.T. Ray and R. Hogg, *J. Colloid Interf. Sci.*, 116, 256–268, 1970.
32. R. Hogg, A.C. Maffei, and D.T. Ray, in *Control'90 — Mineral and Metallurgical Processing*, R.K. Rajamani and J.A. Herbst, Eds., SME, Littleton, CO, 1990, pp. 29–34.
33. D.T. Ray, The Role of Polymers in Flocculation Processes and in the Binding of Compacted Powders, Ph.D. dissertation, The Pennsylvania State University, University Park, PA, 1988.
34. M.J. Vold, *J. Colloid Sci.*, 18, 684–695, 1963.
35. D.N. Sutherland, *J. Colloid Interf. Sci.*, 25, 373–380, 1967.
36. D.N Sutherland and I. Goodarz-Nia, *Chem. Eng. Sci.*, 26, 2071–2085, 1971.
37. P. Meakin, *J. Colloid Interf. Sci.*, 102, 491–512, 1984.
38. A.I. Medalia, *J. Colloid Interf. Sci.*, 24, 393, 1967.
39. B. Koglin, *Powder Technol.*, 17, 219–227, 1977.

40. N. Tambo and Y. Watanabe, *Water Res.*, 13, 409–419 1979.
41. R.C. Klimpel, The Structure of Agglomerates in Flocculated Suspensions, MS thesis, The Pennsylvania State University, University Park, PA, 1984.
42. R.C. Klimpel, C. Dirican, and R. Hogg, *Part. Sci. Technol.*, 4, 45–59, 1986.
43. R.C. Klimpel and R. Hogg, *J. Colloid Interf. Sci.*, 113, 121–131, 1986.
44. R.C. Klimpel and R. Hogg, *Colloids Surf.*, 55, 279–288, 1991.
45. D.A. Weitz and J.S. Huang, in *Kinetics of Aggregation and Gelation*, F. Family and D.P. Landau, Eds., North-Holland, Amsterdam, 1984, pp. 19–28.
46. R.A. Ruehrwein and D.W. Ward, *Soil Sci.*, 73, 485–493, 1952.
47. V.K. La Mer, R.H. Smellie, and P.K. Lee, *J. Colloid Sci.*, 12, 230, 1957.
48. T.W. Healy and V.K. La Mer, *J. Phys. Chem.*, 66, 1835–1838, 1962.
49. J. Gregory, *J. Colloid Interf. Sci.*, 42, 448–456, 1973.
50. G.J. Fleer, M.A. Cohen-Stuart, J.H.M.M. Scheutjens, F. Cosgrove, and B. Vincent, *Polymers at Interfaces*, Chapman & Hall, London, 1993.
51. C. Rattanakawin and R. Hogg, *Colloids Surf.*, 177, 87–98, 2001.
52. C. Rattanakawin, Polymeric Flocculation: Effects of Chemical and Physical Variables on Floc Formation and Growth, Ph.D. dissertation, The Pennsylvania State University, University Park, PA, 2002.
53. X. Yu and P. Somasundaran, *J. Colloid Interf. Sci.*, 178, 770–774, 1996.
54. C. Rattanakawin, Aggregate Size Distributions in Flocculation, M.S. thesis, The Pennsylvania State University, University Park, PA, 1998.
55. R.O. Keys and R. Hogg, AIChE Symposium Series 190, Vol. 75, 1979, pp. 63–72,.
56. R. Hogg, *Colloids Surf.*, 146, 253–263, 1999.
57. L.A. Cutter, *AIChE J.*, 12, 35, 1966.
58. R. Hogg, R.C. Klimpel, and D.T. Ray, *Miner. Metall. Proc.*, 42, 108–113, 1987.
59. R. Hogg, P. Bunnaul, and H. Suharyono, *Miner. Metall. Proc.*, 10, 81–85, 1993.
60. H. Suharyono, Flocculation and Consolidation in Thickening Processes, Ph.D. dissertation, The Pennsylvania State University, University Park, PA, 1996.

61. H. Suharyono and R. Hogg. SME Preprint No 94-231, SME, Littleton, CO, 1994.
62. H. Suharyono and R. Hogg, *Miner. Metall. Proc.*, 13, 501–505, 1996.
63. A.C. Maffei, Scale-Up of Flocculation Processes, M.S. thesis, The Pennsylvania State University, University Park, PA, 1989.
64. B. Fitch, *Trans. AIME*, 223, 129–137, 1962.
65. T. Allen, *Particle Size Measurement*, 5th ed., Chapman & Hall, London, 1997.
66. R. Buscall and L.R. White, *J. Chem. Soc. Faraday Trans. I*, 83, 873–891, 1987.
67. R.H. Weiland, P. Bunnaul, and R. Hogg, *Miner. Metall. Proc.*, 11, 37–40, 1994.
68. H.S. Coe and G.H. Clevenger, *Trans. AIME*, 55, 356–384, 1917.
69. G.S. Kynch, *Trans. Faraday Soc.*, 48, 166–176, 1952.
70. W.B. Talmadge and B. Fitch, *Ind. Eng. Chem.*, 47, 38–41, 1955.
71. F.M. Tiller and Z. Khatib, *J. Colloid Interf. Sci.*, 100, 55–67, 1984.
72. P. Kos, in *Flocculation Sedimentation and Dewatering*, B.J. Moudgil and P. Somasundaran, Eds., Engineering Foundation, New York, 1985, pp. 39–56.
73. J.A. Semon, *Fluid/Part Separ. J.*, 3, 101–109, 1990.

Index

Milton Keynes UK
Ingram Content Group UK Ltd.
UKHW020001071024
449327UK00031B/2606